7th International Conference on Ship and Offshore Technology (ICSOT Indonesia 2021)

Enhancing The Sustainable Use of Ocean-based Resources Through Innovation on Maritime Technology

Surabaya, Indonesia
19 - 20 November 2021

Editors:

Miftachudin
Rikat Eka Prastyawan
Dian Asa Utari
Ayu Nindyapuspa

ISBN: 978-1-7138-8303-6

Printed from e-media with permission by:

Curran Associates, Inc.
57 Morehouse Lane
Red Hook, NY 12571

Some format issues inherent in the e-media version may also appear in this print version.

The Royal Institution of Naval Architects
8-9 Northumberland Street
London, WC2N 5DA
United Kingdom

Phone: 020 7235 4622
Fax: 020 7259 5912

publications@rina.org.uk

Additional copies of this publication are available from:

Curran Associates, Inc.
57 Morehouse Lane
Red Hook, NY 12571 USA
Phone: 845-758-0400
Fax: 845-758-2633
Email: curran@proceedings.com
Web: www.proceedings.com

ISBN : 9781713830368 icsot.ppns.ac.id

ICSOT 2021 INDONESIA

7th International Conference on Ship and Offshore Technology

"Enhancing The Sustainable Use of Ocean-based Resources Through Innovation on Maritime Technology"

19 - 20 November 2021, Surabaya, Indonesia

Conference Proceedings

Organised by :

PPNS Politeknik Perkapalan Negeri Surabaya

The Royal Institution of Naval Architects

Supported by :

Skills for prosperity Indonesia

THE ROYAL INSTITUTION OF NAVAL ARCHITECTS

10 UPPER BELGRAVE STREET, LONDON, SWIX 8BQ Telephone: +44 (0)20 7235 4622

The Royal Institution of Naval Architects 8-9
Northumberland Street

London, WC2N 5DA United
Kingdom

Phone: 020 7235 4622

Fax: 020 7259 5912

publications@rina.org.uk

Table of Contents

Table of Contents

RECONSTRUCTING THE SHIP MOTIONS USING COMBINED MARINE SYSTEMS SIMULATOR TOOLBOX AND A SIX-DOF STEWART PLATFORM

E Yazid, M. Mirdanies, R A Ardiansyah and **Rahmat,** Research Center for Electrical Power and Mechatronics-National Agency for Research and Innovation (BRIN), Indonesia
R Ristiana,Technical Implementation Unit for Instrumentation Development-National Agency for Research and Innovation (BRIN), Indonesia
Y Sulaeman, Research Centre for Electronics and Telecommunication-National Agency for Research and Innovation (BRIN), Indonesia

SUMMARY

Characterization of ship motions is greatly essential in evaluating performance of the ship or a ship mounted subsystems. This paper addresses such a challenge by proposing a methodology on reconstructing the ship motions using combined Marine Systems Simulator (MSS) toolbox and a six-DoF Stewart platform. Given the specifications of the ship, Stewart platform and Inertial Measurement Unit (IMU) system, ship motions are generated and then reconstructed by a six-DoF Stewart platform as a validation. Finding results show that the ship motions are able to be reconstructed and its randomness can be captured with considerable accuracy. However, as the sea state code increases, the accuracy of the proposed method decreases as indicated by the values of sum square error (SSE) of the reconstructed motions. Hence, the proposed methodology can be recommended as ocean waves based ship motions simulator as well as other dynamics system motion simulators.

NOMENCLATURE

X_s	Surge motion (m)
Y_s	Heave motion (m)
Z_s	Sway motion (m)
ψ	Roll motion (degree)
φ	Pitch motion (degree)
ϕ	Yaw motion (degree)
S	Wave spectrum (m s^2)
H_S	Significant wave height (m)
ζ	Wave elevation (m)
k	Wave number(m)
λ	Wave length (m)
T_p	Peak period (s)
ω_0	Frequency at spectral peak (rad/s)
γ	Peakedness parameter
α	Phillip's constant
σ	Shape parameter
ℓ	Stroke (m)

1. INTRODUCTION

Characterization of ship motions is greatly essential in evaluating performance of the ship design and a ship mounted subsystems. To do so, commercial hydrodynamics softwares based numerical methods are common to be utilized. Experimental test either in the wave tank or basin also becomes a favorable validation. However, when such tool and facilities are not available or they cannot be accessed due to the limitation of the test facility, then an alternative solution might be required. Fossen [1] has developed a Marine Systems Simulator (MSS) toolbox to simulate the motions of ships, underwater vehicles, and floating structures. The toolbox also provides library includes guidance, navigation, and control blocks for real-time simulation. Application of MSS toolbox in reconstructing the ship motions can be referred to selected literatures such as [2]-[5]. From those references, prior research with regard to reconstruction of the ship motions is found to be relatively rare. Meanwhile, there have been numerous applications of a six-DoF Stewart platform in reconstructing the motions of dynamics system (e.g. passenger ship[6],active helideck testbed [7],ship motion control [8], high-speed craft simulator [9], ship [10]). However, the existing literatures are still limited to some degrees of freedom, stand-alone use of Stewart Platform and MSS. In particular, to the best of authors's knowledge, combined MSS toolbox and a six-DoF Stewart platform based ship motions is still an open research.As such, this paper proposes enhancements in which the main contribution is listed as,

- To develop a method in reconstructing the motions of ship by using combined MSS toolbox and a six-DoF Stewart platform.
- To investigate the influence of ocean random waves to the performance of proposed method.

This paper is organized as the following: Section 2 presents the system description which covers the modelling of a ship dynamics and concepts of Stewart platform and IMU. Section 3 describes the experimental set-up. Results and discussions are put on Section 4. To end this paper, Section 5 provides the conclusions and recommendations.

2. SYSTEM DESCRIPTION

The motions of a ship in the open sea are illustrated in Figure 1. As can be seen, the ship has two coordinate frames which are inertial and non-inertial frames. The former is denoted by $O_n - X_nY_nZ_n$ which is apart from the ship by a fixed distance, and is taken as a reference coordinate frame. The latter is the ship coordinateframe which is denoted by $O_S - X_SY_SZ_S$, and set to coincide with the principal axes of the ship (surge, sway, and heave). This coordinate is located at the center of gravity of the ship, and affixed to the hull of the ship. Since reconstruction is performed using Stewart platform, then the ship is modeled under zero speed where the ship is not propelled. The purpose is to make the hydrodynamic and the inertial coordinate frames are in a coincidence. It should be noted that dynamics of the ship is predominantly compared to subsystems mounted the board on due to the size of ship is much larger than the size of the subsystems.

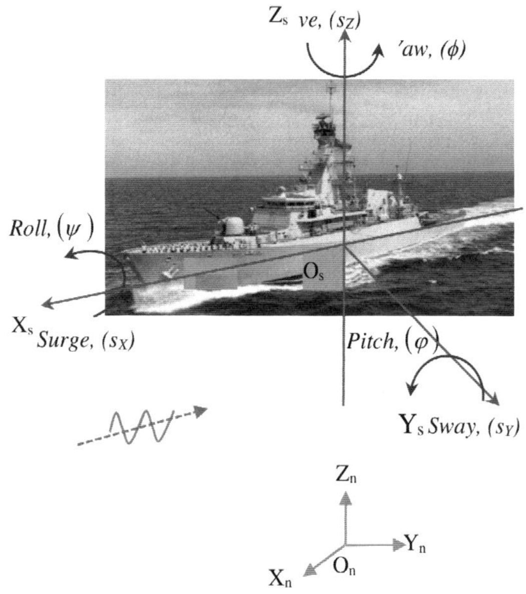

Figure 1: Description of a ship [11].

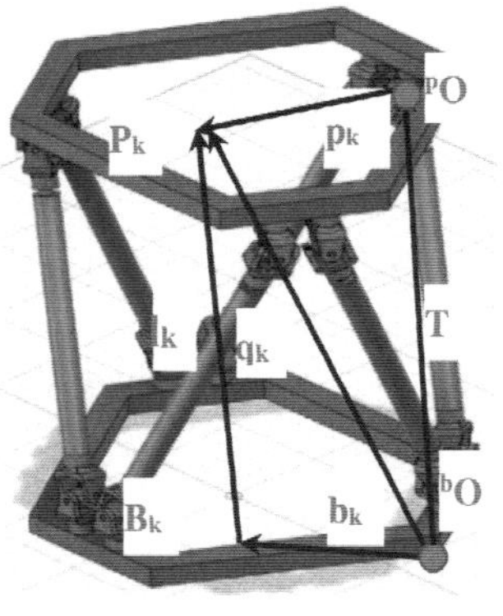

Figure 2: Sizing the stroke of one linear actuator.

2.1 MODEL OF A SHIP IN THE OPEN SEA

As shown by figure 1, modelling of a ship dynamics is performed to compute the surge, heave, and sway displacements, and the roll, pitch and yaw angles, respectively. Those motions are denoted as generalized coordinates of the ship as expressed in the following equation,

$$\eta = \left(X_s, Y_s, Z_s, \psi, \varphi, \phi\right). \tag{1}$$

In this paper, Marine Systems Simulator (MSS) toolbox[11] is employed to generate the ship motions. Generation process is started by defining the sea state code by means of significant wave height (H_S) where the sea state can be varied by adjusting the value of H_S according to the classification of sea state or sea severity [12]-[13]. The value is commonly reflected in several standards ocean energy spectral density formula such as Pierson-Moskowitz (PM) and JONSWAP spectrum. The latter is used instead of the former since it is more versatile to represent the spectral peakedness. JONSWAP spectrum can be defined as in the formula below [14],

$$S(\omega) = \alpha H_s^2 \left(\frac{\omega_0^4}{\omega^5}\right) \exp\left(-1.25\left(\frac{\omega_0}{\omega}\right)^4\right) \gamma^{\exp\left(-\frac{(\omega-\omega_0)^2}{2\sigma^2\omega_0^2}\right)}. \tag{2}$$

As can be seen from equation (2), spectrum is characterized by: H_S denotes significant wave height, ω_0 is frequency at spectral peak, γ represents peakedness parameter and the value is in the range of $[1 \sim 7]$, α denotes Phillip's constant while σ is shape parameter. By adjusting the value of H_S in equation (2), the ship motions in equation (1) will change accordingly. Hence, when the sea state is defined, then the ocean wave height time series may be generated by substituting equation (2) into equation (3) as below,

$$\xi(x, y, t) = \sum_{i=0}^{N} a_i \sin(\omega_i t + \theta_i - k_i x \cos\chi - k_i y \sin\chi). \tag{3}$$

Equation (3) reflects that a random wave at location x, y can be generated by a sum of traveling regular waves, propagating at angle χ with respect to the inertial coordinate frame. Each wave has a random phase θ_i where the value is in the range of $[-\pi, \pi]$, and has a uniform distribution.

To compute the amplitudes of ith wave component in equation (3), equation (4) below may be used by taking the value in equation (2),

$$a_i = \sqrt{2\,S(\omega_i)\Delta\omega}. \tag{4}$$

To obtain the ith wavelength λ_i, the following equation is used,

$$\tanh\left(\frac{2\pi h}{\lambda_i}\right) = \frac{2\pi\lambda_i}{g\,T_p^2}. \tag{5}$$

Equation (5) shows that the wave length is computed iteratively by defining the ocean depth h and the peak period T_p. After it is calculated, the i-th wave number is then calculated by the equation below,

$$k_i = \frac{2\pi}{\lambda_i}. \quad (6)$$

2.2 MODEL OF A SIX-DOF STEWART PLATFORM

After the time series of ship motions are generated as in equation (1) through equations (2)-(5), the next step is to access the Stewart platform. The motions are decomposed into position and orientation vector, and they become desired trajectories for the platform. Inverse kinematics process is performed to size the stroke of the linear actuators. Schematic of a six-DoF Stewart platform with respective coordinate frames is depicted in Figure 2. There are two frames attached on the platform, the first is ${}^{b}O$ put on the circumcenter of the based platform while the second is ${}^{p}O$ put on the circumcenter of the mobile platform. Vector loop equation may be derived as,

$$\ell_k = T + {}^{p}R_b p_k - b_k, \quad k = 1,2,\cdots,6. (7)$$

Term ℓ_k is the length vector of the i-th linear actuator, vector $T = \{X_s \quad Y_s \quad Z_s\}^T$ is the position of the circumcenter of frame ${}^{p}O$ with respect to frame ${}^{b}O$, matrix ${}^{p}R_b$ is the orientation matrix representing the orientation of frame ${}^{p}O$ with respect to frame ${}^{b}O$. Term b_k is the position of all the spherical joints $B_k \ (k = 1,2,\cdots,6)$ on the base platform measured from frame ${}^{b}O$ while p_k is the position of all the spherical joints $P_k \ (k = 1,2,\cdots,6)$ on the mobile platform measured from frame ${}^{p}O$. Thus, stroke of the ith linear actuator is the Euclidean of equation (7) which can be expressed as,

$$\ell_k = \left\| T + {}^{p}R_b p_k - b_k \right\|, \quad k = 1,2,\cdots,6. \quad (8)$$

The values of ℓ_k in equation (8) are taken as desired strokes. The actual values are obtained by accessing the six linear actuators which are represented by six servo motors equipped with control system. General diagram block of Stewart control platform is depicted in Figure 3. From the block, it can be seen that control system receives system input in terms of desired stroke of the ith linear actuator (ℓ_k) obtained from equation (8). Controller has the function to adjust the values of the desired stroke (ℓ_k) with actual (ℓ_{k_a}) so as the system error e can be minimized. Any types of controller such as PD, fuzzy or other adaptive controllers may be applied.

3. EXPERIMENTAL SETUP

Experimental test was performed in the Mechatronics laboratory, Research Centre for Electrical Power and Mechatronics, Indonesian Institute of Sciences.Before it is proceeded, sea state is defined as in Table 1 where the wave spectrum is calculated, and time series of ocean wave height is generated accordingly. Time series of the wave height has duration of 400 s with sampling rate of 0.01 s. It is then fed into the MSS toolbox to produce the ship motions.Flowchart for visualizing the process is depicted in Figure4.

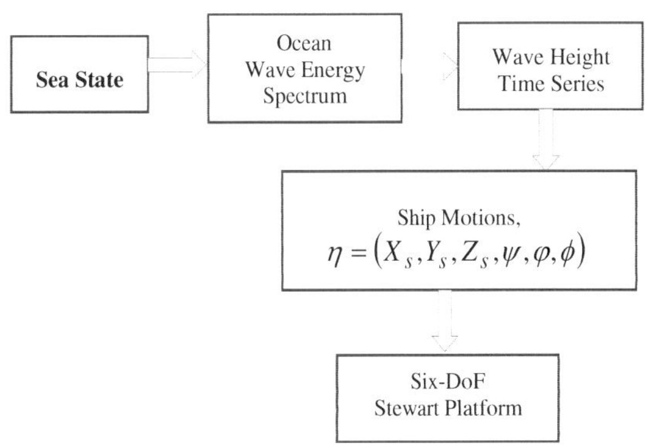

Figure 4: Process flow of reconstruction.

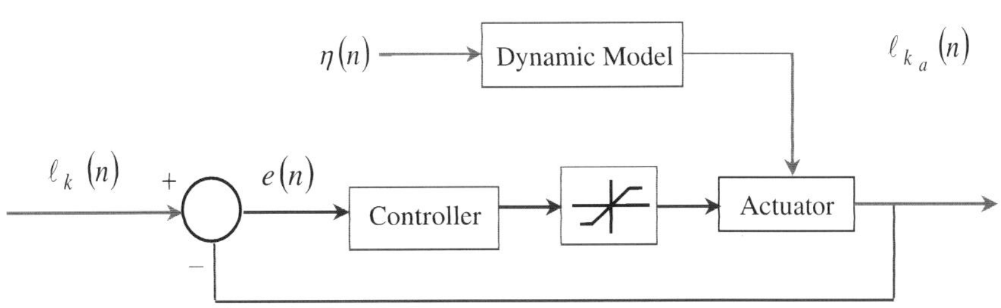

Figure 3: Diagram block of Stewart platform controller.

Experiment wasthen started by moving the platformfollowing the desired ship motions. For such a purpose, the motions were recorded in the control board, and inverse kinematics process was performed. The results were used to manipulate the stroke of the six linear actuators so as to achieve the desired position and orientation. To measure the position and orientation of platform, one IMU system manufactured by Novatel is mounted at the base of platform. To check the influence of ocean random waves induced ship motions to the Stewart platform system, sea state code is defined as in Table 1. It tabulates the sea state code used with its respective peak frequency (f_P) and significant wave height (H_S). The respective time series of surge, heave, sway, roll, pitch and yaw motions are produced by Marine Systems Simulator. Overall time series are then reconstructed by a six-DoF Stewart platform system.

Table 1: Parameters of Ocean Random Waves

Wave Frequency (Hz)	Wave Height (m)	Sea State (Code)
0.83	1.1	Slight (3)
0.77	2.3	Moderate (4)
0.71	3.2	Rough (5)
0.61	4	Very Rough (6)

Table 2:Selectedtechnical specifications of Stewart Platform

Size	1.7 m x 1.5 m x 0.55 m
Payload Maximum	2000 kg
Site Framework	3.2 m
Role Cylinder Stroke	0.6 m
Longitudinal Maximum Stroke	-0.46 m ~ 0.57 m

Table 3: Selectedtechnical specifications of IMU system

Gyroscope	
Input Rate (max)	±150 °/second
Bias Repeatability	0.5 °/second
Bias Instability	3.5 °/hour
Angular Random Walk	0.1 °/√hour
Accelerometer	
Input Rate (max)	±5 g
Bias Repeatability	15 mg
Bias Instability	0.1 mg
Angular Random Walk	0.05 m/s/√hour

In additions, main technical specifications of the platform and the IMU system is tabulated in Table 2 and 3, respectively. The output of the IMU system is then compared to the output of MSS toolbox which an indicator to verify the efficacy of the proposed methodology of this paper. The sum square error (SSE) in every sea state code is compared for further analysis.

4. RESULTS AND DISCUSSIONS

This section presents the results based on equations (1)-(6) and Figures 2-4. Reconstruction of ship motions under sea state (code 4, moderate) is displayed in upper panel of Figure 5 as a sample. As can be seen from the figure, actual ship motions measured by IMU system agree well with the ship motion generated from desired trajectory. This is enhanced by the lower panel of Figure 5. Those results imply that all motions can be reconstructed well.

Table 4:Statistics Properties of Reconstructed Motions (Experiment)

Motion		SSE	Skewness	Kurtosis
Surge	In	0.17	3.38	541.946
	Out	0.17	3.38	541.86
Sway	In	0.057	-0.855	233.15
	Out	0.057	-0.853	232.81
Heave	In	0.049	-1.947	413.20
	Out	0.049	-1.957	412.49
Roll	In	0.018565	1.113	734.85
	Out	0.019113	1.146	736.72
Pitch	In	0.007597	0.062	71.74
	Out	0.008845	0.057	68.421
Yaw	In	0.005172	-8.21	1081.75
	Out	0.006891	-8.87	1071.58

It is also validated by Table 4 where the statistics properties are revealed. There is insignificant difference between the motions get into Stewart platform and are measured by IMU system. To check the effect of ocean random waves to the performance of proposed method, sea state defined in Table 1 is used.

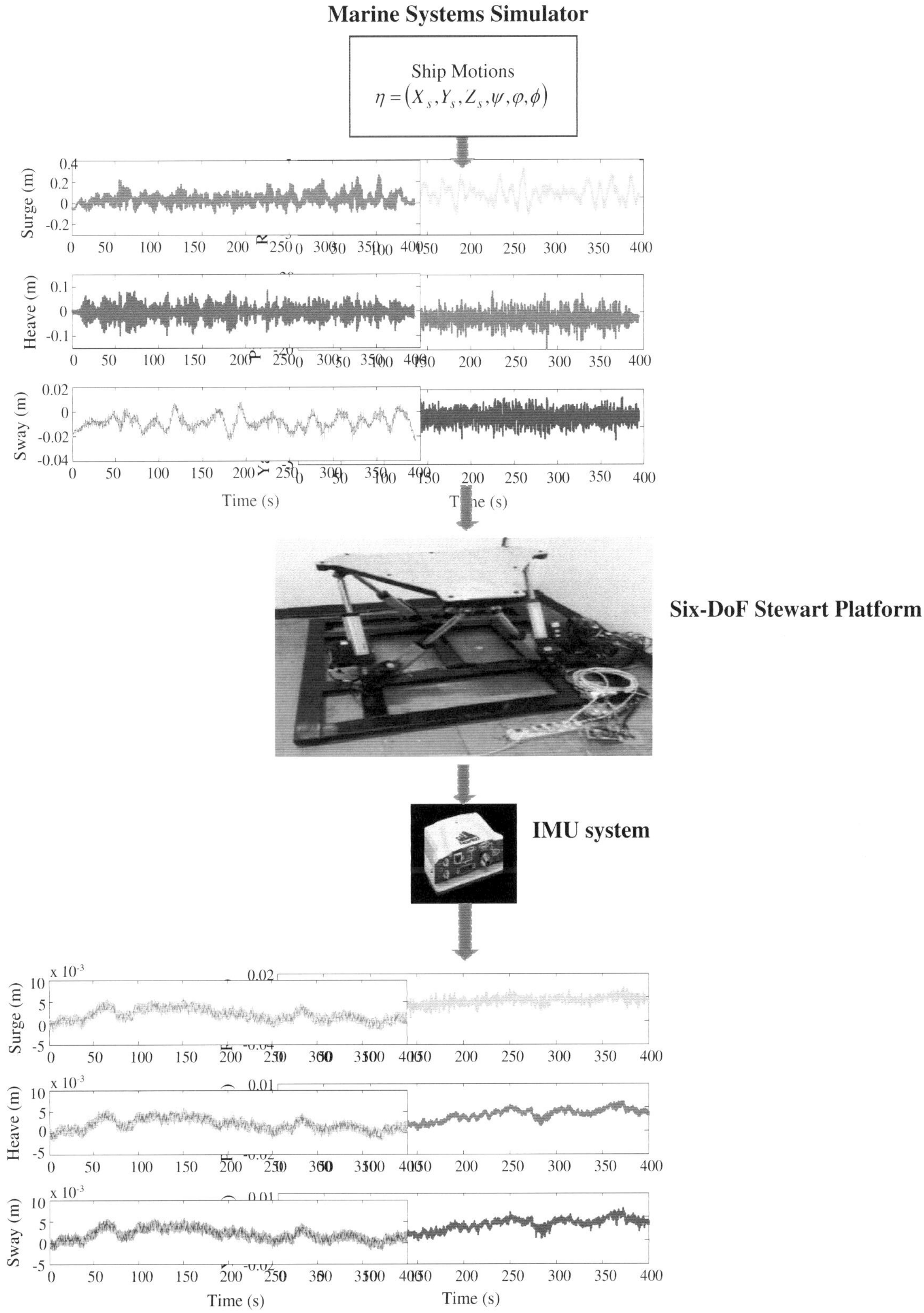

Figure 5: Reconstruction errors.

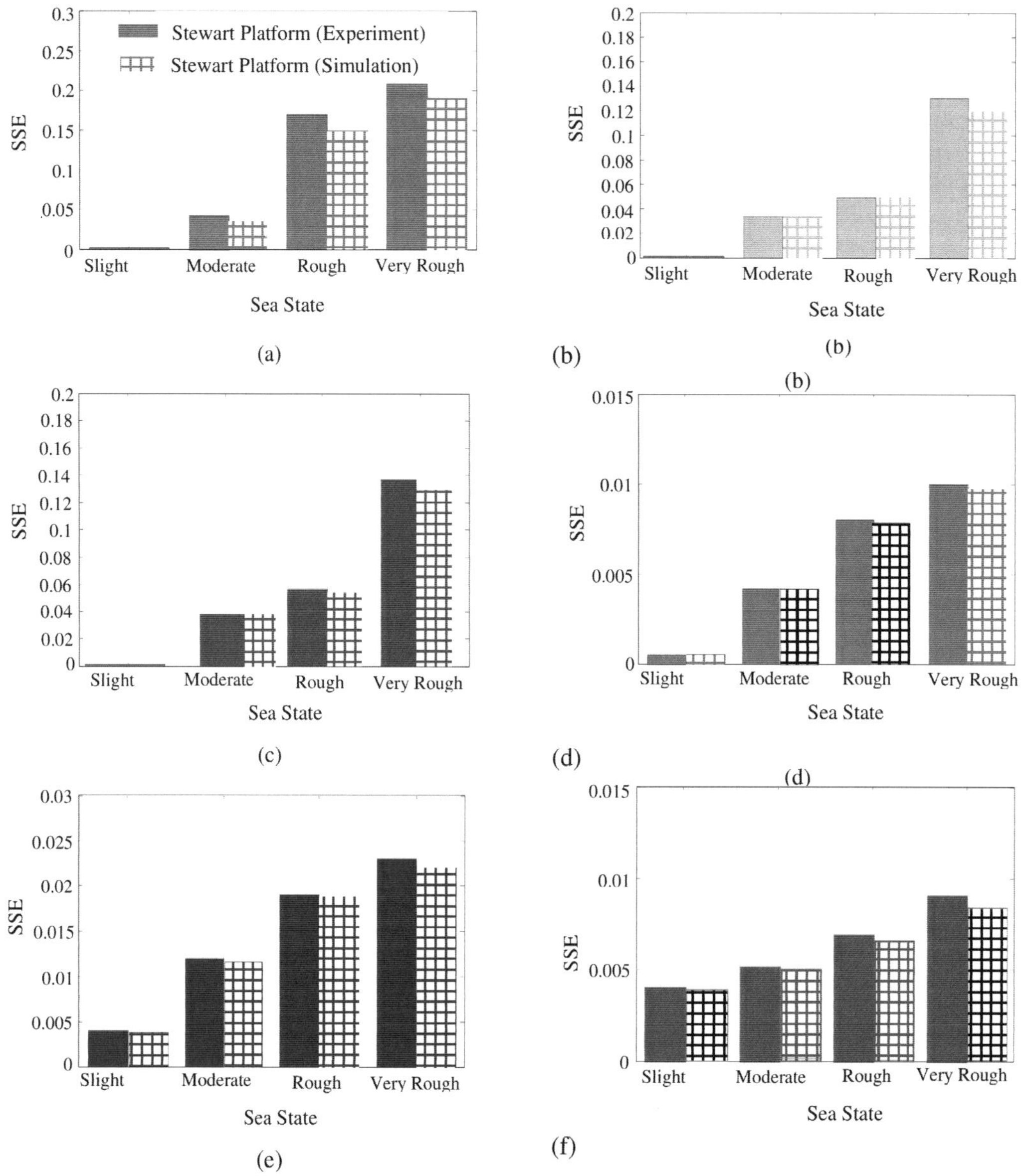

(f)

Figure 6: SSE values of ship motions under influence of sea state code (a) surge (b) heave (c) sway (d) roll (e) pitch (f) yaw

As can be seen from Figure 6 that in the slight sea state, accuracy of reconstruction for all motions is high as indicated by the values of SSE. However, as the sea state code increases, the accuracy is getting lower. The trend seems slight nonlinear. Also, it can be observed that surge motion is found to be the most difficult motion to be reconstructed among the others. This might be due to the in this particular case, amplitudes distribution of surge motion is non-Gaussian type, and it contains nonlinearity of the ship dynamics.

Moreover, compared to pitch and yaw motions, roll motion has lowest reconstruction accuracy. This might be understood since it is a typical motion of common ships. At last, yaw motion is the easiest motion to be reconstructed since it has smaller amplitudes than the others for the underlying case. As benchmark for results in Figures 5-6, reconstruction by Stewart platform via simulation is also performed. The results are compared with experiment as in Figure 6. It is worthy to note that simulation can be carried out using either Eqs. 7-8 and Figure 3 in the form of hand-made codes or MATLAB Toolbox which provides the Simulink ofthe Stewart

Platform. For the latter, parameters in Table 2 and 3 must be defined into the library of respective block.It seems that phenomena in experiment are also found in simulations. However, discrepancy is found in term of amplitude of respective motions. This is to be expected since in simulation process, there are unmodeleddynamics and simplifications such as nonlinearity, friction, and the others.

7. CONCLUDING REMARKS

A method to reconstruct the motions of ship has been developed. The method is combination of MSS toolbox, a six-DoF Stewart platform and IMU system. From finding results, several important conclusions are made, and recommendations are proposed as follows:

- The respective time series of surge, heave, sway, roll, pitch and yaw motions produced by Marine Systems Simulator (MSS) are able to be reconstructed by a six-DoF Stewart platform with IMU system with reasonable accuracy.
- As the sea state code increases, then the sum square error (SSE) also increases for all motions. The trend seems nonlinear.
- Amplitudes distribution of ship motions which leads to nonlinearity and non-Gaussianity properties is able to be captured and reconstructed by using combined MSS toolbox and a six-DoF Stewart Platform.
- The proposed method can be recommended as ocean waves based ship motions simulator as well as other dynamics system motion simulators.
- Ocean random waves used in this paper are still unidirectional long-crested type. Multidirectional short-crested type is recommended as another future work.

8. ACKNOWLEDGEMENTS

The authors are grateful to the Ministry of finance of the Republicof of Indonesia for financial support in terms of LPDP Research Fund under project no. PRJ-92/LPDP/2020 andNational Agency for Research and Innovation (BRIN)for providing research facility.

9. REFERENCES

1. PEREZ, T., SMOGELI, O.N., FOSSEN, T.I., and SORENSEN, A.J., 'An Overview of the Marine Systems Simulator (MSS): A Simulink Toolbox for Marine Control Systems', *Modeling, Identification and Control*, 2006.

2. TU, V.T., HUNG, N.Q., MY, C.A., and HOA, T.N.,'Prediction of Ship Motions on Waves using Marine Systems Simulator', *Journal of Engineering and Applied Sciences*, 2019.

3. Hong, K.S., and Ngo, Q.H., 'Dynamics of the Container Crane on a Mobile Harbor', *Ocean Engineering*, 2012.

4. YAZID, E., NUGRAHA, A.S., SAPUTRA, H.M., and RAHMAT, 'Dynamics of a Ship-Mounted Two-DoF Manipulator: Case: Effect of Sea State to the Manipulator Joint Torque',*Proc. of IEEE Xplore Digital Library*, 2019.

5. YAZID, E., MIDRIEM, M., SAPUTRA, H.M., and RAHMAT, 'Position Control of a Ship-Mounted Two-DoF Manipulator', *Proc. of IEEE Xplore Digital Library*, 2019.

6. RUKAI, Z.,'The Study of 6-DoF Motion Simulation Platform for PassengerShip',*Advances in Engineering Research*, 2017.

7. CAMPOS, A., QUINTERO,J., SALTAREN, N., Ferre, M., and Aracil, R.,'An Active Helideck Testbed for Floating Structuresbased on a Stewart-Gough Platform', *Proc. of International Conference on Intelligent Robots and Systems*, 2008.

8. RIBCZAK, M., and RAK,A.,'Prototyping andSimulation Environment of Ship Motion Control System', *Proc. of International Journal on Marine Navigation and Safety of Sea Transportation*, 2020.

9. REKDALSBAKKEN,W., 'Design and Application of a Motion Platform for a High-Speed CraftSimulator',*Proc. of International Conference on Mechatronics*, 2006.

10. ZHENG, Z., ZHANG, X., ZHANG, J., AND CHANG, Z.,'A Stable Platform to Compensate Motion of Ship Based on Stewart Mechanism', *Proc. of International Conference on Intelligent Robotics and Applications*, 2015.

11. Ship [Internet] [Cited 2020 January 17th], Available from: https://sumselupdate.com/empat-kapal-perang-tni-bantu- pengawasan-natuna/.

12. Tupper, E., Introduction to Naval Architecture, Butterworth-Heinemann (Oxford), 1996.

13. KIM, N.S., 'Extraction of the Second-Order nonlinear Response from Model Test Data in Random Seas and Comparison of the Gaussian and non-Gaussian Models', *Ph. D. Dissertation (America: Texas A&M University)*, 2004.

14. HASSELMAN, K., 'Measurement of Wind-Wave Growth and Swell Decay during the Joint North Sea Wave Project', *(Erganzunscheft: Deutches Hydrographisches Institute)*, 1973.

10. AUTHORS BIOGRAPHY

Edwar Yazidis currently working at Research Center for Electrical Power and Mechatronics, Indonesian Institute of Sciences (LIPI) as a senior researcher. His research interests are in the area of ocean engineering and marine renewable energy.

Midriem Mirdaniesis currently working at Research Center for Electrical Power and Mechatronics, Indonesian Institute of Sciences (LIPI) as a senior researcher. His research interests are in the area of software development, embedded systems, data communication, and computer vision.

Rizqi Andry Ardiansyahis currently working at Research Center for Electrical Power and Mechatronics, Indonesian Institute of Sciences (LIPI) as a young researcher. His researchinterestsare in the area of electronic system design, robotics, control system, and artificial intelligence.

Rina Ristianais currently working atTechnical Implementation Unit for Instrumentation Development, Indonesian Institute of Sciences (LIPI) as a young researcher. His research interests are in the area ofautomation and control instrumentation process equipment and electrical machines & drives, electric transportation, and image processing.

Rahmatis currently working atResearch Center for Electrical Power and Mechatronics, Indonesian Institute of Sciences (LIPI) as a senior technician.His research interestsare in the area ofmechanical system design and manufacture.

YayaSulaemanis currently working at Research Center for Electronics and Telecommunication, Indonesian Institute of Sciences (LIPI) as a young researcher.His research interest is in the area of electronic system design.

PRELIMINARY DESIGN OF DUCTED NOZZLE PROPELLER IN PROPULSION SYSTEM OF CATAMARAN FLAT PLATE BOAT WITH THRUST VECTORING VARIATION

Evaldi Axsendra, Yeddid Yonatan Eka Darma, and **I.G.N.A Satria Prasetya** Ship Manufacturing Engineering Politeknik Negeri Banyuwangi, Indonesia

SUMMARY

The tourism sector is one of the economic supports in Banyuwangi Regency. So with Covid-19 pandemic, tourist visits decreased drastically. To help the problem, they need an innovation that can attract visitors to visit. One solution is a Flat Plate Bottom Glass designed with safety, comfortable, efficient and sophisticated for underwater panoramic tourism facilities. This paper shows the Ducted Nozzle Propeller thrust vectoring ship propulsion innovation. This can be a solution for the thrust to focus and be used entirely to move the ship. Then the propulsion system on the ship becomes effective. The method used in this study used CFDs (Computational Fluid Dynamic). From the simulation results of 12 it's models we know the average thrust increase is as large as 47%.

NOMENCLATURE

[Symbol]	[Definition] [(Unit)]
NCVS	Non Convensioanal Vessel Standart
DWT	Dead Weight Tonnage (ton)
Δ	Displacemet, (ton)
LOA	Length Over All, (m)
LWL	Length Water Line, (m)
LPP	Length per pendicullar, (m)
BM	Moulded Beam, (m)
H	Depth (m)
T	Draugh, (m)
Vs	Voyege speed, (knot)
ρ	Water Density, kg/m3
δ	BlockCoeffisient
ϕ	Prismatic Coeffisient
β	MidshipCoeffisient
s	Wet Surface Area, m
g	Gravitation, (m/s^2)
u	Viscoscity, (shg)
Cf	Drag Resistant Coeffisient
Fn	Froude number
Rn	Reynolt Number
Cr	ResistantCoeffisient
LCB	Longitudinal Center of Buoyancy, (m)
Ca	Adding Resistant Coeffisient
Caa	Air Resistant Coeffisient
Cas	Rudder Resistant Coeffisient
CT	Resistant Coeffisient
Rt	Resistant total(kN)
EHP	Effective Horse Power, (HP or kW)
DHP	Delivery Horse Power, (HP or kW)
THP	Thrust Horse Power, (HP or kW)
SHP	Shaft Horse Power(HP or kW)
BHP scr	Brake Horse Power min, (HP or kW)
BHP mcr	BrakeHorse Power max, (HP or kW)
t	Time, (s)
p	Power, (kW)
N shaft	Number of shaft rotation, (Rpm)
N propller	Number of propeller rotation, (Rpm)
Va	Speed of Advance, (m/s)
Bp	B value
Jb	Ratio advance
Δb	Advance Behind the ShipCoefisien
N	Propeller Rotation (rpm)
A	Blade Propeller Area, (m^2)
r	Moment Arm (m)
Thrust	Force of Push, (kN)
Torque	Force Moment, (kN.m)
V NZT 0,6	Propeller Without Nozzle 0,6 x T
V NZT 0,7	Propeller Without Nozzle 0,7 x T
V XXX XX	Shape of Ducting Thrust Vetoring Nozzle Propeller VariationDegree.

1. INTRODUCTION

Bangsring Beach has the potential of unique marine natural resources and beautiful beaches. The location of Bangsring Beach coast to Tabuhan Island and Menjangan Island makes Bangsring Beach as one of the strategic places as a utilization of natural potential as a tourist destination.But until now the only way to enjoy the beauty of the underwater is by diving or snorkelling. For that, the management need ships that have qualified by safety and comfort standards for tourism facilities on the beach.The ship that we purpose have the design of flat plate Catamaran with glass bottom innovations.

Innovation in the propulsion system is also one of the innovation solutions that make ship performance efficient. The latest technology that can improve the efficiency of ship optimization both in terms of technical and, economical. Ducted Propeller Nozzle is a propeller that has a ducting in the form of foil that surrounds the propeller so that it forms a sheath, tube or nozzle.Ducted Nozzle Propeller make speed of water flow in the tube faster than the flow of water outside the tube resulting in pressure in the tube lower than the pressure outside the tube.This pressure difference results in the addition of thrust or so-called thrust, with the installation of nozzles on the propeller can increase thrust up to 27%. Ducted Nozzle Propeller provides propeller effectiveness at times of high thrust load [1].

In order to support and increase the ship's propulsion campaign, focus the thrust, and the addition of the ship's propulsion, thrust-vectoring is added behind the Ducted Nozzle Propeller. The function of thrust-vectoring is to focus the thrust generated by the propulsion engine to improve the ship's propulsion capabilities. If thrust vectoring is applied to mechanical propulsion it will change the direction of the thrust spread in all directions towards the right. This can reduce losses and power losses by 3% [2].

2. LITERATURE

2.1 BOAT RESISTANT

The resistant of the ship is a fluid force that works on the ship in such a way that it resists movement of the ship. The total resistant of the ship is the number of resistant in the water, namely resistant (RF) and residual resistant (RR) with air resistant (RA). Friction resistance (RF) is a force of friction on a ship caused by friction between the bodyof the ship that is touch in the fluidaround hull [3].

2.2 PROPELLER

Threaded propellers are the most common form of ship propulsion. A threaded propeller has two or more leaves jutting out of the hub or boss. This boss is mounted on a shaft driven by the ship's propulsion engine. The propeller blade can be a part that blends with the hub, or is a part that can be removed from and installed on the hub or is a controllable propeller or a fixed (Fix pitch propeller).

The propellers are generally placed at the lowest possible position at the back of the ship. A propeller must have such a center line (diameter) that if the ship is fully charged the propeller will be adequately immersed so as to avoid as far as possible the phenomenon of airdrawing and propeller driving (racing) when the ship undergoes pitching movements [4]

2.3 PROPELLER HYDRODYNAMICS

In making the basic shape of the propeller requires a hydrodinamis form called Hydrofoil which produces a more lift.Hydrodynamics is an event in which the speed between the top and bottom of the hydrofoils occurs.The fluid that passes through the top of the Hodrofoil travels faster than the fluid that passes through the bottom. This is due to the pressure difference between the upper fluid flow and the lower fluid flow. As we know that the amount of pressure is inversely proportional to the magnitude of the speed.The fluid that passes through the top of the Hydrofoil travels faster than the fluid that passes through the bottom. This is due to the pressure difference between the upper fluid flow and the lower fluid flow. As we know that the amount of pressure is inversely proportional to the magnitude of the speed.So what happens is the flow of fluid through the bottom of the hydrofoil is slower when compared to the top of the hydrofoil. This difference in pressure that occurs is what then eventually gives rise to the phenomenon of lift or lifting style [5].

2.4 DUCTED NOZZLE PROPELLER

Ducted Nozzle Propeller or Nozzle is a propeller protector in the form of a foil-shaped plate. The function of the Nozzle is to increase and concentrate the flow of water flowing into the propeller so as to maximize the water sucked by the propeller. This aims to maximize the work of the propeller so that the thrust generated is also maximal. The resulting force efficiency will be maximized if the water passes through the Nozzle freely without obstruction. This is why as much as possible unung Nozzle should have a free area after body. Not only berfungi to maximize the work of the propeller, nozzle also serves to reduce noise and vibration levels. However, the flow that comes more homogeneously within the Nozzle, the reduction of the difference in central pressure is caused by vibration and cavitation. The combination of propellers inside the Ducted Nozzle Propeller is often referred to as the Ducted Propeller. In principle, nozzles can be used in any type of fast ship such as fast boats, where there is no effect of increased thrust. If the shift in resistance (due to the Nozzle is wider than the increase in thrust, it is a matter of loss of efficiency[1].

2.5 TRUST VECTORING

Thrust vectoring is a tool used to direct the direction of fluid motion and regulate the flow of fluid doringans radiated by the machine. The function of thrust vectoring is able to collect the flow and the missing thrust spread in all directions to be focused to increase thrust to the propulsion equipment. Thrust vectoring is typically used, developed and researched on combat aircraft that are used to regulate air aitan and the direction of power generated by aircraft jet engines. Some are used to regulate the movement or manufer of rockets and missiles to make it easier to hit targets on target and more efficiently. The use of thrust vectoring is in line with the type of machine, usability, machine system and existence[6]. Thrust vectoring is used for mechanical propulsion.Used to organize and direct power in various directions in a focused direction. In fighters and rockets are useful as controllers of their movements such as roll, pitch and yaw. The principle used in the work of thrust vectoring is to manipulate the direction of power flow from the driving engine or what can be called Mechanical Nozzle Manipulation. Thrust Vectoring manipulates the movement of flow that is pressurized physically on the engine output. When this nozzle changes the point of flow movement and has the effect of the flow has the desired direction, it is commonly found in fighter aircraft and rocket propulsion engines. Their experimental data supports this claim. At geometric vector angles up to 25, and through various nozzle pressure ratios, it can be reported to lose a maximum thrust of only 2%. In addition, the experiments suspected that these losses may have been due to miscalculations at the time of the experiment and research.[2]

2.6 THRUST

Thrust is a mechanical force. This is produced through a reaction accelerating the mass of the fluid. The fluid accelerates backwards and the propulsion engine engine accelerates in the opposite direction.To accelerate the fluid, we need some sort of propulsion system. From Newton's second law of motion, we can define force (F) to change the momentum of an object by a change in time (t).Momentum is the mass of an object (m) times the velocity (V). Thus, between twice (t1) and (t2), the force is given by:

$F = ((m \times V)2 - (m \times V)1) / (t2 - t1)$

Force, including thrust, is measured in newton units (symbol: N) based on the International System of Units (SI) and is also used as a unit of measure for the amount required to accelerate 1 kg of mass at a rate of 1 meter per second squared[7]

2.6 TORQUE

Torque is the rotation of a force against a shaft. Torque is measured by the result of force times with the radius of the circle where the force works. It shows on a wheel with the radius r working a Newton force that causes an object to spin at a rate of n spin per second. [8].

It can also be written with the equation:

$$T' = r \times F$$

2.7 FINITE ELEMENT METHODE

Finite element methode (FEM), sometimes referred to as Finnte Element Analysis (FEA), is a computational technique used to obtain approximate solution boundary value problems in a technique. Simply stated, a boundary value problem is a mathematical problem in which one or more varied dependents are able to meet the ubiquitous differential equations in a known independent variable domain and meet certain conditions at the domain boundary. Boundary value issues are also sometimes called field issues. This field is an interesting domain and most often represents physical structure. Element Analysis of fiinite elements helps predict the behaviour of products affected by many physical effects, including:

- Mechanical Stress
- Mechanical Vibration
- Fatigue
- Motion
- Heat Transfer
- Fluid Flow
- Electrostatics
- Plastic Injection Molding [9].

3. METHOD

3.1 RESEARCH METHOD

3.1 (a) COLLECTION AND COMPILATION

Looking for the necessary data and also looking for references to the necessary planning and calculations, namely:

- The Principal Dimension of the ship
- Ship Resistant

The data will be used as a basis for technical modeling ducting thrust vetoring nozzle propeller.

3.1 (b) TECHNICAL MODELING

In the process of technical modeling is divided into two, namely the process of geometry stage and meshing stage. Both of these processes are carried out on ANSYS Workbench on CFD simulation with CFX analysis.

3.1 (c)GEOMETRY STAGE

At this stage geometry is also the creation of a fluid domain where testing. The creation of boundry building on the model that has been designed. Making boundry building by making encoser on geometry. This enclose is formed with desire and adjusted also to the needs of simulation.

3.1 (d)MESHING STAGE

After the fluid domain is formed the next step is to mesh on the models already in the image.

Before starting the meshing first determine the size of the meshing that kiga will use. Keep in mind that the smaller the size of the meshing, the more elements formed so that the time to run will also be longer. Simulation or trial with Computional Fluid Dynamic.

3.1 (e)SET UP STAGE

This stage is something related to parameters with simulation. There are several steps taken at the simulation setup stage by using ANSYS Workbench Software on CFD simulation with CFX analysis

The Rotating Physics Domain Section is the setting of the rotating physics domain of objects. In this simulation, the only part of the propeller.

The stationary Physics Domain Section is the setting of a physics domain of an object that is simulated motionless. In this simulation the imovable parts of The Default Domain and Nozzle.

Boundary is made to know the characteristics of objects and fluids to fit the right conditions. In this simulation process there are boundary parts, including propeller, nozzle and fluid parts that work on the dominant water or on the written Ansys (water).

3.1 (f) SOLVER MANAGER STAGE

Solver manager calculations take from data management, which in this step serves as a file stacker. The number of files is adjusted to the number of time steps in the transient simulation process performed on ducting thrust vectoring nozzle. File results solver manager. Time step part of the sumulation done. Solver Manager of CFX Simulation Is a solver program all parameters that have been entered will be processed in this program. The calculation process must determine the convergence criteria to be calculated.

3.1 (g) EVALUATION STAGE

This stage is the stage of devaluing the results of the design that has been tested before. It concluded whether the Ducted Thrust Vectoring Propeller is fit for use and successfully achieved all the goals that would later be advice for further research development. This evaluation looked at several factors affecting design performance according to the simulation and reported design results.

3.2 PRINCIPAL WORK

The main propulsionmachine to drive the propeller to rotate and work. In properlers flavored by Ducting Thrust Vectoring Nozzle which is able to reduce the amount of thrust value lost by way, the thrust produced by the propeller will be focused to the same purpose as the fliuda flow steering device contained on the back of the propeller tube made of slabs that can be adjusted the output angle of thrust vectoring nozzle.

4. DATA ANALYSIS OF DUCTING THRUST VECTORING NOZZLE

4.1 MAIN DIMENSION OF BOAT

The main dimension is data can use design and calculate of the boat. This data become critical for shape boat in simulating and design application.
This below is main dimension of the boat:

- Type : NCVS catamaran
- DWT : 6,10 ton
- Displasment : 9,60 ton
- LOA : 12 m
- LWL : 12,16 m
- LPP : 11,7 m
- BM : 0,94 m
- H/Depth : 1,85 m
- T/draugh : 0,6 m
- VS : 10 Knot
- δ : 0,758
- ϕ : 0,600
- β : 0,750

4.2 SHIP RESISTANT

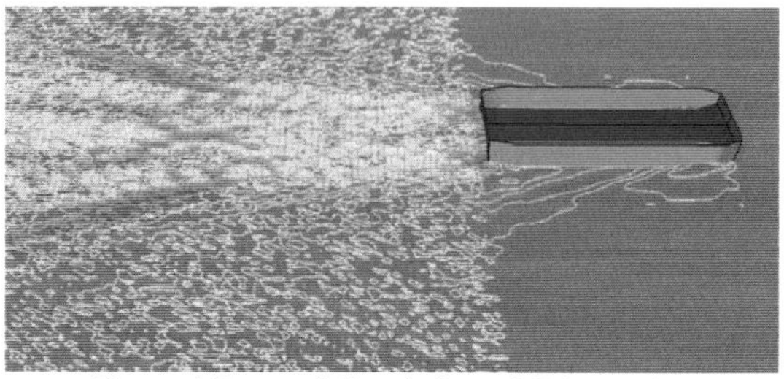

Figure 1.Form of Simulation of Ship Motion

On this ship the ship's resistant are simulated according to ship design data and analysis with maxsurf application by referring to the line plan, and the hydrostatic characteristics of the ship with the form of modeling of tug-of-warulation on the ship.
Drag simulations on ships are used to look for things like the value of ship resistant that occur when the ship moves. The value of ship resistant produced by this ship is shown in the graph of the number of ship resistant values of the ship with the speed on the ship processed data by several methods. According to the results of modeling and simulation of the application of the maxsurf ship's role is 10 Knots with a hull efficiency of 55%:

- Holtrop : 5,3 kN
- Slender Body : 4,5 kN

Average : 4,9kN

4.3 DUCTING THRUST VECTORING NOZZLE

The shape of ducting nozzle trust vectoring is made as simple as possible so that it is easy in the process of manufacturing ducting nozzle trust vectoring that will be paired on a flat plate catamaran. The shape of ducting nozzle trust vectoring is a pipe with a lenght 490 mm and has a diameter adjusted to the minimum diameter of propeller plus clearance radius.
Analysis from all the ship data, the propeller that is as similar as the flat plate catamaran tourism ship with a propeller diameter measuring 0.6 x T, is:
Key Specifications of Propeller:

- Propeller Type = B3 – 50
- D Propeller = 0,380 m
- Blade number = 3
- n (rpm) = 1461,04 rpm
- η propeller = 0,535
- BHP MCR = 17,247 Kw
- η Power = 39 %

Analysis from all the ship data, the propeller that is as similar as the flat plate catamaran tourism ship with a propeller diameter measuring 0.6 x T, is:
Key Specifications of Propeller:

- Propeller type = B3 – 50
- D Propeller = 0,417 m
- Blade number = 3
- n (rpm) = 1200,26 rpm
- η propeller = 0,546
- BHP MCR = 16,2044 Kw

- η Power = 32,9 %

Thrust Vectoring is designed from a plate with a length of 245 mm and has a width of 15 degrees from the total radius of the nozzle that has been made. Ducting Nozzle Trust Vectoring has a variation, the angle of the mouth from thrust vectoring:

- 10 degree
- 20 degree
- 30 degree

The thickness of the nozzle as well as from the Trust Vectoring nozzle designed is 5mm.

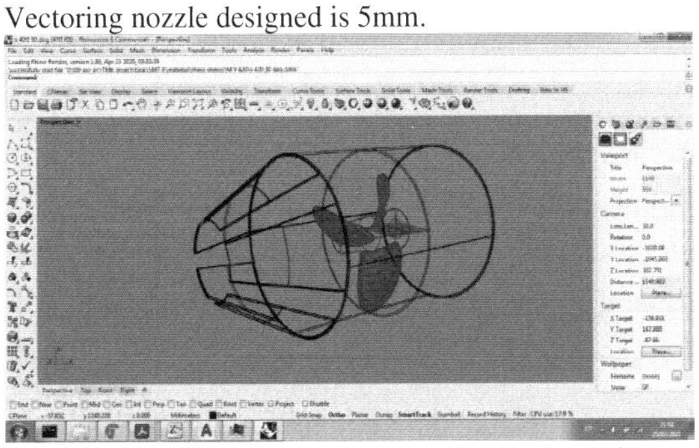

Figure 2.Modeling Nozzle Thrust Vectoring With Propeller

Ducting Nozzle Trust Vectoring design process through AutoCad 2018 and Rhino 6 applications. In the AutoCad application is used for nozzle design and Trust Vectoring.

4.4 DATA ANALYSIS

From the ansys simulation have many information about what simulation thing. In this case, use CFD simulation for ducting thrust vectoring nozzle can get various information like how to fluid flow, number velocity, what pressure happen in ducting thrust vectoring nozzle, and value of its characteristics like value of Thrust and Torque.

From the results of the simulation obtained the value of characteristics,Thrust and Torque. So that from the data can be used for the calculation of the percentage price comparison thrust and torque on each ducting thrust vectoring nozzle. The capture of simulation data is obtained from several value components that need to be taken from the results of the simulation using the Ave function calculator.

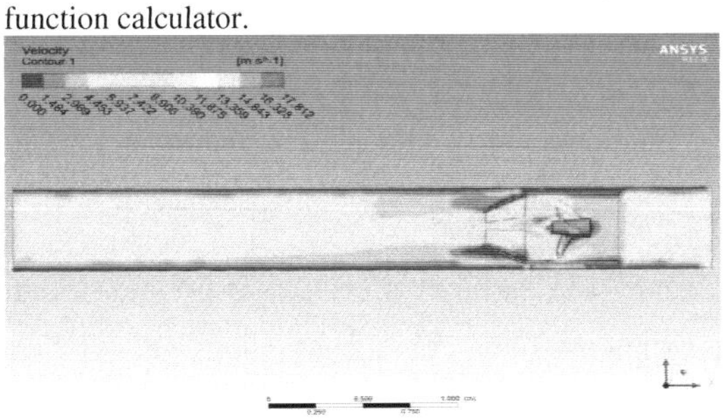

Figure 3.Velocity Ducting Thrust Vectoring Nozzle with Propeller

Figure 3 tell about how to flow direction and velocity quantity effect of propeller adding with Ducting Thrust Vectoring Nozzle. If look the figure can be known about of water velocity. After passing that, the water flow has acceleration and more focus to thrust vectoring nozzle direction. That is shown with green color.

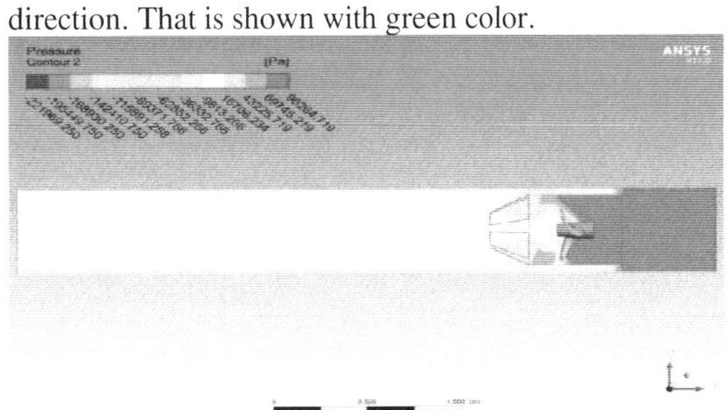

Figure 4.Fluid Pressure Ducting Thrust Vectoring Nozzle with Propeller

How the condition fluid pressure is shown at figure 4. With install it device can be increase number of pressure sourounder it. Especially before or in front Ducting Thrust Vectoring Nozzle with Propeller

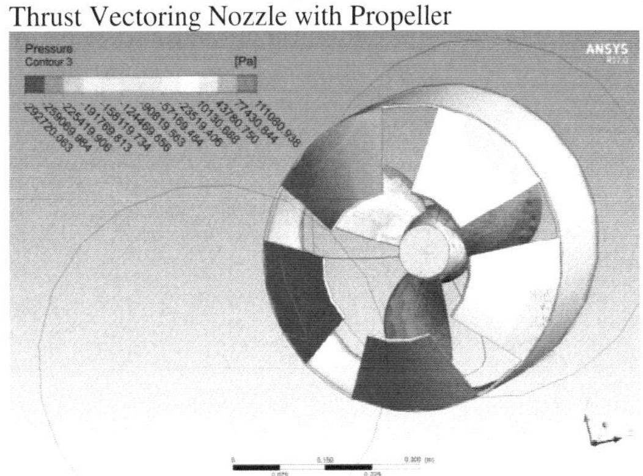

Figure 5.Pressure Ducting Thrust Vectoring Nozzle with Propeller

At the figure 5 will get where location of high pressure at body Ducting Thrust Vectoring Nozzle with Propeller. Follow the figure hight pressure find on top tip surface of propeller.

4.4 (a) THRUST

Thrust is the magnitude of thrust that is welcomed by ducting thrust vectoring nozzle that can push the fludia backwards and can eventually push the ship. This thrust is taken at the force value or force on the propeller rotation to the fluid that can propel the ship. Taken on the X axis in accordance with the direction of fluid movement i.e. moving horizontally or X axis.

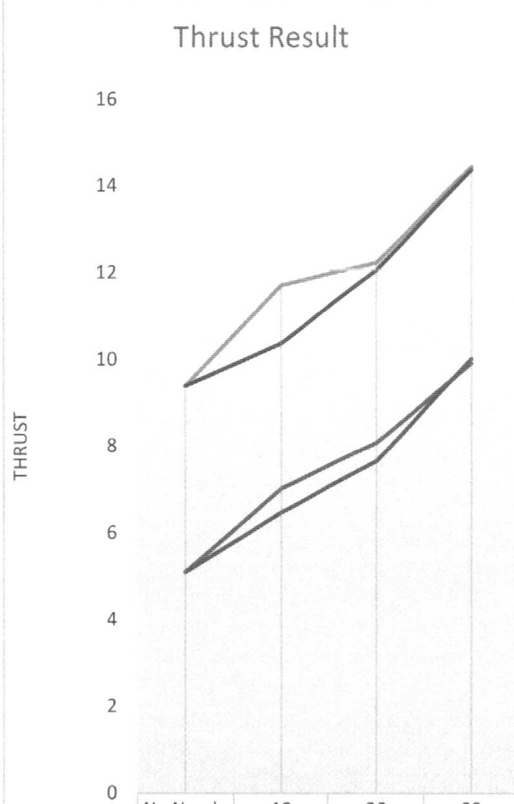

	No Nozzle	10	20	30
V420	5.0939	7.0201	8.0766	9.9112
V490	5.0939	6.4686	7.6553	10.0168
V480	9.3862	11.7103	12.2339	14.4563
V550	9.3862	10.3655	12.06	14.3855

Figure 6. Chart Results from Thrust

If you look at the Thrust graph then all Ducting Thrust Vectoring Nozzle Propeller explores the increase in the value of its thrust. The greater the degree value or nozzle the more it narrow, the lift value shows an increase. The nozzle that has the highest value is V480 30° which is 14.45 kN.

4.4 (b) TORQUE

This torque is taken at the torque value. Taken on the X axis in accordance with the direction of fluid movement i.e. moving horizontally or X axis.

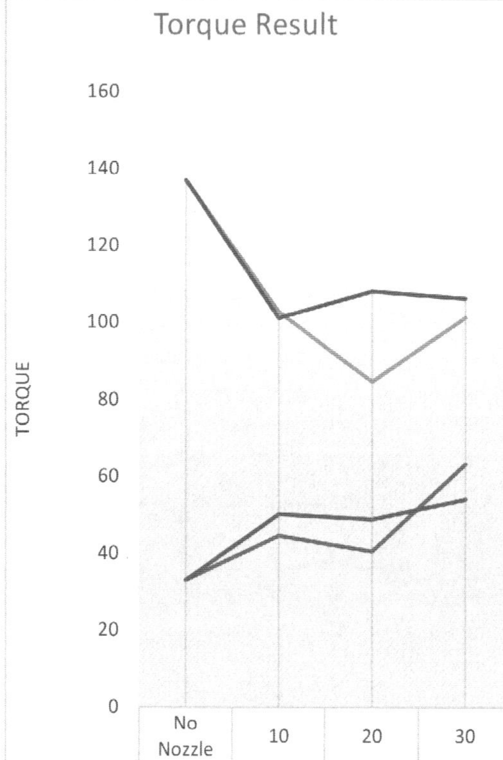

	No Nozzle	10	20	30
V420	33	44.52	40.441	63.178
V490	33	50.16	48.79	54
V480	137	102.8	84.59	101.3
V550	137	101.16	108.17	106.34

Figure 7. Chart results from Torque

If you look at the torque chart then on the nozzle V420 and V490 Ducting Thrust Vectoring Nozzle Propeller explores the increase in the value of the thrust. The greater the amount of the degree value or nozzle the more incup then the value of the elevator shows an increase. But on the nozzle V480 and V550 Ducting Thrust Vectoring Nozzle Propeller explores a decrease in the value of torque. The greater the amount of the degree value or nozzle the more in the bud then the value of the lift is getting down from the propeller without nozzle.

4.5 DATA VALIDATION

In this study to validate the results of the model test, using existing test results software. Validation is used to determine the exact boundary condition to use in boundary conditions when analyzing 12 ducting thrust vectoring nozzle models based on CFD software. Reference the model to be validated using a b-series propeller. The maximal error for validation between CFDs and the test of the magnitude of velocity or the speed of flow of simulation results with Va calculation simulation mergin error from simulation research must be less than 10%. VA simulations were taken on average at boudary outlets.

Table 1.Propeller Data Validation Calculation 0.6 x t

V boatflow	Va simulation	Margin error
5,144 m/s	5,65 m/s	8 %

Table 2.Propeller Data Validation Calculation 0.7 x t

V boatflow	Va simulation	Margin error
5,144 m/s	5,4008 m/s	4,98 %

5. CONCLUSIONS

Based on the experiments and simulations that have been done, it can be concluded as follows:

- This flat plate catamaran ship has a total resistance value of 2.83 kN with a propulsion coefficient of 61%.
- The narrower the clearance between the nozzle wall and propeller, the higher the thrust value
- The more close Thrust Vectoring, the higher the thrust value.
- The thurst value of the 12 simulation Nozzle Thrust Vectoring Propeller is the highest V 480 30° nozzle with a magnitude of 14.45 kN.
- The torque value of the 12 simulation Nozzle torque Vectoring Propeller is the highest V 550 30° nozzle with a magnitude of 108.17 kN.m
- The average thrust increase of all Ducting Thrust Vectoring Nozzles is as large as 47%.After conducting an experiment, the most effective nozzle for the flat plate boat propulsion system is the V 480 30o nozzle with the resulting value of the CFD simulation:
  Thrust = 14,45 kN
  Torque = 101,3 kN.m.

6. ACKNOWLEDGEMENTS

Thanks to Allah SWT who gave rahmad and his guidance. The author can complete research on ducting thrust vetoring nozzle propeller. This research I present to:

- My beloved mother who provides unlimited affection and big support.
- A supervisor who guides me in carrying out the research I do A supervisor who guides me in carrying out the research I do
- All friends of the class of 2017 TMK who always support me.

7. REFERENCES

1. CAHYONO, BAYU SUKMA. Analisa Teknis Penerapan Kort Nozzle untuk Kapal SPOB 4990 Dwt pada Perairan Sungai, tugasakhir Institut Teknologi Sepuluh Nopember, Surabaya, 2016.
2. SCHAEFERMEYER, RYAN M. Aerodynamic Thrust Vectoring For Attitude Control Of A Vertically Thrusting Jet Engine. Master Thesis Utah State University, Logan, Utah, USA, 2011.
3. NUR, MIFTAHUDDIN. DESAIN II – ME091318 Propeller Dan Sistem Perporosan. Institut Teknologi Sepuluh Nopember, Surabaya, 2012.
4. Huda,Nurul, 2013. Analisa pengaruh Energy Saving Device pada Propeller dengan Metode CFD, TugasAkhir, Jurusan Teknik Perkapalan, UNDIP: Semarang
5. LEWIS, E.V. Principles of naval architecture. Edition: 2nd revision" Society of Naval Architects and Marine Engineers. Library of Congress Catalog : United State of America, 1988.
6. EUROPEAN CENTRAL, VERSITA. 2013. A review of thrust-vectoring in support of aV/STOLnon-moving mechanical propulsion system. Research Gate, 2013.
7. NASA, GLENN RESEARCH CENTER. General thrust equalitionN. ational Aeronautics and Space Administration. 2021.
8. TAMPUBOLON, TRENDY S.T. Analisa Pegaruh Reaksi Jangkar Terhadap Kecepatan Dan Torsi Motor Arus Searah Shunt Dan Kompon. Falkutas Teknik, Universitas Sumatera Utara, 2016
9. HUTTON, DAVID V. Fundamental of Finite Element Analysis, Mc Graw Hill, New York, 2004.

8. AUTHORS BIOGRAPHY

Evaldi Axsendra was born in Banyuwangi Regency on September 9, 1998. It's an only child. Starting formal education at SDN 2 Kalirejo then continuing at SMPN 5 Banyuwangi and continuing his education back at SMKN 1 Glagah Banyuwangi by majoring in Electronics with Electronica Audio Video study program with a focus on strengthening signals and audio transmission and graduating in 2017.After graduating directly to study at Politenik Negeri Banyuwangi by pursuing the D4 study program - Ship Manufacturing Engineering by having a Student Identification Number 361721302036. This writer focuses on several activities of both organizations on campus such as IMAM Poliwangi and Poliwangi Hydromodeling Club. The author is also active in the competition related to the field of study studied and Scientific Papers both in the study or other fields.

A NUMERICAL STUDY ON THE EFFECT OF WATER DEPTH ON SHIP HYDRODYNAMIC DERIVATIVES

Jesswin George, Abhishek Raj and **P. Krishnankutty**, Indian Institute of Technology, Madras, India

SUMMARY

Ships behave differently in shallow water when compared to deep water. Added resistance & sluggishness to manoeuvre are the common effects of shallow water conditions. This highlights the necessity of studying ship manoeuvring characteristics in shallow water. In the present study, variation of hydrodynamic derivativeswith respect to water depth is studied. Horizontal planar motion mechanism (HPMM) under different shallow water conditions are numerically simulated using anUnsteady-Reynolds Averaged Navier-Stokes (URANS) based commercial solver.Several empirical formulas are available for predicting the shallow water derivatives, but most of them are for conventional ships. A new regression expression is proposed for estimating the hydrodynamic derivatives based on the simulated results.

NOMENCLATURE

B	Breadth of the ship(m)
C_B	Block coefficient
D	Depth(m)
LBP	Length between perpendicular (m)
m	Mass of the ship (kg)
N_r	First order derivative of yaw moment with respect to yaw rate
N_{rrr}	Third order derivative of yaw moment with respect to yaw rate
$N_{\dot{r}}$	First order derivative of yaw moment with respect to yaw acceleration.
N_v	First order derivative of yaw moment with respect to sway velocity
$N_{\dot{v}}$	First order derivative of yaw moment with respect to sway acceleration
N_{vrr}, N_{vvr}	Third order cross coupled derivative of yaw moment with respect to sway velocity and yaw rate.
N_{vvv}	Third order derivative of yaw moment with respect to sway velocity
r	Yaw rate (rad/s)
$\dot{r}$	Yaw acceleration (rad/s2)
t'	Non-dimensional time
T	Draft of the ship(m)
u	Surge velocity (m/s)
$\dot{u}$	Surge acceleration (m/s2)
v	Swayvelocity (m/s)
$\dot{v}$	Sway acceleration (m/s2)
X_{rr}	Second order derivative of surge force with respect to yaw rate
$X_{\dot{u}}$	First order derivative of surge force with respect to surge acceleration
X_{uu}	Second order derivative of surge force with respect to surge velocity.
X_{vv}	Second order derivative of surge force with respect to sway velocity
X_{vr}	Second order cross coupledderivative of surge force with respect to surge velocity and yaw rate
Y_r	First order derivative of yaw moment with respect to yaw rate
Y_{rrr}	Third order derivative of yaw moment with respect to yaw rate
$Y_{\dot{r}}$	First order derivative of yaw moment with respect to yaw acceleration.
Y_v	First order derivative of yaw moment with respect to sway velocity
$Y_{\dot{v}}$	First order derivative of yaw moment with respect to sway acceleration
Y_{vrr}, Y_{vvr}	Third order cross coupled derivative of yaw moment with respect to sway velocity and yaw rate.
Y_{vvv}	Third order derivative of yaw moment with respect to sway velocity
ψ	Yaw angle (rad)
ω	Angular frequency of HPMM (rad/s)
β	Drift angle (rad)
δ	Rudder angle (rad)
ρ	Density of water (kg m^{-3})

1. INTRODUCTION

Marine traffic in harbours, navigational channels, ports,and coastal water is increasing day by day.Manoeuvring characteristics of shipsdiffer in shallow water when compared with deep water. Different mathematical models are available for predicting ship motions at sea. Most of themuse hydrodynamic derivatives to define the hydrodynamic reaction force acting on the hull. The hydrodynamic reaction force on the hull varies with the pressure variation around the hull. When the ship is in shallow waters, the hydraulic sectional area decreases and the flow of water around its bottom and side get accelerated.This results in the variation of hydrodynamic pressure field across the hull,and thus the derivatives also vary. Hence a qualitative method for interpreting the manoeuvring characteristics in shallow waters is by studying the hydrodynamic derivatives.

The Permanent International Association of Navigation Congress (PIANC) [16] gives proper shallow water classification based on the water depth to draft ratio.

Deep water	$h/T > 3.0$
Medium deep water	$1.5 < h/T < 3.0$
Shallow water	$1.2 < h/T < 1.5$
Very shallow water	$h/T < 1.2$

In an attempt to rationalise the existing mathematical models, Manoeuvring Modelling Group (MMG) model was proposed by Ogawa et. Al. [1]. Hull, rudder and propeller are modelled as three different modules in MMG model. For container ships, Son and Nomoto [2] proposed a nonlinear mathematical model in which rudder derivatives are also incorporated.This model is widely used to simulate ship motions in the horizontal plane. Heave, pitch and roll motions are also included in Son &Nomoto model. A 6-DOF mathematical model for shallow water manoeuvre is established based on captive model tests by Delefortrie et al. [3]. The hydrodynamic reaction forces acting on the hull are modelled as functions of drift, yaw and drift-yaw correlation angles. Several empirical formulas are available to predict the shallow water hydrodynamic derivatives.Clarke [4], Ankudinov [5], Kijima [6], Hirano [7], Kobayashi [8] etc. proposed empirical relations for shallow water derivatives. But most of these equations are based onpotential flow theory and captive model testsofhull formslike tankers &containers.

Many researchers use Reynolds AveragedNavier StokesEquation (RANSE)based solvers to simulate the horizontal planar motion mechanism (HPMM) numerically. Liu et al. [9] did RANSE based simulations of pure sway mode test for DTC hull at $h/T = 1.2$. Pure sway and pure yaw tests for two different Froude numbers using DTC hull were carried out byDeng et al. [10]. For the higher Froude number case, overset mesh method was adopted. He. et al.[11] studied the effect of water depth on hydrodynamic derivatives using pure sway motion. A detailed study of KRISO container ship (KCS) for a range of water depth (h/T = 1.2, ,1.5 and 2) was conducted by Balagopaln et al. [12]. Variation of all derivatives in Son and Nomoto model with respect to water depth was reported.

For the present study, shallow water cases with h/T ratios of 1.1, 1.2, 1.3, 1.4, 1.5 1.8 and deepwater cases will be studied. Son and Nomoto mathematical model is selected for the prediction of manoeuvring characteristics. Benchmark data for HPMM test in shallow water provided by SIMMAN for KCS is used to validatethe numerical setup. And finally, a regression expression will be proposed for the double chine – Inland barge.

2. MATHEMATICAL MODEL

2.1 COORDINATE SYSTEM

Figure 1 shows the coordinate system used for the present study. Earth/Laboratory fixed coordinate system $O_E - X_E Y_E Z_E$, and a body-fixed coordinate system $o_b - x_b y_b z_b$ is defined. For the body fixed coordinate system, the positive x-axis is specified along the ship centreline in the forward direction, positive y-axis along the starboard direction and z-axis towards the bottom of the ship. The ship's velocity along x_b, y_b & z_b axis are defined as *u*, *v* and *w* respectively.

The heading angle (ψ) is defined as the angle between the ship's centreline and laboratory X_E axis. Drift angle(β) is the angle between the ship heading and the resultant velocity (*U*). The angle between the rudder centreline and ship centreline is called the rudder angle δ

.

$$\beta = \tan^{-1}\left[\frac{v}{u}\right] \quad (1)$$

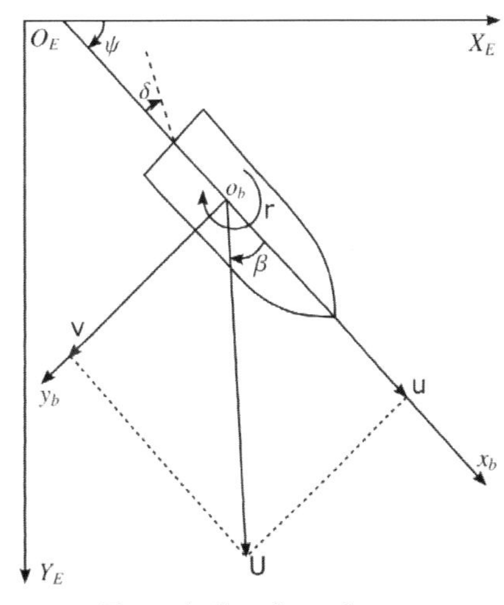

Figure 1: Coordinate System

2.2 SON - NOMOTO MODEL

Equations of motions with surge, sway and yaw degrees of freedomproposed by SON & NOMOTO [2] is used for the present study. Heave, pitch and roll motions are not considered in the current scope. The mathematical model is as follows,

$$\begin{aligned}(m - X_{\dot{u}})\dot{u} - (m - Y_{\dot{v}})vr &= X + \tau_1 \\ (m - Y_{\dot{v}})\dot{v} + (m - X_{\dot{u}})ur - Y_{\dot{r}} &= Y + \tau_2 \\ (I_{zz} - N_{\dot{r}})\dot{r} - N_{\dot{v}}\dot{v} &= N + \tau_3\end{aligned} \quad (2)$$

Where$m$ is the ship's mass.X,Y&Nincludes hydrodynamic reaction forces and moments and τ_1, τ_2, τ_3 are the control inputs.Hydrodynamic reaction forces and moments are defined as,

$$
\begin{aligned}
X &= X(u) + X_{vr}vr + X_{vv}v^2 + X_{rr}r^2 + X_P + X_R + X_{ext} \\
Y &= Y_v v + Y_r r + Y_{vvv}v^3 + Y_{rrr}r^3 + Y_{vvr}v^2 r + Y_{vrr}vr^2 + Y_P + Y_R + Y_{ext} \\
N &= N_v v + N_r r + N_{vvv}v^3 + N_{rrr}r^3 + N_{vvr}v^2 r + N_{vrr}vr^2 + N_P + N_R + N_{ext}
\end{aligned}
\quad (3)
$$

X_P, Y_P & N_P are the forces and moment from the propeller. X_R, Y_R & N_R are the control forces and moment from rudder acting on the hull. Environmental forces and moments are incorporated in X_{ext}, Y_{ext} & N_{ext}. For the present study,the effect propeller and rudder are not modelled.

3. HORIZONTAL PLANAR MOTION MECHANISM (HPMM)

For a surface vessel, hydrodynamic derivatives in the horizontal plane of motion are estimated using Horizontal Planar Motion Mechanism (HPMM). A ship model is towed with a constant carriage velocity in towing tank and oscillated in the transverse direction.Hydrodynamic derivatives are extracted from the measured forces and moments acting on the hull. There are three modes of HPMM.

3.1 PURE SWAY

In this mode of oscillation, ship's centreline is always parallel to the centreline of the towing tank. Heading angle of the ship is kept at zero degrees. All sway dependent derivatives, $X_{vv}, Y_v, Y_{vvv}, Y_{\dot{v}}, N_v, N_{vvv}$ & $N_{\dot{v}}$ are estimated from this mode of HPMM. Figure 2 shows the trajectory and orientation of the ship in the pure sway test case.

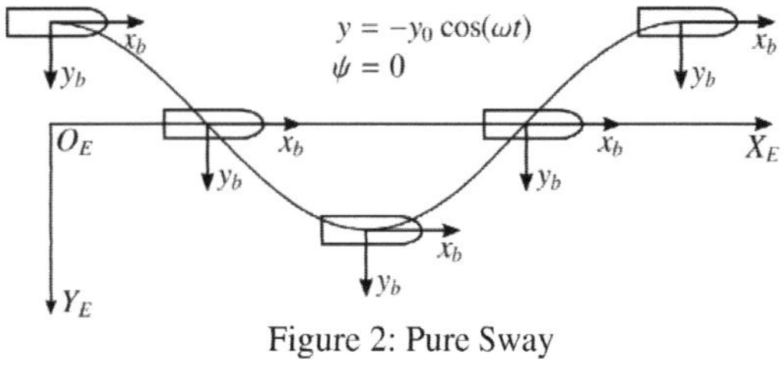

Figure 2: Pure Sway

3.2 PURE YAW

In this mode of oscillation, ship centreline and resultant velocity are always in the same direction. Figure 3 shows the sinusoidal oscillation and orientation of the ship in the pure yaw mode. Sway velocity along the y_b direction is kept zero throughout the test. This eliminates the sway dependent derivatives from the equations of motion. All yaw dependent derivatives, $X_{rr}, Y_r, Y_{rrr}, Y_{\dot{r}}, N_r, N_{rrr}$ & $N_{\dot{r}}$ are estimated from this mode of HPMM.

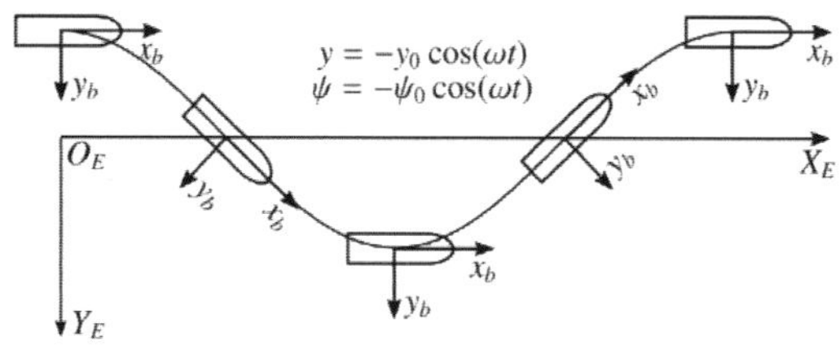

Figure 3: Pure Yaw

3.3 COMBINED SWAY AND YAW

In this mode of oscillation, an additional drift angle (β) is introduced to the pure yaw mode of HPMM. Figure 4 shows the path and orientation of the ship in combined sway and yaw mode. Coupled sway and yaw derivatives, $X_{vr}, Y_{vrr}, Y_{vvr}, N_{vrr}$ & N_{vvr} are estimated from this test.

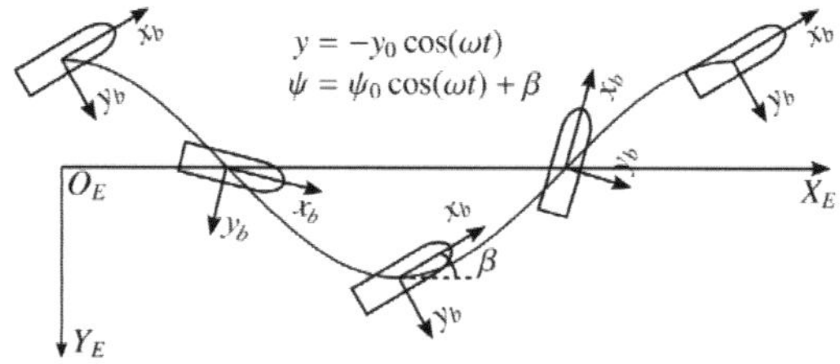

Figure 4: Combined Sway and Yaw

3.4 ESTIMATION OF HYDRODYNAMIC DERIVATIVES

Fourier series analysis is used to estimate the hydrodynamic derivatives from the forces and moments of HPMM. Instantaneous velocity and acceleration is substituted into the equations of motion. Coefficients from the mathematical model and Fourier series are equated to extract the required hydrodynamic derivatives.

4. NUMERICAL METHOD

4.1 SHIPMODEL

An Inland barge with a double chine hull is used for the CFD simulations. Table 1 shows the main particulars of Inland Barge.Figure 5 and 6 shows the perspective view of the inland barge used in CFD simulation.A model scale ratio of 1:17 is chosen. Vessel speed of 4 knots is selected based on PIANC [17] guidelines for vessel speed in navigational areas.

Figure 5: Inland Barge (View 1)

Figure 6: Inland Barge (View 2)

Table 1: Vessel Particulars of Inland Barge

Main Particulars	Full Scale	Model (1:17)
LBP (m)	49.99	3.03
LWL (m)	50.54	2.97
B (m)	9.66	0.57
T (m)	1.30	0.08
Displacement (m^3)	550.47	0.11
C_B	0.877	0.877

4.2 COMPUTATIONAL DOMAIN AND GRID

Dimensions of the computational domain are selected as per the guidelines of ITTC [13]. Along the longitudinal direction,the total length of the domain is fixed as 5.5 times ship length (*LOA*) is selected, in which 3L distance is kept aft of the ship. Three times ship's length is provided on both sides in the transverse direction. For the vertical direction, one ship length is given above the free surface. Below the free surface, domain height is limited by the *h/T* ratios used in different cases.

Unstructured hexahedral mesh is used for the present study. Accurate modelling of the boundary layer is assured using 20 prism layers with a growth rate of 1.2. The first cell thickness of the prism layer was chosen such that a y+ value lies between 0 and 1. Additional volumetric refinements are given near free surface, under keel clearance (UKC), wake regions and bow & aft regions of ship.

k-ω SST, a two-equation eddy viscosity model, is used for modelling turbulence. Other researchers working on similar manoeuvring studies [12] used the same turbulence model. Gravity model and Volume of Fluid (VOF) method are adopted for modelling free surface.Three-dimensional segregated implicit unsteady solver with a second order time discretisation scheme is adopted. The Courant number is kept below 1 in the fluid domain.Hence the time step selected is 0.005s.

4.3 BOUNDARY CONDITION

Boundary conditions are selected such that they mimic the actual physical scenario correctly. In the forward of the ship, aDirichlet boundary condition (velocity inlet) is specified. The inlet velocity is fixed as 0m/s. The aft boundary is kept far away from the ship, and it is approximated that the flow variables do not change in steady-state. Hence the working pressure is imposed on this boundary (Pressure Outlet).The top and sides walls are defined as slip walls. Ship's hull and bottom boundary are treated as a no-slip wall. The velocities at these surfaces are specified as a Dirichlet boundary condition.For the bottom bed, a tangential velocity equal to the speed of the vessel is defined in the opposite direction.

Table 2: HPMM test parameters

Test Case	Sway Amplitude (m)	Carriage Speed (m/s)	Oscillation Frequency (1/s)	Drift Angle (degrees)
Pure sway	0.25	0.499	0.054	-
Pure yaw	0.25	0.499	0.081	-
Combined sway and yaw	0.25	0.499	0.081	3

The Dynamic Fluid Body Interaction (DFBI) module available in StarCCM+is used in the present study. Heave, Pitch and Roll degrees of freedom are restrained. The oscillation frequencies for different modes of HPMM are selected as per ITTC [15] guidelines. The test matrix of HPMM is shown in table 2.

4.4 GRID DEPENDENCE STUDY

A grid dependence study is carried out to ensure that the numerical results are independent of mesh refinement. Three different meshes with a refinement ratio of $\sqrt{2}$, namely coarse, medium and fine,is selected. Peak values of pure sway simulation for *h/T*=1.2 case is compared. Let $S_{i,1}, S_{i,2}, S_{i,3}$ be the peak values for fine, medium and coarse mesh respectively. Grid Convergence ratio *R* is the ratio of change in value between medium-fine $(\varepsilon_{i,21} = S_{i,2} - S_{i,1})$ and coarse-medium $(\varepsilon_{i,32} = S_{i,3} - S_{i,2})$ solutions [14].

$$R_i = \frac{\varepsilon_{i,21}}{\varepsilon_{i,32}} \quad (4)$$

Table 3 shows the magnitude of peak values estimated using three different mesh setups. The grid convergence factor for *X*, *Y*&*N* are 0.54, 0.09& 0.65 respectively,and hence monotonic convergence is observed. For the remaining simulations,the medium mesh will be used.

Table 3: Grid dependence study results

Grid	No. of cells (millions)	*X* (N)	*Y* (N)	*N* (N-m)
Coarse	2.74	-6.6493	15.1375	7.8547
Medium	4.82	-6.3882	15.0237	8.0468
Fine	10.01	-6.2468	15.0132	8.1719

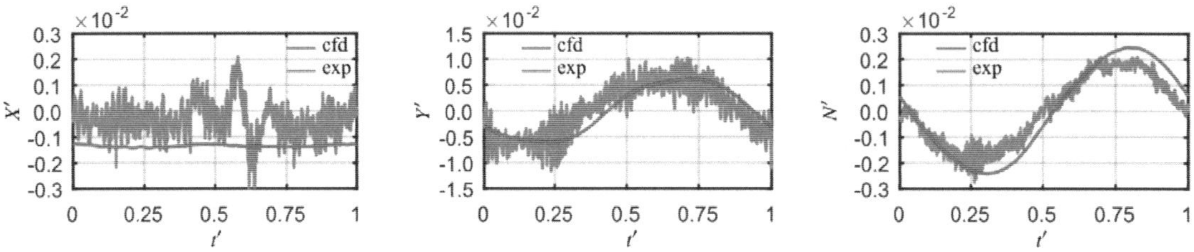

Figure 7: Comparison between CFD and experiment

5. VALIDATION STUDY

Experiment results of HPMM are unavailable for the inland barge used in numerical simulations. Hence experiments results by SIMMAN [18] for KRISO container ships (KCS) is used for validation of numerical setup.A scaled-down model with a scale ratio of 31.599 is used for experiments. Vessel parameters of KCS are shown in table 4. The pure sway test case was carried out with a carriage speed of 0.8007m/sand a water depth to draft ratio of 1.2. The maximum sway amplitude of the test case was 0.525m. The model was free to heave and pitch during the testing. Propeller is rotated at a speed of 3.84rps. However, the propeller was not modelled for the numerical simulation, and the heave, pitch degrees of freedom are disabled.Figure 7 shows the comparison between non-dimensional surge, sway forces, and yaw moment of experiment and numerical simulation.

Table 4: Vessel parameters of KCS

Main Particulars	Full Scale	Model (1:31.599)
LBP (m)	230.0	7.279
LWL (m)	232.5	7.357
B (m)	32.2	1.019
T (m)	10.8	0.143
Displacement (m^3)	52030.0	1.649
C_B	0.651	0.651

Non-dimensional time, forces and moments are calculated using the following relations,

$$X' = \frac{X}{\frac{1}{2}\rho L^2 U^2}$$
$$Y' = \frac{Y}{\frac{1}{2}\rho L^2 U^2} \qquad (5)$$
$$N' = \frac{N}{\frac{1}{2}\rho L^2 U^2}$$
$$t' = \frac{t \cdot \omega}{2\pi}$$

From figure 7, it is evident that the trend of sway force and yaw moment between numerical and experimental results are the same. However, an overestimate of surge force is observed in CFD simulations. To get the correct values of drag, Balagopalan et al. [12] used resistance test to obtain the values of $X_{u|u|}$. However, resistance tests are not in the scope of the present study.Similarly, a slight deviation in the peak values of the yaw moment is visible. This difference may be evolved due to the absence of propeller and heave, pitch motions in numerical simulations. The presence of a propeller in the aft and squat will accelerate the flow around the ship.Hence the hull experiences a higher magnitude of forces and moments on the hull.

6. RESULTS AND DISCUSSION

Pure sway, pure yaw and combined sway & yaw test cases are carried out for six shallow water cases (*h/T* = 1.1, 1.2, 1.3, 1.4, 1.5 and1.8) and a deepwater case. Variation of forces and moments with respect to water depth is shown in figure 8. A steady increase in the magnitude of forces and moments is clearly visible as the water depth decreases.

Figure 9 shows the variation of hydrodynamic derivatives with respect to water depth.An increase in the magnitude of surge derivatives, $X_{vv}, X_{u|u|}, X_{\dot{u}}$ & X_{vr} is seen as we move to lower water depths. A change in signfrom negative to positive is observed at*h/T* =1.3for X_{rr}. The negative sign of $X_{u|u|}$ accounts for the direction of drag. The added mass in the surge direction ($X_{\dot{u}}$) is negative as expected.

The sway derivatives, $Y_v, Y_{vvv}, Y_{\dot{v}}, Y_{vrr}$ & $Y_{\dot{r}}$ increases as we move to lower water depths. On the contrary, a steady decrease in the magnitude of Y_r, Y_{rrr} & Y_{vvr} is observed. Even though the magnitude of Y_{vvr} increases as we move to higher *h/T* ratios, its value for deepwater is lesser than shallow water case. A change in sign from negative to positive is observed for Y_{rrr} as we move to lower *h/T* ratio, and the transition occurs at *h/T* = 1.8. The sign of linear sway derivatives, Y_v & $Y_{\dot{v}}$ are negative as expected.

In the yaw derivatives, except for $N_{\dot{v}}$, a general trend of increase in magnitude is observed as we move to a lower *h/T* ratio.An initial increase in magnitude is observed as we move from deep to shallow water for the coupled derivative $N_{\dot{v}}$. However, from *h/T*=1.5 a steady decrease

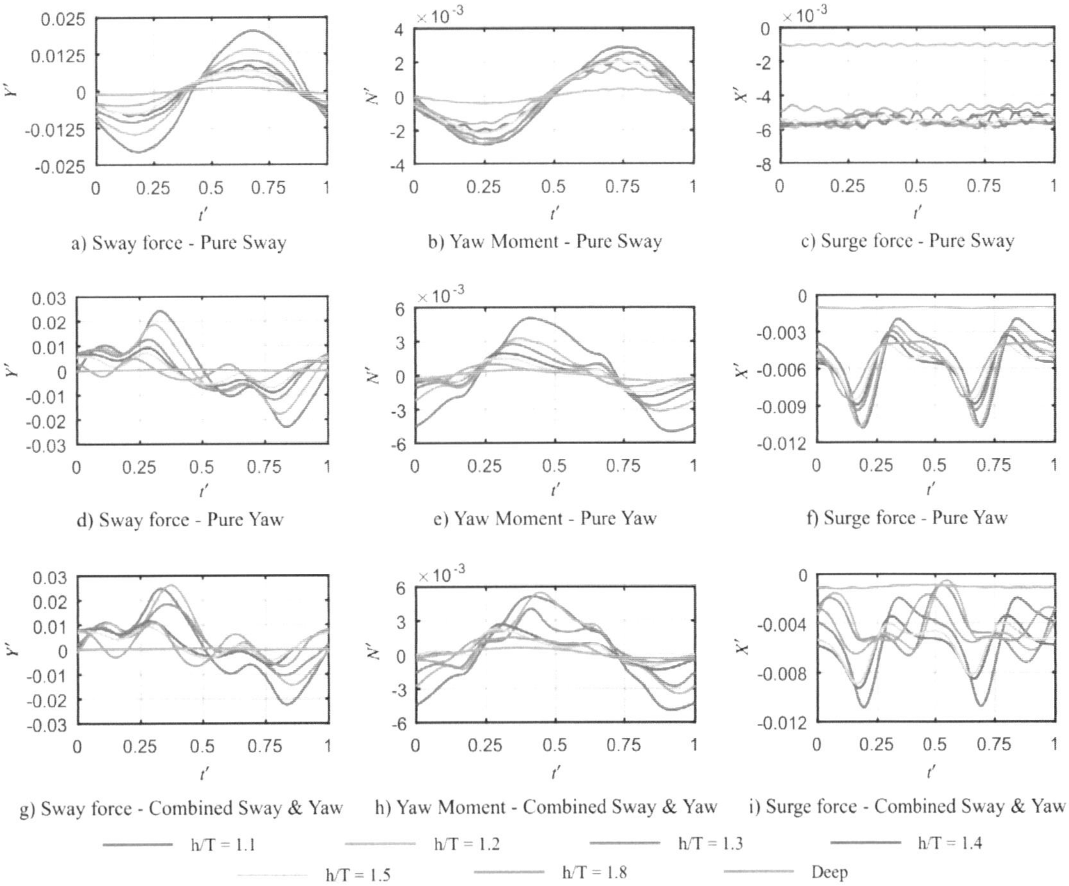

Figure 8: Forces and moments from different test cases of HPMM

in the magnitude of $N_{\dot{v}}$ is visible.Furthermore, the sign of $N_{\dot{v}}$ is positive for *h/T*=1.1 and deepwater cases, whereas, for other water depths, the sign is negative. It is to be noted that the effect of these coupled derivatives in manoeuvring motions will be more negligible when compared to other uncoupled linear derivatives like $N_r, N_{\dot{r}}$ etc. A change in the sign for deepwater case alone is observed for other coupled derivatives N_v, N_{vvr} & N_{rrr} .However, a change in the trend of their magnitude is not visible.

From the calculated hydrodynamic derivatives, a new regression expression is proposed for *h/T* between 1.1 & 1.8. Hydrodynamic derivatives $X_{u|u|}, X_{\dot{u}}, Y_r, Y_{\dot{r}}, Y_{vvr}$, $N_{rrr}, N_{\dot{v}}, N_{rrr}, N_r$ & N_{vrr} are expressed as functions of h/T and other derivatives as functions of $\frac{1}{h/T - 1}$. The proposed regression expressionsareexpressed in terms of x_1 & x_2 ,where, $x_1 = \frac{1}{h/T - 1}$ and $x_2 = h/T$.

Surge derivatives:

$$\begin{aligned} X_{vv} &= -0.003\ln(x) + 2e-05 \\ X_{u|u|} &= 0.002x_1 - 0.0083 \\ X_{\dot{u}} &= -0.0316x_2^2 + 0.1226x_2 - 0.1289 \\ X_{rr} &= -3E-05x_1^2 + 0.0006x_1 - 0.0016 \\ X_{vr} &= -0.0003x_1^2 + 0.0051x_1 + 0.0141 \end{aligned} \tag{6}$$

Sway derivatives:

$$\begin{aligned} Y_v &= 0.0004x_1^2 - 0.014x_1 + 0.0036 \\ Y_{vvv} &= -0.0013x_1^3 + 0.0285x_1^2 - 0.2034x_1 + 0.0817 \\ Y_{\dot{v}} &= -6E-05x_1^3 + 0.0012x_1^2 - 0.0092x_1 - 0.0006 \\ Y_r &= -0.0217x_2 + 0.0439 \\ Y_{rrr} &= 0.0002x_1^3 - 0.0045x_1^2 + 0.0333x_1 - 0.0333 \\ Y_{\dot{r}} &= -0.004\ln(x_1) + 0.0004 \\ Y_{vrr} &= 0.0011x_1^2 - 0.0256x_1^2 - 0.0138 \\ Y_{vvr} &= -0.2967x_2^2 + 0.5104x_2 - 0.2624 \end{aligned} \tag{7}$$

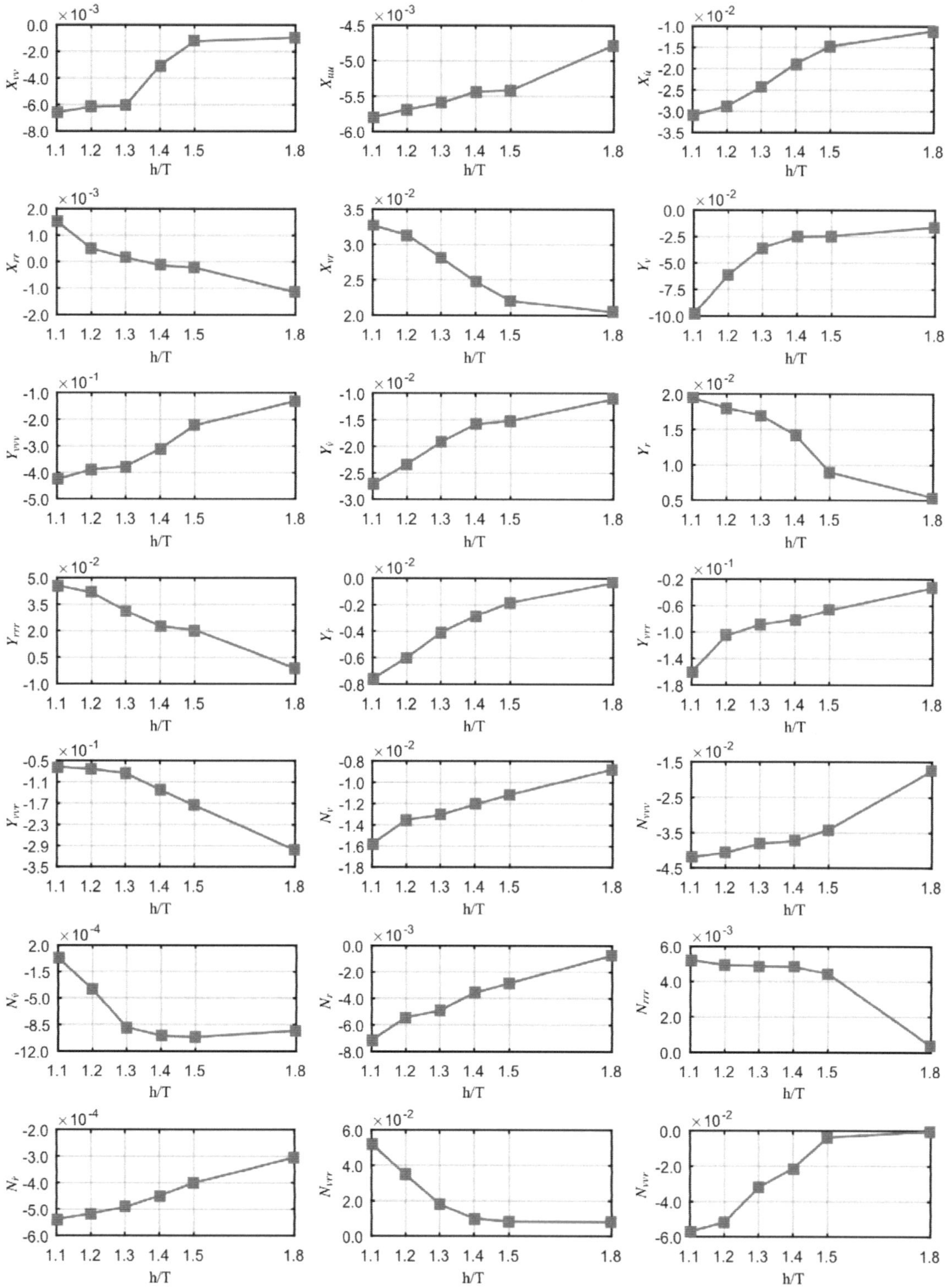

Figure 9: Variation of hydrodynamic derivatives with respect to water depth

Yaw derivatives:

$$
\begin{aligned}
N_v &= -0.0004\ln(x_1) - 0.0082 \\
N_{vvv} &= 0.0463x_2^2 - 0.1003x_2 + 0.0127 \\
N_{\dot{v}} &= -0.0073x_2^3 + 0.0366x_2^2 - 0.0605x_2 - 0.0321 \\
N_r &= -0.003\ln(x_1) - 0.0006 \\
N_{rrr} &= 0.19x_2^5 - 1.35x_2^4 + 3.57x_2^3 - 4.65x_1^2 + 2.97x_1 - 0.74 \\
N_{\dot{r}} &= 0.0003x_2 - 0.0009 \\
N_{vrr} &= -0.2067x_2^3 + 1.0779x_2^2 - 1.8609x_2 - 1.0709 \\
N_{vvr} &= 0.0015x_1^2 - 0.0239x_1 + 0.0318
\end{aligned} \tag{8}
$$

7. CONCLUSIONS

In the present study, numerical HPMM tests are carried out with an inland barge for different water depths. The effect of water depth on hydrodynamic derivatives estimated from the HPMM test is studied. Finally, a new set of regression expressions is proposed to predict the non-dimensional hydrodynamic derivatives.Since only one hull model is used, the application of regression expression to other hull models is limited.Also, the regression expression does not have parameters relating to the shape of the hull. These limitations can be overcome by extending the study to other barge models in the future. And then, a set of generalised expressions for hydrodynamic derivatives can be obtained.

8. ACKNOWLEDGEMENTS

This work is funded from the MOSH project on "Controllability of Ships in Harbours and Navigational Channels".

9. REFERENCES

1. OGAWA, A., KOYAMA, T. and KIJIMA, K., 'MMG report-I, On the Mathematical Model of Ship Maneuvering', *Bull Soc Naval ArchitJpn*, 575, pp. 22-28, 1977.

2. SON, K H. AND NOMOTO, K., 'On the Coupled Motion of Steering and Rolling of a High Speed Container Ship', *Naval Architecture and Ocean Engineering*, 150 (20), pp. 73-83, 1982.

3. DELEFORTRIE, G., ELOOT, K., HOYDON-CK, W.V. AND VANTORRE, M., 'Captive Model Tests Based 6-DOF Shallow Water Manoeuvring Model', *Proceedings of the 4th International Conference of Ship Manoeuvring in Shallow and Confined Water with Special Focus on Ship Bottom Interaction*, pp. 273-286, 2016.

4. CLARKE, D., 'The Shallow Water Effect on Linear Derivatives', *IFAC proceedings*, Vol.30(22), pp 117-122, 1997.

5. ANKUDINOV, V. K.,MILLER, E. R., JAKOBSEN, B.K. and DAGGETTT, L.L., 'Manoeuvring Performance of Tug/Barge Assemblies in Restricted Waterways', *MARSIM &ICMS proceedings*, pp 515-525, 1990.

6. INOUE, S., HIRANO, M., AND KIJIMA, K., 'Hydrodynamic Derivatives on Ship Manoeuvring', *International shipbuilding progress*, 28(321). 1981.

7. HIRANO, M., TAKASHINA, J., MORIYA, S., AND NAKAMURA, Y., 'An Experimental Study on ManeuvringHydrodynamic Forces in Shallow Water', *TWSNA*, 69,pp. 101-110, 1985.

8. KOBAYASHI, E., 'The Development of Practical Simulation System to Evaluate Ship Manoeuvrability in Shallow Water', *PRADSproceedings*,pp. 1.712-1.723, 1995.

9. LIU, Y., ZOU, Z. J. AND ZOU, L., 'RANS-Based Numerical Simulation of Captive Model Tests in Shallow Water for the DTC Container Carrier', *Proceedings of the 4th International Conference on ship manoeuvring in Shallow and Confined Water with Special Focus in Ship Bottom Interaction*, pp. 73-82, 2016.

10. DENG, G., LEROYER, A, GUILMINEAU, E., QUEUTEY, P., VISONNEAU, M., AND WACKERS, J., 'CFD simulations of PMM motion in shallow water for the DTC container ship', *Proceedings of the 4th International Conference on ship manoeuvring in Shallow and Confined Water with Special Focus in Ship Bottom Interaction*, pp. 93-98, 2016.

11. HE, R., ZHANG, Z. Z., WANG, X. Z., AND FENG, D. K., 'Numerical Simulations of the Ship Bottom Interaction of DTC Container Carrier for Different Keel Clearance in Pure Sway Motion', *Proceedings of the 4th International Conference on ship manoeuvring in Shallow and Confined Water with Special Focus in Ship Bottom Interaction*, pp. 65-72, 2016.

12. BALAGOPALAN, A., AND KRISHNAN-KUTTY-, P., 'Manoeuvring Prediction of a Container Ship on Shallow Water using Numerical Planar Motion Mechanism', *Ship technology research*, 2021.

13. Procedures IR. Guideline on Use of RANS Tools for Manoeuvring Prediction'. *Proceedings of 26rd ITTC,* 2011.

14. ITTC, 'Uncertainty Analysis in CFD Verification and Validation Methodology and Procedures', No. 7.5-03-01-01

15. ITTC, 'Captive Model Test Procedures', No. 7.5-02-06-02, 2014

16. PIANC, 'Capability of Ship Manoeuvring Simulation Models for Approach Channels and Fairways in Harbours', *Report of working group no. 20 of permanent technical committee II, Supplement to PIANC bulletin 77*, 1992

17 McBride M, Boll M, Briggs M. Harbour approach channels—Design guidelines. *PIANC Report No. 121,* 2014 Jan.

18 SIMMAN, 2014. Workshop on Verification and Validation of Ship Manoeuvring Simulation Models. https://simman2014.dk/

10. AUTHORS BIOGRAPHY

Jesswin George holds the current position of MS research scholar at Department of Ocean Engineering, Indian Institute of Technology, Madras, India.He is responsible for research on controllability of ships in harbours and navigational channels.

Abhishek Raj holds the current position of project associate at NTCPWC, Indian Institute of Technology, Madras, India. He is responsible for research on controllability of ships in shallow waters.

P. Krishnankutty holds the current position of professor at Department of Ocean Engineering, Indian Institute of Technology, Madras, India. He is responsible for teaching and research on naval architecture and marine hydrodynamics. His major fields of interest are ship manoeuvring and motions, ocean wave structure interaction, ship to ship hydrodynamic interaction and biomimetic propulsion systems.

THE DEVELOPMENT AND CHANLLENGESOF CHINA'S SHIPBUILDING INDUSTRY

Shinjun Ko, American Bureau of Shipping (China) Limited, China, and **Takeshi Shinoda,** Kyushu University, Japan

SUMMARY

China's shipbuilding industry has undergone remarkable progress since 1979 when China adopted the reform and open-up policy. This paper attempts to examine China's shipbuilding industry, analyse its rise and the key factors behind, examine the challenges facing China's shipbuilding industry, and probe the outlook of the industry. The objective is to gain a clearer understanding of the fundamental problems of China's shipbuilding industry and the actions being taken by the industry to overcome the problems. The conclusion it draws is that China's shipbuilding industry has fulfilled its key role in supporting China's continuous development; however how it can continue to flourish in the new digital and green era is yet unclear.

1. INTRODUCTION

The world has seen unprecedented rapid development achieved by China's shipbuilding Industry since China adopted the reform and open-up policy in 1979. In 2010, China surpassed South Korea and became the world's largest shipbuilding nation in terms of newbuilding output DWT. However, China's shipbuilding industry started to decline from 2011 after the 2008 worldwide financial crisis started to undercut export and depress shipping demand. In 2018, South Korea surpassed China and regained the position of the world's largest shipbuilding nation in terms of newbuilding orderbook CGT after 7 years. The rapid development of China's shipbuilding industry as well as the depression it is facing nowadays all result from key factors and have inherent root causes.

This paper introduces and examines China's shipbuilding industry through following sections: the first section introduces the phenomenal rise of China's shipbuilding industry and highlights the key factors that have contributed to the success; the second section identifies basic challenges facing China's shipbuilding industry, analyses the root causes of the problems, and examines how to overcome these obstacles; and the last section examines the current state of China's shipbuilding industry and probes the outlook of the industry under the negative impact of global financial crisis and COVID-19 pandemic.

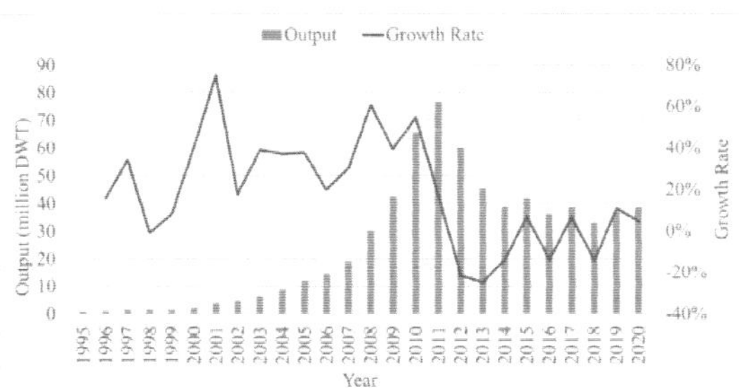

Fig 1: The output and growth rate of China's shipbuilding industry since 1995

This paper concludes that the Chinese government has assigned China's shipbuilding industry a key role in supporting China's continuous development to become a great power. However, to fulfil its role, China's shipbuilding industry will need to overcome the challenges and take decisive actions to accelerate consolidation and reduction of excess capacity and promote transformation and upgrading of the industry into a sustainable and green way of growth.

2. THE RISE OF CHINA'S SHIPBUILDING INDUSTRY

2.1 A BRIEF DEVELOPING HISTORY

Despite the inherent cyclical nature of the shipbuilding industry, China's shipbuilding industry started to grow steadily after China adopted the reform and open-up policy in 1979 and had enjoyed unprecedented rapid growth during the first decade of the 21st century with an astonishing +41.1% annual average growth rate. The output and growth rate of China's shipbuilding industry since 1995 is shown in Fig. 1. The several short periods of the negative growth rate reflect the impacts of the 1997 Asian financial crisis and the 2008 world financial crisis respectively. The output, worldwide share, and rank for China's shipbuilding industry since 1995 is shown in Table 1.

China successfully became the third largest shipbuilding country in 1997 in terms of output following South Korea and Japan and has maintained the number one position in the world in terms of output since 2010.China achieved second place in 2008 taking over Japan in terms of all output, new orders, and orders in hand standing at 28.81 million DWT, 58.18 million DWT, and 204.6 million DWT, respectively. The shares comparison of output, new orders, and orders in hand among South Korea, Japan, and China in 2008 is shown in Fig. 2. China became the number one shipbuilding nation in 2010 in terms of all three indicators, surpassing South Korea for the first time when China delivered 65.6 million DWT inoutput, acquired 75.23 million DWT in new orders, held 195.9 million DWT in orders in hand,

Table 1: The output, worldwide share, and rank for China's Shipbuilding Industry since 1995

Year	Output (Mil. DWT)	Share (%)	World Rank
1995	0.95	4.2	5
1996	1.1	4.3	4
1997	1.48	5.8	3
1998	1.47	5.8	3
1999	1.59	5.7	3
2000	2.24	5.2	3
2001	3.92	5.8	3
2002	4.61	6.6	3
2003	6.41	11.8	3
2004	8.8	14.0	3
2005	12.12	17.0	3
2006	14.52	19.0	3
2007	18.93	23.1	3
2008	30.41	30.7	2
2009	42.43	34.8	2
2010	65.6	43.6	1
2011	76.65	45.1	1
2012	60.21	42.5	1
2013	45.34	41.4	1
2014	39.05	41.7	1
2015	41.84	38.3	1
2016	36.1	35.9	1
2017	38.6	39.4	1
2018	33.1	43.7	1
2019	36.72	37.2	1
2020	38.53	43.1	1

and dominated 43.6% worldwide share.The shares comparison of output, new orders, and orders in hand among South Korea, Japan, and China in 2010 is shown in Fig. 3.China has subsequently retained this leading position so far, even though the leading edge has become very thin.

All data in this paper are collected from CANSI – China Association of the National Shipbuilding Industry, China Shipbuilding Industry statistical yearbook, and Clarkson's.

The leapfrog development of China's shipbuilding industry stemmed from the Chinese government's great ambition to become the world's leading shipbuilding

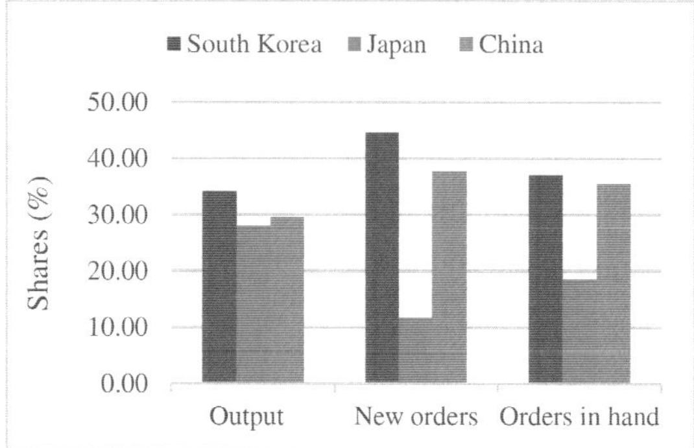

Fig 2: The shares comparison of output, new orders, and orders in hand among South Korea, Japan, and China in 2008

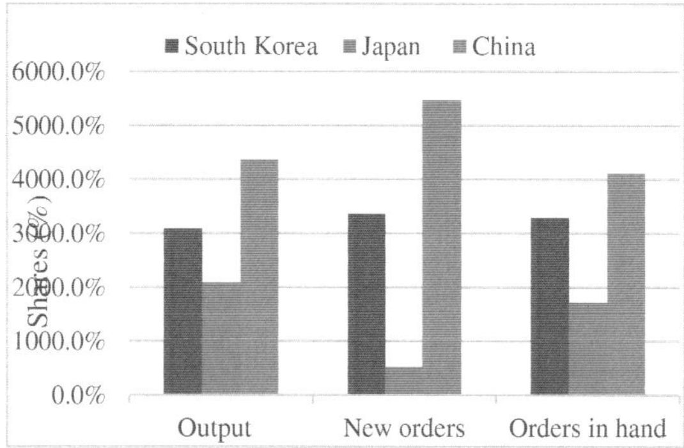

Fig 3: The shares comparison of output, new orders, and orders in hand among South Korea, Japan, and China in 2010

country and was endorsed by the intensive investment on shipbuilding facilities. As a result, China's shipbuilding industry has undergone enormous transformation along the path through repeated corporatization, splitting, newbuilding, upgrading of existing facilities, relocation, foreign investing, privatization, M&A, and consolidation, corresponding to the different eras of development.

China's shipbuilding industry began the process of corporatization in 1982, when the Sixth Ministry of Machine Building was abolished, and its assets were reconfigured into the China State Shipbuilding Corporation (CSSC). Furthermore, in 1999, the CSSC group was divided into two with the new founding of the China Shipbuilding Industry Corporation (CSIC), including all the shipyards in north China with Yangzi River as a border as an effort to encourage the competition. Since then, both CSSC and CSIC have merged and integrated some of their older shipyards. Within CSSC, Hudong merged Zhonghua and Shanghai Edward to form Hudong-Zhong, Jiangnan merged Qiuxin and then relocated to Changxin Island, and Shanghai Shipyard merged with Chengxi Shipyard. Within CSIC, Dalian Shipyard integrated with Dalian New Yard. In 2006, the Chinese government approved the plan to build

Table 2:The Chinese Shipyards' Ranking by Output in 2012

Rank	Shipyard Name	Ownership	CGT (thousand)
1	Shanghai Jiangnan Changxing Heavy Industry	China State Shipbuilding Corporation (CSSC)	1,094
2	Dalian Shipbuilding Industry	China Shipbuilding Industry Corporation (CSIC)	1,010
3	STX Dalian Shipbuilding	Foreign	704
4	Shanghai Waigaoqiao Shipbuilding	CSSC	698
5	Nantong Rongsheng Shipbuilding	Private	616
6	Jinhai Heavy Industry	Private	555
7	New Yangzijiang Shipbuilding	Private	546
8	Nantong COSCO KHI Ship Engineering	Joint Venture	452
9	Hudong-Zhonghua Shipbuilding	CSSC	424
10	Bohai Shipbuilding Heavy Industry	CSIC	416
11	Yangfang Group	Private	413
12	New Times Shipbuilding	Private	405
13	Guangzhou Shipyard International	CSSC	389
14	Tsuneishi Zhoushan Shipbuilding	Foreign	378
15	Hantong Ship Heavy Industry	Joint Venture	365
16	Jinling Shipyard	Sinotrans-CSC	344
17	Yangzhou Dayang Shipbuilding	Private	323
18	Huanghai Shipbuilding	Private	295
19	Shanghai Shipyard	CSSC	289
20	Guangzhou Longxue Shipbuilding	CSSC	288

a number of large shipbuilding facilities and expand shipbuilding capacity in the Bohai Bay region, the Yangtze River Estuary, and Pearl River Estuary to form three large scale modern shipbuilding bases. Since then,CSIC has built Bohai Shipyard, relocated and enlarged Beihai Shipyard and Dalian Shipyard around Bohai Bay region. CSSC has built Waigaoqiao in Pudong, Changxin shipbuilding base in Changxin Island at the YangtzeRiver Estuary, and Longxue Shipbuilding base at the Pearl River Estuary.

Meanwhile, new shipbuilding facilities were burgeoning around China at prime river and costal locations suitable for shipbuilding, ranging from massive shipyards with large docks and advanced building equipment, to small beach yards without proper building equipment with all forms of ownership, such as state-owned, private-owned, foreign-owned, and joint venture, etc.

From the state-owned sector, apart from construction of new mega shipyards by CSSC and CSIC, national shipping companies like COSCO, China Shipping and Changjiang also joined the play by investing shipyard groups. From the private sector, the Yangzijiang Shipbuilding Group invested New Yangzijiang, the New Century Group invested New Times, and the Nantong Rongsheng Heavy Industry was founded as the largest shipyard in the world with four ULCC size docks, followed with Zhoushan Jinhaiwan Shipyard with four different size docks. From the foreign-owned and joint venture sectors, investment from South Korea built new shipbuilding facilities like STX Dalian, DSME Weihai, and Ningbo Samsung; investment from Japan founded Tsuneishi Zhoushan, and Nantong COSCO/KHI Shipbuilding (NACKS), a joint venture between COSCO Shipping and Kawasaki HI of Japan.

The Chinese shipyards' ranking by output in 2012 is shown in Table 2. Eight shipyards including all major shipyards under CSSC and CSIC group made it to the list.

China's shipbuilding industry reached its highest peak in terms of output in 2011 when it delivered 76.65 million DWT with 45.1% worldwide share. There is no doubt that people would want to know how China could make such an achievement of rapid growth within such a short period. In the next section we will analyze the key factors that contributed to the growth of China's shipbuilding industry.

2.2 THE KEY FACTORS THAT CONTRIBUTED TO THE RAPID GROWTH

By looking back and reflecting from available information, historical data, and past experience, we can summarize the key success factors that contributed to the rapid growth of China's shipbuilding industry as follows:

- Flourishing demand due to economic growth

China's economic reform and open-up in 1979 ignited China's remarkable economic growth, which in turn created great demands for energy, raw material, food import, and thus shipping demand. The increasing shipping demand created by China's burgeoning export economy after China's World Trade Organization accession in 2001, accelerated the shipping lines expansion and thus shipbuilding boom. Furthermore, China's shipbuilding industry benefited greatly from the global demand for new ships sparked by the globalization.

- Strong support from the Chinese government

The Chinese government has played a leading role in guiding the development of China's shipbuilding industry. The Chinese central government has regarded the shipbuilding industry strategically important to its defense and economic growth. Driven by the ambitious goal to make China the largest shipbuilding country by 2015, the Chinese government has provided the full support to China's shipbuilding industry. Great efforts have been made by both central and local government at all levels across China to raise the shipbuilding capability to compete globally, especially against the other major shipbuilding countries like Japan and South Korea. A series of favourable policies and guidance have been

Table 3: The China's major shipbuilding policies in different periods

Period	Policies
1980s	Build ships for export
1990s	Chinese owned ships to be Chinese built
2006-2010	11th 5-year plan
2006-2015	Medium- & long-term development plan
2009-2011	Restructuring and revitalization plan
2011-2015	12th 5-year plan
2013-2015	3-year action plan for 12th 5-year plan
2014	Shipbuilding Industry Standard Requirements and White List
2015-2025	Made in China 2025 plan
2016-2020	13th 5-year plan
2016-2020	Action Plan for 13th 5-year plan for Deepening Structural Adjustment and Accelerating Transformation and Upgrading
2019-2021	Action Plan for Promoting Intelligent Transformation of Ship Assembly Construction
2019-2021	Action Plan for Intelligent Ship Development
2019	Special Management Measures for Foreign Investment Access and Special Management Measures for Foreign Investment Access in Free Trade Pilot Zones (Negative List)
	Industrial Investment Encouragement Catalog
	Guiding Catalog for Green Industries

issued by the Chinese central government in differentperiods during the course to support and develop China'sshipbuilding industry to make it self-sufficient as well as globally competitive, as listed in Table 3.

The policy guidance issued by the Chinese government has been playing major roles in supporting the shipbuilding industry's development. However, the effectiveness of those policies and the quantitative contribution of each policy has never been easy to define and evaluate. There lacks direct evidence of whether each policy has brought the expected result. Meanwhile some policies have backfired causing existing problems to worsen or even create new problems.

- Abundant funding and a favorable financingenvironment

China has become the country with the largest foreign exchange reserve in the world through economic growth and international trade over the years. China is also the most favourable destination of foreign direct investment. Taking the advantage of abundant funding, China has invested intensively on the development of the shipbuilding industry. In addition, the reform of China's financial market has provided a favourable financing environment for China's shipbuilding industry. Chinese shipyards enjoyed easier financing and broader financing channels such as the access to global capital markets to raise money. Furthermore, the Chinese government has provided favourable shipyards with direct financial support and preferential tax policies through government subsidies.

- Vast reserves of low-cost labor

The shipbuilding industry, different from other industries, requires intensive labour, capital, and technology at the same time. As a labour-intensive industry, low-cost labour is one of the decisive success factors for shipbuilding industry. China's per capital income is still quite low due to the unaccomplished industrialization, so China can provide the abundant supply of low-cost labour force from vast rural areas. As a result, China's shipbuilding industry maintains a greater labour cost advantage compared to Japan and South Korea due to its much lower salary.

- Huge capacity

China's shipbuilding industry enjoyed the advantage of abundant shipbuilding capacity with the construction of numerous shipbuilding facilities ranging from dozens of mega yards to thousands of small yards along the abundant costal and river sites suitable for shipbuilding. The mega yards bring the economies of scales while smaller yards provide flexible capacities responding to demand variations.

- International cooperation

China, as a latecomer, looked at Japan and South Korea as models for successful paths of development. China's shipbuilding industry actively learned and absorbed shipbuilding experiences and technology from Japan and South Korea through international cooperation and exchanges and achieved leapfrog development.

- Shipbuilding industry's cyclical wave

The world's shipbuilding industry has been shifting from high wage developed countries to low wage developing countries in history. The shift of the world's shipbuilding industry from Japan to South Korea to China coincided with the advent of the cyclical world shipbuilding boom, which contributed to the rise of China's shipbuilding industry.

3. THE CHALLENGES FACING CHINA'S SHIPBUILDING INDUSTRY

The rapid growth of China's shipbuilding industry has been accompanied with inherent problems, which also brought some unwanted problems such as overcapacity resulting from irrational expansion. Some problems and challenges are still facing the industry as of today. Here we would like to take a closer look at those challenges facing China's shipbuilding industry:

- Irrational expansion and serious excess capacity

The booming of China's shipbuilding industry in the first decade of the 21st century resulted in enormous numbers of existing yards being enlarged and new shipyards being constructed without deep thinking and careful planning, including not only the major state-owned CSSC, CSIC yards, quasi-state-owned COSCO, China Shipping yards, but also the foreign and joint venture shipyards, local provincial shipyards, and private owned shipyards. The number of shipyards in China swelled to over 2000 as per one estimation at its peak time, which was too many for people to be able to keep track. The result was low level repetitive construction of new yards, irrational expansion of the capacity, and unreasonable industry structure. Capacities were spread across large numbers of small and medium sized shipyards, including some beach yards without proper building facilities, which was far from achieving the economy of scale and efficiency. China reached an astonishing annual 66 million ton shipbuilding capacity in 2009, which was way beyond the average worldwide annual shipbuilding demand. Excess capacity and significant decline in demand of new ships led to lower utilization, over competition, lower ship prices, less revenue, and/or even loss.

China's shipbuilding industry faced the reality of excess capacity and starts to solve the issue through bankruptcies and closures, mergersand acquisitions, restructuring, consolidating, and creating mega yards to exploit economies of scales.

- Low quality workforce and lack of high-end talents

Even though China has a vast reserve of low-cost labour force from rural areas readily available to supply China's shipbuilding industry, their average education level is quite low. Low skill, a poorly trained workforce and a high turnover rate are common in Chinese's shipyards. The frontline workers' turnover rate can reach 40 percent, as per one statistic, which degrades consistency and quality. Meanwhile Chinese shipyards lack of high-end talents. High level management, technical personnel, and experienced workers are notably insufficient due to unattractive salaries. Low quality workforce and lack of high-end talents are causing low productivity, low quality, and outdated construction methods for China's shipbuilding industry, and it is also one of the reasons for orderbook slippage. Chinese shipyards' average building quality are much lower, and drydock and delivery time are much longer, compared to shipyards in Japan and South Korea.

- Rising production cost

Compared to Japan and South Korea, Chinese shipyards have a much higher raw material and energy consumption due to low management capability, outdated construction methods, obsolete equipment, etc. Meanwhile the price of steel plates as the main raw material for shipbuilding is rising due to the rising of ore and its shipping price and Chinese government's supply structure reform. In addition, Chinese shipyards are losing competitive edge with the wage inflation and currency appreciation. All of these have resulted in a rising production cost.

- Weak design, R&D, and innovation capability

Compared to South Korean and Japanese shipyards, Chinese shipyards still lag on design and R&D capabilities, and capabilities on building high-tech, high value-added products such as Cruise ships, LNG carriers, Ultra large containerships, drill ships, etc. For the LNG carrier, only CSSC group Hudong-Zhonghua shipyard in China has built series of NO96 LNGCs so far. Chinese shipyards' R&D investment is insufficient and has not been spent on innovative technology. Many small and medium sized Chinese shipyards have not developed their own design capability. Most Chinese shipyards are driven by profit to build relatively simple bulk carriers, which have resulted in similar products across different shipyards and more severe competition. Moreover, Chinese shipyards are also way behind Japanese and South Korean shipyards for the application of digital shipbuilding.

- Lagging key marine equipment development

China's shipbuilding industry still lacks the capabilities to produce the key marine equipment and major subsystems. Chinese domestic vendors still concentrate on making low-end equipment, the localization rate of ship equipment is still low, and many key equipment still rely on import. The localization rate for deck machinery, outfitting, and engines are much improved, while the rate for electrical and electronic equipment, especially the rate for communication, navigation, and automation equipment are still very low. Moreover, China-made marine equipment is perceived as low quality and lack competitiveness in the market. The dependence on import of key marine equipment not only hinder the development of China's shipbuilding industry but also kill the price advantage of Chinese shipyards.

- Slowdown of Chinese economic growth

The Chinese economy enjoyed rapid growth with 2-digit GDP growth rate for almost three decades. However, China's economic growth slowed down in recent years with much lower GDP growth rate at around 6.5% after the 2008 global financial crisis sent the world into recession. The shrinking demand on both the oil and raw material import and consumer products export resulted in decline in demand of shipping and thus shipbuilding. In addition, the Sino-US trade friction, the Brexit, and the COVID-19 pandemic has made the situation even worse, undercutting the international trading volume and depressing new shipping demands even further, and causing serious damage to China's shipbuilding industry with the depletion of new orders.

- Depressing financing environment

With the impact of the global financial crisis and the current distressed condition facing the maritime industry, many Chinese financing institutions ceased ship financing, and the remaining ones reduced the ship financing quota while raising the financing bar with stricter examination and higher interest rates. Many Chinese shipyards are now experiencing severe pressure on financial resources due to tightening cash flow, high capital costs, reduced ship prices, reduced outputs, and plummeting new orders as debt and monetary worries made effective financing increasingly difficult. Meanwhile, the orderbook slippage at some Chinese shipyards as some owners giving up or delaying existing orders have put further pressure on the shipyard's tightening cash flow. Chinese shipyards are looking for new financial resources desperately. The situation is more critical among the private sector as Chinese financing institutions almost stopped ship financing to private shipyards, which put them in a disadvantage position in competition with state owned shipyards with financial supports from the government. The financing difficulties have led to bankruptcies of some major Chinese shipyards such as Jiangsu Rongsheng Heavy Industry, Jiangsu Eastern Heavy Industries, Dayang Shipbuilding Co. Ltd, Sinopacific Offshore & Engineering Co. Ltd, Sainty Marine Co. Ltd, Nantong Mingde Heavy Industry, Yangfan Ship Group Co. Ltd, etc, and put some other shipyards on the brink of bankruptcy.

- Difficult to make profit

China's shipbuilding industry faces long existing problems of difficulties on profitability, and the COVID-19 pandemic led to increased difficulty. The pandemic caused the delayed ship delivery resulting in delayed cash return, difficulties on loans renewal, guarantee extension, and other difficulties. The demand for new ships had fallen sharply and the price of new ships decreased continuously. Meanwhile the delay of newbuilding projects and the transportation restriction due to the pandemic caused sharp price increase of raw materials, and the procurement of epidemic prevention materials further added the additional cost to the Chinese shipyards. In addition, Chinese RMB appreciated 6.9% against US dollar in 2020 from the end of 2019, and the average labour cost of Chinese shipyards increased about 15%. The difficulty for Chinese shipyards to make profit becomes more prominent.

In the next section, we want to take a look at how China's shipbuilding industry is doing nowadays under the negative impacts and challenges brought by the financial crisis, trade friction and the COVID-19 pandemic, etc., and where the China's shipbuilding industry is heading.

4. THE CURRENT STATE AND OUTLOOK OF CHINA'S SHIPBUILDING INDUSTRY

The output of China's Shipbuilding industry started to decline after reaching its peak in 2011 due to the negative

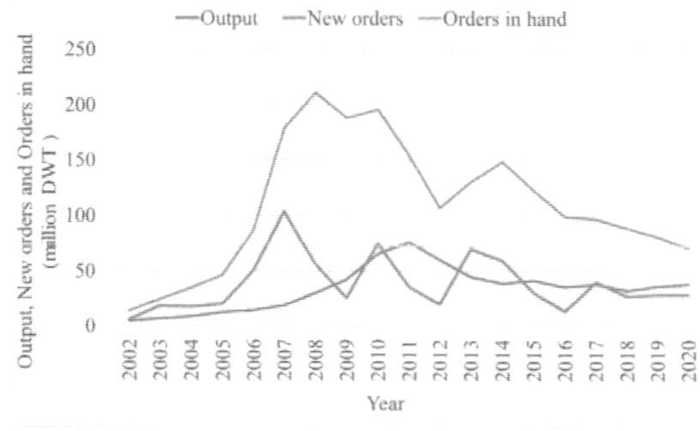

Fig 4: The output, new orders, and orders in hand for China's Shipbuilding Industry since 2002

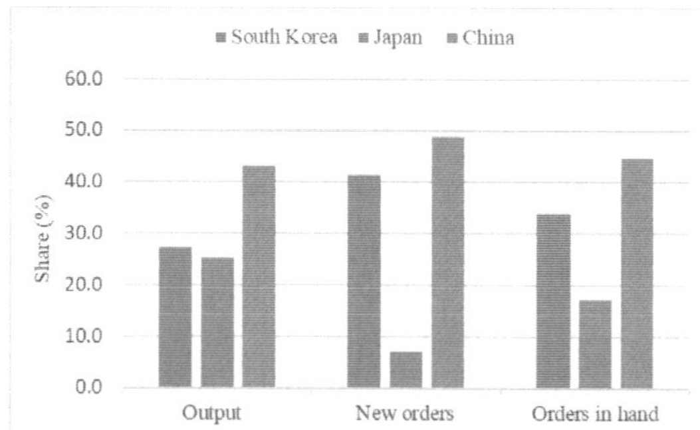

Fig 5: The share of output, new orders, and orders inhand among South Korea, Japan, and China in 2020

impact of the 2008 global financial crisis and other problems. The new orders reached its peak in 2007 while recovered to some extent in 2010, 2013, and 2017 due to the demand created by the Chinese government's various incentive policies. The orders in hand followed the similar movement to the new orders. The output, new orders, and orders in hand for China's shipbuilding industry since 2000 are shown in Fig. 4.Since the global financial crisis, the demand for new ships in the global shipbuilding market has continued to slump, and the zero to negative growth rate of China's shipbuilding industry has become the new normal.

In 2020, China's shipbuilding industry was still at a low point, and the COVID-19 pandemic had brought a significant negative impact on the economy, causing China's GDP recorded an astonishing minus 6.8 % in the first quarter of 2020. However, in the face of severe situation and great difficulties, China proactively promoted prevention and control of COVID-19 pandemic and the resumption of work, and China's shipbuilding industry took decisive measures to overcome the obstacles and difficulties associated with the resumption of work. Chinese shipyards resumed the production comprehensively and reached a production rate of 98%, and the production basically returned to normal condition. Chinese shipyards achieved good results to ensure normal production, ships delivery, new orders completion by using innovative information technology and digital tools such as cloud inspection, cloud contract signing, cloud delivery and other means. The major production and operation indicators for the whole year were better than expected.

In 2020, in the overall downturn of the global shipbuilding market, China's shipbuilding industry had maintained the leading position in the world share. The share of output, new orders, and orders in hand among South Korea, Japan, and China in 2020 is shown in Fig. 5.

Nevertheless, due to the prolonged COVID-19 pandemic and troubled world economy, the trade and shipping demand are expected to drop further. The competition among major shipbuilding countries will become fiercer. Thus, Chinese shipyards will see a further shrink of orders in hand and will face more severe difficulty in operation down the road. Meanwhile China will accelerate the consolidation of shipyards to cope with the fall of the orderbook against overcapacity and promote transformation and upgrading of the shipbuilding industry to develop intelligent and green technology and build high-end ships. Some key measures being taken by the China's shipbuilding industry to tackle the challenges are further examined here.

- Competition & Cooperation

Competition in major shipbuilding countries, especially among China, Korea, and Japan has been mounting especially after the global financial crisis and the COVID-19 pandemic, with the increasing shipbuilding capacity and decreasing shipping demand. Korean shipyards dominate high value-added high-end products such as LNG carriers, ultra large containerships, Japanese shipyards enjoy competitive edge due to strong design capability, advanced building techniques and high efficiency for conventional ships like bulk carriers, while Chinese shipyards still focus on building conventional ship types with advantage of low-cost labor. Nevertheless, Chinese shipyards are trying hard to move upward into building high-end vessels especially for the state-owned shipyards.

With the negative impact of the COVID-19 pandemic, it can be expected that competition in major shipbuilding countries will become fiercer. Chinese shipyards will face greater operating pressures with the difficulty in financing, shrink of new orders, and drop in ship prices.

Competition also brings cooperation. China and Japan have strengthened cooperation to compete jointly against South Korea using Japanese technologies and Chinese cost advantages as weapons. In 2019, China's largest private shipyard group Yangzijiang shipbuilding group joining hand with Mitsui E&S, established a joint venture shipyard Yangzi Mitsui in Taicang, Jiangsu to build bulk carriers to Mitsui designs and aim to undertake LNG carrier orders jointly and build LNG carriers with technologies provided by Mitsui. Meanwhile Kawasaki Heavy Industries ceased its merchant shipbuilding operation in Japan and moved the business to its joint venture shipyards with COSCO Shipping in China, Nantong COSCO KHI (NACK), and Dalian COSCO KHI (DACKS) respectively, including plan to provide LNG/LPG carrier related technologies to undertake LNG/LPG carriers building projects in China. Mitsubishi Heavy Industries also moved part of its marine equipment manufacturing to Chinese company Jiangsu Masada Heavy Industries under licensing.

Domestic cooperation is also on-going. One case is the cooperation among shipbuilding groups and local governments to form shipbuilding alliance, like CMHI group, CIMC group, Shandong Province, Shenzhen City, and State Development & Investment Corporation (SDIC) jointly established China National Offshore Engineering Group to absorb and manage the assets and business related to offshore engineering.

In 2020, China's shipbuilding industry strengthened cooperation in the industrial chain, and actively explored the China's domestic market. Chinese shipyards promoted in-depth cooperation among upstream and downstream industrial chain of the energy trade, transportation and equipment manufacturing, deepened cooperation with domestic financial institution shipowners, domestic trading shipowners, to jointly seize market opportunities, and achieved good results. CSSC signed three 174,000 m3 LNGCs with COSCO Shipping and two 307,000 DWT VLCC with China Merchants Energy Shipping. Many domestic shipping companies placed new orders to promote fleet upgrading and increase the domestic trade competitiveness.

- Consolidation

Since the global financial crisis, the problem of the sharp drop of new ship orders against excess capacity for China's shipbuilding industry has become increasingly prominent. The Chinese government has been continuously promoting organizational restructuring, resource integration and reduction of excess capacity through merges and acquisitions of Chinese state-owned shipyards, and bankruptcy and restructuring of small and medium private shipyards.

In 2015 both CSSC and CSIC groups began to consolidate their member yards. Within the CSSC group, Hudong-Zhonghua Shipbuilding merged Shanghai Shipyard, Guangzhou Shipyard International acquired Longxue Shipbuilding and Huangpu Wenchong Shipbuilding and became known as CSSC Offshore and Marine Engineering (Group) Company Limited. Within the CSIC group, Dalian Shipbuilding Industry merged Tianjin Xingang Shipyard, Bohai Shipyard merged

Shanhaiguan New Shipbuilding, and Qingdao Beihai Shipbuilding merged Qingdao Wuchang Shipbuilding. In 2016, with the merger of COSCO and China Shipping, the COSCO Shipyard Group and the China Shipping Industry restructured into COSCO Shipping Heavy Industry. In 2018, the China Merchants Group merged Sinotrans&CSC group, the China Merchants Industry Holdings, a shipbuilding subsidiary of China Merchants Group, acquired the shipyards under Sinotrans&CSC group such as Jinling Shipyard. In 2019, China Merchants Industry Holdings successively acquired the offshore marine business of CIMC Group and AVIC International including AVIC Weihai, AVIC Dingheng, and Deltamarine ship design company, and formed the third largest shipbuilding group in China. In addition, CSSC and CSIC groups further merged into China Shipbuilding Group, becoming the world's largest shipbuilding group.

Meanwhile, in the private sector, bankruptcy and restructuring accelerated. Large shipyards such as Jiangsu Eastern Heavy Industries, STX Dalian, Nantong Rongsheng, Zhoushan Jinhaiwan, burst one after another, and many small and medium-sized shipyards with small scales, low technology and poor returns just ceased existence without traces. The numbers of shipyards reduced from over 2000 in 2008 to around 120 in 2019.
With the continuous consolidation of shipbuilding capacity, the concentration of China's shipbuilding industry further increased. In 2020, the top 10 Chinese shipbuilding groups accounted for 70.6% output, 74.2% new orders, 68% orders in hand of the whole country, respectively.The Chinese shipyards' ranking by orderbook in 2020 is shown in Table 4. Eight major shipyards under CSSC made it to the list.

It is expected that China will be left with only 20 to 30 shipbuilding companies with active operations when the consolidation of the industry is complete. We can predict that the concentration of China's shipbuilding industry will continue to increase, which will bring monopolistic competition in the future.

- Transformation & Upgrading

China's shipbuilding industry had focused on building simpler, less complex bulk carriers due to the lack of design and R & D capabilities, the lack of building technologies and inclination to earn quick money in the past. To compete globallythe Chinese government has made an ambitious goal to move up the complexity chain and challenge South Korea and Japan by building high-end complex ships such as LNG carries and cruise ships. In September 2019, China Merchants Heavy Industry Jiangsu delivered the first vessel of 4+6 polar expedition cruise ships, which has 135 cabins and can carry 254 people, to Sun Stone Corporation of America. In October 2019, Shanghai Waigaoqiao Shipbuilding Company, a subsidiary of CSSC, started construction of China's first large-scale cruise ship, which can accommodate up to 5,246 passengers and has 2,125 guest rooms, scheduled for delivery in the second half of 2023. The CSSC group signed the contract for the construction of 2+4 135,000 gross tons Vista-class large-scale cruise ships with Carnival Corporation of America and Fincantine Group of Italy in the end of 2018.

Table 4: The Chinese Shipyards' Ranking by Orderbook in 2020

Rank	Shipyard Name	Ownership	CGT (thousand)
1	Jiangnan Shipbuilding	CSSC	1,590
2	Shanghai Waigaoqiao Shipbuilding	CSSC	1,549
3	Guangzhou Shipyard International	CSSC	1,237
4	New Times Shipbuilding	Private	1,180
5	Dalian Shipbuilding Industry	CSSC	929
6	Huangpu Wenchong Shipbuilding	CSSC	885
7	South China Shipyard	Foreign	878
8	Nantong COSCO KHI Ship Engineering	Joint Venture	834
9	Hudong-Zhonghua Shipbuilding	CSSC	811
10	New Yangzijiang Shipbuilding	Private	784
11	Qingdao Beihai Shipbuilding	CSSC	779
12	Yangzi Xingfu Shipbuilding	Private	613
13	Jinling Shipyard Nanjing	CMIH	605
14	Tsuneishi Zhoushan Shipbuilding	Foreign	561
15	Dalian COSCO KHI Ship Engineering	Joint Venture	537
16	China Merchant Heavy Industry Jiangsu	CMIH	536
17	Mawei Shipbuilding	Local Government	501
18	Chengxi Shipyard	CSSC	490
19	Nantong Xiangyu Shipbuilding and Offshore Engineering	Local Government	477
20	Yangzho COSCO Heavy Industry	COSCO Shipping	467

In 2020, China's shipbuilding industry has made new progress and breakthrough on high-tech ship research and development and construction, achieved the remarkable results of transformation and upgrading. CSSC group smoothly delivered 23,000 TEU dual-fuel super-large containerships, green and energy-saving 300,000 DWT VLCC, 18,600 M3 LNG bunkering vessels; received batch new orders of world largest 24,000 TEU ultra-large containerships, 174,000 M3 LNGC, 93,000 M3 VLGC, etc. In April 2020, Hudong-Zhonghua, a subsidiary of CSSC, won the first batch of orders for the large-scale LNG ship project for Qatar Petroleum, which was the largest shipbuilding export contract in China up to now. Hudong-Zhong defeated South Korean major three shipyards (HHI, DSME and SHI) with its strong independent research and development capabilities and the latest design of the "G4 + Changhui" series of 174,000 cubic meters LNG carries. This is a new design launched by Hudong-Zhonghuaon the basis of the successful development of the fourth-generation LNG carrier, absorbing the latest design concepts and technical trends of LNG Carriers in the world. Compared with the same size previous design, its loading capacity increased by 800 cubic meters, and the energy consumption decreased by 8%. This landmark event reflected the rise of the competitiveness of China's shipbuilding industry in the global high-end shipbuilding market.

South Korea had long occupied the world's largest market share of high-tech ships for many years with its first-mover advantage in technology, while China surpassed South Korea for the first time in 2020 in the sectors of ultra-large containerships above 23,000 TEU and VLGC. China had won global new orders for 18 out of 26 ULCS above 23,000 TEU, accounting for 69.1 % of global market share, and 10 out of 19 VLGC, accounting for 52.5% of global market share.

The Chinese government has been promoting the transformation and upgrading of China's shipbuilding industry and intensifying the investment on technology R&D, development of intelligent and green ships and shipbuilding, to create new competitive advantages and strengthen the competitiveness. Due to the continuous efforts, some of the China's technological research has reached the world's advanced level. As one of the examples, Dalian Shipbuilding Industry has developed new designs of 208,000-ton deadweight LNG fuelling bulk carrier and 260,000-ton deadweight LNG fuelling ultra large ore carrier respectively, and both have been approved by American Bureau of Shipping.

With the rapid development of science and technology, and the deepening integration of new generation information and communication technology with shipbuilding technology, the importance of technology becomes more prominent. We can foresee that China will strive to promote the development of the shipbuilding industry towards digitalization, networking, intelligence, and sustainability. The development of energy saving, safe, environment friendly eco-ship design, and intelligent ship systems will be further accelerated.

5. CONCLUSION

In sum, China's shipbuilding industry enjoyed rapid development in the first decade of the 21st century thanks to several key factors such as low-cost labour and huge demand due to the rapid economic growth, while the shipbuilding boom also left some challenges for China's shipbuilding industry such as irrational expansion of the capacity spreading across large numbers of medium and small sized shipyards, lack of own design capability causing similar products in different shipyards and severer competition, insufficient investment in R&D, low-end equipment supply from domestic venders, etc.

In order to overcome the obstacles hindering the progress of China's shipbuilding industry, the Chinese government issued various strategic policy guidance, and promoted restructuring through bankruptcy, mergers and acquisitions to form bigger shipyard groups, particularly among the state-owned shipyards, to strengthen the international competitiveness. However, it is not certain that such big state-owned shipyard groups will become truly competitive in the world market, especially under the current difficult macro situation full of risks and uncertainties resulting from the Sino-US trade friction, the Brexit, and the COVID-19 pandemic. The new difficulties are from the fall of the orderbook against overcapacity, drop in ship price, financing pressure, etc. Nonetheless, China will continue to be a major shipbuilding nation in the world market.

From the examination of China's shipbuilding industry, we draw the conclusion that the Chinese government has assigned China's shipbuilding industry a key role in supporting China's continuous development to become a great power. However, tofulfil its role, China's shipbuilding industry will need to take decisive actions to accelerate consolidation and reduction of excess capacity, and to promote transformation and upgrading of the industry. The question remaining is how China's shipbuilding industry can continue to flourish in the new era of digital, intelligent, sustainable,and green shipbuilding.

6. REFERENCES

1. Ko, Shinjun, and Takeshi Shinoda, 'The State of China Shipbuilding Industry - The Challenges Chinese Shipyards Facing', *Papers of The Japan Society of Naval Architects and Ocean Engineers, Volume 23*, November 2016.

2. Ko, Shinjun, and Takeshi Shinoda, 'An Update of China's Shipbuilding Industry', *Papers of The Japan Society of Naval Architects and Ocean Engineers, Volume 32*, May 2021.

3. Erickson, Andrew S.,*'Chinese Naval Shipbuilding: An ambitious and Uncertain Course'*. Annapolis: Naval Institute Press, 2016.

4. Homepage of China Association of the National Shipbuilding Industry, http://www.cansi.org.cn/

5. China Ship News, *'Review of 2019 reassessment 2020 - where does the world's shipbuilding industry go'*, info.chineseshipping.com.cn/cninfo/News/202002/t20200228_1333082.shtml, 28 February 2020.

6. Jiefang Daily, *'Construction of the first domestically-made large-scale cruise ship costing 5 billion yuan in Shanghai'*,

www.xinhuanet.com/local/2019-10/19/c_1125124583.htm, 19 October 2019.

7. Xinhuanet, '*The first domestic polar expedition cruise ship was successfully delivered*', www.xinhuanet.com/politics/2019-09/06/c_1124969647.htm, 6 September 2019

8. Wang, Li, Industry Watch, '*China's largest shipbuilding export order to date was born, HudongZhonghua won over 20 billion yuan*', www.shobserver.com/wx/detail.do?id=240096, 22 April 2020

7. AUTHORS BIOGRAPHY

Shinjun Ko holds the current position of senior principal surveyor atAmerican Bureau of Shipping. He is responsible for vessels new construction classification matters. His previous experience includes business manager for promoting ABS and getting new orders and district manager for survey operation at ABS.

Takeshi Shinoda is a professor of Kyushu University, Japan. He received D.Eng., M.A. and B.A. in naval architecture from Kyushu University. His main research field is equipment engineering in naval architecture, ship production system, safety engineering, sea transportation including containerized transportation and its facility, and environmental engineering concerning ship.

PROBABILITY ASSESSMENT OF CROSSING SITUATION IN SUNDA STRAIT

F I Prastyasari, Department of Urban and Environmental Engineering, Graduate School of Engineering, Kyushu University, Japan
T Shinoda, Department of Marine Systems Engineering, Faculty of Engineering, Kyushu University, Japan

SUMMARY

The new routeing measures in Sunda Strait, that consists of traffic separation scheme (TSS), precautionary area (PA), and inshore traffic zones (ITZ), have been implemented since July 2020. The Automatic Identification System (AIS) data is utilized to understand the traffic patternin this area. One dataset before and one dataset after the implementation of TSS is then converted into a vessel trip database. Eight vessel courses, two positions, and seven crossing zones are determined from the density plot to categorise the vessel trips. This study introduced a different perspective to assess the crossing situation by using three point-of-views to calculate the probability, namely crossing zone basis, course basis, and vessel type basis. The analysis result shows that the crossing situation is mostly occurred in the precautionary area of the new routeing measure.

1. INTRODUCTION

The Government of Indonesia proposed new routeing measures that consists of two traffic separation schemes (TSS), one precautionary area, and inshore traffic zones, to regulate the traffic in this vicinity. The new routeing measures were proposed to the International Maritime Organization (IMO) due to the high traffic density in Sunda Strait, as the traffic is expected to have a continuous increase in the near future [1]. Another reason is the existence of the Koliot Reefs that is also considered to be dangerous for the vessels passing through this strait due to the risk of grounding. The proposal itself has been adopted by the IMO in the COLREG.2/Circ.74 for the TSS and in the SN.1/Circ.337 for routeing measures other than TSS.

The implementation of the routeing measure in Sunda Straithas four objectives as outlined in the Decree of the Minister of Transportation no. 130 Year 2020 and those are [2]:

a. reduce the number of head-on situations to increase the safety of navigation by separating opposing traffic streams in the area,
b. reduce the grounding accident by keeping all vessel away from the coral reef atolls of Terumbu Koliot,
c. reduce the collision risk between vessels by recommending one precautionary area (PA), and
d. ensure all vessels follow the routes that free from all dangers due to the coral reef atolls of Terumbu Koliot by AToN.

The IMO have adopted the IMO Instruments Implementation Code (III Code) in 2013 and state that the coastal states have obligations and responsibility for the maritime safety, security and protection of the marine environment [3]. In response to that code, Indonesia as a coastal state that have implemented TSS in Sunda Strait has been obliged to ensure that the legislation and guidance related to the TSS is consistently implemented. In addition, a control and monitoring programme shall be carried to be able to make an evaluation about the performance of this TSS and confirm that four points mentioned as the purpose of this TSS is being achieved.

The passage of international cargo vessels and roro/passenger ferries in Sunda Strait cross each other in the precautionary area and may result to a possibility of collision. As it would endanger the human life, the environment, as well as make a great loss to the shipping company, the number of it should be reduced or eliminated, if possible, to create a safe navigation passage not only for the cargo vessels but also the roro/passenger ferries. Hence, serious injuries, fatalities, and other losses that caused by the collision accident could be avoided.

In the safety management, the Heinrich's safety triangle is known to be a triangle that can make a relation between major accidents, minor incidents and near misses. This triangle was theorized by Herbert Heinrich in 1930s and states a ratio between severe injury, minor injury and near miss to be 1:29:300, respectively [4]. A near miss condition is an occurrence that does not result in any injury, loss, or damage but has a potential to do so. While in the case of marine traffic in Sunda Strait, a near miss or crossing situation could also happen due to the crossing traffic.

This paper assesses how probable and how frequent the near miss crossing situation happened in the Sunda Strait. However, looking at the purposes of the TSSthat mainly concern about the head-on collision, this paper attempts to give a different point-of-view for the probability assessment. Instead of head-on situation, a crossing situation is chosen due to several crossing zones are found in the Sunda Strait. The analysis is performed by utilizing the Automatic Identification System (AIS) datasets (DS). There are two DS analysed in this paper. The DS 1 consists of AIS data before the implementation of TSS, specifically on October 8th-14th, 2018 and April 13th-19th, 2019, while the DS 2 contains AIS data of July 1st-7th, 2021, when the TSS has come into force. The paper is organised as follows: AIS data collection, the determination of traffic pattern into courses, steps for constructing a vessel trip database from the AIS, explanation about the theory used to determine the crossing situation and the near miss, and lastly the probability assessment based on three bases: crossing zone basis, course basis, and vessel type basis.

2. AIS DATA COLLECTION

The AIS data collection is conducted twice, before and after the implementation of new routeing measures. Automatic Identification System allows vessels to transmit information about static and dynamic data to other vessels and based stations. The International Maritime Organization (IMO) through the International Convention for the Safety of Life at Sea (SOLAS) regulation V/19 requires all ships to carry the AIS device under the following conditions:

a) Vessels with 300 GT or above that engaged in an international voyage,
b) Vessels 500 GT or above that is engaged in a non-international voyage, and
c) all the passenger vessel regardless the size [5].

The regulation also gives the detail about the information that shall be broadcasted which includes static and dynamic data. The static data shall contain the maritime mobile service identity (MMSI), vessel name, type, callsign, gross tonnage, deadweight tonnage, dimension, flag, etc. While the dynamic data consist of the location (longitude and latitude), timestamp, and the feature of vessel movement (rate of turn, speed over ground, course over ground, and heading). Despite of its potentials, the AIS comes with some limitations, for example, data redundancy, unclear data, or even some dates with no data received at all by the AIS receiver [6].

Two datasets namely DS 1 and DS 2, that have a similar number of trips, were collected for this study. However, both datasets come from a different time interval. The reason of merging the data from two different weeks is because inequal number of trips between each week. The total number of trips extracted from the AIS data in October 2018, April 2019, and July 2021 are 1687, 2041, and 3943, respectively. When the trips from October 2018 and April 2019 are combined, the result is 3728, which is almost the same as the data collected in July 2021. Figure 1(a) shows the data distribution of the DS 1, which is the combination of two weeks AIS data. While Figure 1(b) shows the data distribution of the DS 2. All data used in this study was collected from a company named AISITS. This company is a startup that provides services for vessel tracking and information system based on AIS data.

3. TRAFFIC PATTERN IN SUNDA STRAIT

The AIS data that has been collected and grouped into DS 1 and DS 2, then the traffic pattern is analysed. This is performed by using a software named IALA Waterway Risk Assessment Programme (IWRAP). IALA is short of International Association of Marine Aids to Navigation and Lighthouse Authorities. This software is recommended by IMO to be used for conducting a risk assessment.

In this stage, a traffic pattern needs to be known to categorise each trip made by every vessel. At first, the data is imported to the software and a density plot is generated. This plot shows every point of the AIS data that have been connected to make one perfect trip. The density plot of Sunda Strait area before and after the implementation of TSS is shown in Figure 2 and Figure 3. Eight courses and seven crossing zones are determined based on the density plot of Figure 2. Hence, we would be able to compare those two traffic patterns. Eight courses are coded as C1 until C8, while seven crossing zones are coded as CZ 1 until CZ 7. The importance of course and crossing zone classification is to be used in the development of the vessel trip database later.Each course and crossing zone in the Sunda Strait has its own features and characteristic, which is described in the following paragraphs.

3.1 VESSEL COURSES

Vessels that navigating in Sunda Strait could engage in different routes. This paper distinguished those courses based on the direction and location of the track. The following paragraph briefly describes the feature of all courses in Sunda Strait.

3.1 (a) Course 1 and 2 (C1 and C2)

The Course 1 and 2 are positioned in the IASL 1, which is dedicated for the international merchant vessels navigating in the Indonesia waters. These courses also where the TSS of Sunda Strait is located. Vessels engaged in a route from the Hindia Ocean to the East Asia or in the northbound direction is classed as C1. While C2 is the opposite of C1 which has a southbound direction. The type of vessels that passing these courses are general cargo and bulk carriers. When the TSS has been implemented, the traffic here is separated by a separation zone as shown in Figure 3.

3.1 (b) Course 3 and 4 (C3 and C4)

The new routeing measures in Sunda Strait also including a precautionary area to accommodate the crossing traffic made by roro/passenger ferries cruising round trip from Port of Merak to Port of Bakauheni. Based on an interview with the Officer of Road Transportation Management Center in Banten that was conducted in 2018, the number of trips per day from the Port of Merak and Bakauheni is 100 which is made by 25 vessels.There is no significant difference of the traffic pattern before and after the TSS has been set. However, a precautionary area, that is set in the crossing area between the merchant vessels with the roro/passenger ferries, has a recommendation for the traffic direction for the ferries that should be followed. This recommendation helps regulating the traffic flow of the ferries, which previously was unregulated, and reduce the collision risk.

3.1 (c) Course 5 and 6 (C5 and C6)

A heavy traffic is observed in the eastern part of Sunda Strait and parallel with Course 1 and 2. Similar with C1 and C2, these courses are also dominated by merchantvessels.However, they are mostly engaged in the domesticroute, from the Java Sea to the industrialports close to the Port of Merak. The TSS implementation does not affect the traffic pattern in C5 and C6 as no routeing measure is set in this area.

3.1 (d) Course 7 and 8 (C7 and C8)

The Course 7 and 8 are different from other courses followed by merchant vessels, because these courses span diagonally from the west part of Sumatera Island to the Java Island and return. In addition, C7 and C8 are making intersections with all courses. However, the traffic of C7 and C8 dropped significantly due to the TSS as it would contravene the TSS. The vessels engaged in Course 7 tend to follow the southern TSS first and then sail to the northeast when they are on the precautionary area. The same situation also occurred with the vessels that previously engaged in the Course 8, after the implementation of the TSS they moved their passage tothe northern TSS first and then sail southeast after the PA.

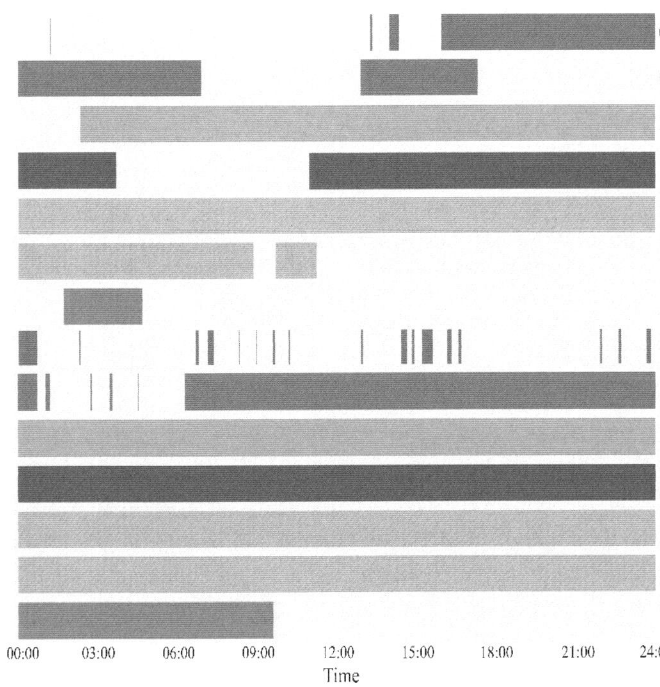

Date	The num. of AIS data	The num. of vessel	The num. of trips	The num. of ferries	The num. of ferry trips
08/10/18	14,816	130	200	56	113
09/10/18	15,085	120	206	52	133
10/10/18	27,782	136	341	55	225
11/10/18	20,351	138	296	55	181
12/10/18	34,413	154	374	61	238
13/10/18	14,360	112	204	52	145
14/10/18	3,404	90	66	43	42
13/04/19	117	26	12	18	11
14/04/19	18,461	135	290	50	192
15/04/19	27,613	139	369	50	226
16/04/19	29,577	147	408	56	247
17/04/19	27,939	144	379	53	250
18/04/19	28,253	147	396	49	254
19/04/19	11,526	113	187	41	113

(a) Data Set 1 (DS 1).

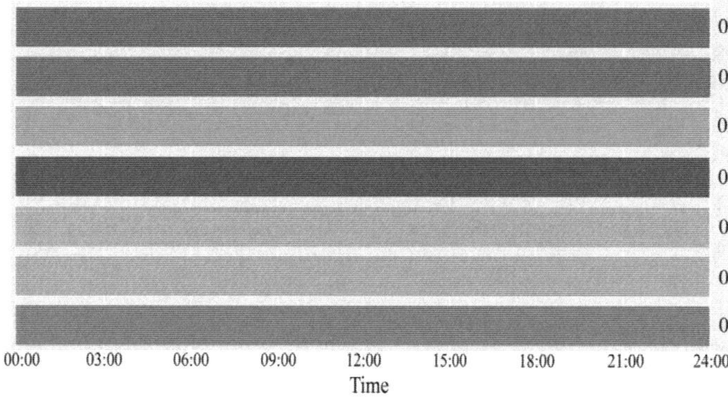

Date	The num. of AIS data	The num. of vessel	The num. of trips	The num. of ferries	The num. of ferry trips
01/07/21	172,764	236	621	48	211
02/07/21	172,158	222	566	48	171
03/07/21	163,942	228	597	50	193
04/07/21	174,162	223	526	48	191
05/07/21	166,623	232	544	48	197
06/07/21	149,064	198	553	49	225
07/07/21	158,069	203	536	47	199

(b) Data Set 2 (DS 2).

Figure 1: Data distribution of DS 1 and DS 2.

However, such vessels are categorised to be contravening the regulation as the southwest of the southern TSS is the inshore traffic zone (ITZ) area, that should be avoided by vessels having length more than 20 meter.

3.1 (e) Course 9 and 10 (C9 and C10)

Both C9 and C10 are not a passage that a vessel could follow, instead two positions that might be performed by such vessel operating nearby the port. The C9 represents the anchoring position outside the port or in the anchorage area, while the vessel labelled with the C10 means that it is inside the port and no movement found. Vessels engaged in C9 and C10 are mostly found in eastern side of C5 and C6 following the anchorage area of every ports.

3.2 CROSSING ZONES

In addition to the vessel courses described in the previous chapter, seven crossing zones (CZ) are also observed in Figure 2. A crossing zone is defined as an area where two or more courses are crossing each other. This area is considered to be the potential area where a crossing situation could happen around the Sunda Strait. The description of all crossing zones is outlined as follows:

3.2 (a) Crossing Zone 1, 2, and 3

As shown in the Figure 2, there are couple of crossing zones located close to the Bakauheni Port, where the vessel navigating in IASL 1 would have a chance to have a crossing situation. The Crossing Zone 1 (CZ 1) is the potential crossing location for vessels navigating in C1 or C2 with other vessels en route in C7 or C8, or even C3 and C4. The Crossing Zone 2 (CZ 2) is positioned in the north of CZ 1 where the traffic in C1 or C2 could have a crossing situation with ferries operating in C3 or C4. While the Crossing Zone 3 (CZ 3) is the intersection area between C3 or C4 with C7 or C8. All type of vessels has a potential to experience a crossing situation in these three CZs.

3.2 (b) Crossing Zone 4 and 5

In addition to the CZ 1,2, and 3, there are two crossing zones located in the east of the Sangiang Island and close to the Merak Port, those are Crossing Zone 4 (CZ 4) and Crossing Zone 5 (CZ 5). Both CZ 4 and CZ 5 are involving the traffic of C5 and C6 with the traffic of ferries in the C3 and C4. The only difference is that the CZ 4 is a crossing location made by ferries from Port of Merak to the west, while the CZ 5 is the opposite.

3.2 (c) Crossing Zone 6 and 7

The last two crossing zones are Crossing Zone 6 (CZ 6) and Crossing Zone 7 (CZ 7). These are made by the extension of C5 and C6 that meet with the extension of C7 and C8. These two are the only CZs that are not involving the crossing traffic of roro/passenger ferries. The difference between these two crossing zones is, the CZ 6 is the crossing location made due to the extension of C5 or C6 that is going to the northwest, while the CZ 7 is made due to the extension of C5 or C6 to the northeast.

4. METHODOLOGY

4.1 VESSEL TRIP DATABASE

A vessel trip database is a collection of data that consist of the processed AIS data and added with information related to the course where the vessel is sailing, and the crossing zones passed by each vessel. The process flow of converting the AIS database into the vessel trip analysis is shown in Figure 4.

Figure 4 shows which data from the AIS are taken to the vessel trip database as well as which data are added to it. A trip count is added to the vessel trip because it gives information about the number of trips a vessel made in a certain time interval. From the number of trips counted, each trip is assigned to a course number. However, one trip can be labelled with more than one crossing zone, because a vessel course may have more than one crossing zone along the way. Thus, all these information are stored for each unique MMSI that represent a particular vessel.

In the process of counting the vessel trips generated by each vessel, the need to validate the result of the IWRAP is emerged. As this software has an ability to connect all coordinates of a vessel, but this feature also has a time constraint (i.e., when the time between two coordinates is exceeding the time limitation, the trip is separated into two different trips). Hence, a visual observation is required to validate it. In addition to that reason, trips made by roro/passenger ferries that is generated by IWRAP can be validated with the schedule released by the operator.

The probability assessment of crossing situation in Sunda Strait is performed with three point-of-views; namely crossing zone basis, course basis, and vessel type basis.Each basis is an information included in the vessel tripdatabaseand it is very interesting to be assessed, as manypoint-of-views would be understood and the TSS can be evaluated on point.

4.2 CROSSING SITUATION AND NEAR MISS

A crossing situation is a condition when two vessels are meeting each other with an angle between 5° to 112.5° or 247.5 to 355° from the own vessel perspective as displayed on Figure 5. In this study, the crossing situation is indicated by the closest point of approach (CPA). The CPA, as shown in Figure 6, pointed out that an own vessel is at its closest location to the target vessel. This indicator has both spatial and temporal features, namely the distance to the closest point of approach (DCPA) and the time to the closest point of approach (TCPA), respectively [7].During this period, the distance between the own vessel and the target vessel reaches the minimum and the time to reach that point is known as the TCPA.

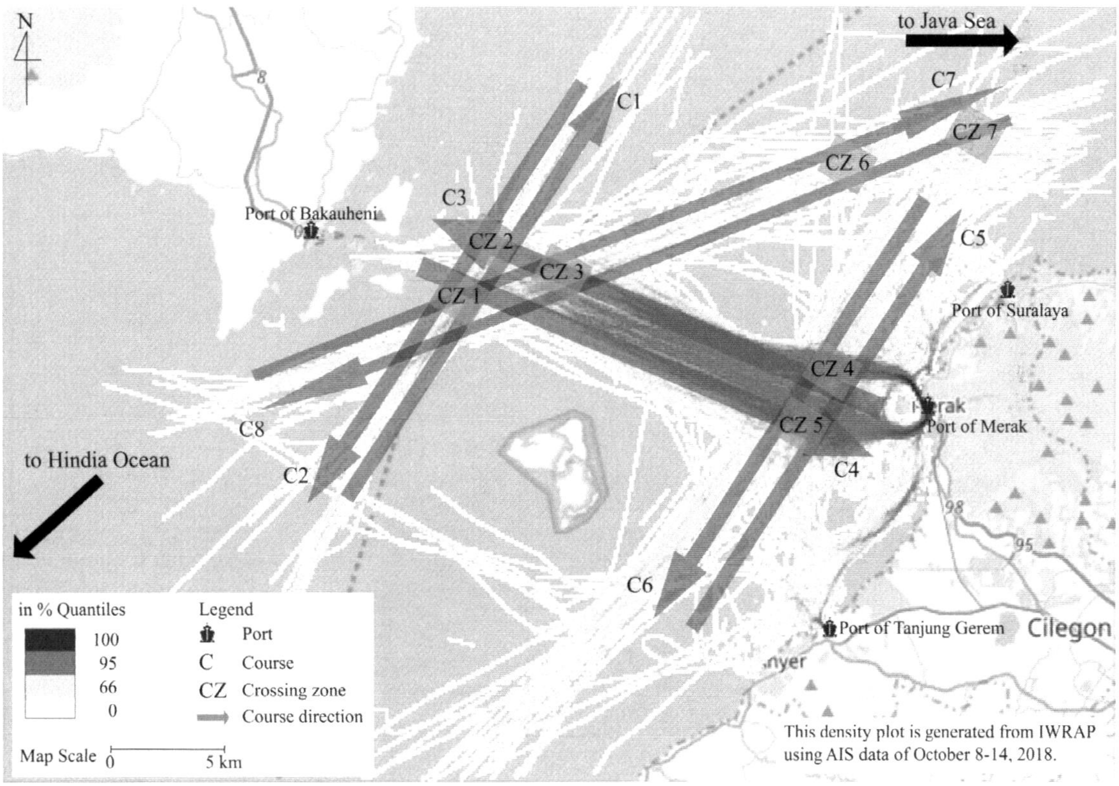

Figure 2: The traffic pattern that represents the DS 1.

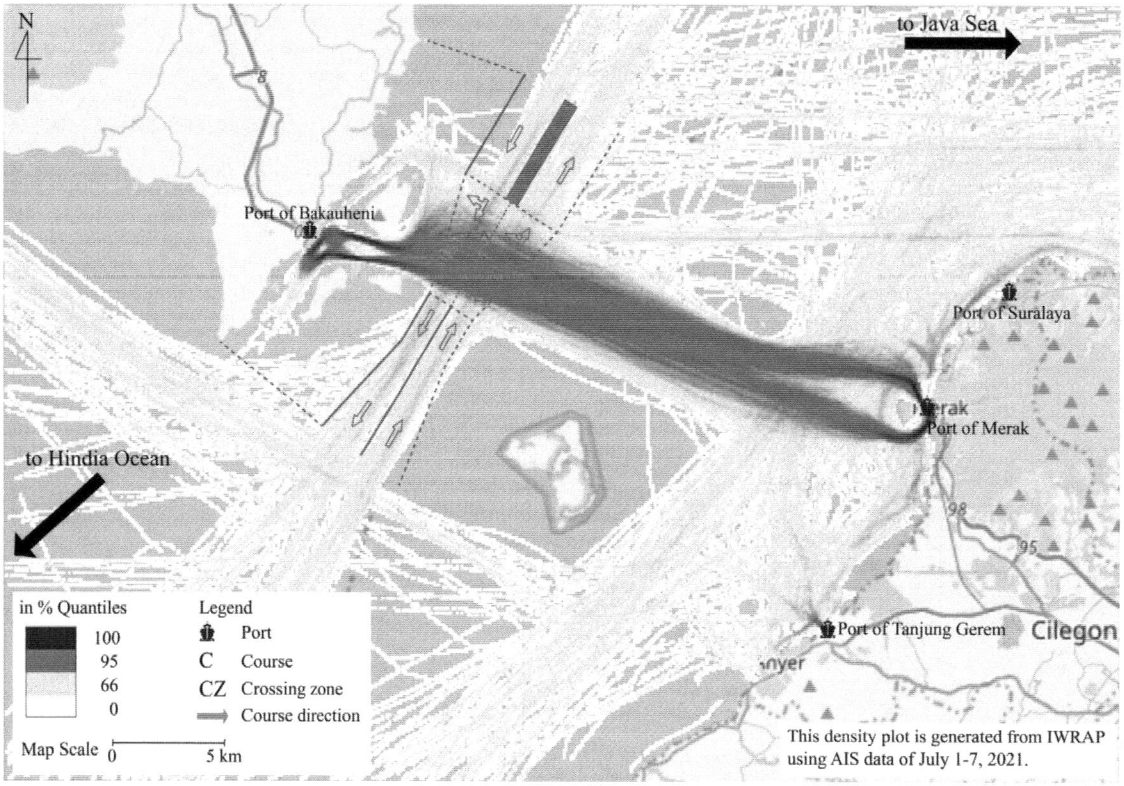

Figure 3: The traffic pattern that represents the DS 2 and the location of TSS Sunda Strait.

The CPA approach is a good methodology to be used in a non-accident probabilistic risk assessment approach as no real accident is involved in the calculation. Several research has discussed the value of CPA based on various circumstances. Research conducted by Goodwin, for example, stated that the safe DCPA is 2.35 nautical miles (nm) for the open sea [8]. Meanwhile, Park in his research found that the DCPA is 0.15 nm and the TCPA is 3 minutes for Korean waters [9]. Furthermore, research by Fukuto and Imazu proposed the value of safe DCPA as 1.0 nm and the safe TCPA is 5 minutes [10].

This study evaluates the DCPA of passing and crossing vessels in Sunda Strait over time during the time intervals previously mentioned. Among those DCPAs that have been proposed by the previous research, an interview with a deck officer of a cargo ship was conducted to get the ideaabout what the crew feels about each DCPA. The result of this interview is that the DCPA from Fukuto and Imazu [10] with the distance of 1.0 nm is taken for categorizing the vessel encounter condition, considering the condition of Sunda Strait that has a passing and crossing traffic. If two vessels are encountering each other and the closest distance between each other is less than 1.0 nm, the encounter is categorized as a crossing situation. The general formula of the DCPA calculation is involving the position and speed of both own vessel and target vessel and the equation is outlined as follows [11]:

$$D_{CPA} = \left| \frac{XV_{ry} - YV_{rx}}{V_r} \right| \quad (1)$$

$$\text{where } V_r = \sqrt{V_{rx}^{\,2} + V_{ry}^{\,2}} \quad (2)$$

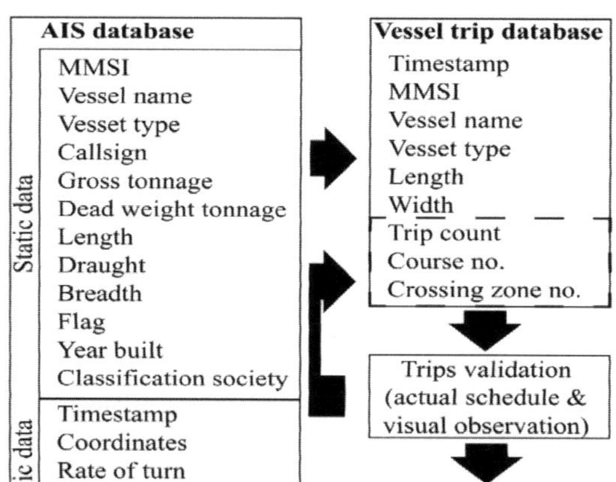

Figure 4: The process flow of generating a vessel trip database and the analysis based on it.

is the relative speed, V_{rx} and V_{ry} are relative speed components of the target vessel in x and y-axis, and all symbol and notation are depicted in Figure 5.

4.3 CROSSING SITUATION DATABASE

The International Regulations for Preventing Collisions at Sea (COLREG) in Rule 15 about crossing situation states the action that shall be taken by the giveway vessels to avoid crossing ahead the other vessel if the circumstances admit.The giveway vessel here is defined as the vessel that encounters another on her starboard side [12].

As defined in the previous chapter for defining the crossing situation, a calculation is performed by IWRAP to find the crossing situation in both datasets. The value of the maximum DCPA is set in the software as 1.0 nm. This is done to give an order to the software to calculate and indicate the encounters that have DCPA as 0.53 nm. Figure 7 shows an example of the crossing situation that has DCPA less than 1.0 nm and indicated as nearmiss situation. The situation occurred during the DS 2, when the TSS has been implemented, and it was located in the Crossing Zone 1.

All crossing situations indicated by IWRAP are collected as a database. This crossing situations database consist of all information related to those situations, namely the date, time, location, name of vessels, MMSI number, type of
vessels, heading, course over ground, speed, course, and role of the vessel based on COLREG Rule 15. Table 1 gives an example of the crossing situation database. This database is later used to carry out the probability assessment.

5. PROBABILITY ASSESSMENT

This paper discusses about the probability assessment to support the obligations for Indonesia to monitor and evaluate the implementation of the new routeing measures. A concept of risk assessment by Kristiansen, [13] is used in this paper to study the risk level of the marine traffic in Sunda Strait before and after the implementation of TSS, can be written as follows:

$$R = PC \tag{3}$$

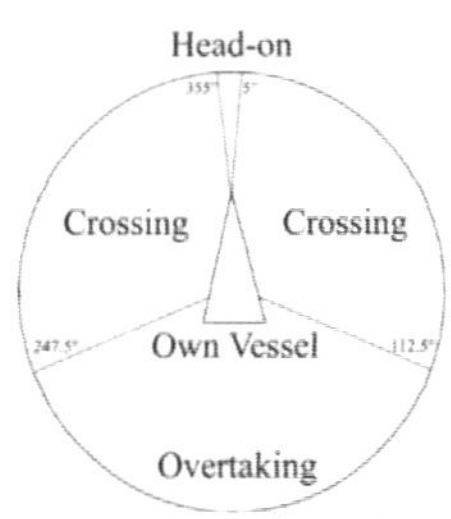

Figure 5: Type of encounters from the own vessel point of view.

In this theory, the risk (R) is affected by both probability (P) and the consequences (C). However, the term of frequency is usually used more frequently to replace theprobability to represent the number of the event in a periodof time (i.e. the number of crossing situations per year) [14]. The guidelines for Formal Safety Assessment (FSA) released by the IMO also explains about the definition of probability. IMO divides probability into two, frequentistic and Bayesian. The first one can also be defined as a relative frequency that can be stated as the ratio of events to the total number of possible events. While the second one is defined as the degree of confidence in the occurrence of an event that can be measured by the scale of 0 until 1 [15].

This paper proposed new perspectives on how to do a probability assessment of the near miss crossing situations in Sunda Strait by looking at three different bases. Thus,the probability can be viewed very specifically for every category. The assessment is utilising the frequentistic approach as outlined in the FSA guidelines by IMO [15] and it gives information about the likelihood of a trip could engage in a near miss trip. This is well estimated by calculating the relative frequency of the crossing situation occurrence in two datasets.

Instead of using the time basis meaning that calculation is based on the duration of the data was collected, the probability assessment is performed based on the number of trips on each dataset. Three bases are introduced here for probability assessment, which after that is developed into annual frequency. Those are crossing zone basis, course basis, and vessel type basis. The assessment taken the near miss information from the crossing situation database that has been developed in the previous chapter, then the number of near misses for each basis is calculated. After that, the probabilities and the annual relativefrequencyare computed by using equations 4-9.

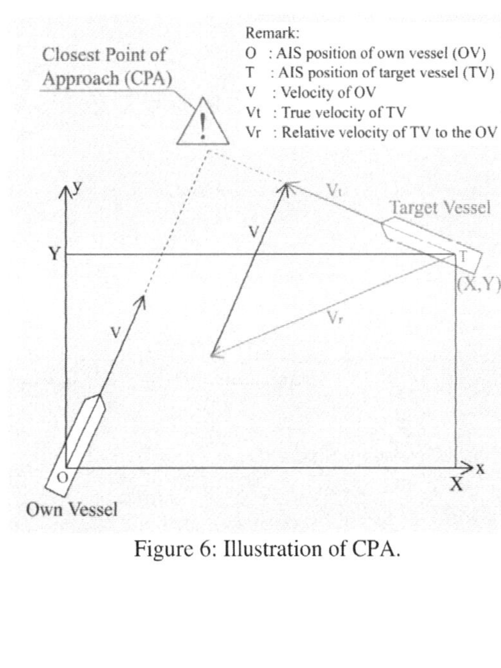

Figure 6: Illustration of CPA.

5.1 PROBABILITY OF CROSSING SITUATIONS

5.1 (a) Crossing Zone Basis

The definition of the crossing situation probability in the crossing zone basis or P_{CZ_i} is how likely a trip that crosses would experience a crossing situation and the following equation is used to computing it.

$$P_{CZ_i} = \frac{n_{CZ_i} \times 2}{t_{CZ_i}} \tag{4}$$

Where n_{CZ_i} is the number of crossing situations in CZ_i and t_{CZ_i} is the number of trips in CZ_i. The number of n_{CZ_i}in Equation 4 is multiplied by 2,becauseone crossing situation in a crossing zoneCZ_i ismade by two trips of two vessels.

5.1 (b) Course Basis

The crossing situation probability under the course basis or P_{C_i}, can be defined as the likelihood of a trip in course C_i to have a crossing situation in that course and the value of P_{C_i} is calculated by using Equation 5.

$$P_{C_i} = \frac{n_{C_i}}{t_{C_i}} \tag{5}$$

Where n_{C_i} and t_{C_i} are the number of crossing situations and the number of trips in each course, respectively.

Table 1: Example of crossing situation database.

Crossing situation ID	DS1 - 1
Date	14/04/2019
Time start (GMT)	14:46:02
Location	CZ 4
MMSI of Vessel A	525012215
MMSI of Vessel B	636014865
Type of Vessel A	Roro/Passenger ferry
Type of Vessel B	General cargo
Course of Vessel A	3
Course of Vessel B	6
Heading of Vessel A	-
Heading of Vessel B	228
COG of Vessel A	334
COG of Vessel B	227
Avg. speed of Vessel A	7.46
Avg. speed of Vessel B	11.13
Role of Vessel A	Give way
Role of Vessel B	Stand on

Theprobabilitycalculation with course basis does not need to be multiplied by 2, because each trip is already counted as 2 trips in a crossing situation.

5.1 (c) Vessel Type Basis

Lastly, the crossing situation probability under the vessel type basis or P_{V_i} can be defined as the chance of a trip made by a type of vessel V_i to have a crossing situation. The equation for calculating this probability is outlined as follow.

$$P_{V_i} = \frac{n_{V_i}}{t_{V_i}} \tag{6}$$

Where n_{V_i} and t_{V_i} are the number of crossing situation and number of trips in V_i, respectively. Similar to the calculation in the Chapter 5.1 (c), the calculation does not need to be doubled for the n_{V_i} as two vessel type is already counted in one crossing situation.

5.2 ANNUAL FREQUENCY OF CROSSING SITUATIONS

The dataset DS 1 and DS 2 only consist of two weeks and one week data, respectively. Thus, to be able to calculate the annual frequency, the number of all possible trips in a year must be estimated. As shown in the Figure 1, the data in the DS 1 is less complete compared to the DS 2, then the DS 2 is taken as the basis for estimating the total possible number of trips per year, N_t, so that the number of trips per year is the same for both dataset and it is calculated as:

$$N_t = \sum_0^i t_{C_i} \tag{7}$$

The definition of annual frequency that is being used in this study is the ratio of crossing situations to the total possible trips in a year. As the data used in this paper is a weekly data, so it is necessary to multiply the result with 52 to annualise the result. The calculation of the annual frequency for every basis are performed by using these following equations:

$$F_{CZ_i} = \frac{P_{CZ_i}}{N_t} \times 52 \tag{8}$$

$$F_{C_i} = \frac{P_{C_i}}{N_t} \times 52 \tag{9}$$

$$F_{V_i} = \frac{P_{V_i}}{N_t} \times 52 \tag{10}$$

Equation 8, 9, and 10 are used to calculate the annual frequency of crossing zone basis, course basis, and vessel type basis.

5.3 FREQUENCY INDEX

To be able to evaluate the risk due to the existance of crossing zones in Sunda Strait, risk acceptance criteria are required. These criteria are written in the International Maritime Organization Maritime Safety Committee (IMO MSC) 72/16 about the Formal Safety Assessment (FSA) decision parameters which include the risk acceptance criteria [15].

This study performed a probability assessment that in the IMO MSC 72/16 is outlined in the Annex 3 as shown in Table 2. The level of frequency or the probability is divided into four levels, frequent, reasonably probable, remote, and extremely remote. The definition of each level is also explained clearly.

5.4 CALCULATION RESULTS

5.4 (a) Crossing Situations Summary

There are two datasets analysed in this paper. The DS 1 consists of two weeks data, ranged from October 8th-14th, 2018 and April 13th-19th, 2019, while the DS 2 contains data from July 1st until 7th, 2021.After carrying the trips validation through visual observation and the conformity with the actual schedule of roro/passenger ferries, the number of trips is 2771 and 2668 for the DS 1 and DS 2, respectively. Furthermore, the total number of crossing situations before the implementation of TSS or the DS 1 is 45, while after the setting of TSS or the DS 2 is 43.

5.4 (b) Probability and Annual Frequency of the Crossing Situation

The probability assessment is divided into three parts based on their basis and they are crossing zone basis, course basis, and vessel trip basis. The result of each basis is shown in Table 3-5, respectively. The results are obtained by following the Equation 4-10.

The traffic in Sunda Strait that passes the CZ 1, CZ 2, and CZ 3 experienced a significant increase. These three crossing zones are located in the precautionary area of the new routeing measure. However, the number of trips that passes the CZ 4, CZ 5, and CZ 6 faced a slight decrease. The implementation of the TSS around CZ 1, CZ 2, and CZ 3 are expected to reduce the number of head-on collisions, instead a higher number of crossing situations are observed in the DS 2, except for CZ 3. Although the number of crossing situations is doubled in DS 2, but the probability in the CZ 1 and CZ 2 have increased 19% and 7% compared to the probability in the DS 1. In addition, the probability of the crossing situation in the CZ 5 also experienced a significant increase and it is expected due to the implementation of the TSS in the western side of the Sunda Strait and that could make some vessels move their passage plan to the eastern part of Sunda Strait to avoid the TSS.

Table 2. The frequency index (FI).

FI	Frequency	Definition	F (per ship year)
7	Frequent	Likely to occur once per month on one ship	10
5	Reasonably probable	Likely to occur once per year in a fleet of 10 ships	1.0E-01
3	Remote	Likely to occur once per year in a fleet of 1000 ships	1.0E-03
1	Extremely remote	Likely to occur once in 10 years in a fleet of 1000 ships	1.0E-05

Table 4 shows the calculation result of the course basis. The number of trips engaged in the Course 1, Course 2, Course 5, Course 6, and Course 7 increased after the implementation of TSS. However, it is also followed by the higher probability of crossing situations, especially in Course 1, Course 2, and Course 4. This result is in a good agreement with the result from the crossing zone basis, because the Course 1 and Course 2 are contributing to the crossing situations occurred in CZ 1 and CZ 2. At the same time, the Course 4 is contributing to the crossing situations happened in the CZ 5.

The calculation results of the vessel type basis are also supporting results of both crossing zone basis and course basis. The probability of crossing situations in the DS 2 is increased compared to DS 1 for a several types of vessels. Those types are container, fishing vessel, general cargo, oil product tanker, roro/passenger ferries, and support vessel. As the types of vessels engaged in the CZ 1, CZ 2, and CZ 5 are cargo vessels and roro/passenger ferries that sailing in the Course 1, Course 2, and Course 4.
The frequency index outlined in the Table 2 is used to classify the annual frequency and proceed any measures to lower it, if necessary. The annual frequency of the crossing situations shown by Table 3-5 are falls somewhat in between FI 1 and FI 5. None of the category fall in the FI 7 or frequent level.

6. CONCLUSION

This studyperformed a probability assessment for crossing situations in Sunda Strait. The significance of this research is to evaluate the crossing situation before and after the implementation of new routeing measures. TheTSS that has been setting is focused to reduce the head-oncollision, while this research offers a different perspective which is targeting the crossing situation. The analysis is performed to two datasets and using three different bases to assess the probability and the annual frequency of crossing situations. There are eight courses (Course 1-8), two positions (Course 9 and 10), seven crossing zones (CZ 1-7), and twelve vessel types are included in this assessment.

The analysis result shows that the crossing situation are mostly occurred in the precautionary area of the new routeing measure. Based on the crossingsituationdatabase, many crossing situations happenedwhen two vessels are contravening the COLREG Rule 15 about the crossing situation. In addition, although several categories

Table 3. The result of probability and annual frequency for the crossing zone basis.

Crossing zone	The num. of trips	The num. of crossings	Probability	Annual frequency
		Dataset DS 1		
CZ 1	432	3	1.39E-02	2.71E-04
CZ 2	234	1	8.55E-03	1.67E-04
CZ 3	510	4	1.57E-02	3.06E-04
CZ 4	1221	29	4.75E-02	9.26E-04
CZ 5	868	7	1.61E-02	3.14E-04
CZ 6	122	0	0.00E+00	0.00E+00
CZ 7	171	0	0.00E+00	0.00E+00
		Dataset DS 2		
CZ 1	845	7	1.66E-02	3.23E-04
CZ 2	876	4	9.13E-03	1.78E-04
CZ 3	796	0	0.00E+00	0.00E+00
CZ 4	1040	19	3.65E-02	7.12E-04
CZ 5	718	12	3.34E-02	6.51E-04
CZ 6	79	0	0.00E+00	0.00E+00
CZ 7	171	0	0.00E+00	0.00E+00

Table 4. The result of probability and annual frequency for the course basis.

Vessel course	The num. of trips	The num. of crossings	Probability	Annual frequency
		Dataset DS 1		
Course 1	44	0	0.00E+00	0.00E+00
Course 2	43	2	4.65E-02	9.07E-04
Course 3	869	32	3.68E-02	7.18E-04
Course 4	882	12	1.36E-02	2.65E-04
Course 5	184	17	9.24E-02	1.80E-03
Course 6	189	19	1.01E-01	1.96E-03
Course 7	64	4	6.25E-02	1.22E-03
Course 8	67	2	2.99E-02	5.82E-04
Course 9	241	0	0.00E+00	0.00E+00
Course 10	188	0	0.00E+00	0.00E+00
		Dataset DS 2		
Course 1	95	5	5.26E-02	1.03E-03
Course 2	97	6	6.19E-02	1.21E-03
Course 3	730	21	2.88E-02	5.61E-04
Course 4	732	20	2.73E-02	5.33E-04
Course 5	272	16	5.88E-02	1.15E-03
Course 6	269	15	5.58E-02	1.09E-03
Course 7	73	1	4.55E-02	8.86E-04
Course 8	57	0	0.00E+00	0.00E+00
Course 9	259	0	0.00E+00	0.00E+00
Course 10	108	0	0.00E+00	0.00E+00

Table 5. The result of probability and annual frequency for the vessel type basis.

Type of vessel	The num. of trips	The num. of crossings	Proba-bility	Annual frequency
		Dataset DS 1		
Bulk carrier	156	15	9.62E-02	1.87E-03
Chemical tanker	31	6	1.94E-01	3.77E-03
Container	21	0	0.00E+00	0.00E+00
Crude oil tanker	23	2	8.70E-02	1.69E-03
Fishing vessel	1	0	0.00E+00	0.00E+00
Gas tanker	40	2	5.00E-02	9.75E-04
General cargo	90	7	7.78E-02	1.52E-03
Oil products tanker	203	9	4.43E-02	8.64E-04
Roro/passenger ferry	1993	45	2.26E-02	4.40E-04
Pleasure vessel	0	0	0.00E+00	0.00E+00
Support vessel	164	0	0.00E+00	0.00E+00
Other vessel	49	2	4.08E-02	7.96E-04
		Dataset DS 2		
Bulk carrier	127	5	3.94E-02	7.67E-04
Chemical tanker	39	1	2.56E-02	5.00E-04
Container	24	2	8.33E-02	1.62E-03
Crude oil tanker	36	0	0.00E+00	0.00E+00
Fishing vessel	29	2	6.90E-02	1.34E-03
Gas tanker	87	3	3.45E-02	6.72E-04
General cargo	113	11	9.73E-02	1.90E-03
Oil products tanker	189	15	7.94E-02	1.55E-04
Roro/passenger ferry	1674	41	2.45E-02	4.77E-04
Pleasure vessel	4	0	0.00E+00	0.00E+00
Support vessel	319	3	9.40E-03	1.83E-04
Other vessel	27	1	3.70E-02	7.22E-04

in each basis experienced an increase on the probability of the crossing situations, but the total annual frequency in the DS 2 is lower than in the DS 1 at all bases. As for the PA, although it has arecommended traffic direction, but there is no regulation that set about which vessel has the priority to pass. The rules said that the vessel should navigating in this area should follow the COLREGS. Hence, a more detailed regulation regarding the PA could be taken as a measure to reduce the crossing situation that leads to a crossing collision and further research can be performed to assess the effectiveness of this new measure.

7. REFERENCES

1. Government of Indonesia, 'Routeing Measures and Mandatory Ship Reporting Systems', *IMO NCSR 6/3/3*, 2018.

2. Ministry of Transportation, 'Ministerial Decree of Transportation Ministry of Republic of Indonesia no. 130', 2020.

3. IMO, 'IMOInstrumentsImplementation Code (III Code)', *A 28/Res.1070*, 2013.

4. HEINRICH, H. W., 'Industrial Accident Prevention. A Scientific Approach', 1941.

5. IMO, 'Regulation V/19: Carriage Requirements for Shipborne Navigational Systems and Equipment', *International Convention for the Safety of Life at Sea (SOLAS)*, 1974.

6. EMMENS, T., AMRIT, C., ABDI A., GHOSH M., 'The Promises and Perils of Automatic Identification System Data', *Expert Systems with Applications vol. 178, p. 114975*, 2021.

7. VAN IPEREN, E., 'Classifying Ship Encounters to Monitor Traffic Safety on the North Sea from AIS Data', *TransNav vol. 9, no. 1, p. 51-58*, 2015.

8. GOODWIN, E. M., 'A Statistical Study of Ship Domains,' *Journal of Navigation vol. 28 no. 3, p. 328-344*, 1975.

9. PARK J. W., 'A Study on the Estimation of a Safe Speed for Prevention of Ship's Collision in VTS Area', *Mokpo Maritime University Korea*, 2014.

10. FUKUTO J. and IMAZU H., 'New Collision Alarm Algorithm using Obstacle Zone by Target (OZT)', *IFAC Proceedings Volumes vol. 46, no. 33, p. 91-96,* 2013.

11. LENART A., 'Approach Parameters in Marine Navigation – Graphical Intrepretations', *TransNav vol. 11, no. 3, p. 521-529*, 2017.

12. IMO, 'Part B: Rule 15', Convention on the International Regulationsfor Preventing Collisions at Sea 1972, 2018.

13. KRISTIANSEN S., 'Maritime Transportation: Safety Management and Risk Analysis', *Routledge*, 2013.

14. ČORIĆ M., MANDŽUKA S., GUDELJ A., LUŠIĆ Z., 'Quantitative Ship Collision Frequency Estimation Models:A Review', *Journal of Marine Science and Engineering*, 2021.

15. IMO, 'Revised Guidelines for Formal Safety Assessment (FSA) for Use in the IMO Rule-Making Process', *MSC-MEPC.2/Circ.12/Rev.2*, 2018.

8. AUTHORS BIOGRAPHY

Fadilla Indrayuni Prastyasariis a doctoral degreestudent in the Department of Urban and Environmental Engineering, Kyushu University, Japan. She received her Master of Science from the Department of Marine Engineering, Newcastle University, UK in 2016. She is also a part of the Center of Excellence of Maritime Safety and Marine Installation in ITS Surabaya, Indonesia.

TakeshiShinodais a professor at Kyushu University, Japan. He received D.Eng., M.A. and B.A. in naval architecture from Kyushu University. His past academic posts include a guest professor at Sao Paulo University, Brazil, a guest lecturer at Hiroshima University, and a part-time professor at Tokyo University. His main research field is equipment engineering in naval architecture, ship production system, safety engineering, sea transportation including containerized transportation and its facility, and environmental engineering concerning ship.

EVALUATION FOR BALLAST TANK DRAINAGE BY TWO-PHASE FLOW MODEL BASED ON HYDRAULICS EXPERIMENT AND SIMULATIONS

G Liu, Department of Urban and Environmental Engineering, Graduate School of Engineering, Kyushu University, Japan
T Shinoda, Department of Marine Systems Engineering, Kyushu University, Japan
T Watanabe, Department of Naval Architecture and Ocean, Graduate School of Engineering, Kyushu University, Japan
H Obata, Outfitting Design Department, Namura Shipbuilding Co., Ltd., Japan
T Nakamori, Hull Design Department, Namura Shipbuilding Co., Ltd., Japan
K Kuroki, Hull Design Department, Hull Design Section, Namura Shipbuilding Co., Ltd., Japan

SUMMARY

This study is utilizing hydraulics experimental and computational fluid dynamics (CFD) methods to verify the flow phenomenon in ballast water during the drainage process and to discuss the drainage efficiency. The model designed for the experiment is a part of ballast tanks of a typical bulk carrier and it is using the scale of 1:10 for the data collection. The experimental results are then validated and verified by conducting a two-phase flow model applied volume of fluid (VOF) simulation. The use of VOF in the ballast tank drainage is approved, so the drainage efficiency of the ballast tank can be predicted. This study also provides an optimization method for a simplified mathematical model, which can improve the agreement between the model and experimental results and create a more efficient yet simpler evaluation method for the configuration of the ballast tank drainage system.

1. INTRODUCTION

The ballast water system provides sufficient seaworthiness conditions in terms of stability, trim strength, and propulsion efficiency for large cargo ships when empty navigation is unavoidable. It is also responsible for adjusting the sailing posture and compensating for stability in some special loading conditions.

At present, there is little research on the efficiency of ballast water drainage during ship loading. The study of ballast water discharge efficiency can help to adjust the ship structure to improve the hull strength, save time for ship loading, and make fuller use of the loading capacity, thereby providing economically added value to shipping companies.

The design of the ballast water system is a systematic engineering problem, which requires comprehensive consideration of various factors such as pumping efficiency, bell-mouth location, drain holes size as well as arrangement, hull structural strength, equipment cost, and so on. Since the ballast tanks are separated by structural members such as longitudinals, transverses, and side girders, and connected by lightening holes, drain holes, etc., the structure is complex. Therefore, most of the configurations about ballast water drainage systems rely on empirical methods, and there is no mature method that can evaluate and predict the configuration of the ballast water system.

The purpose of this research is to develop an experimentally validated computational model for studying the flow phenomena in ballast tanks and evaluating drainage efficiency as a reference for the optimization of the ballast water system design. The ballast tanks of a typical bulk carrier are taken as the research object, and a 1:10 scale model is made to collect experimental data. Then the numerical simulations are performed by the computational fluid dynamics (CFD) method, and the volume of fluid (VOF) method [1] is used to trace the interface of gas-liquid two-phase flow as the water surface in the ballast tank. Then the computing results are compared with experimental results to verify the feasibility of the VOF method to simulate ballast tank drainage. At present, the simplified ballast tank arrangements are carried as the initial study, and the single-longitudinal and single-transverse one-dimensional arrangement model simulations for six compartments are completed. The results are in good agreement with the experimental results, and the VOF method feasibility is preliminarily verified in the simple ballast tank model.

To further explore the reliability of the VOF method used in ballast water discharge, the simulation model is extended to a two-dimensional arrangement. However, along with the complexity of the model, it becomes inevitable that a large amount of computational resources is consumed. To evaluate the ballast water system more economical and efficient, a method of constructing a simple mathematical model is introduced to assist the evaluation and is refined to improve the reliability of the model.

2. MODEL AND EXPERIMENT

In this paper, the area enclosed by the longitudinals and the transverses in the double bottom frame is called a compartment. The coordinate system is established with the bow direction as the positive x-axis direction and the port side direction as the positive y-axis direction.

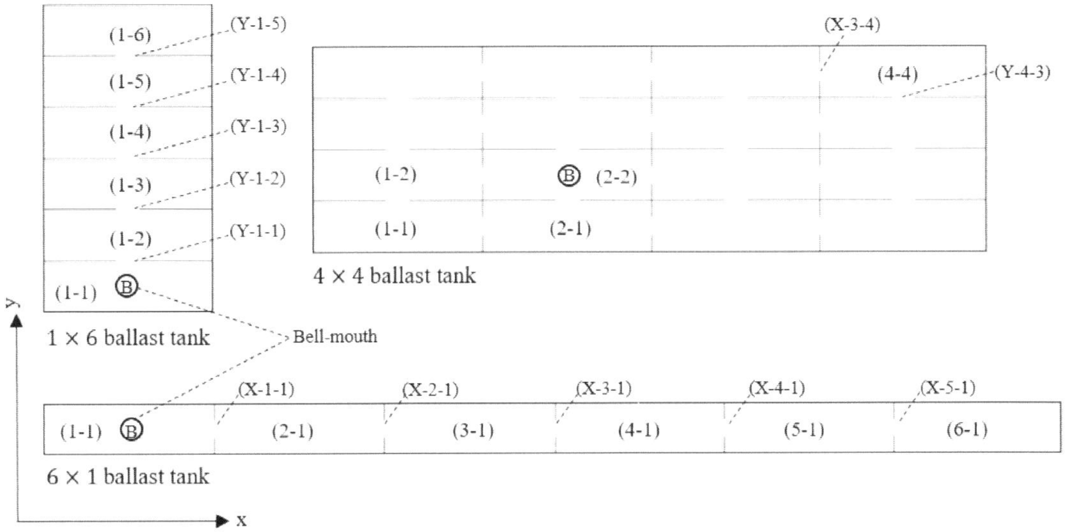

Figure 1. Ballast tank model arrangement and naming method.

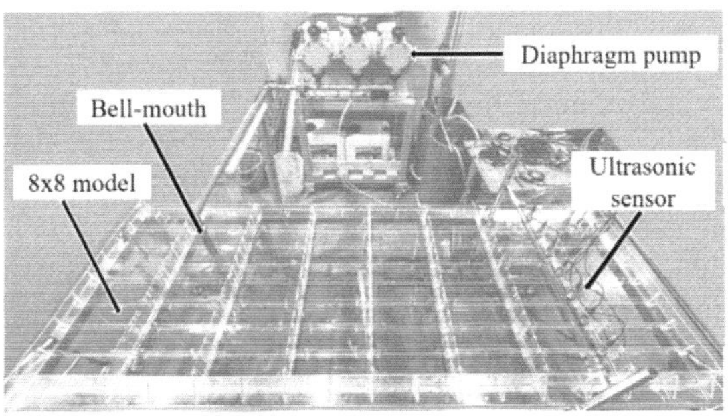

Figure 2. Experimental equipment.

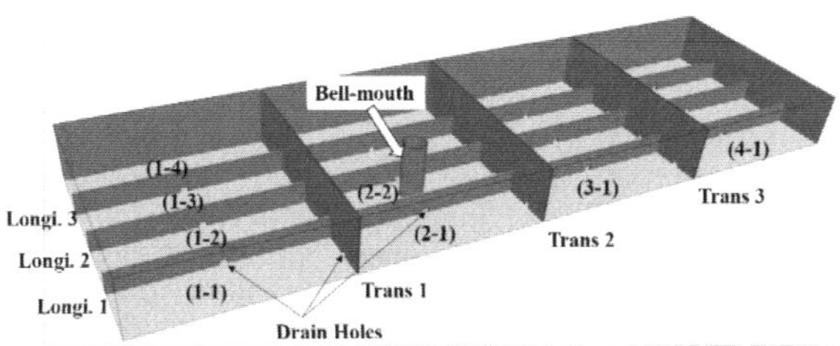

Figure 3. 4×4 ballast tank model.

Table 1. Principal particulars of the 4×4 tank model

Parameters	Dimensions (mm)
Compartment size (L. x B.)	260×77
Longitudinal height	30
Transverse height	80
Plate thickness	2
Drain holes size (B. x H.)	9×6
Bell-mouth external diameter	32
Bell-mouth internal diameter	25

As shown in Figure 1, the arrangement of the ballast tank is described as M × N (M is the amount in the X direction, N is the amount in the Y direction), the compartments are numbered as (m-n) (m is the position in the X direction, n is the position in the Y direction), and drain holes are numbered as (x-m-n) or (y-m-n) (x (y) indicates the axis direction of the drain hole, m-n are the same as the numbering of the compartments).

The hydraulic experimental device of this research consists of a ballast tank model, a ballast pump system, and a set of ultrasonic sensors, as shown in Figure 2.

The 1:10 scale model is designed according to the ballast tank of a typical bulk carrier, assembled with transparent acrylic materials, and made into an opening top ballast tank group composed of 8×8 compartments. The longitudinals, transverses, and drain holes can be freely replaced, which provides conditions for experiments with multiple arrangements of ballast tanks. In order to facilitate the model making and numerical simulation, the detailed structures such as stiffeners and wings are omitted from the model, and the circle holes are simplified to square. It should be noted that the scale of 1:10 refers to the geometric length scale, and the model design does not introduce dimensionless parameters. The refinement of the model and the introduction of dimensionless parameters will be expanded in future research.

Two one-dimensional arrangements of 6×1 and 1×6, which showed in Figure 1, are attempted in the preliminary stage of this study. The CFD simulations obtained high agreements with the experiments and well reflected the characteristics in the water surface descent. In this paper, the one-dimensional arrangement of ballast tanks is extended to two-dimensional, and the 4×4 ballast tank is chosen as the object. Because this arrangement is sufficient to demonstrate the complex flow in the ballast tank during the pumping process, and also economical to consume computational resources in CFD simulation. The 3D model is shown in Figure 3, and the measurements are given in Table 1. As the transverses and side girders are all closed structures connecting the inner bottom and the outer bottom, 80 mm is taken as a suitable height in this model.

The ballast pump system in this study consists of a three-phase diaphragm pump (TACMINA, FXD-FXW-8) and the bell-mouth. The output power of the pump is controlled by the frequency converter. The bell-mouth is simplified to a straight pipe made of vinyl chloride for ease of production and numerical simulation, with measurements as in Table 1, which is set in the compartment (2-2), extending from the top of the tank and the lower edge is placed at a height of 4 mm from the tank bottom. The pumping rate is set to 3.3 L/min, which is equivalent to 200 ton/h in full size and less than 1000 ton/h in actual case, in order to make the numerical simulation easier to converge and to facilitate the observation of flow phenomena.

The water level is measured by using ultrasonic sensors (KEYENCE, FW-02), with a measurement range of 50-200 mm and a response speed of 250 milliseconds. Because the sensor measures the water level at a small area, the measured water level data fluctuates and is difficult to reflect the water surface characteristics in the whole area. Therefore, in the compartment where the water level fluctuates greatly, such as the compartment with the bell-mouth, two measuring positions are placed to reduce the error. In the remaining compartments, only one sensor is set in each for measurement.

In the experiment, the simulated ballast water filled in the model is freshwater instead of saltwater. The ambient temperature is room temperature. The initial water level is 50 mm, which is determined by observing the scales attached to the wall of each compartment. After the water surface is stable, the sensors are first turned on and the data logger is used to record the return voltage value corresponding to the initial water level in each compartment for more than 5 seconds. The pump is then turned on and kept working continuously. When the water surface drops to a maintained level, the final water level of each compartment is observed and recorded through the wall scales, and the recorder of sensors is turned off after storing the final data for about 30 seconds. The collection method of the initial and the final voltage data is to calculate the average value to minimize the measurement error caused by the fluctuation of the water surface. Finally, the initial and final average voltage values correspond to the observation water levels, and the water level during the experiment can be obtained according to the linear relationship between the water level and the voltage data expressed in equation (1).

$$w(t) = \frac{w_f - w_i}{\overline{V_f} - \overline{V_i}} \cdot \left(V(t) - \overline{V_f}\right) + w_f \quad (1)$$

where $V(t)$ and $w(t)$ are the voltage and water level at time t. w_i and w_f are the initial and final observation water level. $\overline{V_i}$ and $\overline{V_f}$ are the initial and final average voltage. Therefore, the water level movement curve in each compartment can be obtained.

3. CFD SIMULATION

The open-source software OpenFOAM (version 5.0) is used for CFD simulation in this study. The mesh-based finite volume method (FVM) is used to solve the governing equations of mass and momentum. The Reynolds-Averaged Navier-Stokes (RANS) standard k-ε model, which is called RAS (Reynolds Averaged Simulation) in OpenFOAM, is used to solve the transient and incompressible problems. The ballast water discharge process is a gas-liquid two-phase flow model consists of air and water, and the interface of two fluids is tracked by using the VOF method. The basic solver of the VOF method in OpenFOAM is interFOAM [2].

In the VOF method, an indicator function is defined to represent the phase fraction of the fluid. For insoluble and incompressible multiphase systems, the transport equation of the indicator function representing one phase is solved with the continuity equation and the momentum equation.

$$\nabla \cdot \mathbf{U} = 0 \quad (2)$$

$$\frac{\partial \rho \mathbf{U}}{\partial t} + (\mathbf{U} \cdot \nabla) \rho \mathbf{U} = -\nabla p + \mu \nabla^2 \mathbf{U} + \mathbf{F} \quad (3)$$

$$\frac{\partial \alpha}{\partial t} + \nabla \cdot (\alpha \mathbf{U}) = 0 \quad (4)$$

where $\mathbf{U}$ is the flow velocity, $\rho$is the fluid density, p is the pressure, μ is the dynamic viscosity, $\mathbf{F}$ is the body forces. In the two-phase system of air and water, the phase fraction α means the cell is filled with water, and α=0 means the cell is filled with air. When α is between 0 and 1, it indicates the mixed state of air and water, and the interface can be approximately considered to be in this interval. In this study, an equivalent surface with α=0.5 is used as the interface between air and water. The VOF method obtains the interface by calculating the phase-field values in each cell and then post-processing without using complex interface tracking algorithms. Although it makes the calculation simple, the numerical discretization of equation (4) generates numerical diffusion at the interface, and a very fine mesh is required to reduce the error which places a higher demand on computing resources. Therefore, the method proposed by Weller [3] is adopted in OpenFOAM, which adds an artificial compression term to equation (4) to compress the phase fraction near the interface, thereby reducing the numerical diffusion and make the interface sharpen. The equation can be written as:

$$\frac{\partial \alpha}{\partial t} + \nabla \cdot (\alpha \mathbf{U}) + \nabla \cdot \left(\alpha (1-\alpha) \mathbf{U}_c\right) = 0 \quad (5)$$

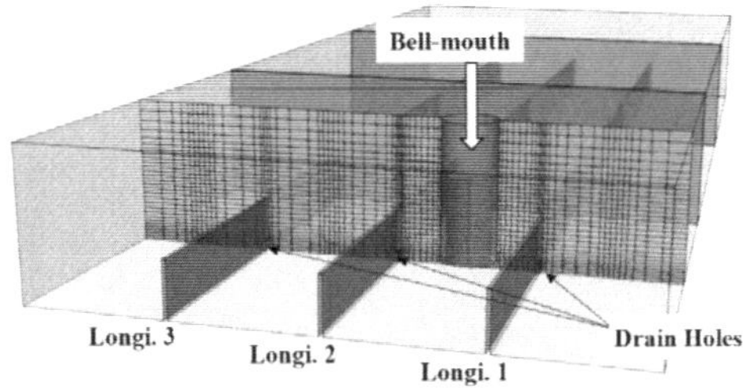

Figure 4. The mesh distribution in 4×4 tank model.

Table 2. Principal parameters of the supercomputer

Equipment name	Fujitsu PRIMERGY CX2550/CX2560 M4
CPU	Intel Xeon Gold 6154 (Skylake-SP) (3.0 GHz (Turbo 3.7 GHz), 18 core) × 2 / node
Theoretical operation performance	3,456 GFLOPS / node (double precision)
Memory	DDR4 192 GB / node
Memory bandwidth	255.9 GB/sec / node

where the last term on the left side is the artificial compression term, in which α - (1 - α) ensures that this term is zero at non-interfaces where α = 0 or 1, and only non-zero at interfaces. $\mathbf{U}_c$ is the compression velocity, which is an artificially added vector used to compress the interface. Weller [3] suggests it can be calculated as:

$$\mathbf{U}_c = c_\alpha |\mathbf{U}| \frac{\nabla\alpha}{|\nabla\alpha|} \tag{6}$$

where the $\frac{\nabla\alpha}{|\nabla\alpha|}$ is the gradient direction of the α at the interface, which is the same as the interface normal, so as to ensure that the compression velocity is perpendicular to the interface and does not produce any bias.

It can be seen that the value of the compression velocity should not be greater than the maximum velocity on the interface, so $c_\alpha|\mathbf{U}|$ is used to limit the maximum value of $\mathbf{U}_c$. c_α is used as a factor to control the compression strength to the interface, which indicates no compression while 0, conservative compression while 1, and enhanced compression while greater than 1. In this work, it is set to 1.

The time increment is controlled by the Courant–Friedrichs–Lewy (CFL) condition and the time step is dynamically adjusted by the automatic time step control function, while the time step is forced to be adjusted to ensure that the data is saved accurately at the specified output time.

The model adopts an unstructured hexahedral mesh, and the mesh distribution of the cross-section at the centreline of the bell-mouth is shown in Figure 4. The mesh is refined in the region of interest, such as near the bell-mouth, drain holes, and frame, to improve resolution. The quantity of cells in this 4×4 ballast tank model is 404,575.

According to the experimental results and numerical simulation experience, 360 seconds is set as the drainage time which is sufficient for obtaining the flow characteristics and saving calculation time.

Since solving transient problems requires a large number of computational resources and the quantity of mesh in this model is relatively large, the computation is carried out using the computer resource offered under the category of General Projects by the Research Institute for Information Technology, Kyushu University. The hardware parameters are shown in Table 2. In order to solve the drainage phenomenon of 4×4 tank model, 288 cores are called for parallel computation, and the calculation time is about 72 hours.

The simulation takes 0.25 seconds as the output time step and extracts the interface height with α=0.5 as the water level in each cell. Then the average water level in each compartment is calculated according to equation (7).

$$H_{mn} = \frac{\sum h_{ij} \times da_{ij}}{\sum da_{ij}} \tag{7}$$

where H_{mn} is the average water level of the compartment (m-n). h_{ij} is the height corresponding to α=0.5 of each cell, da_{ij} is the projected area of each cell perpendicular to the direction of water level. The total volume V_d of discharged water, can be calculated as:

$$V_d = \sum (H_{ini} - H_{mn}) \times A_{mn} \tag{8}$$

where H_{ini} is the initial water level. A_{mn} is the bottom area of each compartment. It is worth mentioning that the water level in a small range is fluctuant in simulation, which reflects the water surface characteristics well. The calculation of the average water level makes the fluctuation to be reduced, and the obtained water level curve is smoother.

4. VALIDATION RESULTS

The comparison between experimental and simulation results is shown in Figure 5. The water level of several compartments of interest is selected from 16 compartments for comparison, and the discharge volumes of the two methods are compared.

From the comparison, the simulation can well show the changing trend of the water level in each compartment. When the water level of each compartment does not drop to 30 mm, they maintain an approximate descent rate.

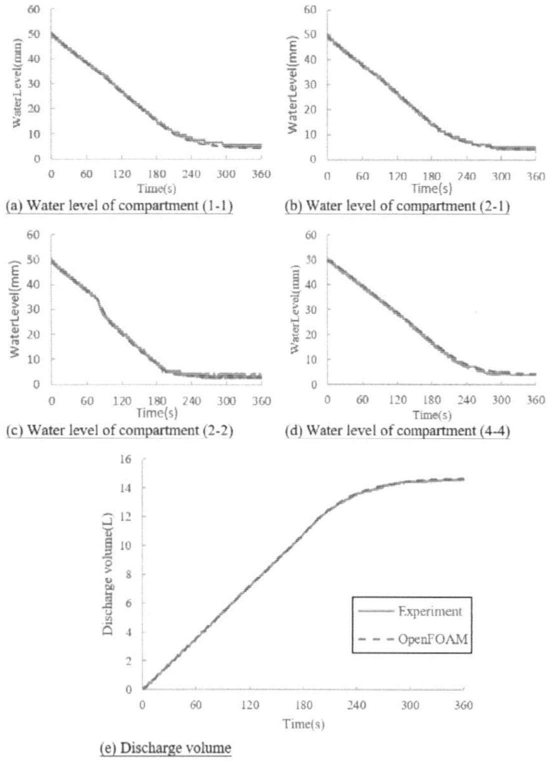

Figure 5. Result comparison for experiment and OpenFOAM of 4×4 tank model.

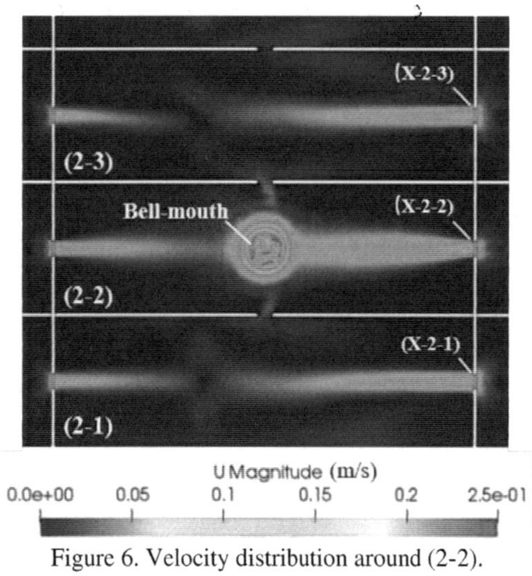

Figure 6. Velocity distribution around (2-2).

When they approach 30 mm, a divergence point appeared due to the influence of longitudinals. After that, they drop individually to the height of about 4 mm and remain stable.

The discharge volume comparison shows the simulation results are in good agreement with the experiment. It shows that the CFD simulation has high credibility in predicting the amount of discharge volume.

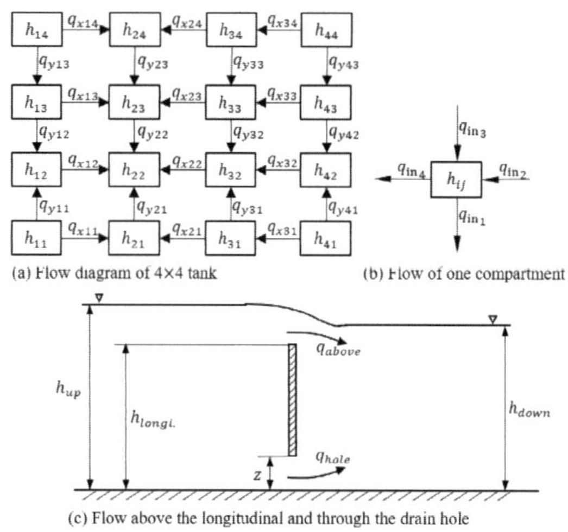

Figure 7. Schematic diagram of simplified model.

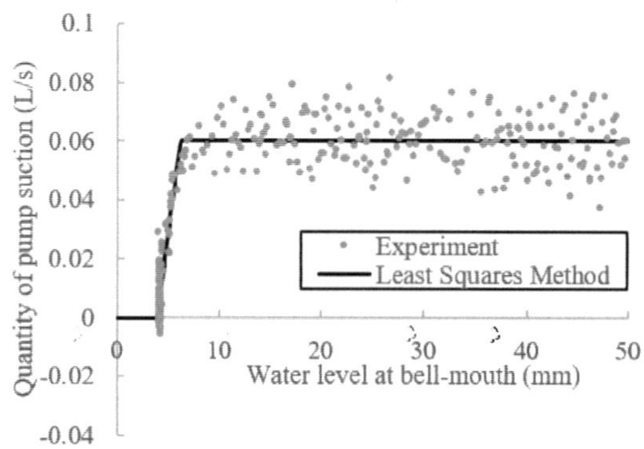

Figure 8. The characteristic curve of the pump.

A horizontal cut plane at a height of 3 mm from the bottom showing the water velocity distribution at 50 seconds is given in Figure 6. At this moment, the water level difference on both sides of the (X-2-2), (X-2-1), and (X-2-3) holes is basically the same, but the flow velocity of the hole (X-2-2) is slightly faster than the other two. This is because the suction of the pump accelerates the flow velocity in the nearby holes, and the influence gradually weakens as the distance from the bell-mouth increases.

5. SIMPLIFIED MATHEMATICAL MODEL

From the CFD simulation of the 4×4 model, the consumption of computing resources is relatively huge. Since the ballast tanks of large cargo ships usually have hundreds of compartments, it is difficult to simulate and evaluate large-scale ballast tanks with CFD methods. Therefore, this article introduces the idea of simplified mathematical model and tries to build an economical and simple way to assist the evaluation.

The flow in ballast tanks can be simplified to a reticulate system [4] as shown in Figure 7(a). For one compartment, only the flow from the adjacent compartments exists, and for the compartment where bell-mouth is located, an additional pumping output is applied. The water level change dh_{mn} per unit time in the compartment (m-n) can be written as:

$$\frac{dh_{mn}}{dt} = (-q_p + \sum q_{in(m,n)_k})\frac{1}{A_{mn}} \quad (9)$$

where the pump flow $q_p = 0$, the formula become the water level change of the no bell-mouth compartment. For $q_{in(m,n)_k}$, the positive value means flowing into the compartment, and negative means flowing out.

The pumping rate is not a fixed value during operation as shown in Figure 8, and it can be approximated as varying with the water depth h_{bm} at bell-mouth. By using the least squares method, the pump characteristic curve can be processed into the form of a function related only to the water level at bell-mouth:

$$q_p = f\left(h_{bm}\right) \quad (10)$$

The flow between adjacent compartments can be divide into the flow above the longitudinals and the flow through the drain holes according to the structural characteristic, and the cross-sectional model is shown in Figure 7(c). It is a classical weir structure with orifice. There are few methods to calculate the total flow in this model [5][6]. The simplest method of them can be assume that the weir and the orifice work independently. The total flow q_{total} between adjacent compartments can be calculated in the form

$$q_{total} = q_{above} + q_{hole} \quad (11)$$

where q_{above} is the flow rate above the longitudinal, and q_{hole} is the flow through the drain hole.

5.1 FLOW ABOVE THE LONGITUDINALS

According to the water level relationship between adjacent compartments, the flow above the longitudinals can be divided into the following three situations.

5.1 (a) Submerged situation

In this case, the upstream and downstream water level are higher than the upper edge of the longitudinal, and the downstream affect the upstream flow. When in a deep submerged state, use equation (12) given by Rajaratnam to calculate the flow rate.

$$q_{above} = c_{above}\sqrt{2g}L(h_{down} - h_{longi.})\sqrt{h_{up} - h_{down}} \quad (12)$$

where c_{above} is a consolidated experimental coefficient, g is gravitational acceleration, L is the length of longitudinal, h_{up} and h_{down} are upstream and downstream water level, $h_{longi.}$ is the height of longitudinal. According to Rajaratnam's suggestion [7], this formula is only accurate when the downstream head $(h_{down} - h_{longi.})$ is greater than 90% of the upstream head $(h_{up} - h_{longi.})$, and c_{above} could be given a constant value of about 0.90. When it is less than 90%, the Villemonte's method [8] using submerged flooding coefficient K is generally utilized.

$$q_{above} = q_{submerged} = Kq_{free} \quad (13)$$

$$K = \left[1 - \left(\frac{q_{down} - q_{longi.}}{q_{up} - q_{longi.}}\right)^{1.5}\right]^{0.385} \quad (14)$$

$$q_{free} = c_{free}\frac{2}{3}\sqrt{2g}L\left(h_{up} - h_{longi.}\right)^{1.5} \quad (15)$$

where equation (14) is the empirical formula of Villemonte's submergence coefficient, and equation (15) is the Kindsvater-Cater formula [9] for the free flow above thin-plate weir. The flow coefficient c_{free} can be calculated according to equation (16), based on the recommendations given by the ISO standard [10].

$$c_{free} = 0.602 + 0.083\frac{h_{up} - h_{longi.}}{h_{longi.}} \quad (16)$$

5.1 (b) Free flow situation

In this case, the downstream water level is lower than the upper edge of the longitudinal, and it is a free flow thin-weir model. The flow rate can be calculated directly according equation (15), and $q_{above} = q_{free}$.

5.1 (c) Non-flow situation

When the upstream and downstream water level are lower than the upper edge of the longitudinal, there is no flow and the flow rate $q_{above} = 0$.

5.2 FLOW THROUGH THE DRAIN HOLE

According to the water level relationship between adjacent compartments, the flow through the drain holes can be divided into the following three situations.

5.2 (a) Fully submerged situation

The water level of upstream and downstream are higher than the upper edge of the drain hole, and the flow is only related to the water level difference on both sides.

$$q_{\text{hole}} = c_{\text{hole}} \sqrt{2g} bz \sqrt{h_{\text{up}} - h_{\text{down}}} \tag{17}$$

where the discharge coefficient $c_{\text{hole}} = 0.65$, b and z are the breadth and height of the drain hole.

5.2 (b) Partially submerged situation

In this case, only downstream water level is lower than the upper edge of the drain hole. According to the general method, the total flow rate is determined by adding the flow of upper portion and lower portion.

$$\begin{aligned} q_{\text{hole}} = c_{\text{hole}} \frac{2}{3} \sqrt{2g} b \left[\left(h_{\text{up}} - h_{\text{down}}\right)^{1.5} - \left(h_{\text{up}} - z\right)^{1.5} \right] \\ + c'_{\text{hole}} \sqrt{2g} b h_{\text{down}} \sqrt{h_{\text{up}} - h_{\text{down}}} \end{aligned} \tag{18}$$

where the first term on the right side is the upper portion flow which is the free flow in large orifice. The second term is the lower portion flow which is in fully submerged orifice case. The discharge coefficient $c'_{\text{hole}} = 0.52$.

5.2 (c) The water level on both sides is lower than the upper edge of the drain hole

This case is modelled on the partially submerged drain hole, and divided into upper and lower portion. The upper portion is regarded as the channel flow, and the lower portion is fully submerged orifice flow.

$$\begin{aligned} q_{\text{hole}} = \frac{1}{n} R^{\frac{2}{3}} I^{0.5} A \\ + c'_{\text{hole}} \sqrt{2g} b h_{\text{down}} \sqrt{h_{\text{up}} - h_{\text{down}}} \end{aligned} \tag{19}$$

$$R = \left[\left(h_{\text{up}} - h_{\text{down}}\right) b / 2 \right] / \left[b + \left(h_{\text{up}} - h_{\text{down}}\right) \right]$$

$$I = \left(h_{\text{up}} - h_{\text{down}}\right) / l$$

Where the first term on the right side is the Manning formula. n is the Gauckler–Manning coefficient, and the acrylic material's n is 0.01, R is the hydraulic radius, I is the linear hydraulic head loss. For the x-direction drain hole, l is the length of the compartment, and for the y-direction drain hole, l is the breadth of the compartment.

5.3 THE PUMP INFLUENCE COEFFICIENT

After analyzing the results of the simulation, it is found that the suction of the pump could increase the flow rate in the drain holes near the bell-mouth as shown in Figure 6. With the hole is farther away, the influence gradually decreases. According to this phenomenon, the flow rate under the influence of the pump suction is obtained by multiplying the influence coefficient c_{p} as:

$$q_{\text{hole-act}} = c_{\text{p}} q_{\text{hole}} \tag{20}$$

where the coefficient is not less than 1. When the coefficient is equal to 1, the drain hole flow is not affected by the pump suction.

$$c_{\text{p}} = q_{\text{hole-act}} / f\left(h_{\text{up}}, h_{\text{down}}\right) \tag{21}$$

Equation (21) is transformed from equation (20). Thus, the coefficient can be obtained using flow rate in drain hole and water level from experimental or simulation results.
Because it is difficult to obtain the flow rate of each hole in the 4×4 ballast tank model which is a two-dimensional arrangement model, the 6×1 and 1×6 tanks, as shown in Figure 1, are first used to obtain the flow rate of drain holes. Then the flow above the longitudinals can be excluded by using the 1×6 tank experimental result with an initial water level of 30 mm. The influence coefficients of each hole under the two arrangements are given with accuracy to two decimal places in Table 3.

It can be seen that in the 6×1 tank model, the flow of the hole (X-5-1) is not affected by the pump, and the distance between the hole and the pump is 1.178 m.

To verify the simplified mathematical model and the pump influence coefficient, the calculating results in the 6×1 tank model are compared with the experiment as shown in Figure 9.

From the comparison, the mathematical model with the addition of the pump influence coefficient is in better agreement with the experimental result than the model with no coefficient.

The discharge volume comparisons of the experiment, simulation, and simplified mathematical model, which use 6×1 and 1×6 tank models with the initial water level of 30 mm, are shown in Figure 10. The simplified mathematical model can well predict the discharge volume of the experiment and the simulation.

The above results show that the introduction of the pump influence coefficient enables the simplified mathematical model to well represent the drainage of one-dimensional arrangement tanks. Compared with the tens of hour computing time of CFD simulation and the requirements of hardware, the code is written according to the simplified mathematical model only takes tens of seconds to obtain the result with a high agreement. However, it is obvious that for two-dimensionally arranged ballast tanks, especially when there are many water holes, the above method to obtain the influence coefficient for each hole is inefficient. Therefore, further research is needed to summarize the variation rule of the influence coefficient.

Table 3(a). Pump influence coefficient on each drain hole in 6x1 tank model

Drain hole	(X-1-1)	(X-2-1)	(X-3-1)	(X-4-1)	(X-5-1)
c_p	1.27	1.24	1.17	1.09	1.00
$L_{p\text{-}h}$ (mm)	130.0	392.0	654.0	916.0	1178.0

$L_{p\text{-}h}$ is the distance from bell-mouth to drain hole

Table 3(b). Pump influence coefficient on each drain hole in 1x6 tank model

Drain hole	(Y-1-1)	(Y-1-2)	(Y-1-3)	(Y-1-4)	(Y-1-5)
c_p	1.29	1.28	1.26	1.25	1.23
$L_{p\text{-}h}$ (mm)	38.5	117.5	196.5	275.5	354.5

$L_{p\text{-}h}$ is the distance from bell-mouth to drain hole

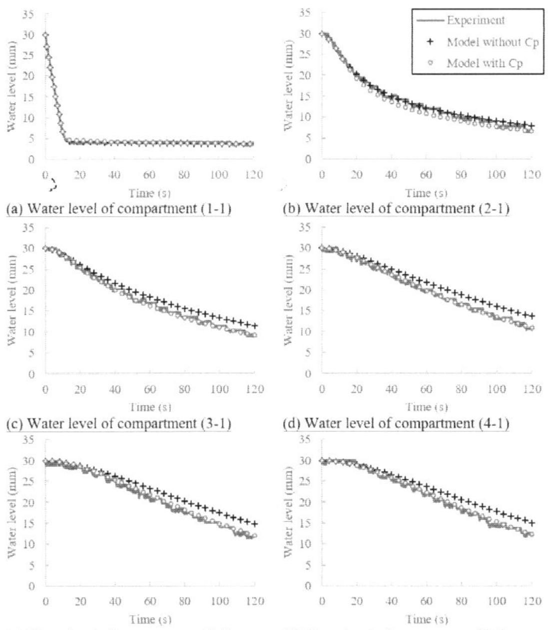

Figure 9. Result comparison for experiment, mathematical model without c_p and model with c_p of 6×1 tank model.

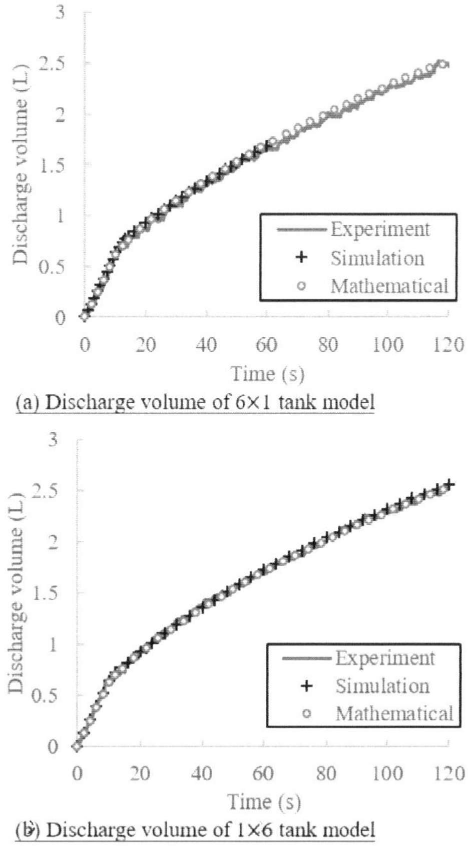

Figure 10. Discharge volume comparison for experiment, simulation, and mathematical method.

6. CONCLUSIONS

In this paper, flow phenomena in a 4×4 ballast tank are verified by hydraulic experiment and CFD simulation. The validation work uses a 1:10 scale model made with a typical bulk carrier ballast tank measurement. The results show that the simulation using the VOF method is in good agreement with the experimental result. In addition, the idea of constructing a simplified mathematical model is introduced, while the influence coefficient of the pump on the flow rate of the drain hole is added. The mathematical model can be in good agreement with the experimental and CFD simulation results under the one-dimensional arrangement, which proves the reliability of the results obtained by the method. The variation rule of the influence coefficient needs more research to be summarized.

The study will also be extended to the simulation of real ships. In addition, the influence of the drain hole size, the bell-mouth location, the pumping rate, and the trim condition on the drainage efficiency of the ballast water system are all topics of future research. This will provide methods and theoretical references for the arrangement and improvement of the ballast water systems, and bring more economic and environmental benefits to shipping.

7. REFERENCES

1. Hirt, C. W., NICHOLS, B. R., 'Volume of Fluid (VOF) Method for the Dynamics of Free Boundaries', *J. Comput. Phys*, 1981. 39(1): 201-25.

2. Greenshields, C. J., 'OpenFOAM User Guide', *Delaware: OpenFOAM Foundation Ltd.*, 2017.p. 60-72.

3. Weller, H., 'A New Approach to VOF-Based Interface Capturing Methods for Incompressible and Compressible Flow', *OpenCFD Ltd. Technical Report TR/HGW/04*, 2008.

4. FUKUCHI, N., SHINODA, T., 'A Practical Method for Analysing the Indeterminable Problems in the Design of Ship Equipment (Part 1) - The System Collecting Oil and Water Through the Tank Bottom Structure', *J. Soc. Nav. Arch. Japan,* 1989.

5. ALTAN-SAKARYA, A.B.,KOKPINAR, M. A., 'Computation of Discharge for Simultaneous Flow Over Weirs and Below Gates (H-weirs)',*Flow Measurement and Instrumentation*, 2013.29: 2-38.

6. NOURI, M., HEMMATI, M., 'Discharge Coefficient in the Combined Weir-Gate Structure', *Flow Measurement and Instrumentation*, 2020.

7. RAJARATNAM, N., MURALIDHAR, D., 'Flow below deeply submerged rectangular weirs', *Journal of Hydraulic Research*, 1969. 7: 355–374.

8. VILLEMONTE, J. R., 'Submerged weir discharge studies',*Engineering news record*, 1947. 866: 54–57.

9. KINDSVATER, C. E., CARTER, R. W. C., 'Discharge Characteristics of Rectangular Thin Plate Weirs', *J. Hydraulics Div. ASCE*, 1957. 83(HY6), 1453/1–1453/36.

10. International Organization for Standardization, 'ISO 1438-2017: Hydrometry — Open Channel Flow Measurement Using Thin-plate Weirs', *ISO: Geneva, Switzerland*, 2017.

8. AUTHORS BIOGRAPHY

G LIU is a doctoral student in the Department of Urban and Environmental Engineering, Kyushu University. In this research, he is mainly responsible for the numerical simulation of ballast tank models on OpenFOAM, and the programming of the simplified mathematical model.

T SHINODA is a professor of Kyushu University, Japan. He received D.Eng., M.A. and B.A. in naval architecture from Kyushu University. His main research field is equipment engineering in naval architecture, ship production system, safety engineering, sea transportation including containerized transportation and its facility, and environmental engineering concerning ship.

T Watanabe is a doctoral student in the Department of Naval Architecture and Ocean Engineering, Kyushu University. In this research, he is mainly responsible for the experiment of ballast water drainage.

H Obata holds the current position of deputy general manager of the outfitting design department at Namura Shipbuilding Co., Ltd. He has outfitting design experience includes drain course arrangement of ship.

T Nakamori, D.Eng. holds the current position of manager of the hull design section at Namura Shipbuilding Co., Ltd. He has structural design experience of the ship includes strength evaluations of the drain hole.

K Kuroki is responsible for hull structural design at Namura Shipbuilding Co., Ltd. He has 12 years of structural design experience of the ship includes strength evaluations of the drain hole.

DEVELOPING A HYBRID VALUE ENGINEERING AND RISK ASSESSMENT (VENRA) FRAMEWORK FOR SHIPBUILDING AND SHIP REPAIR INDUSTRY PERFORMANCE MEASUREMENT

I Baihaqi, I Lazakis, R E Kurt, Department of Naval Architecture, Ocean and Marine Engineering (NAOME), University of Strathclyde, UK

SUMMARY

This paper proposes a novel performance measurementframework to enhance the decision-making process in assessing the ship manufacturing (shipbuilding, ship repair, and ship conversion/modification) industry. Integrated Value Engineering and Risk Assessment (VENRA) approach performance measurement for ship manufacturing is developed to tackle the problem. Thorough literature and critical study of the existing performance measurement for ship manufacturing are reviewed to createnovel VENRA performance measurements. The attributes collected from the literature are grouped into four criteria performance, named VENRA attributes, combining the technical, business, external and safety/risk performance groups. The initial model will be assessed through fuzzy DEMATEL TOPSIS to achieve the weighted score of each attribute. The framework will be applied to three shipyard case studies in Indonesia. The quantitative data (from shipyards) and qualitative data (from expert opinion/judgment) will be included to get the results and the framework's effectiveness.

1. INTRODUCTION

Effective performance measurement seems to be the necessity to achieve more competitive industries. Regardless of its segment, sector, and size, a company cannot become competitive and sustainable unless it mainly accomplishes its value, quality, flexibility, and cost. It means selecting the proper criteria to model performance measurement remains a crucial matter. It shall assist the process of decision-making procedure. The critical factors in performance measurement should be selected to explainwhatcompanies should evaluate [1].

Ship manufacturing industries, including shipbuilding, ship repair and ship conversion, has unique technical and business factors. Technically, it consists of complex production processes, different technology levels, different building strategies, ship-type products and materials, and labours. In business view, it might be similar if compared with general sector industries; however, it has unique business knowledge in terms of finance, payment terms, bank involvement which should integrate with the technical knowledge. With this regard, the performance measurement to assess the competitiveness of ship manufacturing industries should have specific attributes/criteria.

The need for a universal measure of ship manufacturing industryperformance has been recognised for many years. The Organisation for Economic and Co-operative Development (OECD) developed the concept of CGT (Compensated Gross Tonnage) as the global shipbuilding performance measurement. The basic principle is to take the gross tonnage (GT) and multiply it by a compensation factor that "corrects" the GT for differing work content [2]. The correction factors in CGT have changed to consider the changes in ship technology and production, which in this case, the 2007 version of CGT is the latest. CGT as the shipbuilding performance measurement has been developed with different approaches and corrections. Zhangpeng and Flynn[3]proposed a correction of CGT calculation by adding the productive shipyard area, while Bruce[4]proposed a new interpretation in applying the CGT graph to be more helpful. Pires, Lamb and Souza [5]use Data Envelopment Analysis (DEA) and Analytical Hierarchy Process (AHP) for performance measurement benchmarking. Recent developments include[6], who proposed the Balanced Scorecard (BSC) model assessed through a hybrid MCDM approach.

This paper enriches the literature on ship manufacturing performance and competitiveness globally, specifically for Indonesian shipyards with different ship types. The contributions of this paper are as follows. First, it unfolds the ship manufacturing performance measurement methodology, including the key performance indicator collected, providing the gaps of the method used and the selected criteria. Secondly, it purposes VENRA (Value Engineering and Risk Assessment) model applied in selected indicator to assess ship manufacturing performance, combining the technical, business, external and safety/risk grouped attributes. Thirdly, it demonstrates a hybrid MCDM approach through Fuzzy DEMATEL TOPSIS through the VENRA attribute model, allowing practitioners to face complicated issues with multiple contradictory criteria. The model hypothesises that ship manufacturing performance measurement will be more enhanced through the VENRA model attributes assessed by the hybrid MCDM approach and can effectively measure technical, business, external and safety/risk attributes.

The structure of this paper is as follows. Section 1 reviews the development of ship manufacturing performance measurement methodology and the need to pursue enhanced performance indicators. Literature and critical review regarding performance measurement in ship manufacturing, value engineering and risk assessment, fuzzy Multi-Criteria Decision Making (MCDM)is presented in section 2. The VENRA model for performance measurement in shipbuilding assessed through the hybrid MCDM approach is delivered in section 3. Section 4 discusses the case study plan for three shipyards in Indonesia, and section 5 concludes and provides further into future work.

2. LITERATURE REVIEW

2.1 SHIP MANUFACTURING PERFORMANCE MEASUREMENT INDICATORS

The conventional performance measurement in shipbuilding utilises Compensated Gross Tonnage (CGT) to compare one shipyard to other shipyards presented in CGT and cost curves. CGT can be defined as the amount of work necessary to build a given ship. It is measured by multiplying ship tonnage with coefficients, determined according to the type and size of a particular ship [2]. Some researchers utilise CGT in performance measurement in shipbuilding; others criticised the weakness and proposing a new way to represent the use of CGT. Bruce [4] proposed a new interpretation in CGT for shipbuilding performance by adding a correction in interpreting the CGT graph. Zhangpeng and Flynn [3] added the correction of CGT calculation by adding the productive shipyard area factor. Pires, Lamb, and Souza[5] propose a methodology for shipbuilding performance assessment by taking into account the characteristic of the shipyard, the production pattern and the industrial environment of the country region. A Data Envelopment Analysis (DEA) model is proposed to assess the shipyard performance, and an Analytical Hierarchy Process (AHP) method is used to obtain the Industrial environment indicator. Colin and Pinto[7] use several turnover measures, including main physical assets such as dock area, berth length, lifting capacity and the output CGT.

Fareza[8], in his Master Thesis, has developed shipbuilding performance measurement benchmarking using DEA, comparing major shipyards in Asia to acquire the market share by analysing the input and output. Chao and Yeh [9]comparing major shipyards' productivity in China, South Korea, and Japan using the DEA meta-frontier framework.Gavalas, Syriopoulus and Tsatsaronis [6] proposed a performance measurement model to assess the shipbuilding industry's key performance indicators using a Balanced Scorecard (BSC) with the MCDM approach (Fuzzy DEMATEL, Fuzzy ANP, and MOORA).

The application of BSC as the performance measurement has been broadly used in many sectors. However, in BSC, the dimension measured is limited to business process, internal process, learning and growth and customer satisfaction. Concerning the shipbuilding industry case, the attribute can be enhanced more by adding developing own attribute combining business, technical, external and risk/safety factors. In this case, the VENRA attribute model is proposed to tackle the performance measurement attributes gaps.

2.2 VALUE ENGINEERING

Value Engineering (VE) was first created and introduced by Lawrence D. Miles In 1947 named "Techniques of Value Analysis and Value Engineering" at General Electric due to material scarcity during World War [10]. The concept then adopted and implemented multi-discipline, including product design, construction project enhancement, and process enhancement.Society of American Value Engineers) (SAVE) define that VE is a systematic and structured approach for improving projects, products and processes. It analyses and improves manufacturing products and processes, design and construction projects, and business and administrative processes. It also helps achieve the optimum balance between function, performance, quality, safety, and cost, resulting in the maximum value for the project[11]. VE has broadly applied in many sectors and cases. It starts from the construction project, product development, performance-based and decision-making process.

VE is a systematic technical analysis method that targets the lowest lifecycle cost while achieving essential functions for product users. It commonly expresses as the ratio of function to cost [12]:

$$Value = \frac{Function}{Cost}$$

The higher the value, the better the usage of the resources; likewise, a lower value indicates that resources are not efficiently used. There is still an opportunity to improve products or choices by either increasing the functionality or reducing the cost.

The recent paper discusses VE application in the construction project [13], combining the VE and constructability concept in the infrastructure project. Gunarathne et al. [14] propose a framework of integrated sustainability and VE concepts for construction projects, and [15] apply VE techniques in residential buildings. The primary approach used is through the VE-job plan, which uses multi-disciplinary team workshop based, analysing the existing data and purposing the new alternative approach to reduce the cost project without compromising the project quality and performance.

The VE concept is also widely broad applied in product development case studies. Alekseev, Kissin and Gorelov [16]studied the VE implementation for foam-cyclone cleaners by decomposing the product function and cost and altering the model into four options. The analysis includes the hydraulic resistance changes, design complexity, material consumption, dimensions, manufacturing complexity, assembly accuracy, and encrustation sensitivity. The expert evaluation performed the scoring. It shows that VE can help to improve product design while maintaining the techno-economic parameters. A similar approach was also studied by [17] on reducing product cost while maintaining customer satisfaction in global pumping unit production. It tries to change the rotor design as cast slots and reduce the pump body and stator housing weight. Silaskar and Shinde [18]performed the VE concept to find the weight optimisation of the ball valve used hydraulic system by changing the ball valve design and examining it through simulation software to check the tensile stress. Vijayan et al.[19]apply the VE concept to reduce the cost and weight of air suspension products by redesigning the product and analysing the stress and resultant displacement using simulation software. Kalluri and Kodali[20]have implemented VE methodology for hydraulic systems for agricultural and construction equipment by alternating the grade of materials, changing the manufacturing method, and finding possible alternative solutions. Integrated product development method based on VE and design for assembly concept has been proposed by [21], seeking the balance between value and cost of all function of a mechanical subset. Based on the above review, the VE approach has been applied to improve product manufacturing through breakdown function and cost, then purposing the alternative options and involving the expert judgment or simulation software to score or verify the alternative options.

Despite the wide broadly of VE implementation in many sectors, the application of VE in the Marine sector is still few, especially in the shipbuilding or ship repair industry. Garratt [22]applied the VE concept to develop the 33 HP marine industrial engine part, while Tao, Yang and Shi [23]applied the cost control of marine diesel using the VE concept. Hyland [24]implemented the VE concept on designing the division of a large graving dock using slave caissons. Al-Yafei et al.[25]have developed offshore topside facilities manufacturing by alternate the materials using matrix value.VE has been implemented in shipbuilding cases by [26]utilising the ship design value stream. Romano et al.[27]and Formentini and Romano [28]discuss knowledge transfer based on value analysis in a shipbuilding project with the marine piping and electrical cable system case study. Tang and Bittner [29]developed a creative design solution for a marine construction project. The VE's concept has also been applied in ship hull construction by [30]on reducing the rework in the shipbuilding project using matrix and cost-benefit function through the VE workshop. Karami and Olatunji [31] suggested that the marine project's key protocols should be applied to reduce cost and time while optimising the project's quality. From this literature review, the application of VE in shipbuilding and or ship repair is relatively few. It has not been formalised, especially in the shipbuilding and ship repair industries.

In addition, the application of VE as the performance measurement is still few. Lee et al. [32]purpose the performance-based VE application to public highway construction. The project performance criteria are identified and weighted using the AHP method. Through the VE job plan, the alternatives scenario in the creative phase through Function Analysis System Technique (FAST) analysis is developed, and the performance is measured compared to the baseline. The project value is calculated based on performance over the cost in which the higher index has the best alternative / has improved value. Heralova[33]studied the possibility of using VE in a highway construction project. It analyses the possibility of using VE by extracting the benefits and conceiving them into the highway construction project. Anam et al.[34] adopted the [32] approach to implementing the water supply project with the same approach.

The application of VE for the decision-making process has also been widely applied with a combination of other methods. Karunasena and Rajagalgoda [35]examine the existing practices of VE techniques and propose to standardise VE practices with purposed decision-making formula to determine the profitability of VE applications before implementation. A coherent framework for utilising a VE approach to supply chain cost management and overcome the limitation utilising MCDM approach grey multi-criteria decision making presented by [36]. Zhao et al.[37]take the benefit concept of financial and functionality in VE to purpose VE-based facility decision making process by integrating with lifecycle cost analysis.

The literature review about VE shows that:

- VE application can be used in many case studies (product development, construction industry, decision-making and performance measurement)
- The point of VE implements the VE job plan (Information phase, function analysis, creative phase, development phase, implementation phase) and is followed by the decision-making process in the end.
- The main contribution of VE is that it has a concept on how to maintain value and quality and to reduce cost byinvestigating the function of project/product/equipment and provide an alternative to creating the best value for the decision-making process
- The main power also involves a multi-disciplinary team that can fill the knowledge gap of each other decision-makers.
- Many papers also adapt or modify the VE concept into other methods, models, MCDM decision-making tools, and others from the literature review. It shows that the VE can be integrated and modified into other tools to gain the concept's benefits.

2.3 INTEGRATED VALUE ENGINEERING AND RISK ASSESSMENT (VENRA)

The opportunity to integrate VE with formal risk assessment and analysis started in 1993 when a city port authority required a value engineering effort enhanced with a risk assessment application [12]. Other researchers also studied the method integration and explored the advantages. Mootanah [38]interfaced the integrated risk and value management methodologies theoretically and practically in construction project management. It reviews the evolution of value and risk management frameworks and explaining the basis for framework integration. Abd-Karim [39]attempted to apply IRVM (Integrated Risk and Value Management) workshop to integrate risk management and value management qualitatively through four infrastructure projects in the UK. Ahmad et al. [40]explored the potential advantages of integrating RM and VM and the critical success factors (CSF) related to the implemented successfulness of the IRVM workshop. The case study was applied to the combined method following the standard of value management AS 4183:2007 and risk management AS/NZ 4360:2004. Feili et al. [41]used integrated risk management and VE to achieve an optimal approach in developing alternative energy renewable energy projects while reducing the cost and environmental pollution. Pedju and Mawu [42]identify and analyse factors related to the risks and value on the development of the construction projects through the design and engineering phase.The data are collected through questionnaires, interviews, and project's documentation. El Khatib[43]highlighted the concept of project risk management with VE in tendering process. Golabchi et al.[44] represented a model based on the combined VE and risk management to enhance the project cost management process in quantitative analysis of existing risks to improve the project's final cost. Sabzkohi and Pourrostam[45]used integrated VE and risk management method in exterior design project through optimised value index as a measure for decision-making process with clarifying the advantages and disadvantages.

The above literature focuses on integrating risk and value through project management, implemented in general project construction. The method used is primarily analytical; some researchers use quantitative risk analysis based on interviews from expert opinion. However, the integrated VE and risk assessment (VENRA) framework has not been implemented in shipbuilding and ship repair projects. The focus is more on value management, not on VE knowledge. In addition, the implementation of integrated VE and risk assessment has not been attempted for the performance measurement tool. The present paper will adopt the VE concept integrated with risk assessment into the VENRA model to be applied in performance measurement model attributes.

2.4 FUZZY MULTI-CRITERIA DECISION MAKING (MCDM)

In 1965, Zadeh[46] introduced Fuzzy Set Theory (FST) to accommodate the vagueness to answer the question. Since then, various researchers have developed the theory, including [46], who developed fuzzy set theory and its application through disjunction and conjunction. Moreover, Chen and Hwang[47] developed a fuzzy Multi Attributive Decision-Making (MADM) approach and [48] with Fuzzy logic in engineering applications.

FST has been broadly used to enhance the decision-making process in shipbuilding, ship repair and maintenance and the shipyard industry. Guneri et al.[49] and Sahin et al. [50] use a fuzzy AHP shipyard selection based on game theory. There is a broad application of FST in the case of the selection process, such as selecting the best maintenance approach [51], selecting the best propulsion/ manoeuvring system [52], supplier selection in the shipbuilding industry [53], shipping registry selection for existing ship or newly built ship [54], and shipyard facility layout selection [55]. In the performance measurement case, Chou and Liang[56] and Celik et al. [57] use fuzzy to evaluate performance evaluation in shipping companies and docking facilities, respectively. In addition, Freitas et al. [58] demonstrate a fuzzy model for the prioritisation analysis of variable quality performance in shipbuilding. Gayathri et al. [59] and Gavalas et al. [6] use fuzzy MCDM to study port operational and financial performance analysis and assess key performance indicators in the shipbuilding industry, respectively.

Some observations from the literature review concerning shipbuilding performance measurement are summarised as follows:

- The classical MCDM approacheshave their benefits and drawback. The integration of the MCDM model can overcome the limitation of the individual approach and enhance the efficiency of the decision-making process. The combination of MCDM methods has received the most attention recently. However, to the best of the author's knowledge, no study integratesthe MCDM technique of DEMATELand TOPSIS to solve the performance measurement in the manufacturing industry. This study purposes a hybrid Fuzzy DEMATEL TOPSIS approach to determine the importance weights of decision-making criteria and evaluate the performance of different ship manufacturing industries.
- The review also shows that the criteria used are mainly on the business aspect. The technical aspect included is also in the general group, such as steel processing or shipyard facility or capacity. Some of the technical aspects can be derived in more detail to measure the performance measurement. In this study, the criteria are developed based on the VENRA approach by combining technical, business, external and safety/risk criteria, which is part of the VENRA approach.
- The study to model the performance measurement in ship manufacturing by using the MCDM approach is still few. Thus, the paper's contribution is to develop this model based on VENRA by using a novel Fuzzy DEMATEL TOPSIS approach in the ship manufacturing industry.

3. METHODOLOGY

3.1 VENRA MODEL ATTRIBUTES

The developing process of the model starts with collecting attributes related to shipbuilding performance measurement, ship repair, selecting the best docking facility, selecting the best shipyard of shipyard competitiveness in the literature review. The collected attributes were assessed through the VENRA approach by categorising the attributes into four groups: business, technical, external and risk/safety performance, as shown in Figure 1, while the global framework of VENRA through the MCDM approach is shown in Figure 2.

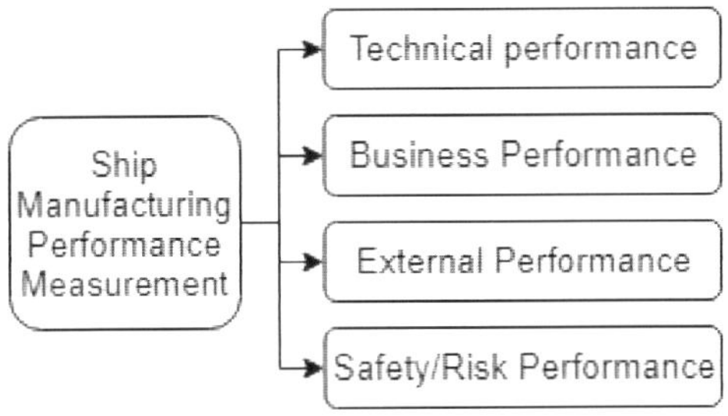

Figure 1. Group attribute for ship manufacturing performance measurement.

3.1(a) Technical Group Performance

It relates to technical things to production processes in ship manufacturing, starting from engineering production to delivery. This attribute group includes performance criteria indicators related to the shipyard capacity, building strategy, workforce, and technology level.

3.1 (b) Business Group Performance

The most important in business group performance is the cost and delivery time of the production process. in addition, shipyard experience and reputation will be included in this group and the organisation and management.

3.1 (c) External Group Performance

The ability to gain external resources is also very crucial to ship manufacturingperformance. It includes vertical integration to gain suppliers, additional workers or equipment, government support and the domestic market.Bruce [60] suggests that the political environment play a vital role in international ship production, such as government support to the shipyard to maintain employability worker European shipyard due to the growth of Asia shipyard.

3.1 (d) Safety/Risk Group Performance

Activity in the shipyard to build or repair a ship is a hazardous activity. This safety/risk performance sometimes have not the priority in performance measurement. Thus the risk/safety group performance is also included to measure the safety operation of the ship manufacturing process. It includes the safety of the workers, the environment, and the risk of product failure. In this case, the safety training induction and the availability of the OHSAS certificate will be included in the criteria. At the same time, the risk of the product will measure the probability of rework in the production process.

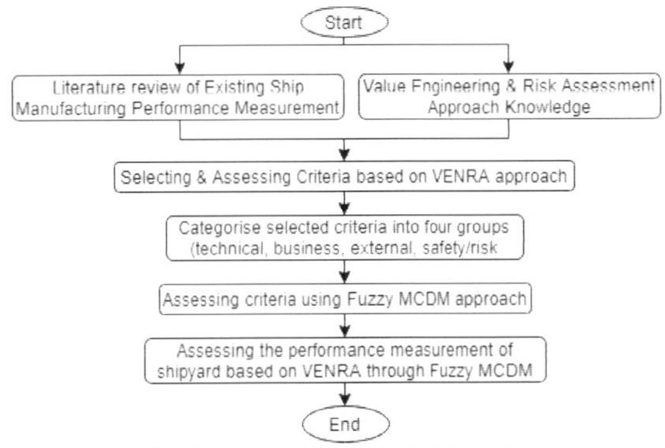

Figure 2. Methodology of VENRA performance measurement for ship manufacturing.

3.2 FUZZY DEMATEL

The performance of ship manufacturing depends on several factors that are either directly or indirectly related. Hence, evaluating ship manufacturing industry performance based on a single factor may not be justifiable. The DEMATEL method (Decision-Making Trial and Evaluation Laboratory) can be used to analyse the relationship amongst multiple complex factors. However, the evaluation criteria in this context may be imperfect or can contain uncertain factors, so the human judgement about preferences can be difficult to estimate by exact numerical values.
The fuzzy DEMATEL method can help to overcome this and has been used by several researchers. This research uses a fuzzy DEMATEL approach to estimate the criteria weights. The relationship between the factors considered is evaluated based on the opinions of experts using a fuzzy scale. Table 1 shows the fuzzy scale mapping the degree of influence with the linguistic terms.

Table 1. linguistic terms for fuzzy DEMATEL evaluation

Linguistic term	Abbreviation	Fuzzy scales
None	N	(0,0,1)
Very Low	VL	(0,0.1,0.2)
Low	L	(0.1,0.2,0.3)
Fairly Low	FL	(0.2,0.3,0.4)
More or less low	ML	(0.3,0.4,0.5)
Medium	M	(0.4,0.5,0.6)
More or less good	MG	(0.5,0.6,0.7)
Fairly good	FG	(0.6,0.7,0.8)
Good	G	(0.7,0.8,0.9)
Very Good	VG	(0.8,0.9,1)
Excellent	E	(0.9,1,1)

Step 1: Obtain a $n \times n$ fuzzy direct-relation matrix $\tilde{A}$from experts, based on pairwise comparisons of the criteria. its elements $\tilde{a}_{ij} = (l_{ij}, m_{ij}, u_{ij})$ represent the degree to which criterion j is affected by criterion i.

Step 2: Determine the normalised fuzzy direct-relation matrix $\tilde{X}$ using equation (1).

$$\tilde{X} = s \times \tilde{A} \quad (1)$$

Where $s = \frac{1}{\max 1 \le i \le n \sum_j^n U_{ij}}$

Step 3: Define three crisp matrices based on $\tilde{X}$, where $\tilde{x}_{ij} = (l_{ij}, m_{ij}, u_{ij})$.

$$X_l = \begin{bmatrix} 0 & l_{12} & \dots & l_{1n} \\ l_{21} & 0 & \dots & l_{2n} \\ \vdots & & & \vdots \\ l_{n1} & l_{n2} & \dots & 0 \end{bmatrix}, X_m = \begin{bmatrix} 0 & m_{12} & \dots & m_{1n} \\ m_{21} & 0 & \dots & m_{2n} \\ \vdots & & & \vdots \\ m_{n1} & m_{n2} & \dots & 0 \end{bmatrix},$$

$$X_u = \begin{bmatrix} 0 & u_{12} & \dots & u_{1n} \\ u_{21} & 0 & \dots & u_{2n} \\ \vdots & & & \vdots \\ u_{n1} & u_{n2} & \dots & 0 \end{bmatrix},$$

Step 4: Obtain the fuzzy total-relation matrix $\tilde{T}$ using equations (2) to (5).

$$\tilde{T} = \tilde{X}(I - \tilde{X})^{-1} \quad (2)$$
$$\text{Matrix}[l'_{ij}] = X_l(I - X_l)^{-1} \quad (3)$$
$$\text{Matrix}[m'_{ij}] = X_m(I - X_m)^{-1} \quad (4)$$
$$\text{Matrix}[u'_{ij}] = X_u(I - X_u)^{-1} \quad (5)$$

$$\tilde{T} = \begin{bmatrix} \tilde{t}_{11} & \tilde{t}_{12} & \dots & \tilde{t}_{1n} \\ \tilde{t}_{21} & \tilde{t}_{22} & \dots & \tilde{t}_{2n} \\ \vdots & & & \vdots \\ \tilde{t}_{n1} & \tilde{t}_{n2} & \dots & \tilde{t}_{nn} \end{bmatrix} \text{where, } \tilde{t}_{ij} = (l'_{ij}, m'_{ij}, u'_{ij})$$

and I is the identity matrix.

Step 5: Defuzzify $\tilde{T}$ using equation (6) and determine the total influence matrix for each set of sub-criteria considered.

$$F(\tilde{t}_{ij}) = \frac{1}{2}\int_0^1 \left(\inf_{x\in\mathcal{R}} \tilde{t}_{ij}^{\alpha} + \sup_{x\in\mathcal{R}} \tilde{t}_{ij}^{\alpha}\right) d\alpha \quad (6)$$

Step 6: compute the row sum *(r)* and the total influence matrix's column sum (c). *r + c* gives the degree to which factor *i* affects or is affected by *j*. A positive value of r-c means that factor i affects other factors and vice versa if it is negative[61].

Step 7: Determine the weights by normalising the *r + c* values.

3.3 TOPSIS

The technique for order of preference by similarity to ideal solution (TOPSIS) assumes that any chosen alternative should be closer to the positive ideal solution and far away from the negativeideal solution. TOPSIS has the following steps [61]:

Step 1: Obtain the decision matrix *D* for evaluation with elements f_{ij}, based on the various criteria considered.

Step 2: Determine the normalised decision matrix with elements r_{ij}.

$$r_{ij} = \frac{f_{ij}}{\sqrt{\sum_j f_{ij}^2}} \quad (7)$$

Step 3: Calculate the weighted normalised decision matrix with elements v_{ij} using the weight $w_i$for each criterion.

$$v_{ij} = w_i . r_{ij} \quad (8)$$

Step 4: Find the ideal and negative-ideal solution using equation (9) and (10), where *I'* is the set of benefit criteria, and *I"* is the set of cost criteria.

$$A^+ = \{v_1^+, \dots, v_n^+\} = \{(maxv_{ij}|i \in I'), (minv_{ij}|i \in I'')\} \quad (9)$$

$$A^- = \{v_1^-, \dots, v_n^-\} = \{(mixv_{ij}|i \in I'), (maxv_{ij}|i \in I'')\} \quad (10)$$

Step 5: Determine the distance from the positive ideal solution $D_j^+$and the distance from the negative-ideal solution $D_j^-$for each alternative *j*.

$$D_j^+ = \sqrt{\sum_i (v_{ij} - v_i^+)^2} \quad (11)$$

$$D_j^- = \sqrt{\sum_i (v_{ij} - v_i^-)^2} \quad (12)$$

Step 6: Evaluate the relative closeness to the ideal solution for each alternative *j* and rank the alternatives.

$$C_j^* = \frac{D_j^-}{(D_j^+ + D_j^-)} \quad (13)$$

4. DISCUSSION & CASE STUDY PLAN

4.1 VENRA TABULATED ATTRIBUTES

Table 2, Table 3, Table 4 and Table 5 showthe collected VENRA attributes categorised into cost/benefit criteria and subjective/objective criteria. The cost criteria mean the criteria will be adverse to the shipyard's performance in conducting the ship manufacturing process, while the benefit criteria increase the performance of each group of criteria. Subjective criteria mean that the assessment process will use expert opinion. In contrast, objective one will include quantitative data and sources through a number of shipyards. Before conducting the shipyard assessment for ship manufacturing, the grouped attributes will be assessed using fuzzy DEMATEL to achieve the weighted criteria and sub-criteria.

Table 2. Technical group performance attributes

No	Criteria	Sub-criteria	Cost/ Benefit	Subjective/ Objective
1	Shipyard Capacity	Total area	Benefit	Objective
		Erection area	Benefit	Objective
		Main Crain Capacity	Benefit	Objective
2	Building Strategy	N/A	Benefit	Objective
3	Workforce	Worker Education level	Benefit	Objective
		Workforce Availability: • Worker availability • Graduate availability	Benefit	Objective
		The average age of worker	Cost	Objective
4	Level of Technology	Fabrication and Assembly • Steel Stockyard and Treatment • Cutting and marking • Sub-assembly • Flat panels • Assembling	Benefit	Objective
		Erection and Outfitting • Erection • Outfitting	Benefit	Objective
		Product and processing	Benefit	Objective

No	Criteria	Sub-criteria	Cost/ Benefit	Subjective/ Objective
		engineering • Ship design (engineering design process) • Production engineering (hardware & software)		
		Layout, material flow, and environment	Benefit	Objective

Table 3. Bussiness group performance attributes

No	Criteria	Sub-criteria	Cost/ Benefit	Subjective/ Objective
1	Ship Manufacturing Cost	Labour cost	Cost	Objective
		Material cost	Cost	Objective
		Overhead cost	Cost	Subjective/ objective
2	Delivery time / speed of process	Keel laying to launching	Cost	Objective / Subjective
		Launching to delivery	Cost	Objective/ subjective
3	Shipyard experience	Build new building	Benefit	Objective
		Ship repair	Benefit	Objective
		Ship Modification/ conversion	Benefit	Objective
4	Organisation and management	Marketing team availability	Benefit	Objective
		Human resources (HRD)	Benefit	Objective
		R&D team availability	Benefit	Objective

Table 4. External group performance attributes

No	Criteria	Sub-criteria	Cost/ Benefit	Subjective/ Objective
1	Vertical integration-the ability to gain:	Suppliers	Benefit	Subjective
		additional workers	Benefit	Subjective
		Equipment	Benefit	Subjective
2	Government support/ subsidies	Domestic market protection	Benefit	Objective
		Investment and research support	Benefit	Objective
		Direct production assistance	Benefit	Objective
		Export credit assistance	Benefit	Objective
		Assistance to customers (eq. tax breaks for shipowners)	Benefit	Objective
		Home credit schemes	Benefit	Objective
		Favourable tax treatment	Benefit	Objective
		Customs import goods policy	Benefit	Subjective/ objective
3	Domestic Market Condition	NA	Benefit	Objective
4	Local Environment	NA	Benefit	Subjective

Table 5. Safety/risk group performance attributes

No	Criteria	Sub-criteria	Cost/ Benefit	Subjective/ Objective
1	Risk of Product	Product rework	Cost	Subjective
		Product failure	Cost	Subjective
2	Risk of personnel	Safety training induction	Benefit	Objective
		OHSAS certificate	Benefit	Objective

4.2 SHIPYARD CASE STUDY

About 245-250 shipyard industries in Indonesia are mainly located in Batam and Java Island, in which about 120 shipyards are associated withIPERINDO (Indonesian Shipyard Association) [62]. The capacity of the newbuilding with the facility of <50.000 DWT has the output of 12 million DWT/year, while the ship repair and maintenance capacity shipyard with <300.000 DWT has the output of 1 million DWT/year. Concerning the activity, the Indonesian national shipyard business industry includes shipbuilding, ship repair and maintenance, ship conversion, offshore industry fabricator, repair and maintenance, component and equipment industry, design and engineering, and ship recycling yard. The experience of the national shipyard includes building bulk carriers(50.000 DWT), ferry RORO(19.000 DWT), chemical tanker (200DWT), LPG carrier, passenger ship, general cargo, container vessel, tanker (30.000DWT), fishing vessel, tugboat, Ancor Handling Tug Supply (AHTS), cement carrier, dredging vessel, and yacht.

Regarding the VENRA attributes, they can be applied in any shipyard to assess the performance measurement, including Indonesian shipyards or other shipyardsworldwide. The authorsinitially plan to apply this framework to three Indonesian shipyards located in Java and Madura Island, as shown in Figure 3.

Figure 3. Location of the shipyards case study plan.

The first shipyard is a state-owned shipyard that has a high level of technology and organisation. With a crane of 300 tons and graving docks up to 200 meters, this shipyard can produce mixed products, including shipbuilding, ship repair, and offshore repair facilities. It is one of the shipyards that have high technology and product quality.

The second shipyard is a private shipyard that has a small area with outstanding human resources and a research and development team. They have experience in building IACS class ships such as harbour tugs. It also has flexibility in design and team.

The third shipyard is located in Madura Island, and it is a private shipyard. They have a large green area that has not been developed yet. Initially, they focused on ship repair for their own group shipping company. Then, they are developing the shipbuilding also for a ferry, general cargo owned by the Indonesian government and ordered from their own group shipping company. In general, this shipyard has an excellent opportunity to involve tender in the domestic market. The overview of the shipyard as the case study is shown in Figure 4, respectively, with the explanation above.

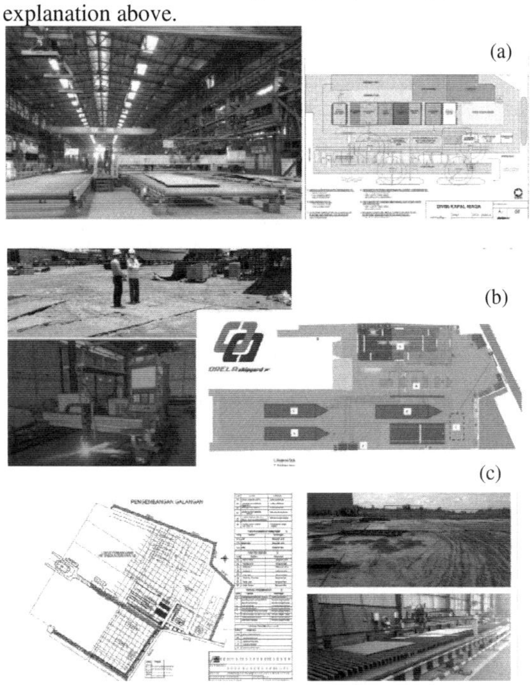

Figure 4. Shipyards case study, the first (a), the second (b), and the third (c).

While the collection and weighting process of the VENRA criteria has been explained before, the next step is to assess the three shipyards case study by collecting the data either in a quantitative or qualitative way. The quantitative data can be directly assessed using the tabulated VENRA attribute and score using the TOPSIS process methodology. The quantitative data can be converted by using expert opinion, which converts the data from qualitative to quantitative data.

5. CONCLUSION

The VENRA model approach has been proposed as criteria for ship manufacturingperformance measurement, combining the technical, business and external value and risk/safety aspect. The criteria assessment is still limited to the literature review and unstructured interviews with expertswithin the maritime domain and shipyards. The fuzzy MCDM approach can be enhancedby studying the benefits and the drawbacks of using Fuzzy MCDM for performance measurement and comparing it with other methods.

Moreover, as the work presented in thispaper is ongoing, further research will include updating and developingthe criteria mentioned above, including feedback and validation by industry experts working in the shipbuilding and ship repair industry. In addition, data related to the three mentioned case studies and shipyards will be collected through a number of sources i.e. annual reports, experts' interviews, and other supporting material in order to apply the VENRA framework.

6. ACKNOWLEDGEMENT

The author sincerely wants to thank the Directorate General of Recourses for Science Technology and Higher Education as BPP-LN "Beasiswa Pendidikan PascasarjanaLuar Negeri", the funding body scholar to pursue a PhD Researcher in the University of Strathclyde, the United Kingdom.

7. REFERENCES

1 J. L. Harbour, *The Basics of Performance Measurement*, vol. 21, no. 2. 1999.

2 OECD, "Compensated Gross Ton (CGT) System," 2007.

3 G. Zhangpeng and M. Flynn, "Productive shipyards," *LLoyds Shipp. Econ.*, pp. 37–38, 2006.

4 G. J. Bruce, "A review of the use of CGT for shipbuilding performance measurement," *2005 SNAME Marit. Technol. Conf. Expo Sh. Prod. Symp.*, vol. 22, no. 2, pp. 99–104, 2006.

5 J. F. Pires, T. Lamb, and C. Souza, "Shipbuilding performance benchmarking," *Int. J. Bus. Perform. Manag.*, vol. 11, no. 3, pp. 216–235, 2009.

6 D. Gavalas, T. Syriopoulos, and M. Tsatsaronis, "Assessing key performance indicators in the shipbuilding industry; an MCDM approach," *Marit. Policy Manag.*, vol. 00, no. 00, pp. 1–29, 2021.

7 E. C. Colin and M. M. O. Pinto, "Benchmarking shipbuilders' turnover of main assets," *J. Sh. Prod.*, vol. 25, no. 4, pp. 175–181, 2009.

8 M. Fareza, "Study of Shipbuilding Competitiveness - Benchmarking analysis as a tool to measure shipyards' competitiveness with a focus on Asian yards," 2020.

9 S. L. Chao and Y. H. Yeh, "Comparing the productivity of major shipyards in China, South Korea, and Japan – an application of a metafrontier framework," *Marit. Bus. Rev.*, vol. 5, no. 2, pp. 193–210, 2020.

10 L. D. Miles, *Techniques of Value Analysis and Engineering*, Third Edit. USA: Eleanor Miles Walker, Lawrence D. Miles Value Foundation, 1989.

11 SAVE International, "Value Standard and Body of Knowledge," 2007.

12 A. Dell'Isola, *Value Engineering: Practical Applications for Design, Construction, Maintenance & Operation*. Kingston - USA: R.S. Means Company, Inc, 1997.

13 S. K. I. Al-Fadhli, "Value Engineering and Constructability Assessment Relating Infrastructure Projects," in *IOP Conference Series: Materials Science and Engineering*, 2020, vol. 737, no. 1.

14 A. S. Gunarathne, N. Zainudeen, C. S. R. Perera, and B. A. K. S. Perera, "A framework of an integrated sustainability and value engineering concepts for construction projects," *Int. J. Constr. Manag.*, vol. 0, no. 0, pp. 1–13, 2020.

15 F. John, A. Reji, A. Vincent, B. Biju, and S. Kuriakose, "Application of Value Engineering Techniques in Residential Building," *Int. J. Res. Appl. Sci. Eng. Technol.*, vol. 9, no. 5, pp. 1193–1195, 2020.

16 N. I. Alekseev, D. A. Kissin, and V. E. Gorelov, "Design improvements in foam-cyclone cleaners by the value engineering method," *Chem. Pet. Eng.*, no. 24, pp. 187–191, 1988.

17 A. Bhosle, A. Sah, and D. K. Shinde, "Application of Value Analysis and Value Engineering for Cost Reduction of Global Pumping Unit," in *Advances in Simulation, Product Design and Development. Lecture Notes on Multidisciplinary Industrial Engineering*, 2020, pp. 663–674.

18 S. M. Silaskar and V. B. Shinde, "Weight Optimization of Valve for Cost Effectiveness: Using Value Analysis," in *Procedia Manufacturing*, 2018, vol. 20, pp. 329–337.

19 R. Vijayan, T. Thanka Geetha, B. Nishanth, M. Tamilarasan, and V. Vijaya Kumar, "Value engineering and value analysis of rear air spring bracket," in *Materials Today: Proceedings*, 2019, vol. 16, pp. 1075–1082.

20 V. Kalluri and R. Kodali, "Component Cost Reduction by Value Engineering: A Case Study," *J. Inst. Eng. Ser. C*, vol. 98, no. 2, pp. 219–226, 2017.

21 P. H. P. Setti, O. Canciglieri Junior, and C. C. A. Estorilio, "Integrated product development method based on Value Engineering and design for assembly concepts," *J. Ind. Inf. Integr.*, vol. 21, no. December 2020, p. 100199, 2021.

22 Garratt G, "Savings from The Use of Value Analysis," *Machy (Lond)*, vol. 115, no. 2972, pp. 682–687, 1969.

23 Y. H. Tao, J. Yang, and J. Shi, "Cost control of the marine diesel based on value engineering," *Appl. Mech. Mater.*, vol. 551, pp. 535–540, 2014.

24 H. L. Hyland, "Value analysis applied to the optimum division of a large graving dock using a slave caisson," *J. Sh. Prod.*, vol. 7, no. 2, pp. 69–76, 1991.

25 E. Al-Yafei, S. Ogunlana, and A. Oyegoke, "Application of value engineering and life cycle costing techniques for offshore topside facility projects: Towards sustainability," *Soc. Pet. Eng. - SPE Kuwait Oil Gas Show Conf. 2017*, 2017.

26 R. L. Storch and M. Williamson, "Technology value analysis of the ship contract design value stream," *IFIP Adv. Inf. Commun. Technol.*, vol. 129, pp. 147–158, 2003.

27 P. Romano, M. Formentini, C. Bandera, and M. Tomasella, "Value analysis as a decision support tool in cruise ship design," *Int. J. Prod. Res.*, vol. 48, no. 23, pp. 6939–6958, 2010.

28 M. Formentini and P. Romano, "Using value analysis to support knowledge transfer in the multi-project setting," *Int. J. Prod. Econ.*, vol. 131, no. 2, pp. 545–560, 2011.

29 P. Tang and R. B. Bittner, "Use of value engineering to develop creative design solutions for marine construction projects," *Pract. Period. Struct. Des. Constr.*, vol. 19, no. 1, pp. 129–136, 2014.

30 T. N. Desai, S. R. Prajapati, and H. R. Patel, "Application of Value Engineering to Rework Reduction in Shipbuilding Project," in *MATEC Web of Conferences 40*, 2016, no. 2, pp. 17–45.

31 H. Karami and O. A. Olatunji, "Key value engineering protocols in marine projects," *Proc. Inst. Civ. Eng. - Manag. Procure. Law*, vol. Vol. 173, no. 1, pp. 21–31, 2020.

32 M. J. Lee, J. K. Lim, and G. Hunter, "Performance-based value engineering application to public highway construction," *KSCE J. Civ. Eng.*, vol. 14, no. 3, pp. 261–271, 2010.

33 R. S. Heralova, "Possibility of Using Value Engineering in Highway Projects," in *Creative Construction Conference*, 2016, vol. 164, no. June, pp. 362–367.

34 S. Anam, E. Pradjoko, and B. Anshari, "Analysis

of infrastructures design at mandalika resort project by using the value engineering," *Int. J. Civ. Eng. Technol.*, vol. 9, no. 5, pp. 626–633, 2018.

35 G. Karunasena and K. Rajagalgoda Gamage, "A decision-making formula for value engineering applications in the Sri Lankan construction industry," *J. Financ. Manag. Prop. Constr.*, vol. 22, no. 1, pp. 77–91, 2017.

36 J. Heidary Dahooie, S. J. Hosseini Dehshiri, A. Banaitis, and A. Binkytė-Vėlienė, "Identifying and prioritizing cost reduction solutions in the supply chain by integrating value engineering and gray multi-criteria decision-making," *Technol. Econ. Dev. Econ.*, vol. 26, no. 6, pp. 1311–1338, 2020.

37 C. Zhao, E. Yang, Y. Nie, and J. D. Russo, "Facility decision making process with modified value engineering approach," *J. Corp. Real Estate*, vol. 21, no. 2, pp. 112–129, 2019.

38 D. P. Mootanah, "Developing an integrated risk and value management framework for construction project management," in *14th Annual ARCOM Conference, 9-11 September 1998, University of Reading. Association of Researchers in Construction Management*, 1998, vol. 2, no. 4, pp. 48–53.

39 S. B. Abd-Karim, D. J. Lowe, H. Abdul-Rahman, C. Wang, I. A. Yahya, and G. Q. Shen, "Integrating risk and value management using IRVM workshops: Case studies in infrastructure projects in UK," *Sci. Res. Essays*, vol. 6, no. 12, pp. 2470–2479, 2011.

40 R. Ahmed, B. John, and C. Nicholas, "Integration of risk management and value management - An Australian case study," *Proc. Int. Conf. Value Eng. Manag. Innov. Value Methodol. ICVEM 2012*, no. Vm, pp. 23–28, 2012.

41 H. R. Feili, S. Nasiri, and N. Akar, "Integrating Risk Management and Value Engineering in the Development of Renewable Energy Project," in *6th Symposium on Advances in Science and Technology*, 2012, pp. 1–10.

42 M. Pedju and E. Mawu, "Considering the integrated value engineering and risk management techniques during the design phase in construction projects – its implications to project objectives (A Case Study on Certain Irrigation Projects, North Sulawesi Province)," *J. Ilm. Media Eng.*, vol. 3, no. 3, pp. 214–226, 2013.

43 D. M. M. El Khatib, "Integrating Project Risk Management and Value Engineering in Tendering Processes," *Int. J. Eng. Res.*, vol. 4, no. 8, pp. 442–445, 2015.

44 M. Golabchi, H. Toosi, and P. Shahsavand, "Quantitative Risk Analysis and Integration of Value Engineering to Increase the Efficiency of Project Management Case Study: Sattar Khan Commercial Centre - Flour Parking - Tehran," *Mediterr. J. Soc. Sci.*, vol. 7, no. 4, 2016.

45 B. A. Sabzkohi and T. Pourrostam, "Integrated Model of Value Engineering and Risk Management Approaches in Empowerment Projects (the Exterior Design)," *Int. J. Adv. Manuf. Technol.*, vol. 3, no. 3, pp. 52–56, 2016.

46 H.-J. (Hans-J. Zimmermann, *Fuzzy set theory-- and its applications*, 2nd rev. e. Boston: Kluwer Academic Publishers, 1991.

47 S.-J. Chen and C.-L. Hwang, "Fuzzy Multiple Attribute Decision Making Methods," in *Fuzzy Multiple Attribute Decision Making: Methods and Applications*, Berlin, Heidelberg: Springer Berlin Heidelberg, 1992, pp. 289–486.

48 T. J. Ross, *Fuzzy logic with engineering applications*, 3rd ed.. Chichester, U.K.: John Wiley, 2010.

49 A. F. Guneri, M. Cengiz, and S. Seker, "A fuzzy ANP approach to shipyard location selection," *Expert Syst. Appl.*, vol. 36, no. 4, pp. 7992–7999, 2009.

50 B. Sahin, D. Yazir, A. Soylu, and T. L. Yip, "Improved fuzzy AHP based game-theoretic model for shipyard selection," *Ocean Eng.*, vol. 233, no. April, p. 109060, 2021.

51 I. Lazakis and A. Ölçer, "Selection of the best maintenance approach in the maritime industry under fuzzy multiple attributive group decision-making environment," *Proc. Inst. Mech. Eng. Part M J. Eng. Marit. Environ.*, vol. 230, no. 2, pp. 297–309, 2016.

52 A. I. Ölçer and A. Y. Odabaşi, "A new fuzzy multiple attributive group decision making methodology and its application to propulsion/manoeuvring system selection problem," *Eur. J. Oper. Res.*, vol. 166, no. 1 SPEC. ISS., pp. 93–114, 2005.

53 X. Ziquan, Y. Jiaqi, M. H. Naseem, X. Zuquan, and L. Xueheng, "Supplier Selection of Shipbuilding Enterprises Based on Intuitionistic Fuzzy Multicriteria Decision," *Math. Probl. Eng.*, vol. 2021, 2021.

54 M. Celik, I. Deha Er, and A. F. Ozok, "Application of fuzzy extended AHP methodology on shipping registry selection: The case of Turkish maritime industry," *Expert Syst. Appl.*, vol. 36, no. 1, pp. 190–198, 2009.

55 V. Dixit, P. Verma, and P. Raj, "Leveraging tacit knowledge for shipyard facility layout selection using fuzzy set theory," *Expert Syst. Appl.*, vol. 158, p. 113423, 2020.

56 T.-Y. Chou and G.-S. Liang, "Application of a fuzzy multi-criteria decision-making model for shipping company performance evaluation," *Marit. Policy Manag.*, vol. 28, no. 4, pp. 375–392, 2001.

57 M. Celik, C. Kahraman, S. Cebi, and I. D. Er, "Fuzzy axiomatic design-based performance evaluation model for docking facilities in shipbuilding industry: The case of Turkish shipyards," *Expert Syst. Appl.*, vol. 36, no. 1, pp. 599–615, 2009.

58 P. de J. Freitas Pinto, C. H. dos Santos Grecco, and C. A. Nunes Cosenza, "Fuzzy Model for the Priorization Analysis of Variable Quality Performance: An Approach in Shipbuilding," *Fuzzy Inf. Eng.*, vol. 12, no. 2, pp. 181–203, 2020.

59 C. Gayathri, V. Kamala, M. S. Gajanand, and S. Yamini, "Analysis of operational and financial performance of ports: an integrated fuzzy DEMATEL-TOPSIS approach," *Benchmarking An Int. J.*, vol. ahead-of-p, no. ahead-of-print, Jan. 2021.

60 G. Bruce, "Competitiveness," in *Shipbuilding Management*, 2021, pp. 23–32.

61 J. K. Chen and I. S. Chen, "Using a novel conjunctive MCDM approach based on DEMATEL, fuzzy ANP, and TOPSIS as an innovation support system for Taiwanese higher education," *Expert Syst. Appl.*, vol. 37, no. 3, pp. 1981–1990, 2010.

62 Kementrian Perindustrian Republik Indonesia and H. Nugroho, "Pengembangan dan Pembinaan Industri Maritim Nasional Dalam Mendukung Penggunaan Produk Dalam Negeri (P3DN)," 2021.

7. AUTHORS BIOGRAPHY

Imam Baihaqi is currently a PhD Researcher at the Department of Naval Architecture, Ocean, and Marine Engineering, University of Strathclyde, United Kingdom. His research interests include shipbuilding manufacturing and productivity, risk analysis in shipbuilding, and marine material experiment. At the same time, he is the lecturer at the Department of Naval Architecture and Shipbuilding Engineering in Institut Teknologi Sepuluh Nopember, Surabaya, Indonesia.

Dr. Iraklis Lazakis, Reader, CEng, MBA, FRINA, FSNAME, FHEA, PhD is a Reader at the Department of Naval Architecture, Ocean, and Marine Engineering at the University of Strathclyde. He has extensive academic and industry experience working in UK and EU research and marine industry projects. His research interests include ship systems condition monitoring, systems maintenance and reliability, ship operations, risk analysis, shipyard manufacturing and productivity, and asset management. He is currently the Chair in the ISSC Committee V.7 - Structural longevity

Dr Rafet Emek Kurt, has principal research interests lie in the field of human factors in the maritime domain. He has investigated the effects of noise and vibration on human performance and developed a human response model for noise and vibration levels on-board ships. His research interests also include design, optimisation, operation of ships, by considering the human factors in the core of the subject. Besides his main research area, he has also been actively working in the area of ship recycling. For 8 years he has been working in this area and contributed the development of tools, models and training tailored for the unique needs of the ship recycling sector. He is also involved in following EU funded ship recycling projects; FP7 DIVEST Project, LLP Ship DIGEST Project and LLP Boat DIGEST Project. He is also actively attending IMO MSC and MEPC Meetings.

SOME REMARKS ON THE APPLICATION OF DEAD SHIP CRITERIA TO AN INDONESIAN RO-RO FERRY

D Paroka, M A Azis, A H Muhammad and **S Rahman,** Hasanuddin University, Indonesia

SUMMARY

This paper discusses thedead ship criteria applied to Indonesian ro-ro ferries. Model experiments consisting of roll decay and roll test in beam seas were conducted to calculate the hydrodynamics factors corresponding to the vulnerability criteria level 1 and level 2 of the dead ship criteria. The results show that the damping factors were smaller than those in the weather criterion. The effective wave slope coefficient was not significantly different from that obtained by the weather criterion when the radius gyration coefficient was the same as that calculated by the formula in the weather criterion. The ships comply with the weather criterion in full loading condition for the hydrodynamics factors determined by the weather criterion and by a model experiment. Inconsistency between the vulnerability criteria level 1 and the vulnerability criteria level 2 was obtained when the scatter wave data of IMO was used to calculate the capsizing index.

1. INTRODUCTION

The International Maritime Organization (IMO) developed Second Generation Intact Stability Criteria (SGISC) to consider ship dynamics on operational safety. The criteria were designed with a multilayer format for five dangerous scenarios of a ship in a seaway. One of the operation scenarios considered in the SGISC was the dead ship criteria. The vulnerability criteria level 1for the dead ship condition was evaluated by using the weather criterion [1] with an extension of the natural roll period to 30 seconds [2]. Here, the damping factors corresponding to the formula to calculate the roll angle due to wave in windward are determined following the values recommended in the weather criterion [1]. Those damping factors were statistically determined based on ships with a breadth-to-draught ratio smaller than 3.5, the block coefficient smaller than 0.7 and the ratio between the vertical centre of gravity and ship draught between 0.7 and 1.5.
The vulnerability criteria level 2 used a capsizing index based on the roll motion of the ship in beam wind and waves [2]. IMO recommended statistical wave data and a formula to calculate the mean wind velocity as a function of significant wave height. The damping coefficients can be calculated by using IKEDA semi-empirical method or model experiment. The effective wave slope coefficient was recommended to calculate using the Froude-Krylov assumption of wave excitation moment with a rectangular shape of the ship cross-section [3]. A ship is vulnerable to level 2 criteria if the capsizing index is larger than 0.06 [4].Here, the capsizing was assumed to occur if the roll angle is larger than the down-flooding angle of the angle of 40 degrees which is one in the less.
The Indonesian ro-ro ferries have different geometric characteristics with the ship used to develop the weather criterion. The breadth-to-draught ratio of the Indonesian ro-ro ferry is mostly larger than 3.50, and the ratio between the vertical centre of gravity and the ship draught is larger than 1.5 [5]. The large ratio between the vertical centre of gravity and the ship draught occurs becausemost Indonesian ro-ro ferries have small draught and all payloads located above the main deck.
The damping factor that corresponds to the breadth-to-draught ratio given in the weather criterion could be overestimated when it is applied to a ship with a breadth-to-draught ratio larger than 3.5. A smaller damping factor due to breadth-to-draught ratio for a ship with a ratio larger than 3.5 has been obtained by Deakin ([6] and IMO [7]. The effective wave slope coefficient of an Indonesian ro-ro ferry could be larger than 1.0 when the formula of weather criterion is used. This large effective wave slope coefficient is obtained because of the large vertical centre of gravity. A smaller effective wave slope coefficient was obtained by model experiment for a large passenger ship and a ro-pax ferry[8], [9]. Paroka[10] and Sato et al.[11] were also obtained a smaller effective wave slope coefficient of ships with a large breadth-to-draught ratio.
The weather criterion's damping factor that corresponds to the bilge keels may also overestimate for a small ship. The bilge keels could reduce the damping coefficient by more than 30 percentages [12]. A more significant reduction of damping coefficient due to bilge keels has been obtained by Gu et al.[13]. Effect of bilge keels on the roll motion of ships with a large breadth-to-draught ratio and a large vertical centre of gravity could be more significant [14].However, the effect of the breadth-to-draught ratio and the vertical centre of gravity was not considered in determining the damping factor corresponding to bilge keels. The damping factor due to bilge keels was determined based on the ratio between the total area of bilge keels and the product of the length of the waterline and ship breadth. A more advanced investigation should also be paid on the wave steepness for a ship with a small natural roll period. A small natural roll period occurs on a small ship, such as the Indonesian ro-ro ferries.
For the vulnerability criteria level 2, the wave characteristics have a significant effect. A different safety level for a ship could be obtained when the ship operated in a different location. Therefore, scatter wave data for each operational location of the Indonesian ro-ro ferries should be developed.

The IMO also recommend the JONSWAP spectrum to model irregular wave during the calculation of the capsizing index. A different wave spectrum may be necessary for a different location. A different capsizing index was obtained for different wave spectrum [15]. IMO recommended the effective wave slope coefficient to calculate by using the Froude-Krylov wave exciting moment with a rectangular assumption of cross-section [3]. Rudakovic et al.[16]estimatedthe effective wave slope coefficient of a river-sea ship with three different methods. They found that the reduced 3 DOF method was the most appropriate for the intact stability assessment of river-sea ships. The Froude-Krylov assumption tends to be conservative when it is applied to the river-sea ship. The effective wave slope coefficient of an Indonesian ro-ro ferry obtained by the Froude-Krylov assumption corresponds to the natural roll period was smaller than that obtained by the weather criterion formula, but it has a good agreement with the result of the model experiment [17].

In order to apply the SGISC to an Indonesian ro-ro ferry, an alternative approach to calculate the damping factors correspond to the formula to calculate the roll angle due to wave in windward should be developed because the values given in the weather criterion were obtained from the ship data with breadth-to-draught ratio smaller than 3.5, the ratio between the vertical centre of gravity and the ship draught smaller than 1.5 and the block coefficient smaller than 0.7. Paroka et al.[18] proposed an alternative method following the three-step procedures of IMO to evaluate ship stability against the weather criterion by model experiment [19]. However, the model experiment is time-consuming and expensive, mainly determining the effective wave slope coefficient for corresponding wave frequency. A numerical method such as computational fluid dynamics (CDF) may become an alternative method because some researches to estimate the damping coefficients and the effective wave slope coefficient using numerical methods have been conducted [16], [20]. The numerical method is also time-consuming and complex froma practical point of view.

This paper discusses the application of the SGISC to evaluate the stability of Indonesian ro-ro ferries in dead ship conditions. Some experimental results correspond to the damping factors, and the effective wave slope coefficient corresponding to the calculation of roll angle to windward due to wave action were presented and compared to the values given in the weather criterion. The differencein scatter wave data and effect to the capsizing index will be presented. The results can be used as consideration when the SGISC is applied to the Indonesian ro-ro ferries. The results could also be used to develop alternative criteria to evaluate the stability of a ship with a breadth-to-draught ratio larger than 3.5 and the ratio between the vertical centre of gravity and the ship draught larger than 1.5 or used for the design of Indonesian ro-ro ferry in the future.

2. SHIP DATA

Two Indonesian ro-ro ferries are used to investigate the hydrodynamics factors corresponding to the weather criterion as well as to analyse the effect of scatter wave data on capsizing index in the vulnerability criteria level 2 for the dead ship condition. The principle dimension and the righting arm curve for full loading conditions are shown in Table 1, Figure 1 and Figure 2, respectively.

Table 1: Principle dimensions of ships

Dimension	Ship 1	Ship2
Length (Lbp)	50.50 m	47.25 m
Breadth (B)	14.00 m	13.00 m
Height (H)	3.80 m	3.45 m
Draught (d)	2.70 m	2.45 m
Block coefficient (Cb)	0.706	0.62
Centre of gravity (KG)	4.72 m	4.00 m

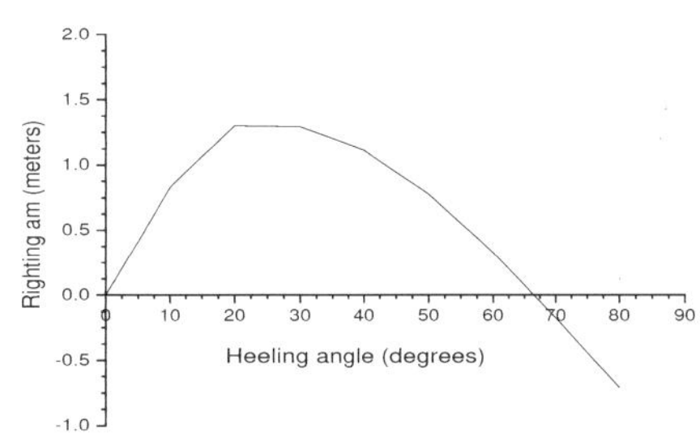

Figure 1: Righting arm of the sample ship 1.

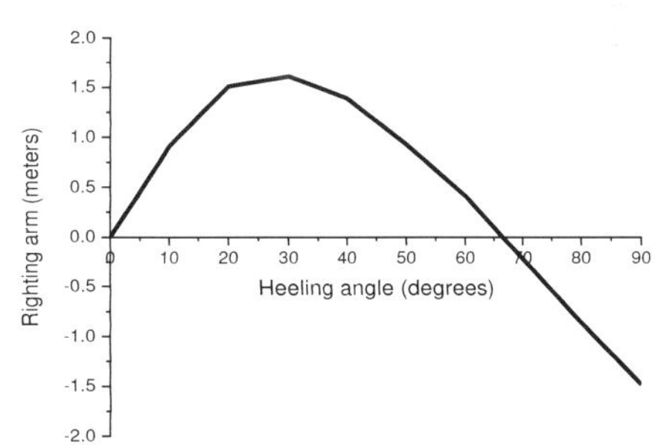

Figure 2: Righting arm of the sample ship 2.

The damping factors correspond to the breadth-to-draught ratio and bilge keels as well as the effective wave slope coefficient,were determined following the three procedures method recommended by IMO [19]. These procedures have been used by Paroka et al.[18], [21] to determine the hydrodynamics factors corresponding to the weather criterion. Here, two model experiments, namely roll decay test and roll test in regular beam waves, were conducted. The model scale of the sample ships was 1 : 40. The radius gyration of the ships was calculated assuming that the heaviest payloads in the vehicle deck located in/near the centreline and the other assumption that the weight distribution follows the radius gyration coefficient the same as that obtained by the formula in the weather criterion. The last assumption results in a larger radius gyration of roll motion compared to the first one. This radius gyration coefficient was larger than the upper limit for the ro-pax ferry recommended by Papanikolaou et al.[22]. However, a significant error of the formula in the weather criterion may occur when it is applied to a ship with a large breadth-to-draught ratio and a larger metacentric height [23].

The roll decay test was conducted with an initial heel angle of 20 degrees and stopped when the amplitude was smaller than 0.5 degrees. The roll test in beam waves was conducted for wave steepness of 0.01, 0.02, 0.03 and 0.04. Five-wave frequencies, namely two frequencies larger and two frequencies smaller than the natural roll frequency for each wave steepness, were used in the roll test. Each test was conducted for 60 seconds to govern a regular roll motion. Wave probes were located before and after the model to measure the actual wave characteristics. The actual wave for calculating the hydrodynamics factors was determined as the average of the wave characteristics measured by those two wave probes. The model experiment was conducted with the model free in sway, heave and roll. A pair of flexible wire ropes was used to restrict the surge, pitch and yaw motions. The tests were conducted for the ships with and without bilge keels to investigate the bilge keels effect.

The obtained damping factors correspond to breadth-to-draught ratio and bilge keels, and the effective wave slope coefficient is compared to those obtained in the weather criterion of IMO or the other calculation method such as the effective wave slope coefficient by using the Froude-Krilov assumption. The hydrodynamics factors are used to evaluate the stability of the subject ship against the vulnerability criteria level 1 and level 2 to understand the effect on the application of SGISC to the Indonesian ro-ro ferries.

3. RESULTS AND DISCUSSION

3.1 EFFECTIVE WAVE SLOPE COEFFICIENT

The effective wave slope coefficient obtained by the model experiment, calculated by the formula in the weather criterion and calculated by the formula proposed by Paroka[10] for the different vertical centres of gravity, are shown in Figure 3. In addition, the effective wave slope coefficient calculated by using the Froude-Krylov assumption of wave excitation moment for the natural roll frequency is also shown in Figure 3.

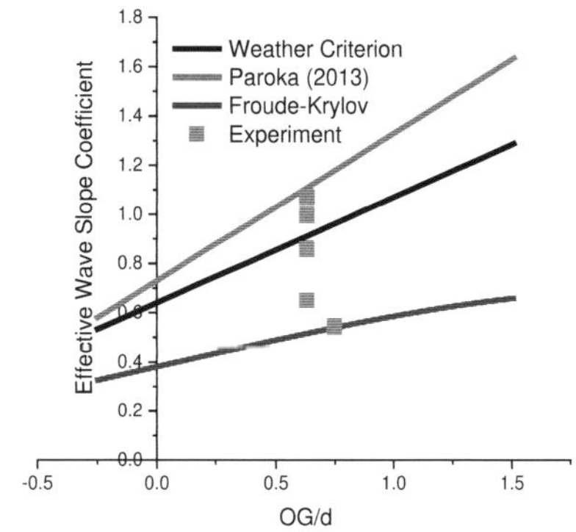

Figure 3: Effective wave slope coefficient.

The formula of weather criterion results in the largest effective wave slope coefficient. A good agreement between the formula of weather criterion and the model experiment was obtained when the radius gyration coefficient of the ship was close to that calculated by using the formula of weather criterion. The effective wave slope coefficient for a smaller radius gyration coefficient is smaller than that for a larger radius of gyration. A good agreement between the result obtained by the formula proposed by Paroka[10] and the result of the model experiment in the case of the smaller radius of gyration when the ship uses bilge keels have been found. The formula proposed by Paroka[10] was statistically determined based on the effective wave slope coefficients of three Indonesian ro-ro ferries obtained by Frank-Closed Fit theory combined with 2D strip theory. The strip theory results did not significantly differfrom the result of the model experiment for the ship with a large breadth-to-draught ratio [11]. For the ship without bilge keels, the Froude-Krylov wave exciting moment results were similar to the result of the model experiment for sample ship 1. Some other researchers also obtained a smaller effective wave slope coefficient than the result of the formula in the weather criterion [8], [9].

The effective wave slope coefficient calculated by the Froude-Krylov wave exciting moment was the smallest result obtained for the Indonesian ro-ro ferry. Rudakovic et al.[16]found that the Froude-Krylov wave exciting moment resulted in a conservative value compared to the other method. The capsizing index in the vulnerability criteria level 2 depends on the effective wave slope coefficient. A larger capsizing index may be obtained when a larger effective wave slope coefficient is used to calculate the roll angle. Following the results shown in Figure 3, the radius gyration of roll or weight distribution against the ship's longitudinal axissignificantlyaffects the effective wave slope coefficient. The heaviest vehicles should be located near the ship centreline and the lightest ones located near to the ship's side to reduce the radius gyration on operational conditions.

3.2 DAMPING FACTOR X_1

The damping factor corresponds to the breadth-to-draught ratio (X_1) significantly affected by the radius gyration of the roll. The model experiment results for the ship with radius gyration coefficientare the same as that calculated by the formula in the weather criterion for the ship with smaller radius gyration is shown in Figure 4.

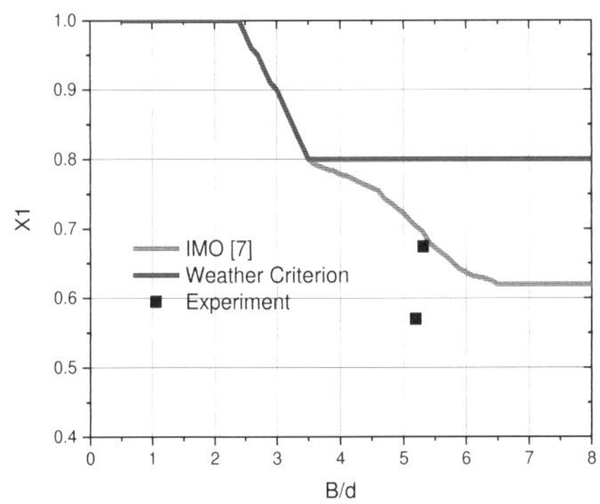

Figure 4: Damping factor (X_1) due to breadth-to-draught ratio.

The red colour shows the damping factor corresponds to the breadth-to-draught ratio proposed by Russia [7] and the blue colour indicates the damping factor based on the weather criterion. The damping factor due to breadth-to-draught ratio obtained by model experiment with the radius gyration coefficient the same as that calculated by the formula in the weather criterion was similar with the value given in the Russian proposal. However, the result of the experiment with a smaller radius gyration coefficient was smaller than the values given in the Russian proposal. The damping factor given in the weather criterion was larger than the model experiment results, even for the radius gyration coefficient the same as the value obtained by the formula in the weather criterion. Deakin [6] also found a similar result forships with a breadth-to-draught ratio larger than 3.5. The effect of the breadth-to-draught ratio on the damping factor in the weather criterion was assumed to be constant for ships with a ratio larger than 3.5. This assumption is because the ship's data were used to determine the damping factor due to the breadth-to-draught ratio havinga ratio smaller than 3.5.

The damping factor corresponding to the breadth-to-draught ratio was determined under the assumption that the damping factor corresponding to the block coefficient was the same as the value given in the weather criterion. Therefore, the damping factor due to breadth-to-draught ratio was smaller than the values given in the weather criterion, mainly for the ship with radius gyration coefficient smaller than the radius gyration coefficient calculated by the formula in the weather criterion.

3.3 DAMPING FACTOR k

The damping factor corresponds to the bilge keels of the sample ships were shown in Figure 5. The damping factor obtained by the model experiment was indicated by rectangular, and those given in the weather criterion were shown as the red line.

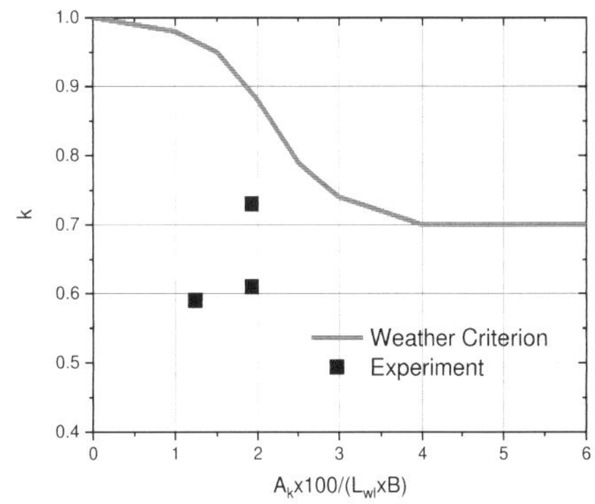

Figure 5: Damping factor due to bilge keels.

The damping factor due to bilge keels was significantly affected by the radius gyration of the roll. The damping factor corresponds to the bilge keels decreasing due to an increase in the radius gyration. The natural roll period of the ship increases when the radius gyration increases. The angular velocity of the roll decreases as the period increase. Therefore the damping coefficient of bilge keels decreases. The damping factor due to bilge keels for the two sample ships was not significantly different for the smaller radius gyration of the roll. The difference occurs due to different ratios between the bilge keels area and the product of the length of the waterline and ship breadth.

The damping factor obtained by the model experiment was smaller than the values given in the weather criterion.The bilge keels could reduce the roll angle by about 30% [12]. The other result shows that the damping moment of the roll could reduce twice due to using the bilge keels depending on the area of bilge keels. For ships with a large breadth and vertical centre of gravity, the bilge keels effect is more significant than the normal ships. Paroka et al.[18] found the reduction of roll angle due to bilge keels of 39% with bilge keels area of 9.0 m^2. Thus, the damping factor corresponding to bilge keels given in the weather criterion seems to be overestimated compared to the reduction of damping moment obtained by some researchers.

3.4 VULNERABILITY CRITERIA

The ratio between area under the righting arm curve from the angle of equilibrium point to the angle of down-flooding or the angle of 40 degrees, which is the smallest and the area under the righting arm curve from the angle of equilibrium point to the heel angle due to wave in windward for several metacentric heights was calculated as the vulnerability criteria level 1 for the dead ship criteria in SGISC. The results are shown in Figure 6.

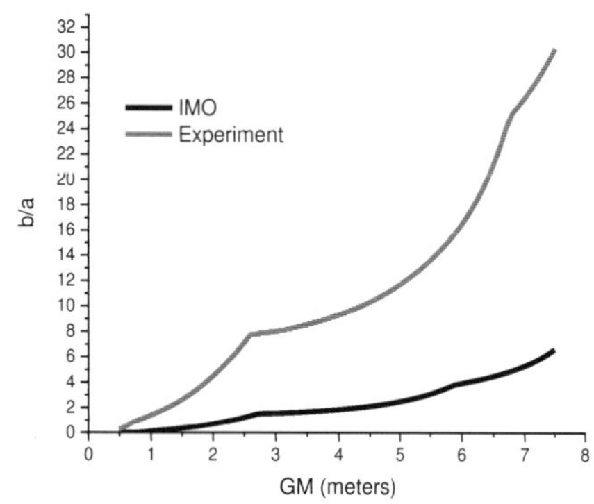

Figure 6: The vulnerability criteria level 1

Here, the b/a was calculated with hydrodynamics factors and the mean wind velocity based on the weather criterion of IMO and based on values obtained by model experiment with wind pressure of 300 Pa. The b/a for metacentric height corresponding to the full loading condition of the subject ship was 2.26 for hydrodynamics factors based on the weather criterion and that of 10.83 for hydrodynamics factors obtained by model experiment. The significant increase of b/a occurs because the damping factors due to the breadth-to-draught ratio and the bilge keels obtained by the model experiment were smaller than those based on values given in the weather criterion. The wind pressure used in the weather criterion was also larger than that used for the damping factors obtained by the model experiment. In the same wind pressure is used, b/a for the damping factor obtained by the model experiment was 7.14. The critical metacentric height was 2.39 meters for the damping factors given in the weather criterion and 0.80 meters when the model experiment determines the damping factors.

These results showed that the subject ship comply with the vulnerability criteria level 1 in full loading condition. The critical metacentric height obtained by using the weather criterion's damping factors seems to be overestimated when the criteria are applied to a ship with a breadth-to-draught ratio larger than 3.5 and a large metacentric height.

The capsizing index based on the scatter wave data of Lombok Strait is shown in Figure 7. Here, the horizontal axis was the b/a values correspond to the hydrodynamics factors obtained by model experiment with wind pressure of 300 Pa. The vertical axis indicates the capsizing index for variation metacentric height as shown in Figure 6. The maximum capsizing index based on the SGISC was 0.06 as stated in SCD 7/WP.6 (Annex-page 16 paragraph 2.2.3) [4].

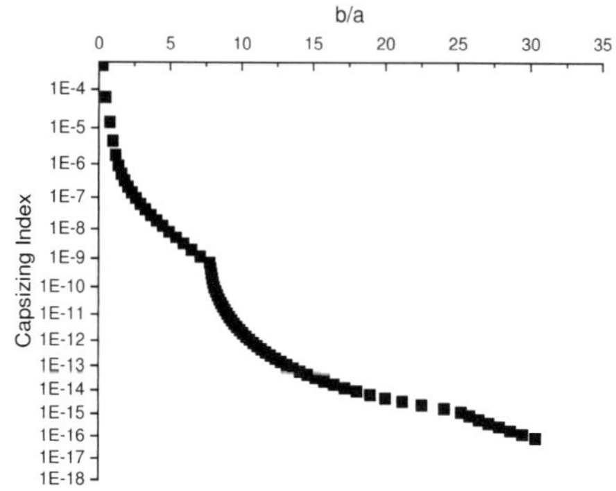

Figure 7: Capsizing index corresponds to b/a

The capsizing index corresponds to the critical metacentric height obtained based on the vulnerability criteria level 1 was 0.106. The metacentric height corresponding to the maximum capsizing index in the SGISC was 2.99 meters. The results showed the inconsistency of the criteria because the vulnerability criteria level 2 did not comply even the ship still complied with the vulnerability criteria level 1. The inconsistency between the vulnerability criteria level 1 and the vulnerability criteria level 2 of dead ship condition in the SGISC has been found for several ship types [2]. However, when the damping factors in the formula to calculate the heel angle due to wave in windward determined by model experiment and the scatter wave data of Lombok Strait were used, the vulnerability criteria level 1 was found to be consistent with the vulnerability criteria level 2. The capsizing index correspond to the critical metacentric height in the vulnerability criteria level 1 was 1.11×10^{-6}. This value is smaller than the maximum capsizing index given in the SGISC.

4. CONCLUSION

The hydrodynamics factors correspond to the formula to calculate the heel angle due to wave in windward consists of the damping factor due to the breadth-to-draught ratio and the damping factors due to the bilge keels have been determined by model experiment for two subject ships. The vulnerability criteria level 1 and the vulnerability criteria level 2 of dead ship condition in SGISC were calculated based on those damping factors and the damping factors given in the weather criterion. Based on the results and discussion, some conclusions can be remarked as follows:

1. The damping factors corresponding to the breadth-to-draught ratio and corresponding to the bilge keels obtained by the model experiment were smaller than those given in the weather criterion.
2. The effective wave slope coefficient obtained by the model experiment agrees with that calculated by the formula in the weather criterion when the radius gyration coefficient was the same as the radius

gyration coefficient determined by the formula in the weather criterion.

3. The vulnerability criteria level 1 and the vulnerability criteria level 2 of the SGISC showed inconsistency when the damping factors in the weather criterion and the scatter wave data recommended by IMO were used. However, the consistency between those two vulnerability levels was obtained when the damping factors obtained by the model experiment and the scatter wave data on Lombok Strait were used.
4. Further investigation of the damping factors is necessary for future work because the damping factors correspond to the block coefficient in the weather criterion used in the present work.

5. ACKNOWLEDGEMENTS

The data presented in this paper is a part of the research supported by Hasanuddin University and the Directorate General of Higher Education with grand number: 752/UN4.22/PT.01.03/2021. The authors would like to express their sincere gratitude to those institutions for their support.

6. REFERENCES

1. IMO, 'The International Code on Intact Stability (IS CODE 2008)', International Maritime Organization, London, 2008.

2. IMO, 'Finalization of Second Generation Intact Stability Criteria: Information Collected by the Correspondence Group on Intact Stability Regarding Second Generation Intact Stability Criteria', SDC 3/INF.10, International Maritime Organization, London, 2015.

3. IMO, 'Development of Second Generation Intact Stability Criteria: Vulnerability Assessment for Dead Ship Stability Failure Mode', SDC 1/INF. 6, International Maritime Organization, London, 2013.

4. IMO, 'Finalization of Second Generation Intact Stability Criteria: Report of the Drafting Group on Intact Stability', SDC 7/WP. 6, International Maritime Organization, London, 2020.

5. PAROKA, D., MUHAMMAD, A.H., RAHMAN, S., 'Estimation of Effective Wave Slope Coefficient of Ships with Large Breadth and Draught Ratio', *Kapal: Journal of Marine Science and Technology*, Vol. 17(1), 2020.

6. DEAKIN, B., 'Evaluation of the Roll Prediction Method in The Weather Criterion', *International Journal of Maritime Engineering*, Vol 150(2), 2008.

7. IMO, 'Severe Wind and Rolling Criterion (Weather Criterion)', Document SLF/46/6/10 Submitted by Russian Federation, International Maritime Organization, London, 2003.

8. FRANCESCUTTON, A., SERRA, A., SCARPA, S., 'A Critical Analysis of Weather Criterion for Intact Stability of Large Passenger Vessels', *Proceedings of 20th International Conference on Ocean, Offshore and Arctic Engineering*, Rio de Janeiro, 2001.

9. ISHIDA, S., TAGUCHI, H., SAWADA, H., 'Contemporary Ideas on Ship Stability and Capsizing in Waves: Evaluation of the Weather Criterion by Experiment and its Effect to the Design of a Ropax Ferry', Springer. London, 2011.

10. PAROKA, D., 'Characteristics of Indonesian Seas: Basic for Development of Local Intact Stability Criteria', JurnalTeknik BKI PROPULSI 1(1), 2014 (in Indonesian).

11. SATO, Y., TAGUCHI, H., UENO, M., SAWADA, H., 'An Experimental Study of Effective Wave Slope Coefficient for Two Dimensional Model', *Proceedings of the 6th Osaka Colloquium on Seakeeping and Stability of Ships*, Osaka, Japan, 2008.

12. FESMAN, E., BAYRAKTAR, D., TAYLAN, M., 'Influence of Damping on the Roll Motion of Ships', *Proceedings of the 2nd International Conference on Marine research and Transportation*, Naples, Italy, 2007.

13. GU, Y., DAY, S., BOULOUGOURIS, E., 'a Study on the Effect of Bilge Keels on Roll Damping Coefficient', *Proceedings of the 12th International Conference on the Stability of Ships and Ocean Vehicles*, Glasgow, UK, 2015.

14. KATAYAMA, T., MATSUOKA, M., IKUSHIMA, K., 'Characteristics of Bilge Keel Damping Component for Shallow Draft', *Proceedings of the 13th International Conference on the Stability of Ships and Ocean Vehicles*, Kobe, Japan., 2018.

15. PAROKA, D., UMEDA, N., 'Analytical Prediction of Capsizing Probability of a Ship in Beam Wind and Waves', *Journal of Ship Research*, Vol. 50(2), SNAME, 2006.

16. RUDAKOVIC, S., BULIAN, G., BACKALOV, I., 'Effective Wave Slope Coefficient of River-Sea Ships', *Ocean Engineering*, Vol. 192, 2019.

17. PAROKA, D., MUHAMMAD, A.H, RAHMAN, S., AZIS, M.A., YUSUF, A., 'Application of the Second Generation Intact Stability Criteria to an

Indonesian Ro-Ro Ferry Supported by Model Experiment', *Proceedings of International Conference on Ship and Offshore Technology*, Semarang, Indonesia, 2019.

18. PAROKA, D., ASRI, S., ROSMANI, HAMZAH, 'Alternative Assessment of Weather Criterion for Ships with Large Breadth and Draught Ratio by a Model Experiment: A Case Study on an Indonesian Ro-Ro Ferry', *International Journal of Maritime Engineering*, Vol. 162, 2020.

19. IMO, 'Interim Guidelines for Alternative Assessment of Weather Criterion', Document MSC.1/Circ.1200, International Maritime Organization, London, 2006.

20. IRKAL, M.A.R., NALLAYARASU, S., BHATTACHARYYA, S.K., 'CFD Approach to Roll Damping of Ship with Bilge Keel with Experimental Validation', *Applied Ocean Research*, Vol. 55, 2015.

21. PAROKA, D., MUHAMMAD, A.H, RAHMAN, S., 'Hydrodynamics Factors Correspond to the Weather Criterion Applied to an Indonesian Ro-Ro Ferry with Different Weight Distribution', *International Journal of Technology*, Vol. 12(1), 2021.

22. PAPANIKOLAOU, A., BOULOUGOURIS, E., SPANOS, D., 'ON THE Roll Radius Gyration of Ro-Ro Passenger Ships', *Proceedings of the 7th International Society of Offshore and Polar Engineers Conference*, Honolulu, USA, 1997.

23. BORISOV, R., LUZYANIN, A., 'An Approach to Assess the Excessive Acceleration Based on Defining Roll Amplitude by Weather Criterion Formula with Modified Applicability Range', *Proceedings of the 12th International Conference on the Stability of Ships and Ocean Vehicles*, Glasgow, UK, 2015.

7. AUTHORS BIOGRAPHY

DaengParoka is a lecturer and a researcher at the Department of Ocean Engineering, Hasanuddin University of Indonesian. His subject is the stability of ships and other floating structures. He has researched probability assessment of ship stability in beam seas and developed an alternative assessment of weather criterion applied to an Indonesian Ro-Ro Ferry.

Muhammad Akbar Azis is a lecturer and researcher at the Department of Naval architecture Hasanuddin University. Recently he worked on a research topic of stability of Indonesian traditional wooden boats.

Andi Haris Muhammad is a lecturer and a researcher at the Department of Marine System Engineering, Hasanuddin University Indonesia. His research topics are ship resistance and ship manoeuvring. Recently he has joined in the research about the stability of Indonesian ro-ro ferries.

Sabaruddin Rahman is a lecturer and researcher at the Department of Ocean Engineering, Hasanuddin University Indonesia. The main topic of his research mainly relates to ocean wave modelling and protection corresponds to the safety of ship operation.

THE USE OF TIDAL ENERGY IN ELECTRICITY GENERATION IN THE CONTEXT OF DECARBONISATION

Y. Petrov, Saint-Petersburg State Marine Technical University, Russian Federation.

SUMMARY

The International Energy Agency's strategy is to reduce carbon dioxide emissions by 2050. Since the energy sector generates most of these emissions, it is this industry that should be affected by the maximum changes. One of the possible ways is the development of hydropower, this article will consider different types of tidal power plants, compare them and conclude that it is advisable to use them as a replacement for the main types of electricity generation. As materials in the study, data from the works of specialists and data from international organizations were taken. The analysis shows the feasibility of various types of tidal power plants to participate in the decarbonization process and reach 'Net Zero' by 2050.

NOMENCLATURE

α_1	Wake velocity coefficient(m s^{-1})
α_2	Correction coefficient(kW)
α_3	Efficiency coefficient (dimensional coefficient)
ρ	Density of water (kg m^{-3})
η	Upstream/downstream water level parameter (m)
g	Acceleration of gravity(m sec^{-2})
h	Local depth (m)
u	Cross-sectional velocity (m s^{-1})
w	Local width (m)
H	Water level difference (m)
A	Impoundment surface area (km^2)
A_s	Cross-sectional flow area (km^2)
B	Ratio $\frac{D}{h}$ (dimensionless coefficient)
C_d	Bed friction coefficient (function)
C_D	Discharge coefficient (function)
C_T	Local thrust coefficient(function)
D	Device diameter (m)
F	Bed friction force (N)
Fr	Frud number (dimensionless coefficient)
P_a	Available power (kW)
P_e	Extracted power (kW)
P_p	Potential power (kW)
Q_T	Flow rate of the turbine (kW h^{-1})
T	Rotating force (N)

1. INTRODUCTION

Nowadays humanity faces different global problems. One of the most important – is global warming. According to the research by World Resource Institute[1] generation of heat and energy is responsible for the most CO2 emissions. And if world community wants to cope with climate change, it is necessary to change the structure of energy producing. There are different decisions- nuclear power plants, usage of solar and wind energy etc. And one of the possible decisions is to use ocean energy – tidal power. It is based on producing electricity from the tides, phenomena caused by gravitational interaction in the Earth-Moon system. For example, a coal based generating plantwith 500 MW energy capacity produces about 3.5 billion kWh per year [2], and tidal power plant can possibly give us 3,800 TWh per year [3]. So, usage of tidal energy can give us a thousand times more watts per yearthan if we were using coal energy. Previous studies mainly considered the issue from a technical point of view, that is, they described only the design of power plants and the amount of energy produced. In this paper, an attempt is made to include in the analysis the practical question of the possibility of including tidal power plants in the global energy network. In addition, previously created works did not include comparisons of various types of power plants with each other, in order to identify the strengths and weaknesses of the objects under consideration.

2. TYPESOFTIDAL ENRGY GENERATORS &PRINCIPLES OFTHEIR WORK

Tidal energy generators can be divided in three groups, by the way how they get tidal energy [4]:

- Tidal stream power plant.
- Tidal range power plant (barrage).
- Tidal range power plant (lagoon).

2.1 TIDAL STREAM ENERGY GENERATION

This type of energy generator uses tidal streams to produce electricity.Water stream gives its kinetic energy of motion to the rotation energy of blades. Then, this energy goes to the electricity motor, which is connected with blades and there the energy of motion turns into electricity.

2.2 TIDAL RANGE ENERGY GENERATION

This type of energy generation uses from the natural rise and fall of tides [5]. The principle of operation of such a power plant is that the water mass is captured behind an artificial structure, and then released to rotate the turbine blades and generate electricity.

3. CHARACTERIZATION

This types of power plants have a number of characteristics:

- Energy producing model and amount of generated energy.
- Construction and maintenance.
- Possible geographic location.

73

3.1 ENERGY PRODUCING MODEL AND AMOUNT OF GENERATED ENERGY.

3.1 (a) Tidal stream type model.

To describe energy efficiency, we can use 2 models. First, is a prediction of available power from the depth-averaged simulation. Second is a prediction based on simplified channel model.

Turbines in the concept of this power plant is situated in the water column. And, these devicesare exposed to frictional forces from interaction with the aqueous medium. The drag resistance on the flow owing to bed friction is given by:

$$F = \frac{1}{2} C_d \rho A_b u|u|$$

Parameters of this equitation ρ, A_b, u are And the coefficient C_d is selected by researcher, it is because bed friction value is highly dependent from chosen model, and is a function of many parameters.

3.1 (a) (a)Inclusion of tidal turbines.

In this model the method of Draper [7] is used. According to this method, turbines are represented as a line discontinuity in elevation. Method represents the relationship between the water level downstream of a homogeneous row of turbines and the water level upstream, and describes 'effluent' representative of the equal and opposite force applied by the turbine to the flow. Denote this force distributed per unit length of the fence as$T$, then it can be shown from continuity and conservation of momentum that the depth change Δh at any point along the fence is given by:

$$\frac{1}{2}\left(\frac{\Delta h}{h}\right)^3 - \frac{3}{2}\left(\frac{\Delta h}{h}\right)^2 + \left[1 - Fr^2\left(1 - \frac{T}{\rho h \bar{u}}\right)\right]\left(\frac{\Delta h}{h}\right) - Fr^2 \frac{T}{\rho h \bar{u}} = 0$$

Where, $Fr = \frac{\bar{u}}{\sqrt{gh}}$ is a local Frud number.

The available power from the generator producing this force, can be estimated using actuator disk theory [8]. If we apply it to the given problem, then we get the following expressions for the T, and for the P_e(total power extracted by the turbines per unit length of fence):

$$T = \frac{BC_T}{2} \rho h u$$

And,

$$P_e = \rho g \bar{u}\left(\frac{\Delta h}{h}\right) h^2 \left(1 - Fr^2 \frac{1 - \left(\frac{1}{2}\right)\left(\frac{\Delta h}{h}\right)}{\left(1 - \left(\frac{\Delta h}{h}\right)\right)^2}\right)$$

Where, the coefficient B is defined as a proportion of the channel swept by turbines, and can be calculated as

$$B = \frac{\pi D^2/4}{Dh} = \frac{\pi D}{4h}$$

$\frac{D}{h}$ is the ratio of device diameter D to local depth h.

and C_T(local trust coefficient for turbine), which can be written functionally as $C_T(Fr, B, \alpha_1)$, $\alpha_1$is the wake velocity coefficient and is defined as the ratio of the velocity in the near wake of the turbine to that immediately upstream of the turbine.

The P_a(available power) is a fraction of the P_e(extracted power). And this value(P_a) can be calculated with satisfy error as follows:

$$P_a \approx \alpha_2\left(1 - \frac{1}{2}\left(\frac{\Delta h}{h}\right)\right)$$

Where, $\alpha_2$is a correction coefficient, which is the ratio of the velocity of the fluid as it passes through the turbine to the velocity upstream.

In the paper [9] numerical implementation and verification of this algorithm into the DG ADCIRC is described.

3.1(a) (b) Simplified channel model.

This model is mostly used to estimate available power in studies, where speed of computation is needed.

To develop a model, we, first turn to Garret& Cummins [10]. Let's assume that tidal flow can be described one-dimensional model satisfying the dynamic equation:

$\frac{\partial u}{\partial t} + u\frac{\partial u}{\partial x} + g\frac{\partial \eta}{\partial x} = -C_d \frac{u|u|}{2h} - F$ (3.1.a.b.a)

Where, u - is the average cross-sectional velocity at an arbitrary location along the channel$x$ – is distance along the channel, t – is time, g – is acceleration due to gravity, η – is the elevation above still water level, C_d - is the bed friction coefficient, F– is a force (per unit mass) parametrizing tidal utilities.

In every moment there are is minimal variation in flow rate along thechannel (for example, PentlandFirth [10]). Integrating equation (3.1.a.b.a) by x gives us:

$g\zeta = c\frac{dQ}{dt} + \left(\lambda_0 + \sum_{i=1}^{N} \frac{C_T B}{2A_i^2}\right) Q|Q|$(3.1.a.b.b)

Where,

$$c = \int_0^l A^{-1} dx$$

And,

$$\lambda_0 = \frac{C_d}{2}\int_0^l h^{-1}A^{-2}dx + A_e^{-2} - A_0^{-2}$$

In equation (3.1.a.b.b), Q(flow rate) can be written functionally, as

$Q(t) = u(x,t)A(x)$,

Where $A(x)$ is the cross-sectional area of the channel ('e' and '0' mean exit and inlet of the channel), B is blockage ratio, C_T is a local thrust coefficient. Following the research [11], the force parameter of the turbines, has been replaced by cumulative force from N rows from homogeneously disturbed tidal turbines, contribution from each of them is

$$F = Tw_i$$

Where, w_i local channel width for ith row. The force T is determined in the same way as mentioned above. The coefficient ζ describes the difference of water level between channel sides, and it can be written as a series of tidal constituents [12]:

$\zeta = a_0 \cos(\omega_0 t) + a_1 \cos(\omega_1 t) + \cdots,$

Phase difference is omitted since it is acceptable when describing the process of generating energy over a long period of time.

So, with only using two constituents to match numerical simulations, and using the non-dimensional parameters:

$$Q' = Q\frac{\omega_0 c}{a_0 g}$$

And,

$$\lambda_0{}' = \lambda_0 \frac{g a_0}{(\omega_0 c)^2}$$

And,

$$t' = t\omega_0$$

We can rewrite (3.1.a.b.a) as

$\cos(t') + k\cos\left(\frac{\omega_1}{\omega_0}\right) = \frac{dQ'}{dt'} + \left(\lambda_0' + \sigma \sum_{i=1}^{N} \frac{C_T B}{2A_i^2}\right) Q'|Q'|$(3.1.a.b.c)

Where, $k = \frac{a_1}{a_0}$;

And, $\sigma = \frac{g a_0}{(\omega_0 c)^2}$

Both of these models are valid to describe the tidal stream power plant.

3.1 (b)Tidal range power plant.

The potential power from tidal range power plant (P_p) can be written using a mathematical expression:

$$P_p \sim AH^2$$

Where, A – impounded surface area, and H – water level difference between downstream and upstream of the impoundment.

To describe small lagoon and barrages ([14], [15], [16]), it is enough to use model, that takes in account tidal conditions, plant operation scheme, and formulae which are represented characteristics of constituent hydraulic structures. This method, which is known as 0D modeling, allows to simulate characteristics of tidal range scheme and gives an informed resource assessment [17]. The model provides for use as input a water level time series, governed by the transient downstream water elevations at the site location.

The common technique of the description – is a usage of the backward-numerical model, with using continuity equation for development. With parameters of upstream $\eta_{up,i}$ and downstream $\eta_{dw,i}$ water level at any point of time t_i, the upstream water level at $t + \delta t$ can be calculated as [18]:

$$\eta_{up,i+1} = \eta_{up,i} + \frac{Q(H_i) + Q_{in,i}}{A(\eta_{up,i})} \Delta t$$

Where, $A(\eta_{up})$ is a wetted surface area of the lagoon, assuming a constant water level of η_{up}. The water level difference H can be defined as $\eta_{up,i} - \eta_{dw,i}$, and feeds into $Q(H)$, which a function for the total dropping contribution from sluice gates and turbines. And, Q_{in} is describe the sum of inflows/outflows through source, which are not the part of the impoundment, for example outflows or rivers. According to Beaker et al. [19], flow through hydraulic structure can be theoretically calculated, as

$$Q = C_D A_s \sqrt{2gH}$$

C_D is a discharge coefficient, and A_s is the cross-sectional flow area. So, the power from the tidal range turbine and given H can be described as:

$P = \rho g Q_T H \alpha_3$,

Where ρ is the fluid density, Q_T flow rate of the turbine, and α_3 is efficiency coefficient, which depends on the turbine.

This method is effective in computation, but it doesn't take into account the influence of hydraulic structure on the localize tidal levels. This assumption can give an over-optimistic results of the research, which don't reflect the real picture. You can see reports about that in the paper by Angeloudis and Falconer [18], and in the work by Yates et al.[20] .

So, to build a model, which can be related to the large barrage, we have to use 2D or 3D modelling

3.1(b) (b) 2D & 3D modelling.

Hydrodynamic simulations of coastal waters can provide valuable insight into resource assessment, the quantification of the potential impacts from planned coastal engineering projects, and the minimization of any detrimental effects through design optimization.

The depth-average(2D) and three-dimensional model(3D) can give a fairly accurate estimate of the energy producing by the power plant [21].

Most of the current researches uses adaptive methods of 2D modeling and shallow water equitation to estimate resources and impacts of designed tidal range power plant. This method is based on the use of the depth-integrated Navier - Stocks equation, with assumption, that the vertical scales are much smaller, than horizontal length scales, and the pressure is close to hydrostatic balance. It can be written in non-conservative and conservative forms:

$$\frac{\partial U}{\partial t} + \frac{\partial E}{\partial x} + \frac{\partial G}{\partial y} = \frac{\partial \tilde{E}}{\partial x} + \frac{\partial \tilde{G}}{\partial y} + S$$

Where, $E$and G are the convective flux vector in the x , y direction, U is the vector of conserved variables, $\tilde{E}$ and $\tilde{G}$ are diffusive vectors in the x and y directions, and S is a source term that includes the effects of bed friction, bed slope and the Coriolis force.

The terms in the equation can be written, as$U = \begin{bmatrix} h \\ hu \\ hv \end{bmatrix}$,

$$E = \begin{bmatrix} hu \\ hu^2 + \frac{1}{2}gh^2 \\ huv \end{bmatrix}, \quad G = \begin{bmatrix} hv \\ huv \\ hv^2 + \frac{1}{2}gh^2 \end{bmatrix}, \tilde{E} = \begin{bmatrix} 0 \\ \tau_{xx} \\ \tau_{xy} \end{bmatrix},$$

$$\ldots \tilde{G} = \begin{bmatrix} 0 \\ \tau_{xy} \\ \tau_{yy} \end{bmatrix}, S = \begin{bmatrix} q_s \\ hfv + gh(S_{\delta x} - S_{fx}) \\ -hfu + gh(S_{by} - S_{fy)} \end{bmatrix}$$

Where, h is total water depth, q_s is the source discharge per unit area, u, v are are the depth-averaged horizontal velocities in the x and y direction. Coefficient f is the part of the Coriolis acceleration, $\tau_{xx}, \tau_{xy}, \tau_{yx}, \tau_{yy}$ are components of the turbulent shear stress over the plane.

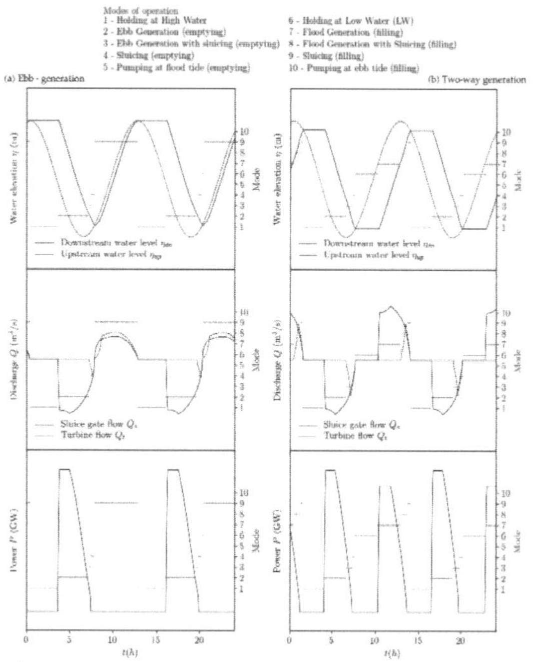

Figure 1: Possible operation modes of tidal range power plants, and two operation strategies [21]. 0D modeling is used and shown as time series of water elevation, flow rate, and power output. (a) ebb-generation on the left side, and (b) two-way generation on the right.
Lagoon type has the same energy producing model.

3.1(c) Amount of generated energy

3.1 (c) (a) Tidal range power plants

First, let's take a look at thetidal range power plants.

Country	Site	Type	Proposed Capacity (GW)	Estimated Annual output (TWh)
Canada	Cobequid	Barrage	5.34	14
India	Gulf of Kutch	Barrage	0.9	1.7
Argentina	San Jose	Barrage	6.8	20
USA	Knik Arm	Barrage	2.9	7.4
UK	Severn	Barrage	8.64	17
Russia	Tugur	Barrage	10	27
South Korea	Garorim	Barrage	0.48	0.53

Table 1 : This table presents several power plants in the tidal range with their energy characteristics. The data was taken from [21].

As we can see from table, we have a wide range of energy capacities (from 0.48 GW to 10 GW) and of estimated annual outputs (from 0.53 TWh to 20 TWh). That means, that we can use this type of power plant for different cases. For example, if we need to organize power supply for a village, we can use tidal range power plant with low powerconsumption and annual production. And if it necessary to supply a big region or industrial area, we can use electricity station with much higher performance. However, it should be emphasized that the construction of such power plants is limited due to geographic reasons, which will be discussed below. Also, we say that in the case of a large distance between the consumer and the producer, there can be energy losses (the world average 8.251% [22]).

3.1 (c) (b) Tidal stream power plants

Country	Site	Type	Proposed Capacity(GW)
UK	West Islay	Farm	0.4
UK	Penthland Firth	Farm	1.9
USA	Cobscook Bay	Farm	0.0075
USA	Western Passage	Farm	0.005
USA	Roosvelet Island	Farm	0.00105

Table 2: This table presents tidal stream power farms with their proposed capacity. This is summarized from these resources ([13], [23], [24], [25], [26]).

As we can see from this table (Table 2), tidal stream power plants have much less capacity, than tidal range power plants (Table 1). If we compare the tables, the advantage of tidal range power plants over tidal stream farmsin the energy factor becomes obvious. In addition to a higher energy capacity value, we also see a wider spread of this value, which indicates greater flexibility in choosing in the choice of the supplied object.

3.2 CONSTRUCTION AND MAINTENANCE

3.2 (a) Tidal stream power plant

As we mentioned above, this type of power generating facility is mainly presented in the form of farms. That is, the facility is a set of turbines, device for setting on the seabed and a device for storing energy and transferring it to a consumer.So, the installation scheme is quite simple. It is necessary to prepare the seabed by installing the foundation for the turbine there, after which the turbine itself is installed using a crane or manipulator device. Everything ends with the connection of the turbine to the general power grid.The complexity of the installation process lies in the fact that it is necessary to correctly place the turbines, since the efficiency of electricity production depends on the correct location.Poor weather conditions, inappropriate bottom contours, and imperfect installation equipment can cause errors in the placement of turbines. In order to neutralize the impact of the above factors, it is necessary to conduct a thorough hydrological reconnaissance of the station location, constant monitoring of meteorological conditions, as well as monitoring the health of the equipment.All these actions increase the final cost of the project.
Maintenance is reduced to maintaining the operability of turbines and parts of the power grid (cables, wires, etc.), as well as protecting them and their foundations from biofouling. Various methods are used to simplify the maintenance process as well as to make it cheaper. For example, changes in the design of turbines, consisting in the creation of lifting devices that lift the turbine blades

for their maintenance, or the use of special materials to prevent biofouling.

3.2 (b) Tidal range power plant

Tidal range facility is a complete hydrological structure. It is consists of several main parts ([27], [28]) :
Embankments are the foundation of any tidal station. They are artificial structures that form a closed contour of the reservoir, they should also have a maximum length with the maximum possible area of a closed surface in plan. Minimization of changes in the natural flow becomes a prerequisite for their design.
The turbines are located in special places made in the embankment. They convert the potential energy of a change in the water level into rotational energy, and then with the help of generators into electricity.
Openings are equipped with control and slice gates. This is necessary to ensure the transmission of flows at a given time with minimal disruption.
Locks built into the structure allow to ensure the safety of the passage of ships through the impoundment.
As we can see from the description, the tidal range power plant requires large investments for construction. First of all, for the zero cycle (embankment construction),in addition to this, it is necessary to correctly position the openings and locks, since if they are incorrectly located, the entire work of the power plant can be disrupted.
Maintenance of such power plants requires maintaining the integrity of the embankment, maintaining the operability of the turbines, as well as protecting all components of the structure from biofouling.
We can say that the simplicity of the structure of the tidal stream power plant(farm) allows you to spend much less money on construction and maintenance. However, a tidal power plant is a more stable structure, due to its complexity, the same quality will explain the higher complexity of servicing such facilities.
3.3 Possible geographic location.
The concept of tidal power generating systems requires access to the resource right at the location of the power plant, it is obvious that such a resource is powerful tidal phenomena.

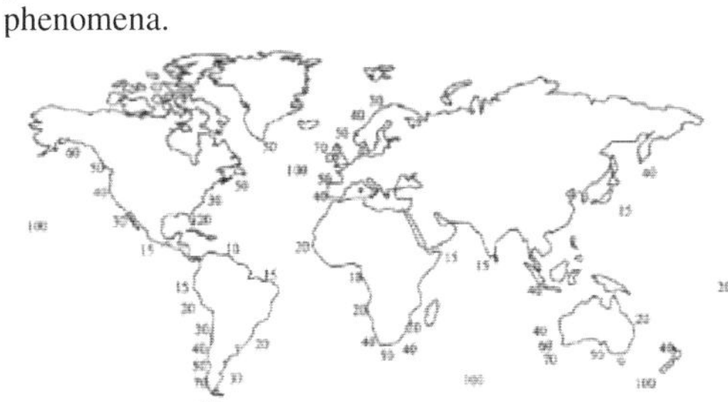

Figure 2: A Map that represents the power density of a wave, measured in kilowatts (kW) per meter (m) of wave front [29].

If we analyze map, we see that the regions richest in hydrological resources are the southern tip of South America and Australia, the northeastern part of the Atlantic Ocean.This means that it is in these parts of the Earth that the installation of tidal power plants will be most effective. However, it should be noted that the greatest concentration of the required resources is located in the equator region, so the states located there can use this fact to diversify their energy industry. For example, combined power plant (solar photovatic and micro-hydropower) can produce energy for 962 households in Indonesia [30].In addition, hydrological resources can make a great contribution to countries with a large number of available fossil fuels, for example, in Nigeria, hydropower can provide 30% of all electricity generation [31].However, the development of hydropower is typical not only for developing economies, but also for leading countries, we see examples of such projects, for example, in the USA and Great Britain(Table 1, Table 2). Nevertheless, the distribution of tidal stations is limited by natural factors, primarily the relief factor, both of the seabed (tidal energy farms) and land (tidal range power plants). Also important is the question of the availability of a consumer of the generated electricity, and here some problems arise.
According to the World Bank research [32] the highest population density coincides with the highest wave density only in Europe.From this, we can conclude that the construction of power plants in the most convenient areas in terms of resource availability is impractical, since there is a consumer who needs a large amount of generated electricity. But, on the other hand, a high population density corresponds to a large amount of smaller tidal resources. This allows us to make the assumption that in such regions it would be more rational to build power plants of low power, but in a large number.
Thus, the strategy for choosing thelocation of a particular type of power plant, its capacity, method of placement depends on many factors from the characteristics of the area to economic feasibility.

4. COMPARISION

Let's make a short interim conclusion, which is a comparison of the two types of power plants, in the form of a table.

	Tidal Range Power Plant	Tidal Stream Power Plant
High Energy Efficiency	+	-
Ease of Construction & Maintenance	-	+
Flexibility of Geographic Location	Equal	Equal

5. CONCLUSION

Based on the analysis carried out, we can say that the use of electric spaces based on tidal energy allows us to realize a large supply of hydrological resources available on our planet. A wide range of possible energy production, as well as the possibility of choosing different design options for power plants, allows you to create a unique solution for the selected tasks.However, in general, we can talk about the prevalence of advantages over disadvantages and the expediency of using tidal power plants as instruments for decarbonization and enhancing development sustainability. Other studies show similar results in assessing the energy efficiency of various types of tidal power plants. However, it should be said that different methods of assessing the electricity produced are used in the works and this must be taken into account when using their results. As recommendations for future research work in this area, it is possible to indicate the development of projects for the optimization of existing models of power plants, as well as complexes of renewable energy sources, which will include tidal power generation systems

6. REFERENCES.

1 https://www.wri.org/insights/4-charts-explain-greenhouse-gas-emissions-countries-and-sectors

2 https://www.mcginley.co.uk/news/how-much-of-each-energy-source-does-it-take-to-power-your-home/bp254/

3 https://www.britannica.com/science/tidal-power

4 https://www.nationalgeographic.org/encyclopedia/tidal-energy/

5 http://www.tidallagoonpower.com/tidal-technology/what-is-a-tidal-lagoon/

6 VENNEL, R., 'Tuning turbines in a tidal channel', Journal of Fluid Mechanics, 2010.

7 DRAPER, S. et al., 'Modelling tidal energy extraction in a depth-averaged coastal domain', IET Renewable Power Generation,2010.

8 HOULSBY,GT. et al., 'Application of linear momentum actuator disc theory to open channel flow'. Technical report no. 2296-08, University of Oxford, Oxford, UK, 2008.

9 SERHADLIOĞLU, S. et al., 'Assessment of tidal stream energy resources in the UK using a discontinuous Galerkin finite element scheme'., 17th Int. Conf. on Finite Elements in Flow Problems, San Diego, CA, 2013.

10 DRAPER, S. et al., 'The extractable power from the Pentland Firth. Renew. Energy',

11 VENNELL, R,'Tuning tidal turbines in-concert to maximise farm efficiency', Journal of Fluid Mechanics, 2011.

12 GARRET C. andCUMMINSP.,'The power potential of tidal currents in channels', Proceeding of the Royal Society a mathematical, physical and engineering sciences, 2005.

13 ADCOCK, T., et al., 'The available power from tidal stream turbines in the Pentland Firth', Proceeding of the Royal Society a mathematical, physical and engineering sciences, 2013.

14 YATES, N., et al.,'Appraising the extractable tidal energy resource of the UK's western coastal waters',Philosophical Transaction of the Royal Society a mathematical, physical and engineering sciences, 1985, 2013.

15 BURROWS, R., et al.'Tidal energy potential in UK waters', Proceedings of the Institution of Civil Engineers-maritime Engineering, 2009.

16 BURROWS, R., et al., 'The tidal range energy potential of the West Coast of the United Kingdom', Applied Ocean Research, 2009.

17 ANGELOUDIS, A. andFALCONER, R.A.,'Sensitivity of tidal lagoon and barrage hydrodynamic impacts and energy outputs to operational characteristics'Renewable Energy, 2017.

18 ANGELOUDIS, A., et al.'Numerical model simulations for optimisation of tidal lagoon schemes'Applied Energy, 2016.

19 BAKER C. and LEACH P.,'Tidal Lagoon Power Generation Scheme in Swansea Bay: a Report on Behalf of the Department of Trade and Industry and the Welsh Development' AgencyDTI, 2006.

20 YATES, N.,et al.'The energy gains realisable through pumping for tidal range energy schemes',Renewable Energy, 2013.

21 NEIL, S.P.,et al., 'Tidal range energy resource and optimization – Past perspectives and future challenges', Renewable Energy, 2018.

22 https://data.worldbank.org/indicator/EG.ELC.LOSS.ZS

23 https://tethys.pnnl.gov/project-sites/western-passage-tidal-energy-project

24 https://tethys.pnnl.gov/project-sites/cobscook-bay-tidal-energy-project

25 https://tethys.pnnl.gov/project-sites/roosevelt-island-tidal-energy-rite-project-pilot

26 https://www.power-technology.com/projects/west-islay-tidal-farm-scotland/

27 KADIRI, M., et al., 'A review of the potential water quality impacts of tidal renewable energy systems', Renewable Sustainable Energy Reviews., 2012.

28 BAKER, C. and LEACH, P., 'Tidal Lagoon Power Generation Scheme in Swansea Bay: a Report on Behalf of the Department of Trade and Industry and the Welsh Development Agency', DTI, 2006.

29 W. THROPE, T., 'An Overview of Wave Energy Technology: Status, Performance and Costs', Proceedings of a Conference on Wave Power: Moving towards Commercial Viability, 1999.

30 SYAHPUTRA, R., and SOESANTI, I.,'Renewable energy systems based on micro-hydro and solar photovoltaic for rural areas: A case study in Yogyakarta, Indonesia.',Energy Reports, 2021.

31 OHUNAKIN, O.S., 'Energy utilization and renewable energy sources in Nigeria', Journal of Engineering and Applied Science, 2010

32 https://data.worldbank.org/indicator/EN.POP.DNST?type=points&view=map&year=2018

6. AUTHOR BIGRAPHY

Yegor Petrov holds the current position of student at the St.-Petersburg State Marine Technical University. His previous experience includes presentation during the IMarEST Annual Conference 2021.

BIG DATA ANALYSIS FOR CONTAINER STAYING TIME PROBLEM OF CONTAINER STORAGE YARD

T N Mathias and **T Shinoda,** Kyushu University, JP

SUMMARY

The increasing demand for goods transportation inherent the risks of the terminal congestion, consequently decreasing the operational efficiency. Maritime transportation containerization has forced container terminals to carefully evaluate how to manage their yard operations. Nowadays, with the Internet of Things (IoT) and the development of data technologies the complexity of data within and around the terminal has been increased, and a lack of data-driven approaches in this matter can still be identified. Using data mining techniques, some effects that are producing inefficiency might be discovered and fixed. This paper presents a data-mining approach to extract useful information from the terminal daily operations report. Additionally, it is shown the container's length of stay of a Straddle Carrier system container terminal.

1. INTRODUCTION

Recently, industries are paying attention to the usage of their resources much more than in the past. Their equipment, land size, buildings, financial power, and human resources need to be managed very carefully to avoid wastes. With the growth of information technology, data is a new kind of resource that is added to this list. Data became one of the most important features in the modern world, not only in the heavy industries but also to understand consumption behaviour or marketing planning. With the implementation of the internet of things (IoT), companies have been changing the way of dealing with operations, using the data from their operations to increase efficiency and consequently profit.

Transforming the traditional way of interpreting data to a more advanced analytics technology may be the main usage of the Internet of Things (IoT). New technologies have always been used for economic improvements, and the IoT is no different. Researchers in various business shows the annual impact of IoT in the business would be an average of \$4.4 trillion, with respectively lower and upper boundaries of \$2.7 and \$6.2 trillion [1][2].

Several decades of computing hardware development created the possibility to store large amounts of data and via combining with the IoT, much more data can be generated and stored to further analyze or even for real-time analysis. The current methods of handling data cannot be applied while dealing with large-scale data. Moreover, the nature of business has been changed because of data science technologies, and many companies exist based on insights and solutions that only can be gathered through a big data approach [3][4].

Mainly used as a storage location (land size) and for the handling cargo equipment, the physical resource of a container terminal is mainly used as storage. By handling the containers as efficiently as possible the terminal can increase their throughput and so their profit, the smart use of their resources can be considered vital to maintaining competitive business in the market. The measurement of the efficient use of a container storage yard is related to operation with a high throughput in a small area, which means the trade-off between land and handing operations must be studied and improved constantly, that includes, for example, the numbers of cargo handling equipment, the container volume handled, operating status, and energy consumption.

It is important to analyze the current cargo handling situation to grasp the operational status of physical resources, but sometimes it is not possible to obtain cargo handling data to evaluate specific points of the operations due to the format in which the data is collected by the designed system. Then it is necessary to create a methodology to extract useful information from the available data. Therefore, we introduce a method of evaluating the data generated by daily S/C operations when handling containers in the terminal yard.

This study proposes a data mining method to show the container's length of stay in the yard and the panorama of a container terminal operational efficiency indicators. The data was collected from the Kashii Park Container Terminal at Hakata Port, a medium-sized terminal in Japan.Started operation in 1994, equipped with four over-Panamax gantry cranes on a quay with a total length of 600 m and a water depth of 13 m, and a Straddle Carrier (S/C) cargo handling method (6 out of 17 hybrid types) at Hakata Port, handle about 40% of the container operations (about 320,000 TEU, 2014 results).Kashii Park is located about 15 minutes to the Kyushu Expressway "Fukuoka Interchange" and Fukuoka Airport, making it a highly convenient container terminal.

2. THEORETICAL BACKGROUND

Information technologies are being used from many different fields to improve operations efficiency and to gain productivity regarding daily management. Container terminals are not different, computing systems are implemented to guarantee better outcomes.

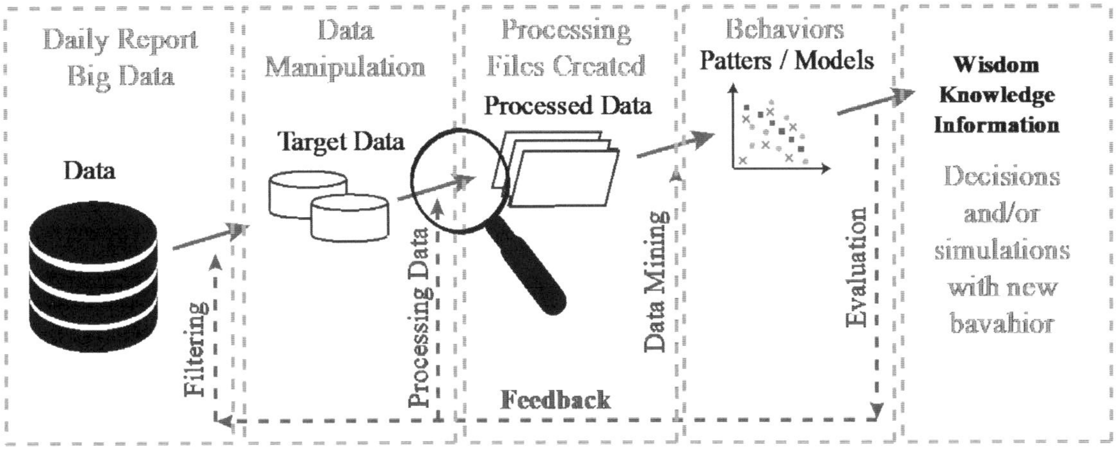

Figure 1. Knowledge Discovery from Database (KDD) applied on Kashii Park Container Terminal.

Automated terminals are the state of art regarding goods transshipment between transportation modes. But, to introduce a high level of automation, the investment necessary is equally high. Then, many terminals around the globe adopt a semi or no automated system, but there is still a large amount of data generated from the terminal daily operation.

2.1 BIG DATA ANALYSIS

Big data is addressed to solve distinguished problems where traditional data analysis is not enough. More specifically, big data deals with different requirements such as multiple datasets, hidden information, and large amounts of unstructured data. Furthermore, big data methodologies are not limited by statistical approaches, new analytics algorithms can and should be applied when dealing with a large amount of data [5].

One well-known definition of big data is "datasets that could not be perceived, acquired, manages, and processed by traditional IT within a tolerable time" comparatively, the Apache Hadoop definition in 2010 says that big data are "datasets which could not be captured, managed, and processed by general computers within an acceptable scope" [6].

The characteristics that define big data can be compiled on the three Vs model: Volume, Velocity, and Variety. Additionally, "big data technologies describe a new generation of technologies and architectures, designed to economically extract value from very large volumes of a wide variety of data, by enabling the high-velocity capture, discovery, and/or analysis." Then, the definition gained another V, Volume, Velocity, Variety, and Value [7][8][9]. Veracity was also added to this definition.

- Volume: refers to the size of the datasets that need to be processed and analyzed. The total volume of data requires different technology or methods than traditional. In other words, this means that the data is too large to be processed with regular commercial software.
- Velocity: refers to how fast the data is generated.
- Variety: is related to the sources of the data. It is usually multiple sources and one out of three types: structured, semi-structured, or unstructured data. Variety often makes the database larger because of its multiple source characteristic.
- Value: is the usefulness of the data.
- Veracity: refers to the quality or fidelity of the data. This item may add a new processing stage to remove invalid data or noise. Noise is data that cannot be translated into information.

Figure 1 shows the KDD process scheme representation where each step is linked by a process to create knowledge. First, from the big data, a target is selected, this phase is usually known as pre-processing, processing, and transformation of the data are needed to change the data to a more comprehensive format for the next phase. After that, data mining algorithms play a role, finding different patterns, behaviour, or new methods to be categorized,interpreted, and evaluated by a specialist, followed by generating wisdom and knowledge information. Note that every process is composed of a loop between any two steps, and it is common to have multiple interactions, feedbacks, and loops between them.

Data mining can be considered a step in the KDD process to apply different methods and algorithms to generate new models over the data as an outcome. Then, Knowledge Discovery in Database (KDD) was highlighted to emphasize that extracting knowledge out of the data is the main goal of data-driven analysis. [9][10].

There are endless possibilities when combining big data analysis, and a data mining approach to a daily report container terminal operation to evaluate all and each resource in the container yard operation [11].

3.2 CONTAINER TERMINAL OPERATIONS

A container terminal is defined as a node within the road, rail, inland waterway, and maritime traffic networks. Containers are transferred from and to, each of the different transportation modes. Figure 2 shows the four

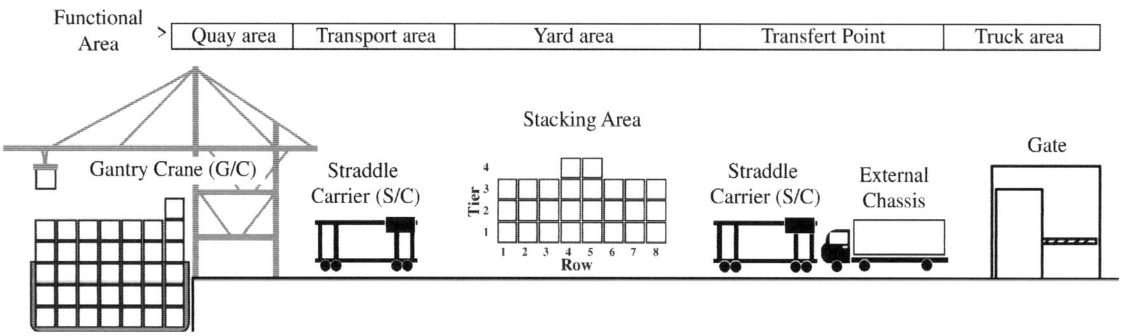

Figure 2. Schematic layout of container terminal with Straddle Carrier System

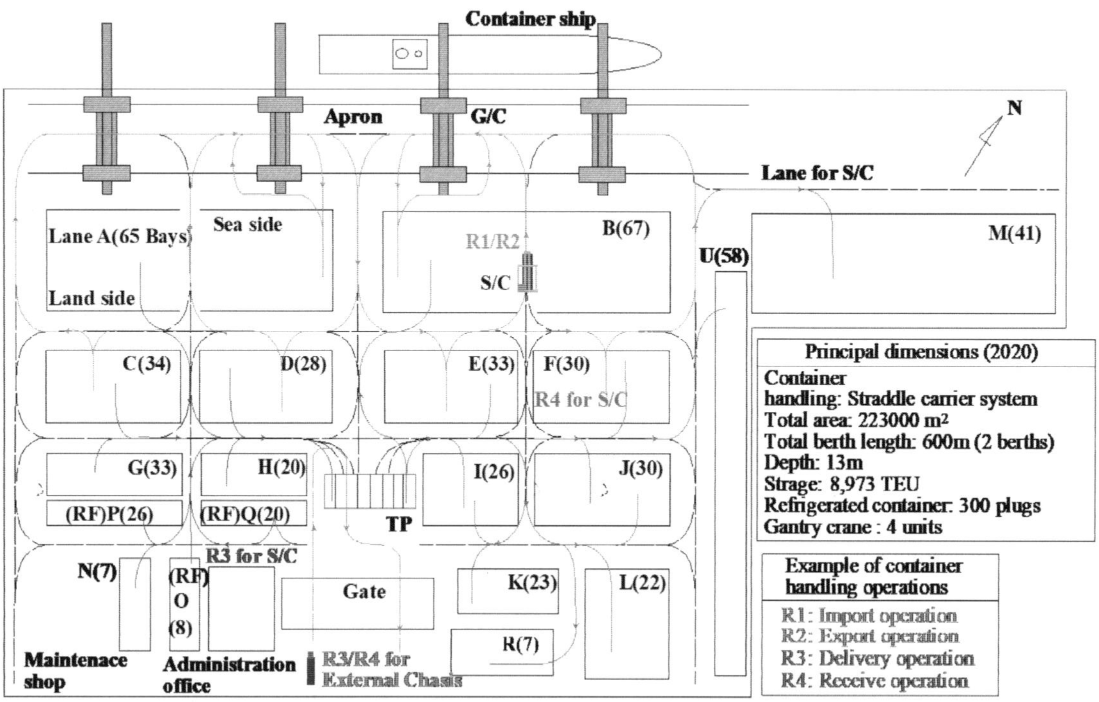

Figure 3. Kashii Park Container Terminal Layout

operational areas in a container terminal: Quay area, designated for mooring the container vessels; Transport, to move containers within the terminal; Container yard to store all containers when necessary; and the trucks and trains which serve as the integration between the terminal and external modes. Each area has its sort of equipment and degree of complexity, the container flow goes to both directions, and is mostly operated by gantry cranes (G/C), straddle carriers (S/C), and the Transfer Point (T/P)to ling external trucks (chassis) to S/C [1].

Handling container cargo is the most important function of a container terminal. Container terminal managers need to be careful to make efficient container transportation from the vessel to the trucks (import and delivery), also the other way from trucks to the vessel (receive and export) to make the operation as efficient as possible [13].

Figure 3 represents the Kashii Park Container Terminal layout operations. The movement flow works from the apron to the yard for Import (R1) and Export (R2) taking the container from and to the storage stacking and moving from and to the container ships. Then, Delivery (R3) and Receive (R4) moves containers inward and outward the terminal. Note that, in this case, the operation needs to transfer the containers to external trucks through the Transfer Points (T/P). Also, it's important to highlight that the external truck movements can also combine receive (R4) and Delivery (R3).

The daily report operational database is shown in figure 4. The Terminal Operational System (TOS) takes every movement done with a container is recorded into the database. Each record has the characteristics of the respective operation, such as date, company, and operator responsible to get the task done, equipment and locations designated for each container storage place, container unique identification (IMO) and its physical characteristics, and then, the timing composition. Placing time stamp moment when the order arrived in the TOS,

年月日	ふ頭	荷役機器	作業会社	作業者	荷役作業	GC	コンテナ番号	Sz	Tp	Ht	外来ヘッドID	コンテナ移動 FROM	コンテナ移動 TO	蔵置アドレス	ゲート受付時刻	作業完了時刻	待ち時間
20190601	香椎	S35	Cop	Wk1	動		C***0*****1	20	RF	96		HKK	M17GH3		09:50:42	10:35:30	44:48
20190601	香椎	S35	Cop	Wk1	リ		K***3*****2	20	DR	86		F02 F3	F02 C1	F02 D3	10:33:32	10:38:23	04:51
20190601	香椎	S35	Cop	Wk1	出		R***7*****3	40	DR	86		F02 F2	TP01		10:33:32	10:39:45	06:13
20190601	香椎	S35	Cop	Wk2	動		K***6*****4	40	RF	96	J*8**	HKK	M17EF1		09:50:42	10:43:00	52:18
20190601	香椎	S35	Cop	Wk2	動		G***5*****7	40	RF	96		HKK	M17EF2		09:50:42	10:47:03	56:21
20190601	香椎	S35	Cop	Wk2	カ		U***4*****3	40	DR	96		YOKO	B22AB1		08:26:32	10:50:07	143:35
20190601	香椎	S35	Cop	Wk2	入		E***2*****0	40	DR	96		TP01	B19EF1		10:39:58	10:52:30	12:32
...							...					...			...		

Figure 4. Kashii Park Container Terminal Layout

Table 1. Types of container cargo handling operation.

Data Type	Abbr. in paper	Name of operation	Directionof Movement	Detail
揚	Import	Import	Inward	Unloading container from ship to importing container from ship to storage lane through internal chassis.
出	Deliv	Delivery	Outward	Delivering container from storage lane to external chassis.
リ	RH.Dl	Rehandle for delivery	Inside of storage lane	Re-handling container for delivery. It is instructed automatically by the system, closest place, and lowest tier.
積	Export	Export	Outward	Loading container to the ship to export a container from storage lane to ship through internal chassis.
入	Recv	Receive	Inward	Receiving container from external chassis to storage lane.
動	RHSf	Re-handling for shift	Inside of storage lane	Re-handling container with SO, convenience for operations, or protection for high wind. It is instructed by the yard operator, YO, closest place, not above today's container, and lowest tier.
無	RH.Dr	Rehandle by the driver	Inside of storage lane	Re-handling by Equipment driver when they see dangerous stacking.
仮	Tp	Temporary	Vessel Operations	Tentative storage containers from ship to rehandle containers inside ship's cargo hold. It is paired operation with a return operation.
戻	Rt	Return	Vessel Operations	It is paired operation with a temporary operation.
カ	Es	Escape	Inward/Outward	Escaping empty container from VP (vanpool) to storage lane (or reverse).

and later when the equipment operator finished the task. From the database give, it is possible to find not only the basic Import, Export, Receive, and Delivery operations, but also a few more movements done by the container yard operator. A detailed description of each movement item seemed in the database is given in table 1. First, due to the location of this terminal, the movement's name is given in Japanese, in this paper it is using an abbreviation of it. An entire name of the operation is sown as well as its flow directions, and finally a brief description. Inward and Outward flows are referred to when containers are taking in and out of the terminal, whether from the vessels or tucks. When rehandling and/or marshaling is needed, a movement inside of the storage lane is found. Also, if it is necessary to reposition a container that is already on board, vessels operations of Temporary and Return are done by the Gantry Crane (G/C). Lastly, Escape operations dealing with empty containers moving from or to the vanpool area.

3. METHODOLOGY

The concepts of the data value chain, data mining, big data analysis, and statistics were applied to develop a procedure to extract, understand and evaluate the container terminal daily report data to support decision-makers to take the best actions possible when planning the daily yard operations. Figure 5 shows the flow chart of this paper methodology that uses Knowledge Discover from Database (KDD) to extract useful information out of the terminal daily operational data. The Kashii container terminal data report was used to verify and validate the methodology effectiveness, concepts of Big Data value chain, data mining, data analysis is part of the core of this process[14][15][16].

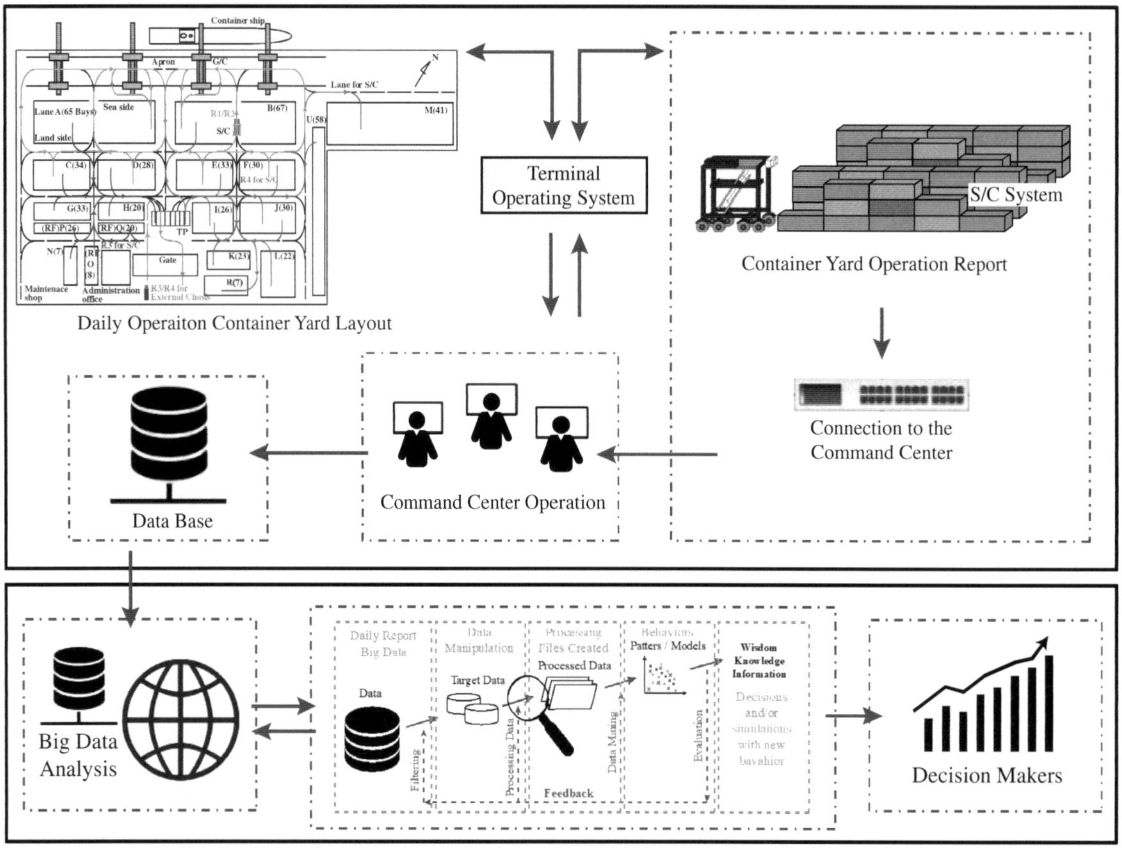

Figure 5. Big data analysis and datamining methodology.

We divided this methodology into two parts. First, data collection plays its role in recording each container's operational data which is taken from the moment the order arrives until the end of the movement itself when the operator press a button to record the time stamp. The TOS through the Command Centre Operation is responsible for the addressing system, manage the yard capacity and operation, and sending each job information to the equipment operator that is responsible for the task. Note that the operator may need to change the address according to the terminal condition, weather, or experience. If any change is made, the data is manually changed to the newest location, but the timing process remains the same. The data from Kashii Park is the record of each S/C operation completed in the yard, but it is also possible to verify other equipment during the operation. The external chassis number (C/Y), G/C, S/C, TP, Quay, company, and driver responsible for that operation are also taken in the same record. It means that by using this proposed methodology, it is not only possible to extract information from S/C, but it also to estimate all resources used in the terminal yard.

Then, the second block represents the analysis, starting with the database of the daily operation; using appropriate approaches to extract information and analyze the data, and the key performance indicators can be calculated and compared. It is in this step that the core of this methodology is applied, mining the data to find out unknown patterns and behaviourto propose operational changes. [16][17]

All information collected from the S/C daily work report is gathered; the analysis is based on the turnover container's movements, which is the composition of all movements each container had during its time in the terminal, also how many times the containers visited the yard during one year. Also, the terminal throughput, yard and TP utilization, G/Cs productivity, C/Y and external chassis, average stacking tear, container status, etc can be extracted from it [18][19][20].

4. RESULTS AND DATA ANALYSIS

A comprehensive description of operational behaviour is mandatory to be examined. In this study, the Kashii Park data of approximately 600,000 container handling operations from April 1st, 2019 to March 31st, 2020. It was developed Python modules to extract useful information from data. Several relations between equipment operations, storage positioning, etc were seen from the data analysis, particularly when the equipment operator change the address based on his experience.

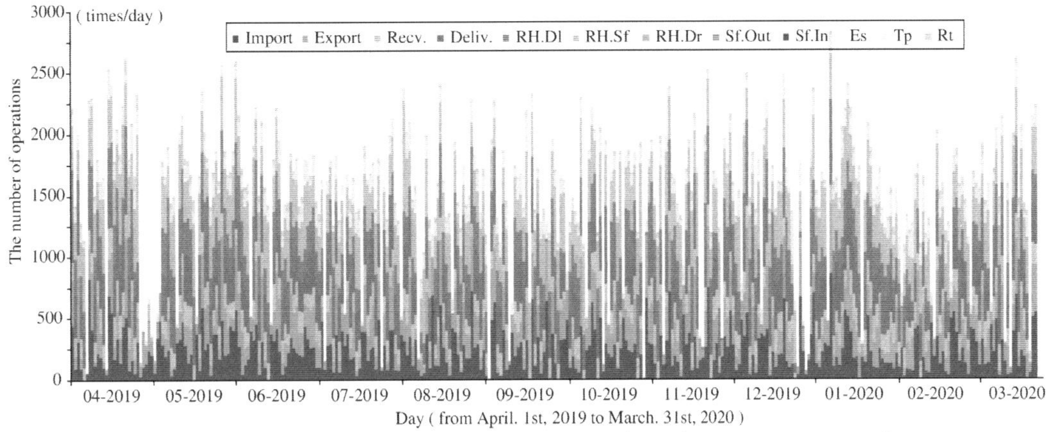

Figure 6. Container terminals types of movements throughout a one-year operation

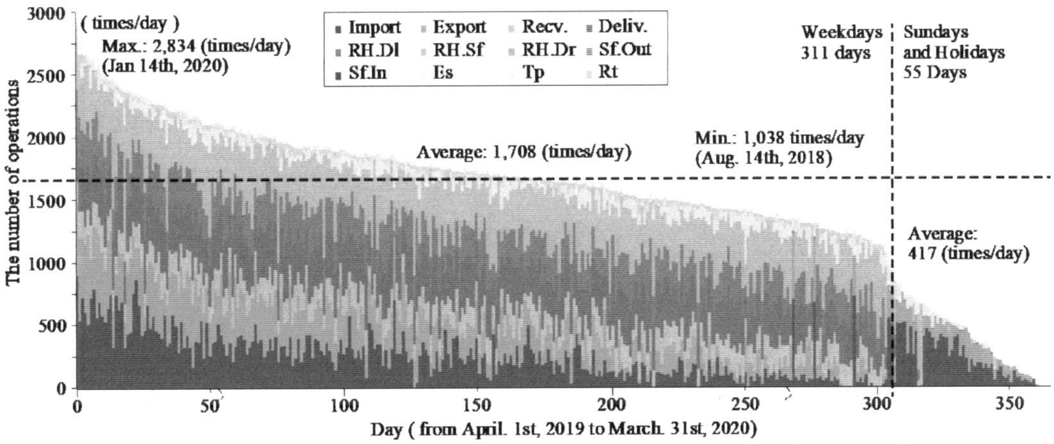

Figure 7. The daily number of operations is sorted by the highest to the lowest number

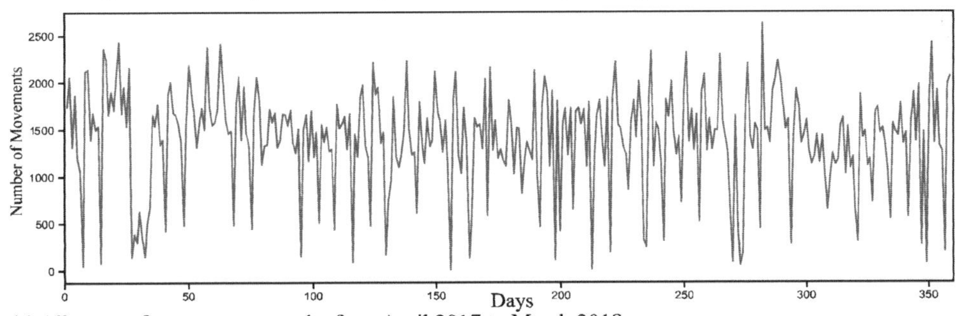

(a) All types of movements per day, from April 2017 to March 2018

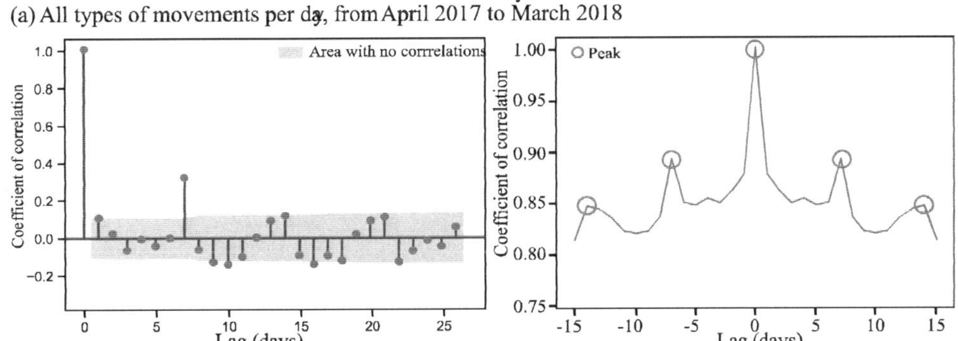

(b) Correlogram (ACF)

(c) Autocorrelation plot

Figure 8. Container terminals types of movements throughout a one-year operation

To comprehend the system, initially, it was analyzed the daily number of operations grouped by its handling movement type, as mentioned before. Figure 7 shows the breakdown of each cargo classification. There is a large fluctuation in the amount of container handling per day throughout the year. It is possible to see that there is a big number of shift operations, that is due to repositioning containers from a different area to inside the terminal yard.

Figure 6shows the data sorted by decrescent order of the total handling cargo volume. There are 311 working days are and 55 holidays, a total of 366 days due to leap year. The first, feature to notice is about lack of external chassis during holidays, but there are still ship operations running. An average of 417 containers moved during no working days, 4 times lower than weekdays. Then, during normal days, peak and valley operational days also have about 2.7 times differences between themselves, being 2,834 times/day and 1,038 times/day, respectively, with an average of 1,708 times/day.

The movements analysis is important to have an overview of how the terminal resources are utilized. If it is considered the terminal was sized to operate in its full capacity, more than 90% of the time there will be idle equipment, or non-essential operations being conducted. That represents damage to the operation's efficiency. Another feature that can be detected is related to seasonality; the first trimester had a higher movements rate than others but following a cyclic behaviour returning to the peak on the following year.

Figure 8 (a) shows the total movements per day for one year, starting day 1 and finishing day 366. Figure 8 (b) uses the Auto Correlation Function (ACF) to check randomness in the data set. This is determined by computing autocorrelations for data values at varying time lags which is the calculated correlation with the observed time-series with its previous step. If random, the autocorrelations should be near zero for all time-lag. Otherwise, then one or more of the autocorrelations will be significantly non-zero. Also known as correlogram, the autocorrelation of this time series is done using a daily lag. The confidence interval is drawn in the light-blue shade area with a 95% interval. Furthermore, when the variable fits the Gaussian distribution, Pearson's correlation coefficient can be used to calculate the correlation between those values. The autocorrelation at lag zero is always equal to 1 because represents the autocorrelation between the lag and itself. Each spike that rises above or falls below zero can be considered statistically significant. Additionally, Figure 8 (c) computes the autocorrelation of the movements within 15 days positive and negative, a weekly periodicity of the data. It means that every seven days the same behaviour tends to repeat, just in a different proportion [19][20].

A different aspect worth evaluating is how long the container stays in the terminal. Here forth, this timing will be referred to as turn-over movement. To take the turn-over movement time for each container, data processing is required, all possible combinations were processed and evaluated to take the turn-over time. Figure 9 illustrates the turn-over histogram for import and delivery and receives and export operations. There is an average of 9.4 days to be delivered, and 6.6 days to be exported. Container owners might be using the terminal as an extension of their storage, damaging the terminal operation. Then, if the storage period of the container is restricted, it may lead to the improvement of the terminal throughput and the reduction of the storage area, possibly dropping the waste of resources.

To understand the effect longer staying time containers have on terminals, it was evaluated the turn-over movement regarding the number of rehandles necessary to make before the containers leave the terminals. First, it was taken all those three different types of rehandling and summed up, to assure that only the rehandling operations would be shown, records with no repositioning movements were deleted from this dataset. Then, a histogram for 1, 2, 3, and 4 rehandling movements was built to show their behaviour. The time figures are showing only 30 daysbecause after being so long in the terminal, eventually, the container will hit the 1st tier and so be placed on the ground, then no more rehandling is necessary. It is vital to highlight that the yard operators have the autonomy to reposition containers, which means that not always when the container is placed on 1st tier it will remain there, the driver might move it to a different position (or tier) if he thinks it is needed. Then, a policy to refrain the operator to change containers positions when it is not extremely necessary might help to decrease the total number of rehandling in the terminal.

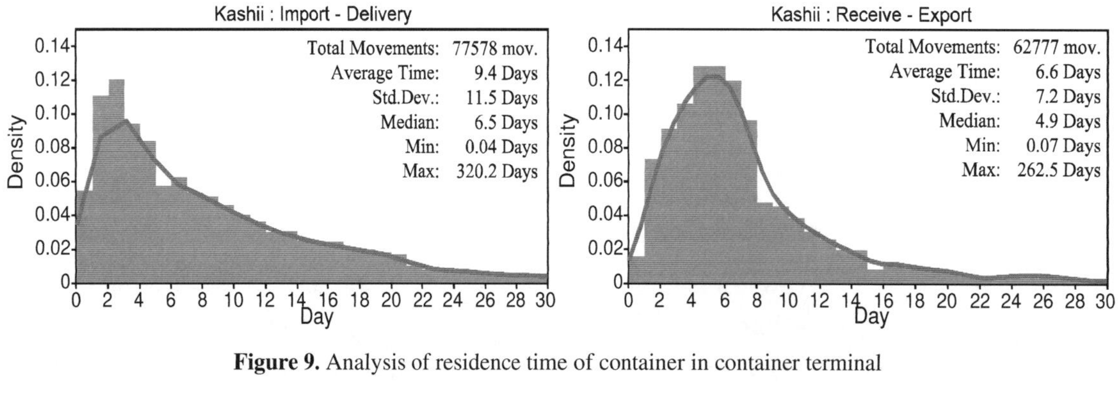

Figure 9. Analysis of residence time of container in container terminal

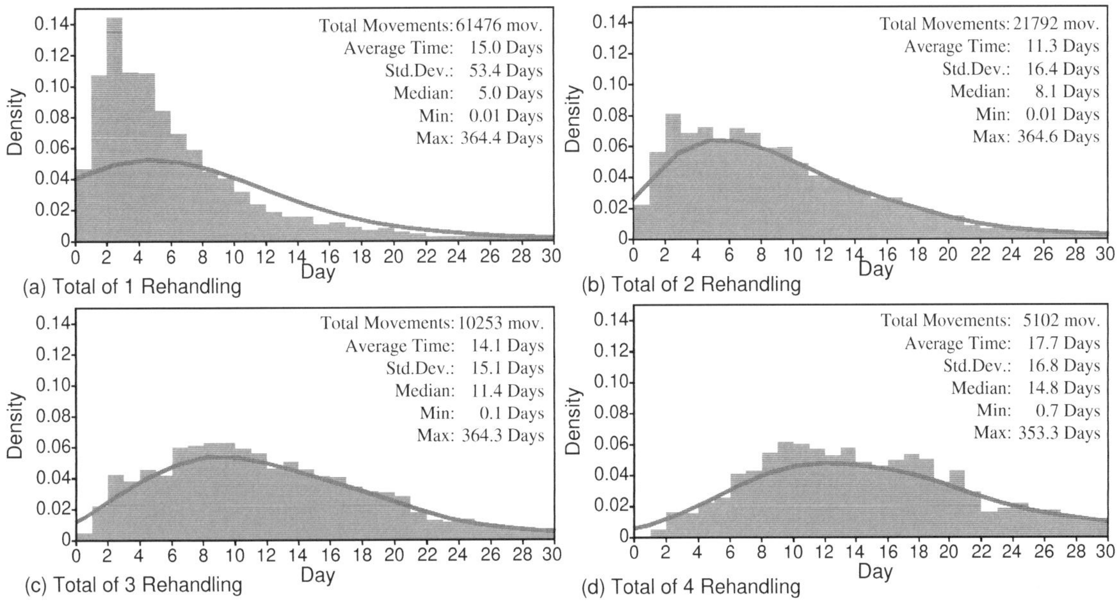

Figure 10. The number of rehandling operations compared to the time spent in the terminal

Figures 10 show the different behaviour when rehandling happens. Note that the kernel density estimation (KDE), which is a non-parametric way to estimate the probability density function, shows that when adding rehandling movement, the probability curve gets flattened, resembling a normal distribution curve, then to a uniform distribution. The average turn-over time in the terminal gets higher as the number of rehandling also increases, on the other hand, the efficiency and probability of higher throughput decrease because the terminal is being used as a warehouse, not a modal transshipment facility.

3. CONCLUSIONS, LIMITATIONS, AND FUTURE RESEARCH

This paper uses a methodology of datamining and big data analysis framework to extract information and evaluate a container terminal database daily report. It was used datasets from an S/C transportation system of Kashii Park Container Terminal. It was shown examples of how powerful this methodology could be to take useful information out of the daily operations records, regarding the container's length of stay in the container storage yard.

Daily movements were presented to understand the operationalbehaviour during oneyear. The analysis presented suggests that the terminal equipment was oversized during the design; the terminal can handle peak days. However, on days with a high number of movements that are not frequent, then many important resources are underutilized or being used when it is not necessary.

Also, it was demonstrated how long containers are left in the terminal. Some container owners might be using the terminal yard as an extension of their storage, increasing the terminal handling work. Taking the container out of the terminal faster would help to optimize the yard, reducing the total number of operations. Therefore, reducing the number of rehandling and the average tier in the terminal yard bays.

From another perspective, the terminal has still space for improvement and it is not saturated, in other words, it might be possible to achieve better efficiency adjusting operational metrics that are dragging down the terminal key performance indicators, and consequently the throughput.

It was mentioned before that the TOS was designed to control the terminal operations; therefore, it is not the scope of the system to provide multiple parameters necessary to conduct this research, and it is an important limitation to be considered.

For further studies it is suggested to use a bigger database, multiple years would confirm trends and might show different hidden patterns.

8. ACKNOWLEDGEMENTS

We would like to show our gratitude to Kyushu University and Hakata Port Terminal Co., Ltd. both for supporting us, including measurements and data, on which this study was based.

9. REFERENCES

1. MOHAMMADI, M., AL-FUQAHA, A., SOROUR, S., GUIZANI, M. 'Deep Learning for IoT Big Data and Streaming Analytics: A Survey', IEEE Communications Surveys & Tutorials, 4 (3), 2932-2960, 2018.
2. MANYIKA, J., CHUI, M., BUGHIN, J., DOBBS, R., BISSON, P., MARRS, A. 'Disruptive Technologies. Advances That Will Transform Life', Business and the Global Economy, 180, 2013.
3. FISHER, D., DeLINE, R., CZERWINSKI, M., DRUCKER, S. 'Interactions with big data analytics',Interactions, 19 (3), 50-59, 2012.
4. TSAI, C. W,. LAI, C. F., CHAO, H. C., VASILAKOS, A. V. 'Big data analytics: a survey. Journal of Big Data', 2(1), 1-32, 2015.
5. ERL, T., KHATTAK, W., BUHLER, P. 'Big data fundamentals: concepts, drivers & techniques', Prentice Hall, 2015.
6. CHEN, M., MAO, S., LIU, Y. 'Big Data: A Survey', Mobile network and applications, 19, 171-209, 2014.
7. LANEY, D. '3-d data management: controlling data volume, velocity and variety', META Group Research Note, 2001.
8. GANTZ, J., REINSEL, D. 'Extracting value from chaos', IDC iView, 1–12, 2011.
9. FAYYAD, U. PIATETSKY-SHAPIRO, G., SMYTH. P. 'From Data Mining to Knowledge Discovery in Databases', American Association for Artificial Intelligence, 7(3), 37-54, 1996.
10. PIATETSKY-SHAPIRO, G. 'Knowledge Discovery in Real Databases: A Report on the IJCAI-89 Workshop', AI Magazine, 11(5), 68–70, 1991.
11. NOVAES MATHIAS, T., SHINODA, T., HANGGA, P., INUTSUKA, H. 'Big Data Approach to Identify the Waste Management of Container Terminal Resources', Asian Transport Studies, 5(4), 653–678, 2019.
12. MEISEL, F. 'Seaside Operation Planning in Container Terminals', Contribution to Management Science. Physica-Verlah, HD, 2009.
13. HANGGA, P., SHINODA, T. 'Big Data Utilization and Analysis to Improve Operational Performance of Hybrid Straddle Carrier in Container Terminal', Asian Transport Studies, 5(4), 679–693, 2019.
14. BRUCE, P., BRUCE, A., Gedex, Peter. 'Practical Statistics for Data Scientists', Second Edition O'Reilly, 2020.
15. WILEY, J. 'Data Science & Big Data Analytics: Discovering, Visualizing and Presenting Data', EMC Education Services, 2015.
16. McKINNEY, W. 'Python for Data Analysis', O'Reilly, 2018.
17. TAMHANE, A. C., DUNLOP, D.'Statistics and Data Analysis from Elementary to Intermediate', Prentice-Hall, 2000.
18. WICKHAM, H.,GROLEMUD, G. 'R for Data Science', O-Reilly, 2016.
19. HATCHER, L.,'Advanced Statistics in Research: Reading, Understanding, and Writing Up Data Analysis Results', Shadow Finch Media, 2013.
20. COWPERTWAIT, P. S. P., METCALFE A. V. 'Introductory Time Series with R', Springer, 2009.

9. AUTHORS BIOGRAPHY

Tiago Novaes Mathias holds the current position of assistant professor at Kyushu University. He is responsible for research technologies based on Big Data Analysis for Container Terminal yard operations. His previous experience includes simulation projects for: container handling operations, logistics for soybean harvest, sugar cane transportation, iron ore yard operation, etc.

Takeshi Shinoda holds the current position of Professor at Kyushu University. He is responsible for developed a decision support system for eco-friendly container terminal and presented an in-depth energy analysis for a single hybrid straddle carrier to identify energy consumption sources and clarify energy reduction potentials by focusing on detailed horizontal and vertical motions of the carrier based on coordinated Big Data.

HULL VANE INVESTIGATION WITH NACA 4412 ON VARIATION ANGLE OF ATTACK AND POSITION

W Sulistyawati and **P J Suranto,** Naval Architecture Engineering, Universitas Pembangunan Nasional Veteran Jakarta, Indonesia
N Utami, Undergraduate student of Naval Architecture Engineering, Universitas Pembangunan Nasional Veteran Jakarta, Indonesia

SUMMARY

Fuel efficiency is the goal of every ship, the accuracy of the motor selection, propulsion efficiency, as well as in the hull design for reducing drag. There have been many innovations in ship hull forms, such as the bulbous bow, transom steering, changing shape and becoming a multihull. This study adopted the NACA 4412 experimental as a Hull Vane®that other researcher carried out and perform with CFD simulations, then compared them with experimental results. The suitability of the trend can be further developed by varying the angle of attack from NACA and changing its position to the hull. The angle of attack variations on 1°, 2°, 3°, and positions change to the ship draft and AP-after perpendicular. Changes in the angle and position of NACA were investigated of their effect on the resistance component. From the analysis of the wave resistance, the most magnitude resistance decrease founded at the NACA position at a distance of 40 mm Trailing Edge of ship Draft and 85 mm Trailing Edge of AP. The lowest wave resistance was generating by the angle of attack of 3°, with a significant decrease of 19% compared to 1°. The more considerable lift drag was obtained after Fn 0.3 from NACA at an angle of 3° and the smallest at 2°. This investigative study was presented in graphs and data as well as conclusions and discussion of the results.

NOMENCLATURE

A_{VT}	Ship's projected face area above the waterline
C_A	Correlation allowance
C_{AA}	Coefficient of air resistance
C_T	Coefficient of total resistance
C_P	Coefficient of pressure resistance
C_R	Coefficient of residuary resistance
C_V	Coefficient of viscous resistance
C_W	Coefficient of wave-making resistance
D_{HV}	Hull Vane's drag force vector
F_{HV}	Hull Vane's force vector
FX_{HV}	Hull Vane's X component of lift force vector
FZ_{HV}	Hull Vane's Z component of drag force vector
L_{HV}	Hull Vane's lift force vector
R_T	Total resistance (N)
R_P	Pressureresistance (N)
R_V	Viscous resistance (N)
R_W	Wave-making resistance (N)
$(1+k)$	Form factor

1. INTRODUCTION

Several studies have been conducted to save energy to reduce fuel consumption and the environmental impact of ship CO2 emissions. Installing a Hull Vane®/foil at the ship's stern is one method of reducing drag. The lifting force of the Hull Vane® produces additional thrust, which reduces drag. Lifting force effectively produces trim and bending moment, reducing wave formation in the stern area due to the negative pressure on the top of the foil. Several investigations have shown that Hull Vane® can reduce fuel consumption as in the result of Uithof et al. (2016), Andrew et al. (2015), and Hou et al. (2020). The placement of the hull vane provided a resistance reduction of 8.05% at Fn 0.7 and up to 14.33% at Fn 0.2 (Andrew et al., 2015). Even a study by Celik et al. (2021), Hull Vane® can reduce drag by up to 31% compared to a bare hull.

The Hull Vane is appropriate for vessels with relatively large displacements and a more comprehensive range of operating speeds (Bouckaert, 2018). The Hull Vane® position is also critical in preventing flow interactions that induce wave-making between the stern shell and the Hull Vane®. The correct position of the Hull Vane® can result in good lift and greater thrust. Bouckaert et al. (2016) showed that vertical acceleration on the stern area decreased resistance by 13.1%. According to Uithof et al. (2016), changing the position of Hull Vane® in the vertical direction had a minor effect on total resistance. Suastika et al. (2019) find the most advantageous position of the Hull Vane® is the position of the leading edge two chord lengths behind the transom. Hou et al. (2020) discovered that the angle of attack has a more significant influence than the location of the Hull Vane®.

According to Van Oossanen (1992) and Uithof et al. (2014), the most efficient Hull Vane®/foil used as a Hull Vane® is NACA 4412. The influential position of the foil is located at the steepest point of the wave system interacting behind the ship's stern. The purpose of this study is to investigate by Computational fluid dynamics (CFD) the effect of angle variation on the most effective NACA hydrofoil in terms of resistance reduction. The NACA as Rhodes St. Gense 32 by Hou et al. (2020) that modification from NACA 4412 was used in this work. The Hull Vane® at the vertical position of 40 mm from trailing edge to draft and longitudinal 85 mm trailing edge to after perpendicular (AP). The various angles attack of 1, 2, and 3 degrees.

2. RESISTANCE COMPONENTS

The measurement of resistance components based on CFD is expressed as a measurement of the model's total force, which is expressed by the equation:

$$R_T = R_V + R_P \quad (1)$$

where R_V is the viscous resistance and R_P, the pressure resistance. According to the ITTC-78 method, the total resistance components should be expressed as follows:

$$C_T = (1+k)C_F + C_R + C_A + C_{AA} \quad (2)$$

where *(1+k)* is the form factor determined from the low-speed resistance test (Fr 0.12-0.24), and C_F is the coefficient of frictional resistance calculated according to ITTC 1957:

$$C_F = 0.075 \Big/ (\log_{10} Rn - 2)^2 \quad (3)$$

C_R is the residuary resistance coefficient, C_A is the correlation allowance, and C_{AA} is the air resistance coefficient, that are given by:

$$C_R = C_T - (1+k)C_F \quad (4)$$

$$C_A = (5.68 - 0.6 \log Re) \times 10^{-3} \quad (5)$$

$$C_{AA} = A_{VT} \Big/ 1000S \quad (6)$$

where A_{VT} is the ship's projected face area above the waterline. Then, the total resistance coefficient,C_T and wave resistance coefficientC_W can be determined by the formula:

$$C_T = R_T \Big/ \tfrac{1}{2} \rho S V^2 \quad (7)$$

$$C_W = R_W \Big/ \tfrac{1}{2} \rho S V^2 \quad (8)$$

where ρ, water density, S, wet surface area, and V, speed.

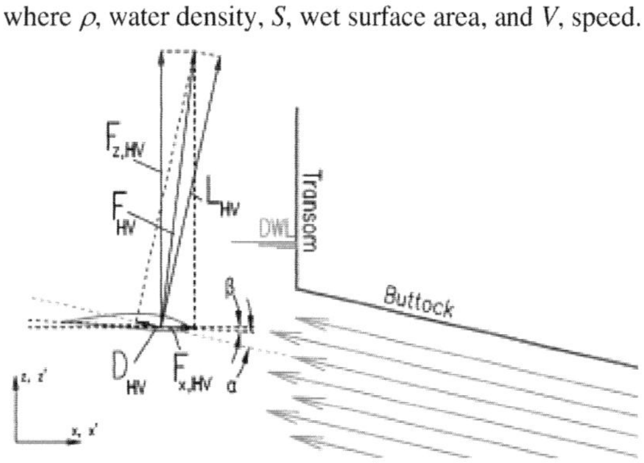

Figure 1: The forces schematic on the Hull Vane®(Uithof et al., 2014)

In the direction of flow, the foil effect as a Hull Vane® provides lift and drag force, and the combination of the two produces thrust, illustrated in Figure 1. The lift vector's x-component and the drag force vector's z-component can be written as:

$$\vec{L}_{HV} + \vec{D}_{HV} = \vec{F}_{HV} = \vec{F}x_{HV} + \vec{F}z_{HV} \quad (9)$$

where LHV is the lift force vector, DHV is the drag force vector, that given by equations:*L*

$$L_{HV} = \tfrac{1}{2} \rho V^2 S C_L \quad (10)$$

$$D_{HV} = \tfrac{1}{2} \rho V^2 S C_D \quad (11)$$

3. CFD SETTING

3.1 HULL MODELING

The experimental model used in this investigation was DTMB 5415 by Hongbo Hou et al. (2020), with the dimension parameters mentioned in Table 1 and the body plan displayed in Figure 2.

Parameter	Dimension	
LPP	2.784	m
Lwl	2.788	m
Bwl	0.374	m
T	0.121	m
Displacement	0.066	m^3
Wetted Surface	1.2668	m^2
CB	0.507	
CM	0.821	
LCB (%Lpp)	-0.683	

Table 1: Dimensions of the model

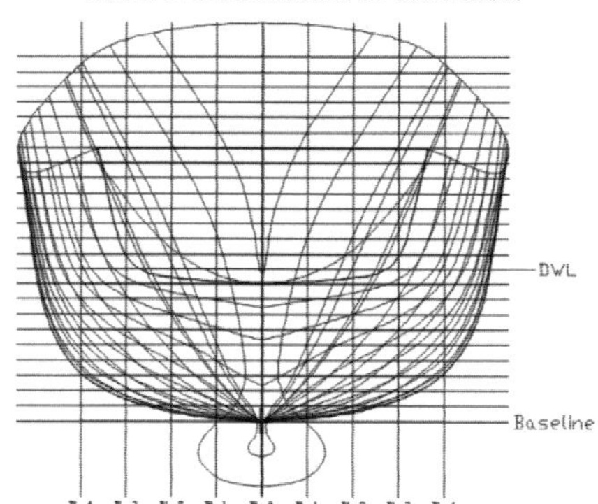

Figure 2:The body plan of model

Parameter	Unit	
Chord (c)	m	0.055
Span	m	0.27

Table 2: specification of NACA

The NACA specification is shown in Table 2 as model Hongbo Hou et al. (2020)shown in Figure 3. The vertical position of 40 mm from trailing edge to draft and longitudinal 85 mm trailing edge to after perpendicular (AP) as in Figure 4. The various angles of attack of 1, 2, and 3 degrees.

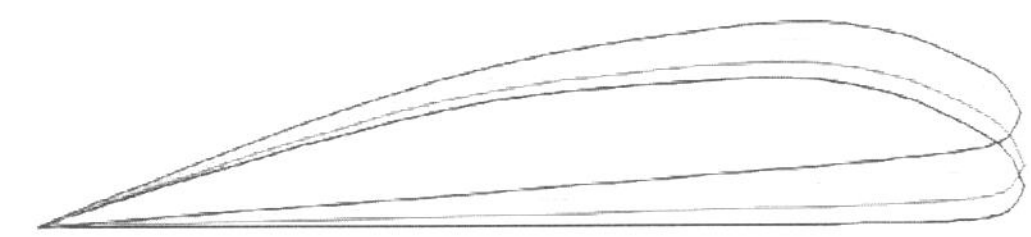

Figure 3: NACA model of Hongbo Hou et al. at various angles of attack of 1°, 2°, and 3°

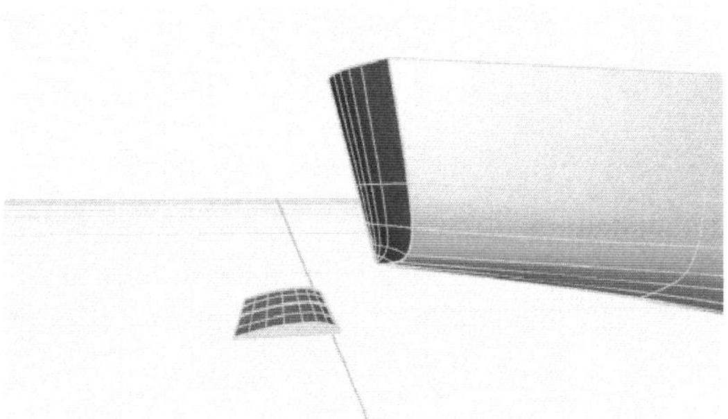

Figure 4: Foil position

3.2 BOUNDARY

The model domain boundaries refer to the 2011 ITTC rules, as shown in Figure 5. The inlet is set at 1 *Lpp* from the bow, while the outlet is 2 *Lpp* from the stern. The side is at 1 *Lpp* from the symmetry plane, the bottom 1 *Lpp* from the keel, and the top 0.5 *Lpp* from the model's keel.

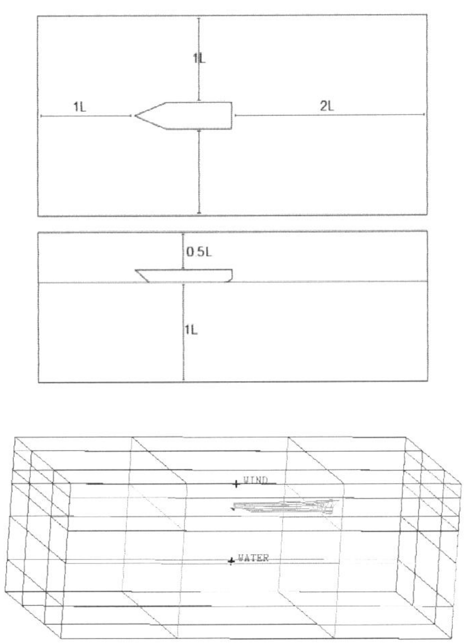

Figure 5: Boundary illustration

3.3 MESHING

The initial process in this simulation uses ICEM with an iterative meshing process to produce convergent values. The analysis of the resistance solution was verified using element number convergence on meshing and numerical uncertainty evaluation. The mesh processing repeated up to seven times with element sizes for both the boundary and hull in 1.0 to 0.05. Convergence is obtained when the value of the force difference is not significant, as shown in Figure 6. The physical domain in the free surface model for the domain boundary was defined: mesh tetrahedral, mixer; physics domain: homogeneous water/air multiphase, K-e, automatic wall function, and standard free surface model; inlet: volume fraction, turbulence intensity 0.05; an outlet, relative pressure; wall, free slip condition; top, Opening, relative pressure 0 Pa; hull and bottom, a wall with a no-slip condition; solver settings with convergence criteria on residuary type RMS, Target: 10-e4; and multiphase control used Volume fraction coupling. All models run in six times computations with increasing elements differently to get a convergent result up to 17 million elements, as shown in Figure 7.

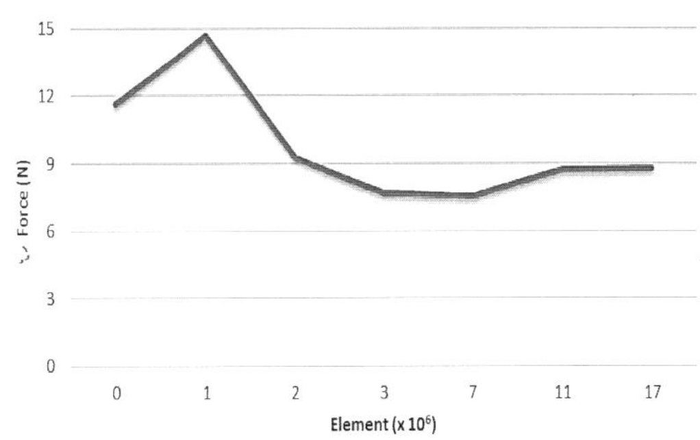

Figure 6: Convergence of model

Figure 7: Meshing results

4. HULL VANE® INVESTIGATION

Calculations using ANSYS CFX against NACA installations with different angles of attack are identified in the total resistance, viscous, and form factor by the Prohaska method to estimate the wave resistance. Figures 8-11 shows a comparison of the total resistance, viscous and wave of NACA with variations in the angle of attack: 1°, 2° and 3°. The total resistance, *RT*, (Figure 8) at Fn 0.28–0.4, findings of the three adjustments in the angle attack of NACA did not reveal a meaningful difference. With the increase in speed, NACA's third angle of attack provides a decrease. It provides the lowest resistance by an angle of attack of 3° at Fn 0.6. NACA 3° revealed a 2% reduction compared to 1° and 18% from 2°. Figure 9 compares simulation and test from Hongbo Hou et al. (2020). The disparity between the simulation and the test is significant, but it offers a relatively appropriate trend line.

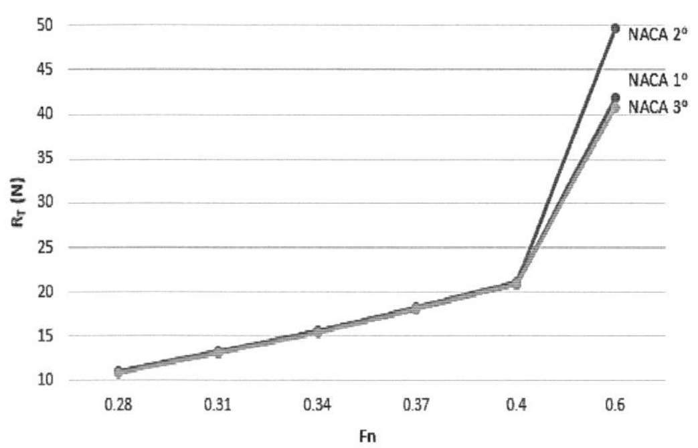

Figure 8: Total resistance R_T

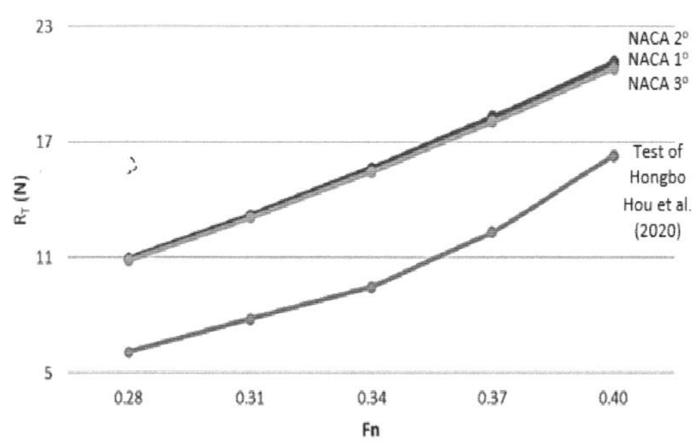

Figure 9: Comparison Total resistance $R_T$between CFD simulation and test of Hongbo Hou et al. (2020)

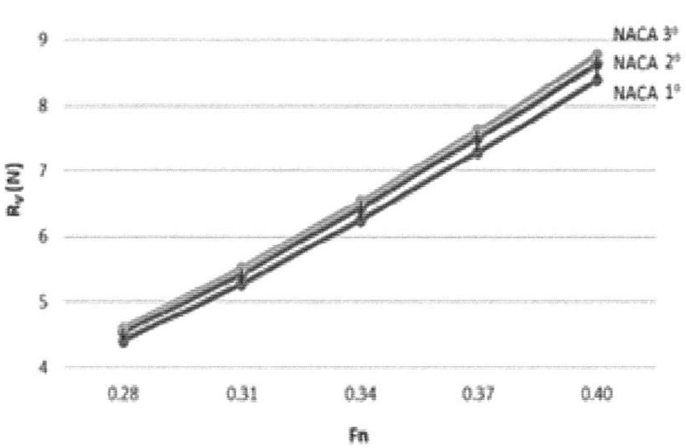

Figure 10: Viscous resistance R_V

Viscous pressure resistance increases (Figure 10) with increasing Froude number and degree of attack angle. The angle of attack 1°-2° has a 2.85 % average deviation, 2°-3°has a 1.6 %, and 1°-3° has a 4.4 %. The biggest viscosity is produced by NACA 3, which is possible because the higher angle of attack underwater increases the wet area, which increases friction. The value of the resulting wave resistance is inversely proportional to the slope of the Angle of Attack, as seen in Figure 11. Compared to 2° and 3°, the Angle of Attack 1° delivers a substantial increase in wave resistance compared to the increase in speed. At Fn 0.6, there is a rather considerable variation. The average deviation of the angle of attack 1° -2° is 12.8%, the angle of attack 2°-3° is 8.9%, and the angle of attack 1° -3° is up to 20.5%. It is achievable because the 3° angle of attack minimizes the flow of waves from the hull's stern.

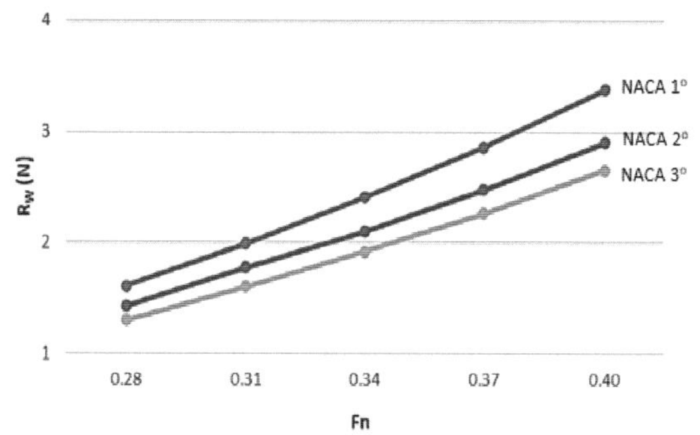

Figure 11: Wave resistance R_W

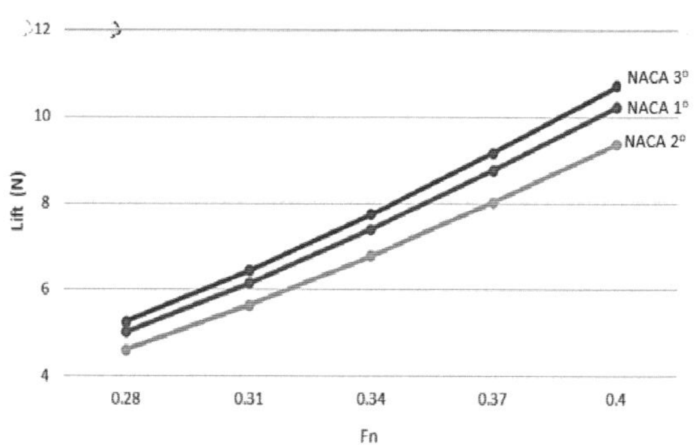

Figure 12: Lift force

Foil behind the stern to produce a lift or thrust effect to provide additional propulsion force. The foil thrust effect will be generated when the vertical lifting force syncs with the lifting speed.Changes in the NACA angle have a substantial impact on the generated lift. The greater the lift created by the NACA, the lower the resistance. NACA's vertical lift may exert negative pressure on the foil's top surface, minimizing wave generation. From Figure 12, the lift force of the three NACAs shows an increase in line with changes in velocity. The angle of attack 3° produces the most significant lift force with a deviation of 12.5% from 2° and 4.5% from 1°. It shows that the correct placement and angle of attack will affect the possibility of producing a more significant lift force.

5. CONCLUSIONS

The effect of the hull vane's angle of attack at a distance of 40 mm trailing edge of ship draft and 85 mm trailing edge of AP on the resistance component was explored in this study. Moreover, it was compared to other researchers' studies at the foil position with the most significant decrease in resistance. The simulation of total resistance with a 3° angle of attack resulted in a 2% reduction in resistance. The researcher's test data, which was used as secondary data, was reduced by 3.7%. It demonstrated that the simulation followed the ITTC's Practical Guidelines for Ship CFD. The study found that increasing the angle of attack to 3° creates a more significant lift, minimizes the wave's additional resistance, and lowers total resistance.The lowest wave resistance is generated at a 3° degree angle of attack, which is a considerable reduction of 19% compared to a 1°. At an angle of 3°, a greater lift drag is obtained after Fn 0.3 of foil, and the smallest at 2°.

ACKNOWLEDGEMENTS

As an undergraduate student of Naval Engineering, the third author acknowledges financial supports from the Faculty of Engineering of Universitas Pembangunan Nasional Veteran Jakarta. All authors sincerely thank the support of the Dean of the Faculty of Engineering Universitas Pembangunan Nasional Veteran Jakarta.

REFERENCES

1. UITHOF K, B.B., VAN OOSSANEN, P.G., MOERKE, N.,'Hull Vane ESD on DTMB5415 Destroyer – A CFD Analysis of the Effect on Resistance',*MAST Conference 2016, Amsterdam, Netherlands, pp. 21–23*, 2016.

2. ANDREWS, I., AVALA, V.K., SAHOO, P.K., RAMAKRISHNAN, S., 'Resistance characteristics for high-speed hull forms with vanes',*SNAME 13th International Conference on Fast Sea Transportation,*September 2015.

3. HOU, H., KRAJEWSKI, M., ILTER, Y.K., DAY, S., ATLAR, M., SHI, W., 'An experimental investigation of the impact of retrofitting an underwater stern foil on the resistance and motion', *Ocean Engineering*, *205*, *p.107290*,2020.

4. ÇELIK, C., DANIŞMAN, D.B., KHAN, S., KAKLIS, P., 'A reduced order data-driven method for resistance prediction and shape optimization of hull vane', *Ocean Engineering*, *235*, *p.109406*, 2021.

5. BOUCKAERT, B.,'An underwater spoiler on a warship: why, when and how?', *Zeszyty Naukowe Akademii Marynarki Wojennej*, *59*, 2018.

6. BOUCKAERT, B., UITHOF, K., MOERKE, N., VAN OOSSANEN, P.G.,'Hull Vane on 108m Holland-Class OPVs: Effects on Fuel Consumption and Seakeeping', *Proceeding of MAST Conference*, 2015.

7. UITHOF, K., HAGEMEISTER, N., BOUCKAERT, B., VAN OOSSANEN, P., MOERKE, N.,'A systematic comparison of the influence of the Hull Vane®, interceptors, trim wedges, and ballasting on the performance of the 50m AMECRC series# 13 patrol vessel' *Proc. Warship*, 2016.

8. SUASTIKA, K., HIDAYAT, A., RIYADI, S.,'Effects of the application of a stern foil on ship resistance: A case study of an Orela crew boat', *Int. J. Tech*, *8(7), pp.1266-1275,*2017.

9. UITHOF, K., VAN OOSSANEN, P., MOERKE, N., VAN OOSSANEN, P.G., ZAAIJER, K.S.,'An update on the development of the Hull Vane', *9th International Conference on High-Performance Marine Vehicles (HIPER) Athens*, 2014.

10. ITTC,'Recommended Procedures and Guidelines Practical Guidelines for Ship CFD Applications', *7.5-03-02-03*, 2011.

AUTHORS BIOGRAPHY

Wiwin Sulistyawati is a lecturer with a research focus on ship hydrodynamics. Before joining the Faculty of Engineering, Universitas Pembangunan Nasional Veteran Jakarta, she was a ship design consultant. She holds a Doctorate in Mechanical Engineering.

Purwo Joko Suranto is a lecturer with research focusses on ship design and hydrodynamics. Currently, he is head of a Naval laboratory at the Faculty of Engineering, Universitas Pembangunan Nasional Veteran Jakarta. He holds a Master's degree in Naval Architecture Engineering.

Nurhayati Utami is an undergraduate student of Naval Architecture Engineering at Faculty of Engineering, Universitas Pembangunan Nasional Veteran Jakarta. And this research is part of her thesis.

REAL TIME NOTIFICATION FOR AUTOPILOT SHIP DATA COMMUNICATION BASED ON LORA AND MQTT

Afif Zuhri Arfianto, Lilik Subiyanto, Dimas Pristo and **Yusril Ihda Imamuddin,** Dept. of Marine Electrical Engineering, Politeknik Perkapalan Negeri Surabaya, Indonesia

SUMMARY

Autopilot is a system control application on moving vehicles which serves to stabilize the direction of motion in the desired time and the programmed path's direction. The autopilot control system is controlled by a series of microcontrollers and GPS that has the function to be able to determine and the desired object. A control system is needed to control the boat to its destination. This system is broadly divided into two parts: the control station part and the autopilot unmanned ship part. These two parts communicate with each other using LoRa device. The data sent from the control station to the autopilot unmanned ship is target latitude and longitude coordinates. The autopilot unmanned ship will send feedback in the form of boat latitude and longitude coordinates so that users can find out the current location of the boat. The control station will send feedback from the boat to a database that can be monitored via the website to find out the boat's current condition on the website. The multi interface communication module was designed, implemented and evaluated, which was to transfer low-power, long-range large amount of data using MQTT and LoRa.

1. INTRODUCTION

In this era of globalization, the development of technologies that apply electronic sciences and control systems, especially microcontrollers, is increasingly popular. A large number can know this of equipment that implements a mix of science and systems control, one of which is shipping [1]. In shipping, there is much work that has difficulty in terms of supervision and observation of the work.Soit is necessary to system that can assist in the work of the company. One such system is an auto control system for pilots who can significantly assist with work on board to drive the boat. In such systems, the autopilot control system is controlled by a series of microcontrollers and GPS that has the function to be able to determine and the desired object. This control system is designed with electrical system controls that utilize microcontrollers, RSSI and GPS as control media [2][3][4][5][6]. With input data derived from GPS and the microcontroller will be sent to drivers and motors for the boat's propeller until it reaches and if there is a barrier in front of the boat, the boat will stop automatically. This research will use small boats as testing media. The pilot's auto control system uses GPS to be tested directly in the water reservoir[7]. The data used in this GPS program is the data settings and real-time data, where data sets are data the destination of the boat stops and the real-time data is the actual position of the boat. With both data above, if the real-time data is the same as the data set, the boat will stop, and if real-time data is not the same as the data set, then the boat will continue to run and look for the location of the position of the data setting or destination [4].

Autopilot is a system control application on moving vehicles such as airplanes, helicopters, or boat, which serves to stabilize the direction of motion in the desired time and the programmed path's direction [8]. The development of autopilot prototypes is used for testing new control algorithms and the reliability of other electronic components such as sensors and microprocessors [9]. In rocket control, the control system is used to guide the rocket in the desired direction. The rocket's track can be read from the inertia sensor or additional from GPS data[10]. Changes in the motion of the rocket's path by changing the direction of the rocket motor's thrust or by rotating a few degrees of the tilt of the rocket's fins by utilizing the principle of aerodynamic forces[11]. To develop this system, using a rate gyroscope to calculate angles and a stable control actuator is an essential part of the system.

The autopilot system is a mechanical, electrical, or hydraulic system that guides a vehicle without human intervention [12]. Generally, the autopilot is associated with an airplane [13]. However, the autopilot is also used on boats with the same term. A navigation system and control system guide the movement of the boat in autopilot mode [4]. The boat navigation system on autopilot uses the waypoint method. The navigation system will guide the autopilot boat to move systematically following the waypoints that have been set at the ground control station. In the movement of an autopilot boat, external forces such as wind and waves and internal forces such as an imbalance in the rotation of the main propulsion speed will influence its movement. Therefore, a control system is needed to control the boat to its destination. The relatively accurate system dynamics model affects the steering performance of the autopilot system. The system in question can be considered a boat with an actuator rudder, which experiences external disturbances, as shown in Fig 1.

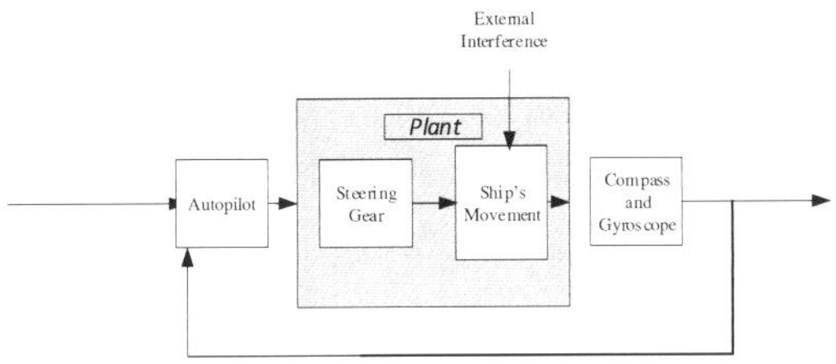

Fig 1 : Autopilot System Diagram Blocks

LoRa module is a device with a wireless frequency or radiofrequency with a long-distance using low power consumption. LoRa module can be used to control and monitor something with radiofrequency. Such as monitoring mountain activities, drones, electricity, agriculture, and others. LoRa (Long Range) is a unique and extraordinary modulation format created by Semtech. The resulting modulation uses FM modulation. LoRa is a digital wireless data communication technology patented and developed by Cyclo from Grenoble, France, and acquired by Semtech in 2012

LoRa is a long-range wireless communication protocol that can compete with other Low-Power Wide-Area Network (LPWAN) such as Narrowband IoT (NB IoT) or LTE Cat M1. Compared to this technology, LoRa has a very long-range, which exceeds 10 km, but the resulting data rate is low. The SX1276 / 77/78/79 transceiver is a LoRa® long-range wireless transceiver module that provides Ultra-long spread spectrum communication with a high degree of resistance to low power consumption interventions. Using the LoRa modulation technique patented by Semtech, the module can run with a sensitivity up to -148 dBm and function at a low cost. This LoRa module has advantages, such as selectivity and blocking capabilities, which are much better than conventional modulation systems, making this module one of the transceiver devices superior in range, resistance to intervention, and power consumption. However, the disadvantage of LoRa itself is that the data speed is below 50 kbps and because LoRa has a duty cycle and other weaknesses, in practice, this technology is suitable for applications where the delay in sending data is not taken into account. RFM95 is a LoRa type widely circulating in the market and uses the LoRa SX1276 chip with a frequency of 915 MHz [14] .The specifications and forms of the Arduino LoRa module are shown in Table 1.

Table 1:LoRa RFM95 915Mhz Specifications

Parameter	Specification
Frequency	915 MHz
Chips	SX 1278
Supply Voltage	1.8 - 3.3 V
Data Transfer Speed	1.2 – 300 kbps
Data	SPI data Interface
Sensitivity	-139 dBm
Parameter	Specification
Modulation	LoRa

2. METHODOLOGY

2.1 LORA RFM95

LoRa (Long Range) is a wireless frequency device or long-distance radio frequency using low power consumption. With this LoRa module, users can control and monitor something utilizing a radio frequency network. Such as monitoring mountain, forest, sea activities, drones, electricity, agriculture and others. LoRa (Long Range) is a unique and fantastic modulation format created by Semtech. The frequency value in LoRa varies according to the region. In Asia, the frequency used is 433 MHz; in Europe is 868 MHz, while in North America is 915 MHz.

Table2 :Communication Technology Parameters Comparison

Technology	Distance	Maximum Rate	Power Consumption
Bluetooth	10 m	2 MB/s	Low
WiFi	0-60 m	54 MB/s	High
RFID	0-100 m	10 KB/s	Low
Zigbee	0-1500 m	250 KB/s	Low
LORA	0-15 km	600 KB/s	Low

Table 2shows that communication technology uses LoRa has a pretty long range compared to other communication technologies and has a low power consumption [16].

2.2 MQTT (MESSAGE QUEUING TELEMETRY TRANSPORT)

MQTT (Message Queuing Telemetry Transport) protocol is a protocol that runs on the TCP / IP stack and is designed specifically for the machine to machine that does not have a particular IP address, as well as Arduino, Raspi, or other devices. The MQTT work system uses the Publish and Subscribe data system. The device will connect to a broker. The broker's function in MQTT is to handle the Publish and Subscribe data from various devices linked to a server with a particular IP address.
Some examples of existing brokers are Mosquitto, HiveMQ, and Mosca. Publish is a way for a device to send its data to subscribers, which is the input occupied by the sensor. Subscribe is a way for a device to receive various kinds of data from publishers. Subscribers can be in the form of web sensor monitoring applications and others. This subscriber will later receive data from the publisher through an intermediary broker MQTT.
MQTT publishers can send topic-based messages in the form of strings to MQTT brokers. MQTT subscribers can subscribe to the topics they want and what users want to the MQTT broker. After being subscribed, the MQTT broker can convey messages from the MQTT publisher to the MQTT subscriber. Through the master broker, the MQTT broker can deliver the communication between MQTT publishers and subscribers.

According to Figure 2the simple MQTT system has three essential parts: Publisher, Broker, and Subscriber. Each section has its task:

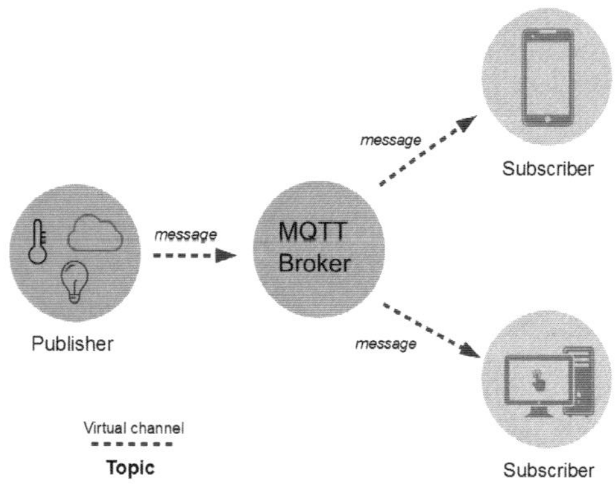

Fig 2 :MQTT System

- Publisher Block

In the publisher block, there are sensors as input to be sent; for example, there is sensor A. Each sensor will connect to a gateway such as an Ethernet shield, ESP8266, or device that can connect to MQTT Broker. Between gateways and sensors, there must be Arduino, RaspberryPi, NodeMCU, or the like. The task of this publisher block is quite simple, i.e., sending data taken from sensor A to an MQTT broker.

- MQTT Broker

MQTT Broker has an address that publishers and subscribers can access. The job of this MQTT Broker is transaction liaison or as a data intermediary between publishers and subscribers. MQTT broker will recognize data through a grouping or called "Topic". For example, a publisher sends sensor data A, with topic data 1. Then one day, if a subscriber subscribes to the same "topic," the subscriber will receive sensor data A from the publisher.

- Block Subscribers

The block subscriber is in charge of subscribing data on topic data 1. After getting data in the form of sensor values A from the publisher, it can display these data on the localhost web or process it into a structured real-time system monitoring without refreshing the browser.

2.3 HARDWARE DESIGN

The hardware design process begins by assembling all the required components, including NodeMCU and RFM95. NodeMCU acted as a data processor programmed through the Arduino IDE and received through the LoRa RFM95 module. Then the data is connected to the MQTT broker for the webserver to monitor or detect fish via the web. The following is the wiring of NodeMCU with LoRa.

Figure 3 is a design of NodeMCU and LoRa with a pin connection as a tool to receive information to be displayed on the web and telegram notifications. The pins used for the connection shows in table 3

2.4 DETERMINING THE LORA RFM95 ANTENNA

For the antenna design to function optimally, calculations are carried out using an antenna calculator to determine how high is the antenna of LoRa RFM95 required.

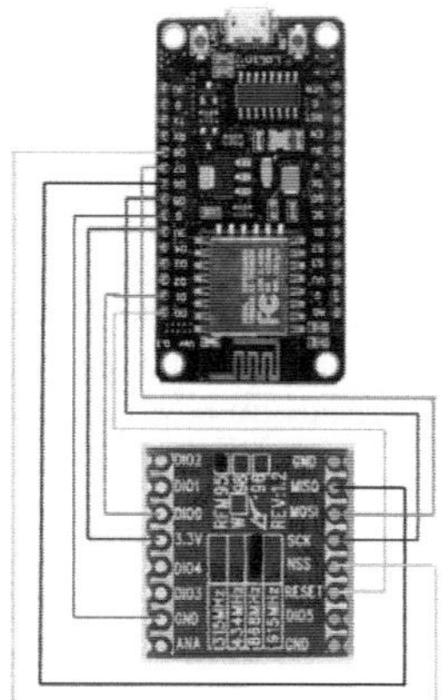

Fig3 :Wiring NodeMCU LoRa

Tabel 3 :NodeMCU Pin Connection with LoRa

NodeMCU Esp8266	LoRa RFM95
Gnd	Gnd
3 V	3.3 V
D0	Rst
D1	D10
D5	Sck
D6	Miso
D7	Mosi
D8	Nss

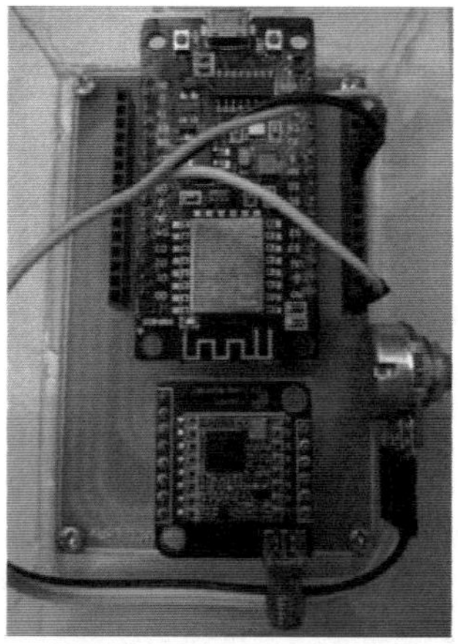

Fig 4 : NodeMCU hardware with LoRa

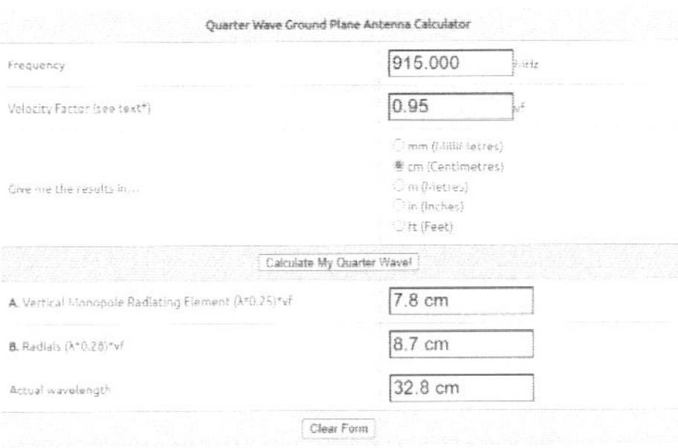

Fig 5 : LoRA Antenna Calculator

Figure 5 is the result of the antenna required calculation for LoRa RFM95, with a specification of 915 MHz, it shows that what is needed is a 7.8 cm antenna.

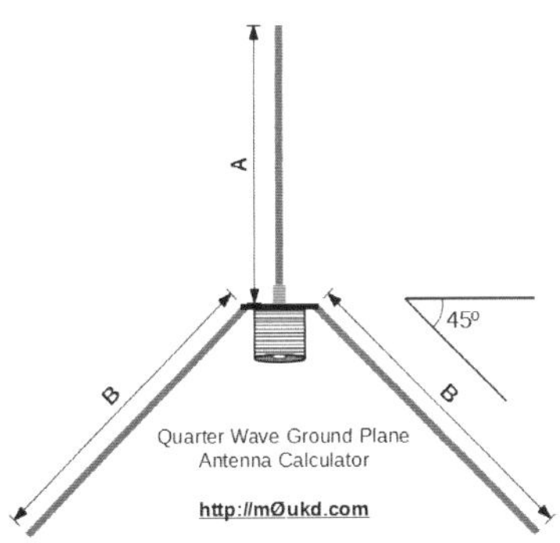

Fig 6 :Illustration of LoRa Antenna

2.5 DATA COMMUNICATION TX SENDER AND RX RECEIVER

The data used for communication between LoRa is using 4 LoRa. Including 3 LoRa transmitters with a potentiometer as fish detection inputs and 1 LoRa receiver that has received analog data programmed between LoRa TX and RX. Also using a simulation of playing minimum-maximum on the value settings of a pre-programmed potentiometer that is done by initializing the potentio program. If the lowest or minimum reading is less than 700, it is simulated that there are no fish. Furthermore, when the potentiometer is played at maximum or the data reading is more than 700, it is estimated that there are fish in the area. First, sensor readings use 1, 2, and 3 potentiometer transmitters as input for sending data readings to the LoRa receiver, which will receive data 1 from transmitter 1, data 2 from transmitter 2, and data 3 from transmitter 3.

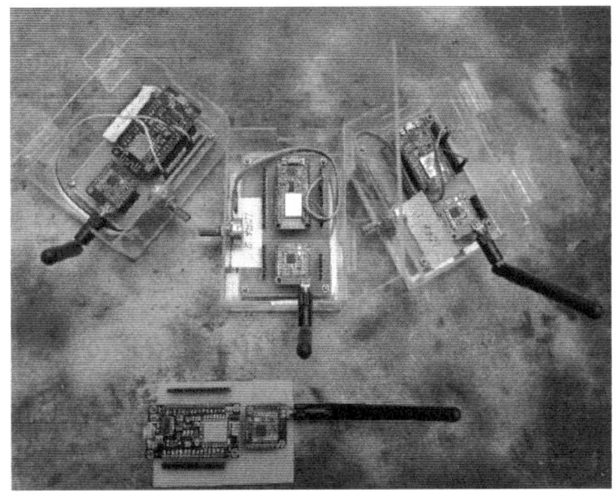

Fig 7 : LoRA Transmitter 1, 2,3 and Reciever

3. RESULTS AND DISCUSSION

LoRa Communication Distance Testing. This test was performed using an 8 cm antenna, also called a 3 dB antenna, and was tested in open conditions in Mejoyo village, Bangsal districts, Mojokerto, East Java. The test location is an open and flat area without trees or high-rise buildings. GPS assistance is used as a distance measurement. The result can be seen in an online map by testing 0xBB, 0xCC, and 0xDD (LoRa's sender). While 0xAA is a LoRa receiver. The test was performed in ten attempts. The purpose of this test is to determine how far the LoRa signal can communicate. Testing is done from sending to receiving data with a certain distance variation and delay to the recipient.

Based on the testing results of the LoRa communication range with an open condition using an 8 cm vertical antenna or 3 dB antenna above, the author discovered that the antenna's type affects the quality of transmission and the maximum range that LoRa can cover. In this test the distance difference only affects the Receive Signal Strength Indicator or (RSSI)'s value. This RSSI is an indicator of the signal strength received by a wireless device. The farther the distance, the smaller the RSSI value, or the more negative the value, indicating a bad signal. Whereas, the closer the distance, the RSSI value will be greater or more positive, which means a good signal.

The test is carried out starting in an open ricefield area. Table 4 shows that at a distance of 10 meters, the RSSI value is 60. Meanwhile, at a distance of 450 meters, the RSSI value is -96. So to receive data, LoRa is only capable of reaching 450 meters. More than 450 meters, LoRa has difficulty communicating and can no longer connect.

Table 4 :LoRa Communication Distance Testing

Distance (m)	Data Received	RSSI Value (dBm)	Delay (s)
10	√	-60	0.4
50	√	-80	0.3
100	√	-96	0.4
300	√	-98	0.4
450	√	-96 Unstable	0.5

500	X	Not Connected	X
700	X	Not Connected	X
800	X	Not Connected	X
>805	X	Not Connected	X

4. CONCLUSION

To universally supply autopilot unmanned ship data communication, various sensor data with versatile application and transmission technology for large amount of data are needed. In this paper, the multi interface communication module was designed, implemented and evaluated, which was to transfer low-power, long-range large amount of data using MQTT and LoRa. The MQTT was able to communicate high transmission rate and long range while Lora was able to communicate low-power and long-range. The usability of this module in applicable area would be further studied after building the test bed using real sensor.

5. ACKNOWLEDGEMENTS

This research was supported by Directorate General of Research and Development, Ministry of Research, Technology, and Higher Education, Indonesia

6. REFERENCES

1. T. Porathe, Å. Hoem, Ø. J. Rødseth, K. Fjørtoft, and S. O. Johnsen, "At least as safe as manned shipping? Autonomous shipping, safety and 'human error,'" Saf. Reliab. Soc. a Chang. World. Proc. ESREL 2018, June17-21, 2018, Trondheim, Norw., 2018.

2. A. Z. Arfianto et al., "Unmanned Vehicle UsingReceived Signal Strength Indicator (RSSI) in Instant Beverage Industry," in 2019 International Conference on Advanced Mechatronics, Intelligent Manufacture and Industrial Automation (ICAMIMIA), 2019, pp. 340–343, doi: 10.1109/ICAMIMIA47173.2019.9223401.

3. M. Rifai, A. Z. Arfianto, and A. T. Muzazanah, "Radio Frequency-based Smart Control for Lighting in Public Service," in Journal of Physics: Conference Series, 2020, vol. 1595, no. 1, doi: 10.1088/1742-6596/1595/1/012026.

4. M. B. Rahmat, A. Z. Arfianto, T. B. Santoso, T. Santoso, and N. Gunantara, "Development of Autopilot Unmanned Smartboat Vehicle (AUSV) Based on Fishing Zone Prediction Map," in Journal of Physics: Conference Series, 2020, vol. 1595, no. 1, doi: 10.1088/1742-6596/1595/1/012036.

5. B. E. Putra et al., "Multi Automated Guided Vehicle (AGV) cardboard carrier using wireless communication," IOP Conf. Ser. Earth Environ. Sci., vol. 340, p. 012009, Oct. 2019, doi: 10.1088/1755-1315/340/1/012009

6. R. A. Atmoko, D. Yang, M. I. Abas, A. Z. Arfianto, and R. Rahim, "Teleoperation cloud industrial robot using XMPP protocol," Int. J. Recent Technol. Eng., vol. 8, pp. 6280–6284, 2019.

7. T. A. Putra et al., "KOMUNIKASI DATA BLUETOOTH UNTUK PERANGKAT INFORMASI PERSEBARAN IKAN (PORTABLE VIRTUAL ASSISTANT) PADA KAPAL NELAYAN TRADISIONAL," J. Teknol. Marit., vol. 1, no. 2, pp. 45–52, 2018.

8. L. R. Ribeiro and N. M. F. Oliveira, "UAV autopilot controllers test platform using Matlab/Simulink and X-Plane," in 2010 IEEE Frontiers in Education Conference (FIE), 2010, pp. S2H-1.

9. E. Capello, G. Guglieri, and G. Ristorto, "Guidance and control algorithms for mini UAV autopilots," Aircr. Eng. Aerosp. Technol., 2017.

10. S. Melnychuk, W. Foote, Z. Q. Zhao, and F. C. Moore, "Target acquisition and tracking system." Google Patents, Oct. 22, 2009.

11. T. Li, A. M. Esteban, and S. Zhang, "Enhanced disturbance rejection control based test rocket control system design and validation," ISA Trans., vol. 84, pp. 31–42, 2019.

12. C. Yan, W. Xu, and J. Liu, "Can you trust autonomous vehicles: Contactless attacks against sensors of self-driving vehicle," DEF CON, vol. 24, no. 8, p. 109, 2016.

13. M. Liu, G. K. Egan, and F. Santoso, "Modeling, autopilot design, and field tuning of a UAV with minimum control surfaces," IEEE Trans. Control Syst. Technol., vol. 23, no. 6, pp. 2353–2360, 2015.

14. R. K. Kodali, K. Y. Borra, S. S. GN, and H. J. Domma, "An IoT based smart parking system using LoRa," in 2018 International Conference on Cyber-EnabledDistributed Computing and Knowledge Discovery (CyberC), 2018, pp. 151–1513.

15. A. M. Yousuf, E. M. Rochester, B. Ousat, and M. Ghaderi, "Throughput, coverage and scalability of LoRa LPWAN for internet of things," in 2018 IEEE/ACM 26th International Symposium on Quality of Service (IWQoS), 2018, pp. 1–10.

16. M. K. Hasin, "DETEKSI LOKASI PESEBARAN IKAN PADA PETA DIGITAL UNTUK PORTABLE VIRTUAL ASSISTANCE (PVA) NELAYAN TRADISIONAL," 2017.

7. AUTHORS BIOGRAPHY

Lilik Subiyanto is lecturer in the Department of Automation Engineering, Politeknik Perkapalan Negeri Surabaya. In this research, he is mainly responsible for the experiment of LoRA for communication between the control station and the autopilot unmanned ship

Afif Zuhri Arfianto is lecturer in the Department of AutomationEngineering, Politeknik Perkapalan Negeri Surabaya. In this research, he is mainly responsible for multi interface communication module LoRA and MQTT was designed, implemented and evaluated

Dimas P. Riananda is lecturer in the Department ofMarine Electrical Engineering, Politeknik Perkapalan Negeri Surabaya. In this research, he is mainly responsible for electrical controls systemwas designed, implemented and evaluated

Yusril Ihda Imamuddin is student in the Department ofMarine Electrical Engineering, Politeknik Perkapalan Negeri Surabaya. In this research, he is mainly responsible for lora communication distance testing.

NUMERICAL SIMULATION OF SLOSHING IN THE PRISMATIC TANK WITH VERTICAL BAFFLE USING SMOOTHED PARTICLE HYDRODYNAMICS

H Atthariq, A Trimulyono, D Chrismianto, Department of Naval Architecture, Diponegoro University, Indonesia

SUMMARY

The capacity of LNG carriers is increasing year by year due to the demand for liquified natural gas (LNG). One of the types of LNG carriers is membrane type that the tank is prismatic. A natural phenomenon in a liquid carrier such as LNG is sloshing. Sloshing is a violent phenomenon in fluid dynamics caused by excitation force. The present study will conduct sloshing with a baffle shape to reduce the free surface inside the tank and impact pressure. In this study, a single vertical baffle and a double vertical baffle are used to reduce sloshing. In the present study, the meshless approach, so-called Smoothed particle hydrodynamics (SPH), is used to reproduce sloshing in the prismatic tank. The three-dimensional prismatic tank was used to capture sloshing with the baffle. Regular roll motion with one pressure sensor uses to validate the SPH results. The result shows that vertical baffle effectively reduced fluid movement inside the tank and impact pressure. As a result, the hydrodynamic force and moment decreased. The ratio of the baffle and the water depth 0.9 shows has significant influence to sloshing.

1. INTRODUCTION

Sloshing on an LNG ship is a dangerous phenomenon caused by a tank encountering an external oscillation. If the tank moves very energetic, it will be dangerous when the value is close to the natural frequency [1]. Violent sloshing could happen and cause damage to the tank due to impact pressure caused by violent fluid inside the tank with the wall.

One of the ways to reduce sloshing is to use additional blocks in the tank. These blocks are called baffles or internal blocks. The configuration of the baffles is needed properly well designed to get an effective effect reduce fluid movement. Baffle with proper height will reduce fluid movement. As a result, it reduces the impact pressure [2,3]. It was showed the ratio of baffles height and water depth is essential. The quantity of baffles affects the reduction in wave height and pressure on the tank wall [4]. Additional factors such as slosh amplitude, frequency, liquid level, and tank geometry shall be accounted [5].

Computational fluid dynamics (CFD) is a widely known approach to calculating fluid problems. One of the methods is Mesh-based CFD that uses an Eulerian scheme to solve Navier-Stokes equations. The finite element method was used to solve sloshing problems on a three-dimensional tank [6]. Kim et al. [7] calculate impact load on sloshing problems using the finite difference method.

The sloshing phenomenon is related to free-surface flow. As a result, the meshless method has the advantage for sloshing cases. One of the methods is smoothed particle hydrodynamics (SPH), also known as mesh-free computational fluid dynamics (CFD). This method is often applied to sloshing problems because the particle methods are fully Langrangian. The first research related to the use of the SPH method on free surface flow was carried out by Monaghan [8].

The use of the SPH method in recent years is increasing. SPH method is used to calculate obstacles in the rectangular tank [9]. SPH application for sloshing was carried out for prismatic tank with low pass filter to reduce pressure noise [10]. Long duration sloshing of single-phase and two-phase SPH was used to calculate the impact pressure [11]. The results were showed SPH has fairly agreement with the experiment, although computation time was costly. Long duration sloshing in ractangular tank in single-phase was carried out by Green and Peiró[12]. Moreover, long duration of sloshing using the SPH approach in different tank shapes was successfully carried out in three-dimesion [13]. It was shown that SPH has the advantage to overcome sloshing phenomena because of the nature of the meshless approach that SPH easily captures large deformation.

The present study aims to know the vertical baffle's effect on sloshing in the prismatic tank. The variations of baffles use single and double vertical baffles. Roll motion with one pressure gauge near the free surface is used, and the tank filling ratio is 25%. It shows that the filling ratio of baffle height and water depth significantly affects reducing the sloshing phenomenon.

2. PRISMATIC TANK DIMENSION

In this study, we use a prismatic tank from a previous study that has been successfully conducted in the experiment facility in Tokyo [14]. Figure 1. shows the sketch of three dimensions of the prismatic tank that will use for SPH computation. The detail is shown in Table 2, which explains the principal dimension of the prismatic tank used in the present study. Figure 1. shows the movement tank is roll motion.

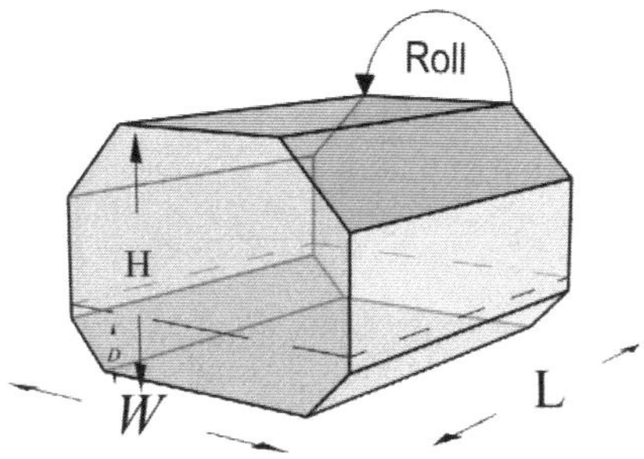

Figure 1. Prismatic Tank

Table 2. Principal Dimension of Prismatic Tank

Dimension of the Prismatic Tank			
Length	(L)	0.21	m
Width	(W)	0.38	m
Height	(H)	0.30	m
Draught	(D)	0.0525	m

3. BAFFLE CONFIGURATION

The tank model based on the prismatic tank is shown in the reference [14]. The variations single and double vertical baffle are made in the present study. The baffle thickness in this study is 0.006 meters. The ratio of the height of the baffle and water depth is found using the following equation.

$$\frac{h_b}{d} = 0.9 \quad (1)$$

Where h_b is the height of the baffle, d is the height of the liquid.It shows that the optimal ratio between the baffle's height and the liquid's height was 0.9 to avoid fluid run-up to the top of the tank [2].Meanwhile, the position of baffles evenly will reduce the impact pressure on the tank wall [4]. The position on a single vertical baffle was in the mid of the tank. The baffles are placed distributively on the left and right sides of the tank wall (see **Figure 4.**). **Figure 2**. shows the geometry of the tank model without baffle and **Figure 3**. shows a single baffle. In this study, the filling ratio tank is 25%, with external oscillation motion rolling.

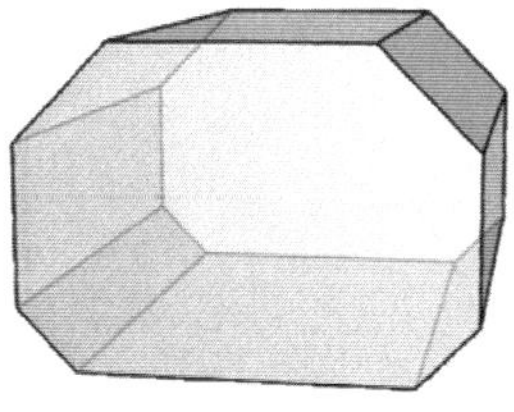

Figure 2. Prismatic tank without baffles.

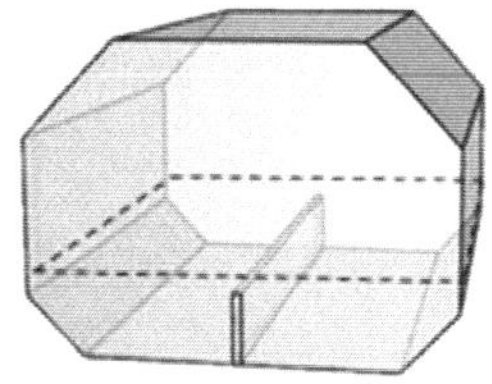

Figure 3. Prismatic tank with single vertical baffle.

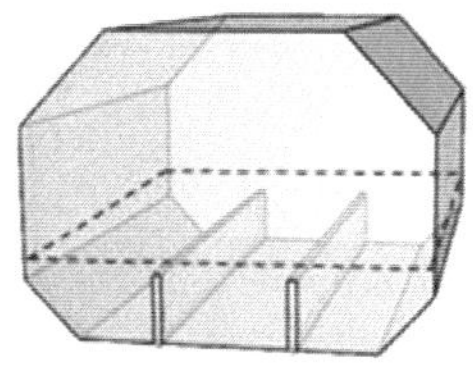

Figure 4. Prismatic tank with double vertical baffle.

4. EXTERNAL OSCILLATION AND SENSOR

The external oscillation force was based on reference [14]. **Figure 5** depicts the movement of the tank. The roll motion is regular motion with a time simulation is 30 second. Table 3 shows the frequency and amplitude of oscillation. There is a pressure gauge to measure impact pressure near the free surface. **Figure 6** shows the position of the pressure gauge. The pressure gauge was set in the mid of the tank.

Table 3. External Force

External Force	Frequency (Hz)	Amplitude (°)
Roll	1.04	8.66

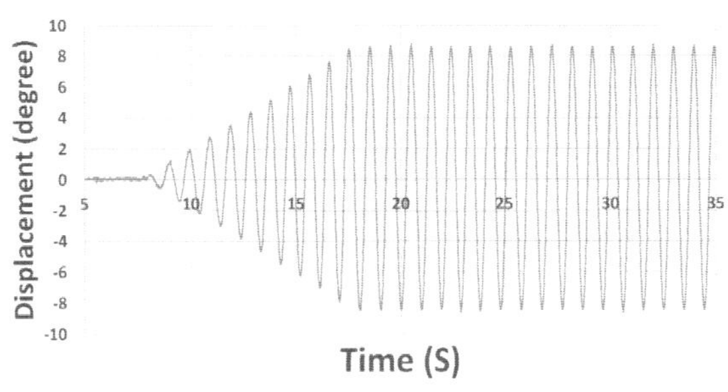

Figure 5. Time history of tank motion

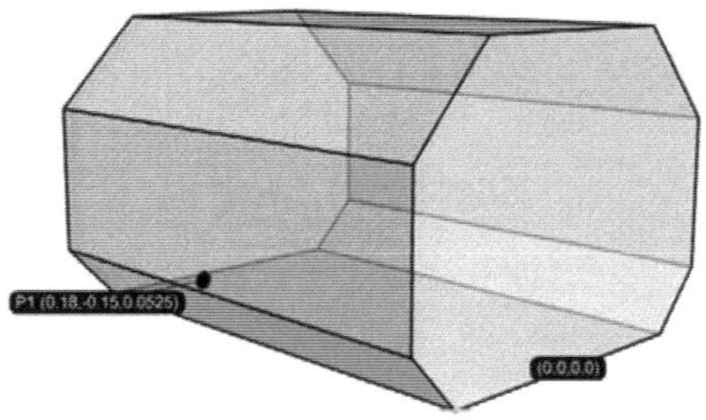

Figure 6. Pressure Gauge location.

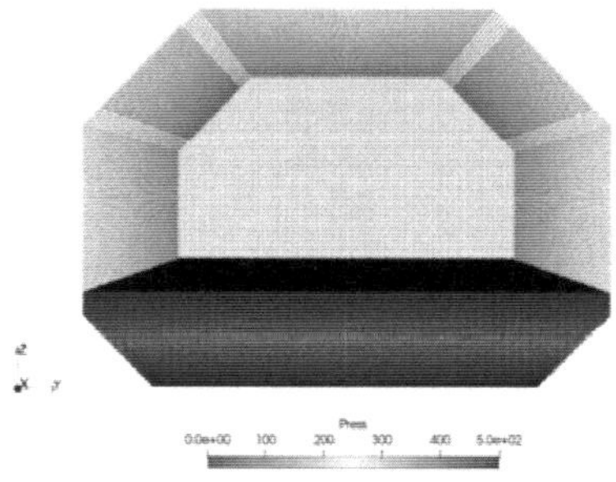

Figure 7. Hydrostatic pressure

5. SMOOTHED PARTICLE HYDRODYNAMICS (SPH)

Smoothed Particle Hydrodynamics (SPH) is a meshless and Lagrangian approach. The method is discretized the computation domain using point or particle. This method was first applied to the astrophysical problem. This study uses an open-source SPH solver DualSPHysics version 5.0 [15] to overcome sloshing. DualSPHysics is software based on C++ and CUDA programming languages. In the DualSPHysics XML programming structure, there is a command structure consisting of a casedef and an execution program. Casedef consists of several parts, which is constantsdef, i.e., constant required in simulation, mkconfig is a label configuration between fluid and boundaries. Creation of geometry particle to determine the location of the object, geometry mainlist to create a tank model, and fluid settings. Moreover, a motion is a label about movement performed at the tanks. After working on casedef, settings are carried out to execution to set the parameters in the DualSPHysics application.

6. RESULTS

a. Hydrostatic Pressure

In the rest conditions, the tank continues to experience hydrostatic pressure from the static fluid. The hydrostatic pressure is compared with the analytical solution. In order to determine the hydrostatic pressure accuracy that occurs in rest conditions, the following pressure equation can be used.

$$P = \rho \,.\, g \,.\, h \tag{2}$$

The hydrostatic pressure calculation was carried out at a filling ratio of 25% (0.0525 m). The hydrostatics pressure in the SPH simulation shows in **Figure 7**.

b. Hydrodynamic Pressure

Figure 8. shows the comparison between the experimental results and SPH simulation. The red line is the impact pressure of the experiment, green is the tank with a single baffle, and the purple line is a tank with two baffles. The mean peak of the experiment is 498.51 Pa, and the mean peak of SPH without baffle is 645.33 Pa. Moreover, the mean peak of SPH simulation with a single baffle is 141.2 Pa. The mean of peak SPH simulation with double baffles is 100.615 Pa. The results showed that the difference between SPH simulation and experiment is 22%. Meanwhile, baffles can reduce the impact pressure up to 78% using a single baffle and 85% using a double baffle.

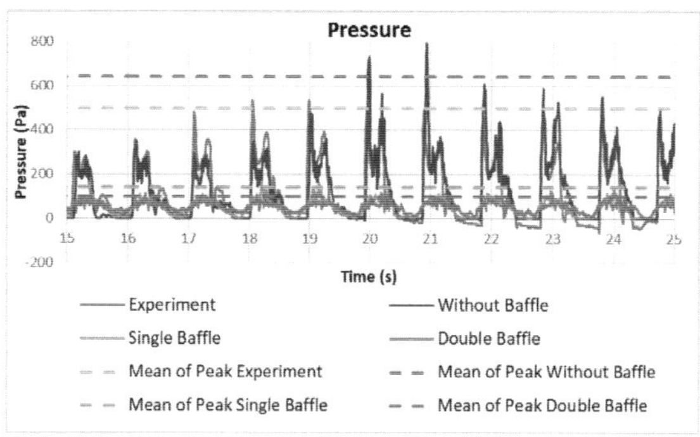

Figure 8. Hydrodynamic pressure without and with baffle.

c. Water Level

Figure 9. shows the difference in water level with or without baffle. The water level without baffles is 0.0891 m. Using single baffles is 0.073 m, and double baffles are 0.0674 m, respectively. The use of baffles can reduce baffle height up to 18% for the single baffle and 24.3% for double baffles.

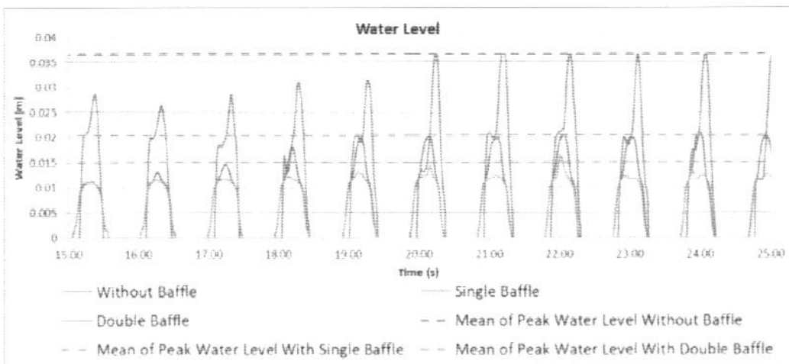

Figure 9. Water height of water inside of tank without and with baffle.

d. Free surface deformation

In this paper, we use advanced visualization of the SPH model using blender [16]. Time simulation 22.06 s had used to see the baffle effect to free surface deformation inside the tank. Time simulation 22.06 s is the highest point of water that runs up to the top of the tank in the original shape (see **Figure 10**).

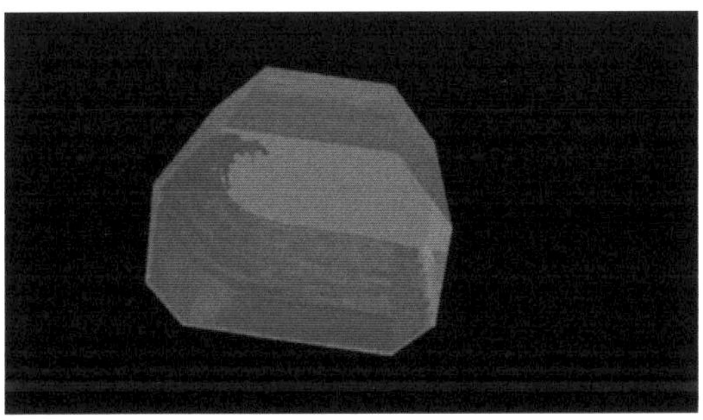

Figure 10. Prismatic tank without baffle

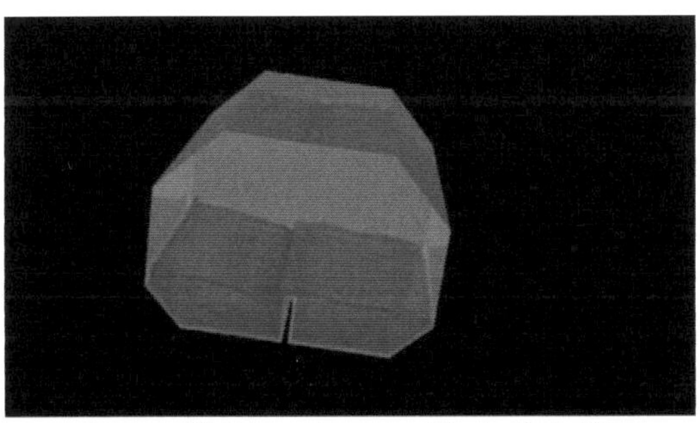

Figure 11. Prismatic tank with single baffle

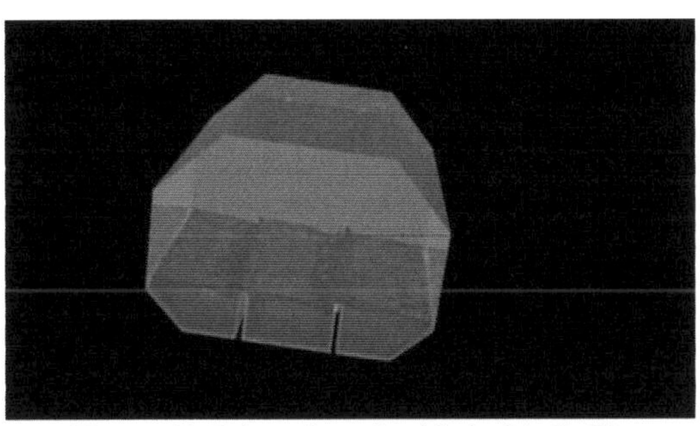

Figure 12. Prismatic tank with double Baffle

Figure 11. Shows the free surface deformation using a single baffle. It depicts vertical baffle effectively reducing the run-up of water as a consequence. The impact pressure has decreased. A similar result has shown in **Figure 12.**

e. Hydrodynamics Force

Figure 13. shows hydrodynamics forces that occur during sloshing. The hydrodynamics force without baffles shows the mean of peak force 15.451 N. For single baffle is 12.65 N. The double baffle is 12.78 N. The use of baffles can reduce the hydrodynamics force a difference of 18% for a single baffle. The double baffle is 17.2%, respectively.

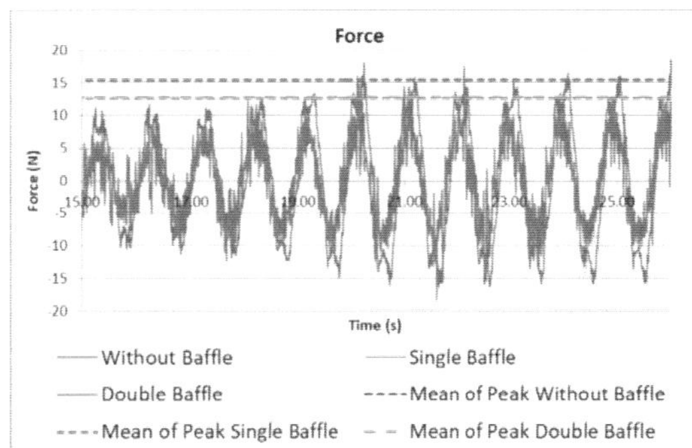

Figure 13. Hydrodynamics forcewithoutand with baffles

f. Hydrodynamics Moment

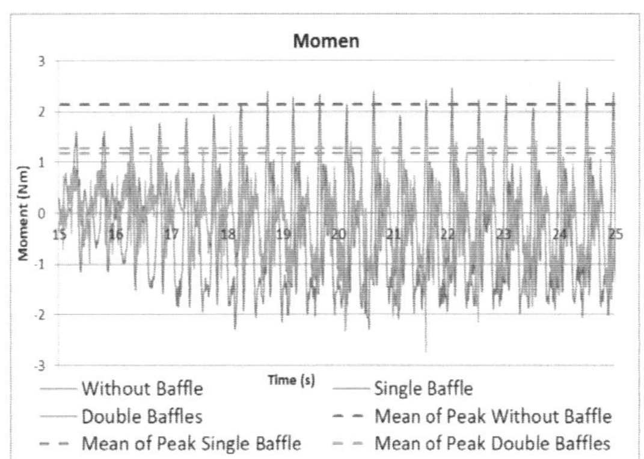

Figure 13. Hydrodynamics moment of without and using baffle

Figure 13. shows a graph of moments that occur during sloshing. The graph of moments without baffles reached its mean peak at 2.146 Nm, while for single baffle is 1.177, and for double baffles is 1.276 Nm. The use of baffles effectively decreases moments during sloshing, up to 45.1% for the single baffle and 40.4% for double baffles.

7. CONCLUSION

The present study shows that the results of hydrostatic pressure have reasonably good accuracy. In addition, for dynamic pressure, the accuracy is acceptable compared with experimental results. The use of vertical baffle in the tank has a significant effect on the sloshing phenomenon. The baffle can reduce the dynamic pressure by 85%, hydrodynamic forces are 18%, and hydrodynamics moment is 45.1%. The use of vertical baffles reduces fluid movement effectively in the sloshing phenomenon. Meanwhile, the comparison of the effects of single and double vertical baffles does not have a significant difference. Future works of sloshing in two-phase SPH need to be conducted to know the effect of compressibility of air into impact pressure [17].

8. ACKNOWLEDGEMENT

The authors wish to thank the ship hydrodynamics laboratory for the computer facilities and conference fee.

9. REFERENCES

1. R. Ibrahim, *Liquid Sloshing Dynamics: Theory and Applications*. Cambridge University Press, 2005.

2. J. H. Jung, H. S. Yoon, C. Y. Lee, and S. C. Shin, "Effect of the vertical baffle height on the liquid sloshing in a three-dimensional rectangular tank," Ocean Engineering, Vol. 44, pp. 79–89, 2012.
3. Q. Jin, J. Xin, F. Shi, and F. Shi, "Parametric studies on sloshing in a threedimensional prismatic tank with different water depths, excitation frequencies, and baffle heights by a Cartesian grid method," Int. J. Nav. Archit. Ocean Eng., 2021.

4. C. Ma, C. Xiong, and G. Ma, "Numerical study on suppressing violent transient sloshing with single and double vertical baffles," Ocean Engineering, Vol. 223, pp. 108557, 2021.

5. NASA, "Slosh suppression," *Nasa Sp-8031*. p. 36, 1969.

6. G. X. Wu, Q. W. Ma, and R. Eatock Taylor, "Numerical simulation of sloshing waves in a 3D tank based on a finite element method," *Appl. Ocean Res.*, vol. 20, no. 6, pp. 337–355, 1998.

7. Y. Kim, Y. S. Shin, and K. H. Lee, "Numerical study on slosh-induced impact pressures on three-dimensional prismatic tanks," *Appl. Ocean Res.*, vol. 26, no. 5, pp. 213–226, 2004.

8. J. J. Monaghan, "Simulating Free Flows with SPH." Journal of Computational Physics, Vol. 110, pp. 399-406, 1994

9. Syamsuri, M. J. Chern, and N. Vaziri, "SPH model for interaction of sloshing wave with obstacle in shallow water tank," *J. King Saud Univ. - Eng. Sci.*, 2020.

10. A. Trimulyono, Samuel, and M. Iqbal, "Sloshing Simulation of Single-Phase and Two-Phase SPH using DualSPHysics" Kapal: Jurnal Ilmu Pengetahuan dan Teknologi Kelautan, Vol. 17, No. 2, pp. 50-57, 2020.

11. A. Trimulyono, D. Chrismianto, Samuel, and M. H. Aslami, "Single-phase and Two-phase Smoothed Particle Hydrodynamics for Sloshing in the Low Filling Ratio of the Prismatic Tank" International Journal of Enginerring Transactions B: Applications, Vol. 34, No. 5, pp. 1345–1351, 2021.

12. M. Green, and j. Peiró, "Long duration SPH simulations of sloshing in tanks with a low fill ratio and high stretching" Computer and Fluids, Vol. 174, pp. 179–199, 2018.

13. M. Green, Y. Zhou, J. M. Dominguez, M. G. Gesteira and J. Peiro, "Smooth particle hydrodynamics simulations of long-duration violent three-dimensional sloshing in tanks" Ocean Engineering, Vol. 229, 2021.

14. A. Trimulyono, H. Hashimoto, and A. Matsuda, "Experimental validation of single- and two-phase smoothed particle hydrodynamics on sloshing in a prismatic tank," Journal of Marine Science and. Engineering, vol. 7, No. 8, 2019.

15. J. M. Dominguez, G. Fourtakas, C. Altomare, R. B. Canelas, A. Tafuni, O. Garcia-Feal, I Martinez-Estevez, A. Mokos, R. Vacondio, A. J. C. Crespo, B. D. Rogers, P. K. Stansby, M. Gomez-Gesteira,"DualSPHysics: from fluid dynamics to multiphysics problems"Computational Particle Mechanics, March, 2021.

16. O. Garcia-Feal, A. J. C. Crespo, and M. Gomez-Gesteira, "VisualSPHysics: advanced fluid visualization for SPH models," Computational Particle Mechanics, January, 2021.

17. A. Trimulyono, D. Chrismianto, and M. Iqbal, "Two-phase SPH for sloshing simulation in prismatic tank" 2ndInternational Conference on Mechanical Engineering Research and Application (iCOMERA 2020), Malang, 7-9 October, 2020.

10. AUTHORS BIOGRAPHY

H Atthariq, Undergraduate student at Department of Naval Architecture, Diponegoro University (UNDIP). He research interest is smoothed particle hydrodynamics (SPH) and computational fluid dynamics (CFD).

A Trimulyono, Assistant Profesor and head of ship hydrodinamics laboratory at department of naval architecture Diponegoro University (UNDIP), He teachs hydrodynamics, fluid mechanics and computational fluid dynamics (CFD).

D Chrismianto, Assistant Profesor and head ofship design and computer aided design laboratory at Department of Naval Architecture Diponegoro University (UNDIP). He teachs hydrodynamics, ship design and CAD.

BENCHMARK TESTS OF FINE^{TM}/MARINE CFD CODE FOR THE CALCULATION OF SHIP RESISTANCE AT HIGH FROUDE NUMBERS

I K Suastika, A F Fauzi, A S Saputra, I K A P Utama, Department of Naval Architecture, Institut Teknologi Sepuluh Nopember, Surabaya, Indonesia
A Firdhaus, Department of Systems and Naval Mechatronic Engineering, National Cheng Kung University, Taiwan
Sutardi, Department of Mechanical Engineering, Institut Teknologi Sepuluh Nopember, Surabaya, Indonesia
M Hariadi, Department of Computer Engineering, Institut Teknologi Sepuluh Nopember, Surabaya, Indonesia

SUMMARY

Benchmark tests of FINE^{TM}/ Marine CFD code were performed to verify the ship-resistance calculation results at high Froude numbers. The CFD code solves the Reynolds-averaged Navier-Stokes (RANS) equations with modelling of the free surface effects, that is, the generation of waves due to the ship movement on the water surface. A high-speed OSV boat was considered at three speeds (17, 20 and 26 knots), which correspond to the Froude numbers based on the volume of displacement Fr_V = 1.29, 1.52, and 1.98.The CFD results show a good agreement with the results of Savitsky's formula and the numerical results obtained from Maxsurf Modeller. Further, a reasonable agreement was found between the CFD results and available experimental data. Flow visualisations show that the wave pattern behind the boat consists of transverse and divergent waves. As the boat's speed increases, the wave height increases, and the wave pattern becomes more dominated by the divergent waves. These observations are consistent with the classical analytical predictions.

NOMENCLATURE

[Symbol]	[Definition] [(Unit)]
ρ	Density of water (kgm^{-3})
ω	Turbulence dissipation rate (m^{2}s^{-3})
τ_{ij}	Turbulence shear stress (Nm^{-2})
θ	Running trim (deg)
c_i	Volume fraction of fluid i (-)
e	Percent error or percent difference (%)
g	Gravitational acceleration (m s^{-2})
h	Water depth (m)
k	Turbulence kinetic energy (m^{2}s^{-2})
n	Run number (-)
$\mathbf{n}$	Outward normal unit vector (-)
p	Pressure (N m^{-2} or Pa)
t	time (s)
B	Breadth of the ship (m)
C_B	Block coefficient (-)
C_M	Midship coefficient (-)
C_W	Water plane coefficient (-)
Fr_h	Froude number based on the water depth(-)
Fr_L	Froude number based on the length at the waterline of the ship LWL (-)
Fr_V	Froude number based on the volume of displacement of the ship (-)
H	Height of the ship (m)
LOA	Length overall (m)
LBP	Length between the perpendiculars (m)
LCB	Longitudinal center of buoyancy (m)
LCF	Longitudinal center of floatation (m)
LWL	Length at the water line (m)
N	Number of cells (-)
R_T	Total ship resistance (N or kN)
S	Control surface (m^2)
T	Draft of the ship (m)
$\mathbf{U}$	Velocity field (m s^{-1})
$\mathbf{U}_d$	Velocity of the control volume (m s^{-1})
V	Control volume (m^3)
V_s	Ship speed (m s^{-1})
V_{max}	Maximum ship speed (knots or ms^{-1})
WSA	Wetted surface area (m^2)

1. INTRODUCTION

Benchmark tests are commonly carried out to verify the results of a CFD code, such as verification of results of a flow solver, boundary conditions, discretisation methods, numerical schemes, or turbulence models. Larsson et al. [1]reported CFD benchmark tests where three modern hull forms were considered, to verify results of turbulence modelling and wave prediction. Their study indicates that free-surface waves may now be well predicted, also away from the ship hull, and full-scale viscous flows can be computed without numerical difficulties.

Considering the ship resistance, numerical modelling of the free-surface boundary, that is, the generation of waves due to the vessel movement on the water surface, has been the topic of many ongoing studies. It is closely related to the vessel's wave-making resistance. This wave-making resistance has a much more significant contribution to the total ship resistance for high-speed crafts than low-speedvessels. Most industrial CFD codes utilise an interface capturing method for the free surface modelling, such as the volume of fluid (VoF) method [2]. The other important components of the total ship resistance, in addition to the wave-making resistance, are the friction and viscous pressure resistance [3].

More recently, Chiroşcă and Rusu [4]performed CFD benchmark tests by comparing the numerical results obtained from NUMECA FINE™/Marine, ANSYS CFD and SHIPFLOW with experimental data. They utilisedthe Duisburg Test Case (DTC)of a typical container ship in their studyfor benchmarking and validation of the numerical methods. The hull geometry and model-test results of ship resistanceare publicly available. A good agreement was found between the towing test results and the computation results. Further, amongst the three CFD codes considered, the SHIPFLOW performed best for the case considered, with a mean error of 2.011% between the numerical and experimental results.

Benchmark tests of FINE™/MarineCFD code were also reported by Firdhaus et al. [5]for ship resistance calculations at a low Froude number. They utilised the DTMB 5415 navy surface combatant ship model in their study, which advanced in calm water at Froude number (based on the water line length of the ship)$\mathrm{Fr}_L = 0.248$. They reported a good agreement between the CFD results and experimental data. Further, the FINE™/Marine results were found to be consistent with the results of earlier numerical studies[1, 6, 7].

Considering the FINE™/MarineCFD code, ithas been utilised in many earlier studies of high-speed crafts, particularly in the application of hydrofoils as Hull Vane®,[1][8, 9] or hydrofoil applications on catamaran ships. Further, most CFD benchmark tests on ship flows available in the literature reported only the verification and validation of CFD results for ships moving at relatively low Froude numbers[1, 4, 5]. Because of the above reasons, benchmark tests of theFINE™/MarineCFD code are deemed necessary for the ship resistance calculations at high Froudenumbers,that is, for the cases where the displacement Froude numbers are larger than one($\mathrm{Fr}_V > 1$). Accordingly, the purpose of this study is to perform benchmark tests of FINE™/Marine CFD code for the ship resistance calculationsat high Froude numbers.

2. METHOD

A high-speed semi planing OSV boat (crew boat) is considered in this study, which is a 31.20 m long boat with a hard chine. Its maximum speed is 27 knots. The same boat has been utilised in earlier studies reported by Suastika et al. [8] and Riyadi and Suastika [9]. The main particulars of the vessel are summarised in Table 1.

Three high Froude-number speeds are considered in this study, namely, 17, 20 and 26 knots, which correspond to the displacement Froude numbers $\mathrm{Fr}_V = 1.29$, 1.52, and 1.98. Further, to gain more insight into the high Froude number results, it is helpful to contrast these results with the results of a reference case at a low Froude number.

[1]Hull Vane® was invented by van Oossanen in 1992 and patented in 2002.

To this end, a ship speed of 11 knots, corresponding to $\mathrm{Fr}_V = 0.837$, was chosen as a reference case. Table 2 summarises the ship speeds considered in this study with their corresponding length and displacement Froude numbers, denoted as Fr_L and Fr_V, respectively. The LWL and volume of displacement of the boat vary with varying ship speed due to planning of the hull and they were calculated with the aid of Maxsurf Modeller [10].

Table 1: Main particulars of the vessel.

Length Overall (LOA)	31.20 m
Length between the Perpendiculars (LBP)	28.80 m
Breadth (B)	6.80 m
Depth (H)	2.75 m
Draft (T)	1.40 m
Maximum Speed ($V_{\max}$)	27 kn
Displacement (Δ)	104.68 t

Table 2: Ship speeds considered in this study and their corresponding length and displacement Froude numbers,denoted as $\mathrm{Fr}_L$and Fr_V, respectively.

Ship speed V_s [knots]	Fr_L	Fr_V
11	0.340	0.837
17	0.524	1.29
20	0.617	1.52
26	0.801	1.98

The FINE™/Marine results are verified by usingthe results of Savitsky's empirical formula[11] and the numerical results obtained from the Maxsurf Modeller [10], which are also based on Savitsky [11]. A comparison is also made between the CFD results and the experimental data of Riyadi and Suastika [9]. Further, the resulting wave patterns are verified with the analytical predictions described by Faltinsen [12].

3. FINE™/MARINE CFD FLOW SOLVER

The FINE™/MarineCFD code utilises the ISIS-CFD flow solver, developed by Ecole Centrale de Nantes and CNRS[13, 14].Theflow solver is based onthe second-order accurate finite volume method[15].The solver solves the Reynolds-averaged Navier-Stokes (RANS) equations with appropriate boundary conditions and appropriate turbulence modelling.The free surface boundary is modelled with an interface capturing method utilising the volume of fluid (VoF) method[2].Further, the turbulence model utilised in this study is the k-ω SST turbulence model[16], which gives accurate predictions of the onset and the amount of flow separation [17].

The mass, momentum and volume fraction equations for incompressible unsteady flows are given in integral form in Equations (1), (2) and (3), respectively, as follows:

$$\frac{\partial}{\partial t}\int_V \rho \mathrm{d}V + \int_S \rho(\mathbf{U}-\mathbf{U}_d)\cdot \mathbf{n}\mathrm{d}S = 0 \qquad (1)$$

$$\frac{\partial}{\partial t}\int_V \rho U_i \mathrm{d}V + \int_S \rho U_i(\mathbf{U}-\mathbf{U}_d)\cdot \mathbf{n}\mathrm{d}S = \int_S \left(\tau_{ij}I_j - pI_i\right)\cdot \mathbf{n}\mathrm{d}S + \int_V \rho g_i \,\mathrm{d}V \qquad (2)$$

$$\frac{\partial}{\partial t}\int_V c_i \mathrm{d}V + \int_S c_i(\mathbf{U}-\mathbf{U}_d)\cdot \mathbf{n}\mathrm{d}S = 0 \qquad (3)$$

In Equations (1-3), **U** is the velocity field relative to a stationary observer,$\mathbf{U}_d$ is the velocity of the control volume*V*relative to the stationary observer, *S* is the control surface with an outward normal vector **n**, ρ is the mass density of the fluid, *p* is pressure,*g* is the gravitational acceleration and *t* is time. Further τ_{ij} are the Reynolds stresses,I_i is an identity tensor and c_i is a volume fraction, representing the presence of fluid *i* (c_i = 1) or its absence (c_i = 0). Equation (3) is the transport equation representing a model for the free surface boundary utilising the volume of fluid (VoF) method [2].

4. CFD SIMULATIONS

The procedure for the CFD simulations is described in this section.The procedure consists of numerical modelling of the boat's hull geometry, definition of the computational domain and its meshing, definition of the boundary conditions and grid-independence tests.

In earlier studies,Suastika et al. [8]and Riyadi and Suastika [9]reported a comparison between the results of FINE™/Marine and experimental data at high Froude numbers. They reported relatively large discrepancies between the CFD results and experimental data of total ship resistance (for the case of boat without vane, Case 0) at some Froude numbers, reaching a maximum value of 10.88% [8]. Therefore, their Case-0 CFD simulations are re-done in this study for the relevant ship speeds.

4.1 GEOMETRICAL MODELLING

The geometry of the hull was modelled with the aid of the Maxsurf 3D modeller[10]. The hull geometry is shown in Figure 1. To ensure that the numerical model represents the prototype accurately, the hydrostatic characteristics of the model were verified with those of the prototype. Table 3 summarises the results of the verification. As shown in Table 3, the differences for all parameters considered between the prototype and the numerical model are less than two percent, indicating accurate geometrical modelling results.

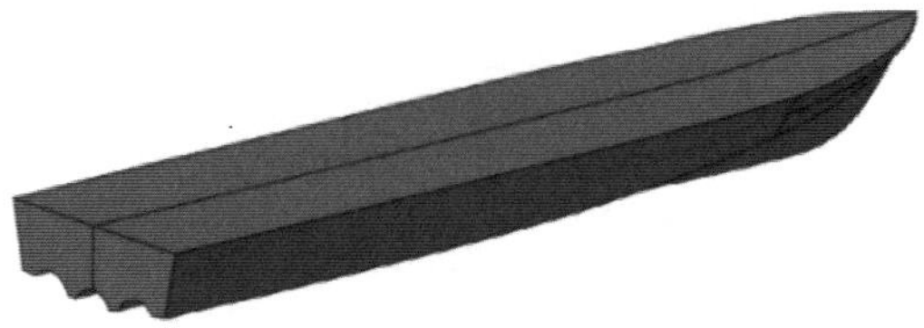

Figure 1: Hull geometry of the OSV boat considered in this study.

Table 3:Comparison of the ship hull hydrostatic characteristics between the prototype and the numerical model.

Parameter	Prototype	CFD model	Percent error [%]
Displacement [ton]	104.68	105.90	1.17
Wetted surface area WSA [m^2]	177.13	177.48	0.196
Block coefficient C_B	0.409	0.415	1.47
Midship coefficient C_M	0.470	0.479	1.89
Waterplane coefficient C_W	0.830	0.822	-0.964
Longitudinal center of buoyancy LCB, measured from AP [m]	11.79	11.73	-0.483
Longitudinal center of floatation LCF, measured from AP [m]	11.44	11.44	0.035

4.2 COMPUTATIONAL DOMAIN

Figure 2 shows a mesh of the computational domain with the boat in it,which is heading to the right. A multiblock-structured grid was utilised in the spatial discretisation of the computational domain [18]. The governing equations, boundary conditions and the turbulence model equations were discretised using a second-order accurate finite volume method.

The locations of the boundaries of the computational domain areas follows [19].The inlet was located at 1.5*L* upstream from the ship, the outlet at 3*L* downstream from the ship and the side walls at 1.5 *L* aside from the ship, where *L* is the ship's overall length (LOA). Further, the bottom wall was located at 1.5 *L* below the ship and the top wall was located at 1.0 *L* above the ship.

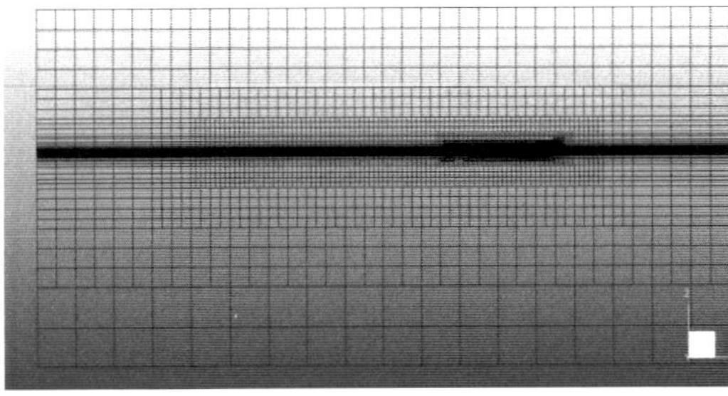

Figure 2: Side view of the discretised computational domain utilising multiblock-structured grid, showing the boat heading to the right.

4.3 BOUNDARY CONDITIONS

The boundary conditions are as follows. On the inlet and side walls,the flow velocities were defined as far-field free stream velocity. On the outlet and the bottom wall, the pressure was prescribed. The boundary condition on the ship hull was defined as no slip where a wall function was utilised.The free surface was resolved utilising the method of volume of fluid (VoF) [2]. The heave and pitch motions were resolved in the simulations.

4.4 GRID-INDEPENDENCE TESTS

Grid-independence tests were done to determine the optimum number of cells used in the final setting of the simulations and to check the convergence of the numerical calculations. The reference case with ship speed of 11 knots, corresponding to Fr_V = 0.837, was considered in the grid-independence tests. In these tests the number of cells was increased by a factor of approximately two in two subsequent simulations. The total ship resistance was determinedfrom each simulation and the percent error between two subsequent simulations was calculated, which is denoted as $e_{n+1,n}$, where n = 1, 2, … is the run number.

Table 4: Total ship resistance R_T calculated using an increasing number of cells in the simulation for ship speed of 11 knots (Fr_V = 0.837).

Run number n	Number of cells N	Total resistance R_T [N]	Percent error $e_{n+1,n}$ [%]
1	637,183	27,762	
2	1,223,562	25,763	-7.76
3	2,294,875	25,121	-2.56
4	4,701,551	24,787	-1.35

Table 4 summarises results of the grid-independence tests for this reference case. As shown in Table 4, the total ship resistance decreases monotonically with increasing number of cells used in the simulations. The total ship resistance is expected to reach an asymptotic value for a very large number of cells (the number of cells N tends to infinity), which is approximately 24.79 kN in this case. Further, the monotonically decrease of the percent error $e_{n+1,n}$ as the number of cells increases indicates that the numerical scheme is consistent and convergent. The value of the total ship resistance of 24.79 kN obtained from the CFD simulations is closed to the experimental data of 24.44 kN reported by Suastika et al. [8], which differs only 1.43%. The ship resistance at this speed cannot be calculated using the method of Savitsky [11] because the displacement Froude number Fr_V is less than one.

Figure 3 shows visualisations of the wave pattern generated by the boat for this reference case with ship speed of 11 knots (Fr_L = 0.340; Fr_V = 0.837). The wave pattern consists oftransverse waves propagated in the direction of the ship course and divergent waves propagated in the direction making an angle with the ship course. The ratio of water depth to ship length h/L = 1.5 and for values of h/L larger than approximately 0.4, there is very small effect of the water depth [12]. The Froude number based on the water depth, $\mathrm{Fr}_h = V_s/\sqrt{gh}$ = 0.264, indicating a subcritical speed. The ratio of the length of the transverse waves to the ship length is estimated as $\lambda_T/L = 2\pi Fr_L^2$= 0.726 and the ratio of the length of the divergent waves to the ship length at the minimum angle of propagation relative to the ship course is estimated as $\lambda_D/L = (4/3)\pi Fr_L^2$ = 0.484. The Kelvin angle of the wash in infinite water depth is 19.47° [12].

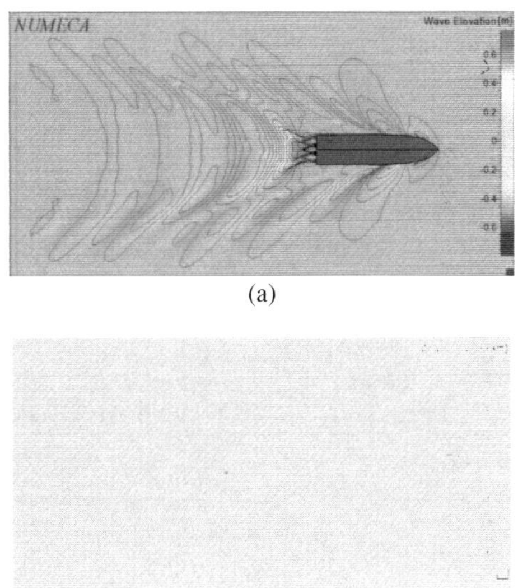

(a)

(b)

Figure 3: Wave pattern at 11-knot ship speed(Fr_L = 0.34; Fr_V = 0.837; Fr_h = 0.264): (a) top view and (b) side view.

5. RESULTS AND DISCUSSION

Results of the ship resistance calculations utilising FINE^{TM}/Marine are discussed in this section where they are compared with calculation results based on Savitsky

[11] and the experimental data of Riyadi and Suastika [9].

Table 5 summarises results of the total ship resistance, obtained from FINE™/Marine CFD simulations, Savitsky's empirical formula [11] and numerical method implemented in Maxsurf Modeller, which is also based on Savitsky [11]. The percent differences from the three methods are tabulated in Table 6. Table 6 shows that the differences between the results of FINE™/Marine and Savitsky's empirical formula are relatively small, with a maximum value of 3.54%. In addition, the differences between the results of FINE™/Marine and Maxsurf Modeller [10] can reach a maximum value of 4.92%. Table 5 shows that the results from the Maxsurf Modeller overestimate the results of Savitsky's empirical formula.

A comparison between the FINE™/Marine results and the experimental data of Riyadi and Suastika [9], summarised in Table 7, shows that the difference between the CFD results and experimental data can reach 5.05%. An agreement with a percent difference of approximately five and less is considered as reasonable.

Table 5: Total ship resistance obtained from FINE™/Marine,Savitsky's empirical formula[11] and Maxsurf Modeller[10].

Speed [kn]	Total ship resistance [kN]		
	FINE™/ Marine	Savitsky'sfor mula[11]	Maxsurf Modeller[10]
17	59.04	57.02	62.10
20	72.33	71.31	71.90
26	92.25	89.61	93.20

Table 6: Percent differences between the results ofFINE™/Marine and Savitsky's formula[11],and between the results of FINE™/Marine and Maxsurf modeller[10].

Speed [kn]	FINE™/Marine-Savitsky's formula[11] [%]	FINE™/Marine-Maxsurf Modeller[10] [%]
17	3.54	-4.92
20	1.43	0.598
26	2.94	-1.01

Table 7: Total ship resistance obtained from FINE™/Marine and the experimental data of Riyadi and Suastika [9].

Speed [kn]	Total ship resistance [kN]		Percent difference [%]
	FINE™/ Marine	Experiment [9]	
17	59.04	-	-
20	72.33	72.11	0.305
26	92.25	97.16	-5.05

Table 8 summarises the results of the wetted surface area (WSA) and running trim of the vessel as function of speed obtained from the CFD simulations. A negative running trim indicates a bow-up trim. This is consistent with the right-hand rule of the Cartesian coordinate system shown in Figure 2, where the x-axis points to the right and the z-axis positive upward. Table 8 shows that the WSA increases with increasing speed. Further, the bow-up trim also increases with increasing ship speed, which is ascribed to planing of the hull.

Table 8: Wetted surface area (WSA) and running trim of the ship obtained from FINE™/Marine CFD simulations.

Speed [kn]	WSA [m^2]	Running trimθ [deg]
17	188.26	-0.859
20	188.99	-1.38
26	192.34	-1.53

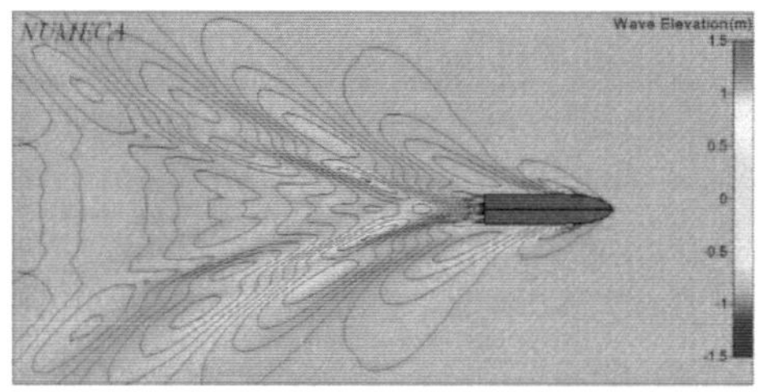

(a)

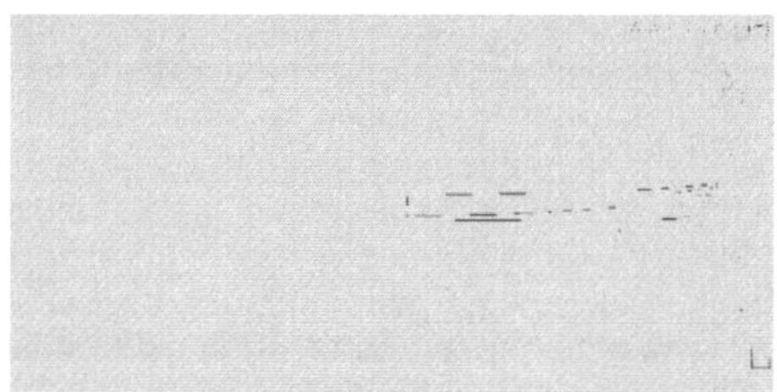

(b)

Figure 4: Wave pattern at 20-knot ship speed(Fr_L = 0.617; Fr_V = 1.52; Fr_h = 0.480): (a) top view and (b) side view.

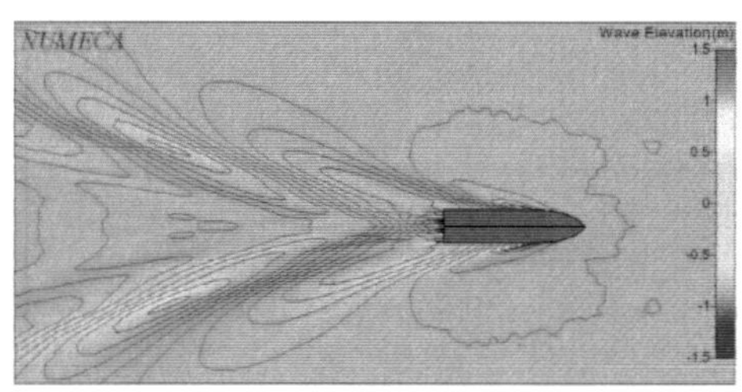

(a)

(b)

Figure 5: Wave pattern at 26-knot ship speed (Fr_L = 0.801; Fr_V = 1.98; Fr_h = 0.624): (a) top view and (b) side view.

The wave patterns at high Froude numbers, as shown in Figures 4 and 5, arerather different from that at a low Froude number shown in Figure 3. The depth Froude number $\mathrm{Fr}_h = V_s/\sqrt{gh}$ for the 20-knot speed is 0.480 while that for the 26-knot speed is 0.624. Both are less than one, indicating subcritical speed. For values of Fr_h less than approximately 0.7, the Kelvin angle is approximately equal to the deep-water Kelvin angle, namely, 19.47°. Figures 4 and 5 show that, as the ship speed increases, the wave height increases, and the wave pattern becomes more dominated by the divergent waves.These observations are consistent with the analytical predictions described by Faltinsen [12].

6. CONCLUSIONS

Benchmark tests of FINE^{TM}/Marine CFD code were done for the calculations of ship resistance at high Froude numbers (Fr_V> 1). A high-speed 31.20 m long OSV boat was considered in the study. The differences between the FINE^{TM}/Marine results and the results of Savitsky's empirical formula[11]are relatively small, with a maximum value of 3.54%. In addition, the differences between the FINE^{TM}/Marine results and the numerical results from the Maxsurf Modeller [10]can reach 4.92% and between the FINE^{TM}/Marine results and the experimental data of Riyadi and Suastika [9] can reach 5.05%.An agreement with a percent difference of approximately five and less is considered as reasonable. Flow visualisations show that the wave pattern behind the boat consists of transverse and divergent waves. As the ship speed increases, the waves generated by the boat become higher and the wave pattern becomes more dominated by the divergent waves(in the high Froude-number regime). These observations areconsistent with the classical analytical predictions[12].

7. ACKNOWLEDGEMENTS

This research project was supported by the National Research and Innovation Agency (BRIN) of the Republic of Indonesia under the Grant Penelitian Dasar Unggulan Perguruan Tinggi (PDUPT) with contract no. 887/PKS/ITS/2021.

8. REFERENCES

1. L. LARSSON, F. STERN, AND V. BERTRAM, "Benchmarking of computational fluid dynamics for ship flows: The Gothenburg 2000 workshop," Journal of Ship Research, vol. 47, no. 1, 2003, doi: 10.5957/jsr.2003.47.1.63.

2. C. W. HIRT AND B. D. NICHOLS, "Volume of fluid (VOF) method for the dynamics of free boundaries," Journal of Computational Physics, vol. 39, no. 1, 1981, doi: 10.1016/0021-9991(81)90145-5.

3. A. F. MOLLAND, S. R. TURNOCK, AND D. A. HUDSON, Ship Resistance and Propulsion. 2011. doi: 10.1017/cbo9780511974113.

4. A. M. CHIROȘCĂ AND L. RUSU, "Comparison between model test and three CFD studies for a benchmark container ship," Journal of Marine Science and Engineering, vol. 9, no. 1, 2021, doi: 10.3390/jmse9010062.

5. A. FIRDHAUS, I. K. SUASTIKA, K. KIRYANTO, AND S. SAMUEL, "Benchmark Study of FINETM/Marine CFD Code for the Calculation of Ship Resistance," Kapal: Jurnal Ilmu Pengetahuan dan Teknologi Kelautan, vol. 18, no. 2, 2021, doi: 10.14710/kapal.v18i2.39727.

6. Y. AHMED, J. M. A. FONFACH, AND C. GUEDES SOARES, "Numerical simulation for the flow around the hull of the DTMB model 5415 at different speeds," International Review of Mechanical Engineering, vol. 4, pp. 957–964, 2010.

7. Y. M. AHMED AND C. GUEDES SOARES, "Simulation of the flow around the surface combatant DTMB model 5415 at different speeds," in 13th Congress of Intl. Maritime Assoc. of Mediterranean IMAM, Oct. 2009, p. [Online]. Available: https://www.researchgate.net/publication/263126369

8. I. K. SUASTIKA, A. FIRDHAUS, R. AKBAR, W. D. ARYAWAN, I. K. A. P. UTAMA, S. RIYADI, AND B. GANAPATHISUBRAMANI, "Experimental and numerical study of ship resistance due to variation of hull vane® positioning in the longitudinal direction," in Proceedings of International Conference on Ship and Offshore Technology, 2019, pp. 5–14.

9. S. RIYADI AND K. SUASTIKA, "Experimental and numerical study of high froude-number resistance of ship utilizing a hull

vane®: A case study of a hard-chine crew boat," CFD Letters, vol. 12, no. 2, pp. 95–105, 2020.

10. MAXSURF, "MAXSURF Naval Architecture Software," https://maxsurf.net/, Jun. 25, 2021.
11. D. SAVITSKY, "Hydrodynamic Design of Planing Hulls," Marine Technology and SNAME News, vol. 1, no. 04, 1964, doi: 10.5957/mt1.1964.1.4.71.
12. O. M. FALTINSEN, Hydrodynamics of High-speed Marine Vehicles, Cambridge University Press, 2005.
13. E. GUILMINEAU, G. B. DENG, P. QUEUTEY, M. VISONNEAU, AND J. WACKERS, "Assessment of turbulence models for flow around three-dimensional geometries," Notes on Numerical Fluid Mechanics and Multidisciplinary Design, vol. 137, 2018, doi: 10.1007/978-3-319-70031-1_21.
14. E. GUILMINEAU AND F. CHOMETON, "Effect of side wind on a simplified car model: Experimental and numerical analysis," Journal of Fluids Engineering, Transactions of the ASME, vol. 131, no. 2, 2009, doi: 10.1115/1.3063648.
15. Z. LI ET AL., "Comparison of wave modeling methods in CFD solvers for ocean engineering applications," Ocean Engineering, vol. 188, 2019, doi: 10.1016/j.oceaneng.2019.106237.
16. F. R. MENTER, "Two-equation eddy-viscosity turbulence models for engineering applications," AIAA Journal, vol. 32, no. 8, 1994, doi: 10.2514/3.12149.
17. J. E. BARDINA, P. G. HUANG, AND T. J. COAKLEY, "Turbulence Modeling Validation, Testing, and Development," Nasa Technical Memorandum, no. 110446, 1997.
18. J. H. FERZIGER AND M. PERIĆ, Computational Methods for Fluid Dynamics. 2002. doi: 10.1007/978-3-642-56026-2.
19. H. K. VERSTEEG AND W. MALALASEKERA, An Introduction to Computational Fluid Dynamics: The Finite Volume Method. Harlow, UK: Longman Scientific, 2007.

9. AUTHORS BIOGRAPHY

I K Suastika received the doctoral degree from Delft University of Technology, The Netherlands. He holds the current position of professor in the field of hydrodynamics at the Department of Naval Architecture, Institut Teknologi Sepuluh Nopember, Surabaya, Indonesia. He is responsible for the teaching and supervising students in the fields of hydrodynamics and naval architecture. His previous experience includesstudies in wave-current interactions including wave blocking, boundary layer flows and application of vanes as energy saving devices to single-hull and catamaran vessels.

A F Fauzi was formerly an undergraduate student at the Department of Naval Architecture, Institut Teknologi Sepuluh Nopember, Surabaya, Indonesia, and earned his B.Eng.-degree in 2020.

A S Saputra was formerly an undergraduate student at the Department of Naval Architecture, Institut Teknologi Sepuluh Nopember, Surabaya, Indonesia, and earned his B.Eng.-degree in 2020.

I K A P Utama is professor in ship hydrodynamics at the Department of Naval Architecture, Institut Teknologi Sepuluh Nopember, Surabaya, Indonesia. He is a fellow and the Vice President for the Asia Region of the Royal Institution of Naval Architects (RINA). He is also a fellow of the Indonesian Academy of Sciences (AIPI). His research interests include ship resistance and propulsion, experimental and computational fluid dynamics, marine renewable energy, and maritime education.

A Firdhaus is currentlya PhD student at the Department of Systems and Naval Mechatronic Engineering, National Cheng Kung University, Taiwan. He earned his M.Eng.-degree from the Department of Naval Architecture, Institut Teknologi Sepuluh Nopember, Surabaya, Indonesia in 2020. His research interests include effects of the application of Hull Vane® on high-speed crafts and application of vanes on catamarans as energy saving device.

Sutardi received the doctoral degree from Memorial University of Newfoundland, St. Johns, Canada. He holds the current position of professor in the field of fluid mechanics at the Department of Mechanical Engineering, Institut Teknologi Sepuluh Nopember, Surabaya, Indonesia. His research interest includes turbulent fluid flow, fluid flow visualisation, and viscous drag reduction in turbulent boundary layers. He is responsible for several fundamental and advanced fluid courses such as fluid mechanics, turbulent flows, and viscous fluid flow.

M Hariadi received the B.E. degree in Electrical Engineering Department of Institut Teknologi Sepuluh Nopember, Surabaya, Indonesia, in 1995. He received both M.Sc. and Ph. D. degrees from the Graduate School of Information Science Tohoku University Japan, in 2003 and 2006 respectively. Currently, he is the staff of Computer Engineering Department of Institut Teknologi Sepuluh Nopember, Surabaya, Indonesia. He is the project leader in joint research with PREDICT JICA project Japan and WINDS project Japan. His research interest is in video and image processing, big data and computational intelligent. He is a member of IEEE, and member of IEICE.

AN INVESTIGATION INTO THE JUSTIFICATION OF THE SERVICE SPEED OF RO-RO FERRY WITH BLOCK COEFFICIENT 0.8 BASED ON THE RESISTANCE AND SEAKEEPING PERFORMANCE

R C Ariesta, **M H N Aliffrananda, S Riyadi,** and **I K A P Utama** Department of Naval Architecture, Institut Teknologi Sepuluh Nopember, Surabaya, Indonesia

SUMMARY

Determination of service speed has always become a primary concern when designing a ship. However, a ship designer may not consider its effect on the use of energy and seakeeping performance of the vessel. The current paper discusses the determination of the actual and proper speed of a ro-ro ferry, operated between Tanjung Perak (Surabaya) and Lembar (Lombok), based on resistance estimation and seakeeping performance. The investigation is carried out using commercial software Maxsurf^{TM} and further compared with computational fluid dynamics (CFD) approach to capture the effect of detailed wave drag, which is not quantified by the design software. The discussion and analysis include the determination of optimum speed based on the size of main engine and the seakeeping performance of the ship covering the aspects of heave, pitch, and roll.

NOMENCLATURE

[Symbol]	[Definition] [(Unit)]
$(1+k)$	form factor
ζ_{k0}	ship amplitude
ζ_0	wave amplitude
C_T	non-dimensional form coefficient
Hs	wave height *(m)*
γ	peak enhancement factor
μ	heading angle *(^{o})*
m_0	area under spectrum curve
ρ	density (kg/m^3)
S	wetted surface area(m^2)
$S_{(JWP)}$	JONSWAP Spectrum(m^2/(rad/s))
$S_{(PM)}$	Pierson-Moskowitz Spectrum in (m^2/(rad/s))
$S_{\zeta r}(\omega)$	spectra response (m^2/s)
$S_{\zeta}(\omega)$	wave spectra(m^2/s)
RAO (ω)	RAO transfer function
RAO	Response Amplitude Operator
RMS	Root Mean Square*(^{o})*
R_T	total resistance (kN)
R_F	frictional resistance(kN)
R_R	residual resistance(kN)
R_{VP}	viscous pressure resistance(kN)
R_W	wave-making resistance (kN)
τ	shape parameter
ω	wave frequency (rad/s)
v	speed (knot)

1. INTRODUCTION

In ship design process, two factors are always considered i.e., technical constraints and economic criteria. Technical restrictionsprovide challenges for engineers to fulfilaspects such as adequate speed, good stability, and optimum seakeeping performance. Meanwhile, economic factorscorrelate to the cost of fuel consumption, maintenance of ship, tax of service at ports, etc.Among others, the primary aspects of ship operation include optimum resistance and powering and hence fuel consumption and emission, and ship performance and comfortability. Moreover, ships operated at certain service speed in oblique wave condition may cause uncomfortable to passengers attributed to excessive heave and roll s.

The choice of hull form together with hydrostatics characteristics for passenger ships such as Ro-Ro vessel are of importance[1]. Ro-Ro vessel which carries passengers, cars, coaches, and trucks across islandswith adequate distance should haveoptimum resistance and seakeeping performance for saving energy and reliable performance. The speed of ship is not only fast enough, but also having good stability and dynamic performances, namely heave, pitch, and roll s [2]. Those aspects are considered important as the rate of Ro-Roaccidents in Indonesia are quite high according to the National Committee for Safety Transportation (KNKT)[3].

In order to attract and promote economic growth beyond Java Island, the direct operation of ships from East Java to Lombok Island has increased considerably in the last 2 years. A Ro-Ro vessel which can carry 242 passengers, 40 trucks, and 82 cars is planned to operate from Surabaya (East Java) to Lembar (Lombok). The vessel is designed to operate at sea state 3 which has significant wave height between 0.75m and 1.25m [4]. In order to obtain adequate speed, but with low engine size, several improvements are made out on the existing ship including introducing wider bilge radius and generating better underwater hull-form (i.e., streamflow)along the bow of vessel. Those are conducted as to reduce resistance and increase efficiency and well as improve the seakeeping performance [5].

Therefore, this paper aims to investigate and justify the appropriate service speed of the Ro-Ro vessel based on the resistance and sufficient seakeeping performance in terms of the response of heaving, pitching, and rollingof the ship. The study was carried out using a commercial ship design software called Maxsurf^{TM} and justified with computational fluid dynamics (CFD) method. Comparative analysis with published data was also conducted to support the findings.

2. RO-RO VESSEL MODEL

The investigation is conducted by improving the existing hull of the Ro-Ro vessel and two modifications of hull are made. The existing model has block coefficient (C_B) 0.8 (Model 1) and its modification has C_B 0.794 (Model 2) and C_B 0.790 (Model 3), respectively.Principal particulars of the vessel are given in Table 1.

Table 1 Principal Particulars of the Ro-Ro Vessel

Hull	LOA	LPP	B	H
Existing (1)	106.43	100.00	19.70	5.00
Modification (2)	106.43	100.00	19.70	5.00
Modification (3)	106.43	100.00	19.70	5.00

Model 1 has a rectangular hull, does not have a bilge radius and maximum displacement. Model 2 has a bilge radius of 0.5 m and Model 3 has a bilge radius of 0.75 m. Those models are tested at two draughts: 3.3 m and 3.8 m. Details of models can be found in Table 2.

Table 2 Characteristics of C_B and displacement model

Hull	C_B	Displacement (ton)		C_B Deviation Based on 0.8 (%)
		Draft 3.3 (m)	Draft 3.8 (m)	
Model 1	0.802	5413	6384	0.3
Model 2	0.794	5345	6324	0.8
Model 3	0.790	5320	6298	1.3

Optimum hulls have been generated after investigation of changing different hull form parameters such as C_B and displacement that value must be around 0.8 and ±5000 ton. It found the optimum hull Ro-Ro that is shown in Figure 1. In modifying the hull, significant changes to the ship's construction will incur high costs, hence, to avoid this, changes are only made to the bilge and bottom shape of the hull. Then this will reduce the value of the displacement, which indirectly reduces the area of the submerged area. Therefore, the resistance value can be reduced so that the most optimal hull shape is obtained. Moreover, changes to the body plan are the hallmark of a ship, therefore needs to be considered in the shipbuilding process.

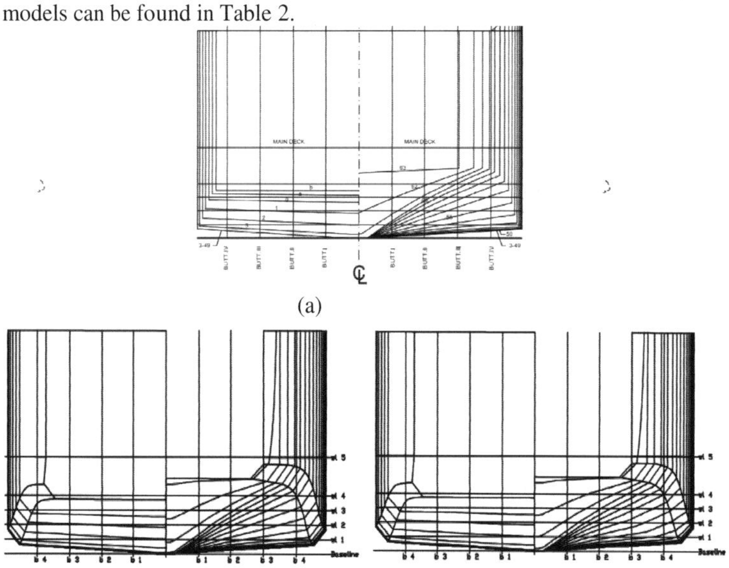

(a) (b) (c)

Figure 1. Types of hull modification; (a) Existing Ship, (b) Model 2 and (c) Model 3

3. RESISTANCE ANALYSIS

The total resistance (R_T) consists of the total frictional resistance (R_F) and residual resistance (R_R)[5]. Friction resistance is a force working tangentially to the hull surface and as an accumulation of shear stress induced by the viscosity of a fluid. In addition, the residual resistance component is influenced by pressure. The normal force acts perpendicular to the hull and produces viscous pressure resistance (R_{VP}) and wave-making resistance (R_W). If a hull is fully submerged in a fluid, then the wave-making resistance is equal to zero. However, the viscosity of the water affects the amount of pressure resistance experienced by the ship [2]. The resistance components are formulated with equations(1) and (2):

$$R_T = R_F + R_R + R_A \tag{1}$$

$$R_T = R_F + R_{VP} + R_W + R_A \tag{2}$$

Several components of the total resistance ship are able to represent in the non-dimensional form coefficient (C_T) summarized (3), formulated from ρ is the density of the fluid, S is wetted surface area, and v is the velocity of hull:

$$C_T = \frac{R_T}{0{,}5\rho S V^2} \quad (3)$$

equation (3) can be rewritten as a non-dimensional form (4):

$$C_T = C_F + C_{VP} + C_W + C_A \quad (4)$$

Moreover, the parts of appendages resistance are constructed by the underwater components such as bilge keels, propeller, skeg, and rudder.In this paper, the appendage resistance is assumed to be equal to zero. Therefore, the total coefficient can be expressed as (5) and (6):

$$C_D = C_F + C_{VP} + C_W \quad (5)$$

$$C_D = (1 + k)C_F + C_W \quad (6)$$

where (1+k) is a form factor to generate in the extrapolation process between a hull model. It represents the viscous pressure resistance that reason of pressure on a hull is influenced by the own hull form. While the models in this study will conduct an analysis using Maxsurf™ and CFD. Maxsurf™ and Computational Fluid Dynamics (CFD) solver method compare use for resistance analysis. Resistance estimates the value of using two numerical software, namely (1) Maxsurf™ and (2) CFD is used to predict resistance just in calm water while wave conditions using CFD. There is a parameter that cannot define using Maxsurf™. Therefore, subsequent analysis was carried out using CFD packages. In all of the simulation steps, several stages must be done: During the pre-processing, model geometry must be clean bodies and need to determine the volume of fluid exposures into the discrete mesh. Also, the physical and boundary condition is defined. [6]. The simulation using both packages of software is done by test the existing model of the Ro-Ro vessel. The simulation obtained in the ship with drafts 3.3 m and 3.8m, the percentage difference in resistance for CFD and Maxsurf™ is about 11% and 22%, respectively.This condition is influenced by the wake that is exposed to the surface of the hull. Thus, further analysis is carried out with CFD.

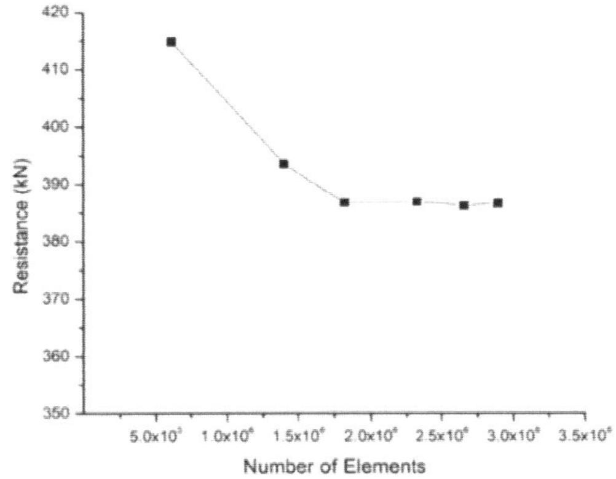

Figure 2. Grid Independency study

Figure 2 Illustrates the estimates of the number of elements used, which isperformed from the grid independence study. A total of grid approaches better for the model reaches approximately two million elements. The grid of the independence study gained 0.07%. Figure 3 presents the motion of a ship in the water produces two types of flow patterns, forms a wave pattern around the hull surface and a turbulent flow along the hull and is denser at the stern and it is called the wake. A Wake phenomenon is formed from the submerged hull into the deep of water. Moreover, the depth of the hull influences the energy absorbed through the increase of speed affects the pattern of wake pressure added resistance in the ship stern. The phenomena of wake from the Ro-Ro vessel in draft 3.3 m and 3.8 m are different when the draft is deeper; the wake appears significant from the surfaces of hull. That concludes the hull form influenced the flow throughout the ship below the waterlines.

Furthermore, analysis of resistances is generated in numerous variations that are 10 knots, 13 knots, and 15 knots. Range of speed observations is set to obtain the amount of decreasing energy in modification of a hull. Uncertainty issues in predicting ship resistance are influenced by the following factors such as ocean current, wind speed, and wave interaction. That included into the resistance estimation but extremely difficult to count separately. Therefore, it is recommended to provide an additional margin of 10-30% for the previously shipping routes to reduce the uncertainty of the prediction considering the many factors that can affect the calculation results.

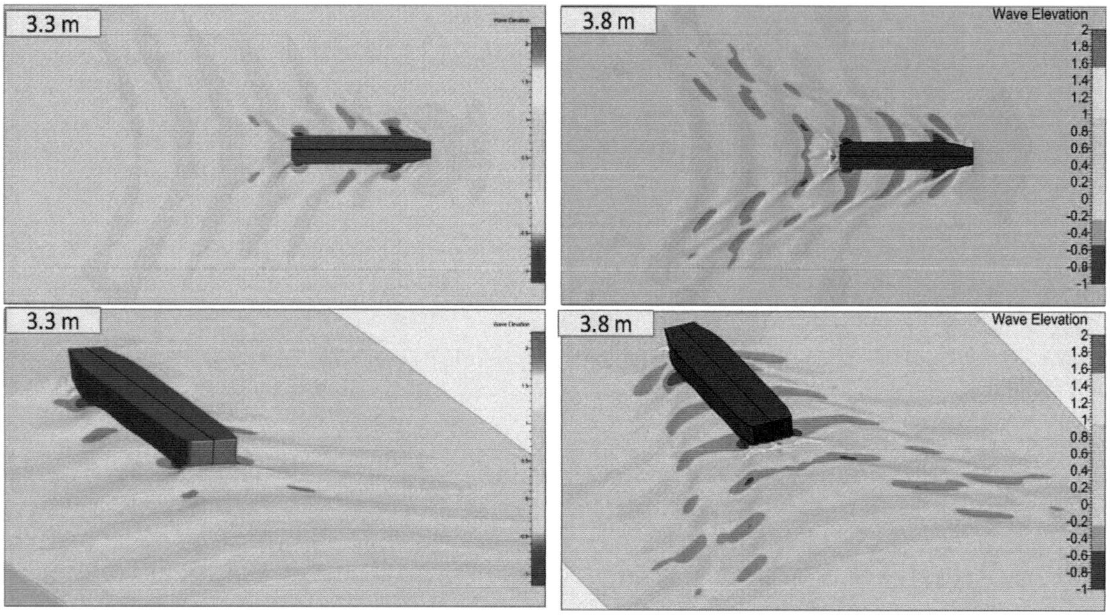

Figure 3. Ship wake phenomenon with draft 3.3 m and 3.8 m

Furthermore, from the calculation of resistance that has been carried out, it is included with the addition of a margin of conditions for new shipping routes [5]. Figure 4 showed the results of the prediction of resistance in the three variations of speed. Existing hulls producemore drag force than the modified hulls. The average resistance reduction in the modified hull in calm and wavy waters and the draft at 3.3 m and 3.8 m, respectively, is about 10 %.The calculation of the prediction of the ship's main engine power requirement is obtained from the predicted value of the ship's resistance that has been obtained previously. However, arranging the main engine power is first calculated related to the effective power required to move the ship according to the operating speed. The value of EHP (effective horsepower) has the same trend as the value of the ship's resistance because the value of EHP is a function of the speed and resistance of the ship[5]. Furthermore, power conversion from EHP to BHP (brake horsepower) requires calculating the efficiency of the ship's equipment, including, among others, the hull, propeller, and rudder. Therefore, the magnitude of the efficiency ranges from 50-65%[2].

Hence, the analysis efficiency taken based on sea trial data on the operating route as a reference for calculating BHP's efficiency value show that the efficiency average is 30% with a machine workload of 85%. The conversion of the results of the calculation of resistance to BHP in calm water and wave condition is presented in Table 3.

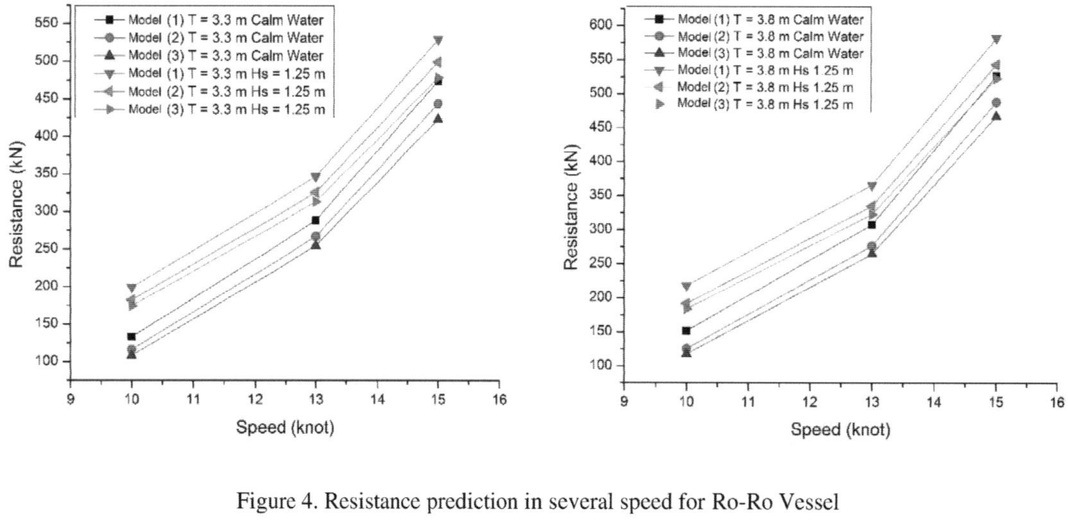

Figure 4. Resistance prediction in several speed for Ro-Ro Vessel

Table 3. Comparison data of power requirements of each model (HP)

Draft (m)	Speed (knots)	Calm Water			Wave Condition Hs = 1.25 m		
		Model 1	Model 2	Model 3	Model 1	Model 2	Model 3
3.3	10	2970	2601	2427	4447	4083	3912
	13	8374	7757	7399	10071	9454	9097
	15	15883	14885	14186	17736	16738	16040
3.8	10	3390	2812	2638	4868	4294	4123
	13	8932	8036	7678	10629	9733	9376
	15	17661	16364	15666	19514	18217	17519

Based on Table 3, improving the bilge radius and streamline of the bow can reduce the power needed. It is assumed that if the Ro-Ro ship is given an engine limit of 2 x 3200 HP in each variation of the hull model. Thus, the most optimal model is obtained when using modifications to the hull. Before the modification, the Ro-Ro ship was only able to run at a speed of 9.8 knots. In contrast, after making changes to model 1 and model 2, the speed increase was obtained. Ro-Ro ships can increase the Model 3 by 11.5 knots, assuming the same engine is used. Hence, the percentage increase in speed is 17% by improving the bilge and bow sections. Details of the speed increase can be seen in Figure 5. The green area shows the service speed that can be reached by Ro-Ro vessel.

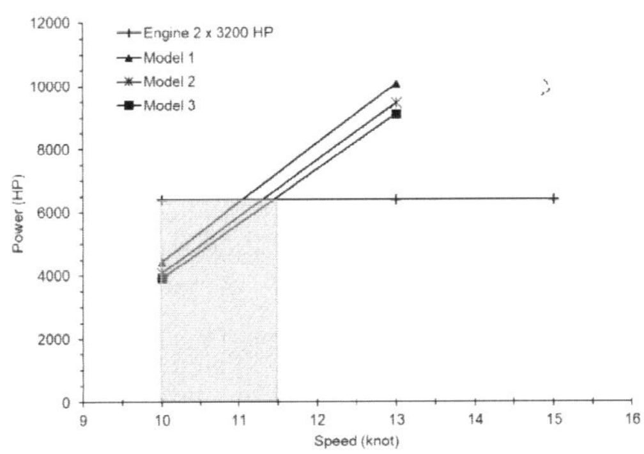

Figure 5. Service speed range

4. SEAKEEPING ANALYSIS

External forces influence seakeeping performance of the ship due to water conditions. The ship's occurs at six degrees of freedom consisting of surge, sway, heave, which are translational s, then roll, pitch, and yaw are rotational motions[7]. The motion is symbolized by space coordinates that allow the ship to rise, heel, trim, and submerge on the bow and stern.

Analysis related to shipping is conducted numerically using CFD approach. Seakeeping investigation on irregular waves according to the wave energy distribution recommended by ITTC [8]. Ship's performance analysis at sea is investigated to measure the ability of ship to hold on dangerous situation during operation[9]. Ship coordinates systems in this study are used by sets the earth fixed coordinate systems by the axis O-XYZ or body-fixed coordinate system[10]. In formulaeof the ship in the frequency-domain defined the O-XYZ coordinates as a fixed coordinate system in space. The formulaein the frequency domain show the space fixed following o-XYZ coordinates.

Afterward, that position related to the vessel passes through the waterline and cross-section of the centre of gravity and motion of the ship. Regarding the direction of the different axes, it isexplained that the bow in the direction of the OX axis corresponds to the pitch motion, then sideways is represented by OY associated with rollalong with the OZ axis moves vertically up and down. The ship hull is assumed to move forward direct OX at speed, and the heading angle was μ. The true mass, position of the center of gravity, and gyration of radius in the longitudinal and transversal directions following the data on the full-scale ship model should be used. In the absence of exactness knowledge, hence a formula of roll radius of gyration is 0.35B to 0.45, and both pitch and yaw are 0.25 Lpp generally used[11].The illustration of ship coordinates is shown in Figure 6.

Determination of response on the ship can be analogous to a black box with an input filter and an output. However, black boxes in the ship work by arranging like waves as input, ship as a filter, and as output or response[12]. According to ship response, it is needed to calculate into several stages as:

i) Response Amplitude Operator (RAO) is the transfer of the response function of the force due to waves on the dynamic response of the ship in the form of frequency. RAO is also known as the transfer function. RAO graph generated from the frequency parameters as the abscissa and the ratio of the amplitudes of in a mode at specific coordinates. According to[7], translational RAO is calculated as follows (Equation (7)):

$$RAO\ (\omega) = \frac{\zeta_{ko}(\omega)}{\zeta_0\ (\omega)} \qquad (7)$$

Where $\zeta_{ko}(\omega)$ is ship amplitude and $\zeta_0\ (\omega)$ is wave amplitude. While the rotational in RAO is a comparison between the of the rotational amplitude and the slope of the wave, which is obtained by multiplying the wavenumber by the wave amplitude can be obtained by Equations (8) and (9):

$$RAO\ (\omega) = \frac{\zeta_{ko}(\omega)}{\zeta_0\ (\omega)} \qquad (8)$$

$$RAO\ (\omega) = \frac{\zeta_{k0}(\omega)}{\left(\frac{\omega^2}{g}\right)\zeta_0} \qquad (9)$$

ii) Wave Spectra is influenced by two parameters: significant wave height (Hs) and zero up crossing period (Tz).The spectrum able to represented by JONSWAP spectrum. Based on data collected at the Joint North Sea Wave Observation Project or called JONSWAP. the application of wave spectra requires a wave spectrum to obtain it in this analysis.The wave spectrum used in the analysis is the Pierson-Moskowitz. The Pierson Moskowitz spectrum is formulated in Equation (10) [13]:

$$S_{(PM)} = \frac{Hs^2}{4\pi}\left(\frac{2\pi}{T_Z}\right)\omega^{-5} exp\left[-\frac{1}{\pi}\left(\frac{2\pi}{T_Z}\right)^4 \omega^4\right] \qquad (10)$$

$S_{(PM)}$ is defined as Pierson-Moskowitz Spectrum in $m^2/(rad/s)$ and ω is the wave frequency rad/s. Nowadays, JONSWAP spectral is commonly used in the analysis and design of offshore platforms that exist in Indonesia. However, some researchers, that recommended the Indonesia waters using small ranges, about 2.0 to 2.5.Therefore, this is to decrease the major of energy occurred by only some wave frequencies. The formulating presented by Equation (11):

$$S_{(JWP)} = S_{(PM)}\gamma^r \qquad (11)$$

$S_{(JWP)}$ is defined as JONSWAP spectrum, where γ is a peak enhancement factor in Indonesia the value around 2 to 2.5, r is formulated by (12):

$$r = exp\left[\frac{(\omega - \omega_0)^2}{2\tau\ \omega_0^2}\right] \qquad (12)$$

Where τ represent shape parameter and ω_0 is wave peak frequency, with 0.07 if $\omega \leq \omega_0$ and 0.09 if $\omega \geq \omega_0$. In the analysis JONSWAP Wave Spectrum applied with the height wave in Surabaya – Lombok assumed in sea state 3 condition, the height of wave is between 0.5m and 1.25m.

iii) Response spectra are the interaction effect between added mass, potential damping, and external force, which is analysed on the ship's motion in the regular wave. The random response is transformed from the wave spectrum to the response spectrum to obtain an energy density response on the ship due to the wave load. The response spectra is calculated using Equation (13):

$$S_{\zeta r}(\omega) = [RAO\ (\omega)]^2 x\ S_\zeta(\omega) \qquad (13)$$

Where $S_{\zeta r}(\omega)$is spectra response (m^2/s), $S_\zeta(\omega)$ is wave spectra (m^2/s), and RAO (ω)is transfer function.

iv) Therefore, the response of the Ro-Ro vessel can be obtained by multiplying the RAO and wave spectrum. Also using statistical processing to obtain the Root Mean Square (RMS) of ship's mode, that constructs from heave, pitch, and roll. The estimation of ship is summarized in Equation (14) [7]:

$$RMS = \sqrt{m_0} \qquad (14)$$

m_0 is under spectrum response curve are that reached from the numerical method. Area under spectrum curve can be formulated in the Equation (15)[7]:

$$m_0 = \int_0^\infty \omega_e{}^0 S_\zeta d_\omega \qquad (15)$$

Therefore, the Ro-Ro vessel response was obtained from the equation for the two modes of for the ship in modification (2) and modification (3) with speedsfrom 8 knots until 15 knots. The study of seakeeping results in simulation RMS responses of the heave, pitch, and roll modes of the ship with a significant wave height of 1.25m with differences in ship drafts through 3.3m and 3.8m, respectively. The simulation also requires recognition of the path of the incoming wave considering that the difference in the direction of wave can induce various responses to the ship's .

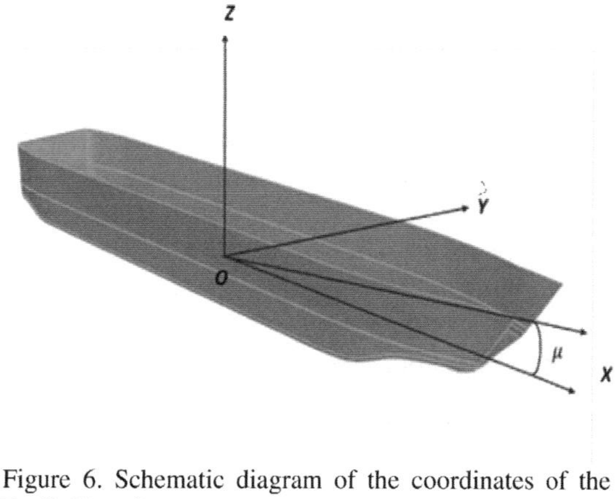

Figure 6. Schematic diagram of the coordinates of the Ro-RoVessel

A polar diagram describes the prediction response due to the uncertainty of environmental conditions on the vessel. Changes in the environment, the principle implies wave height, will increase the potential of the ship's response. In this analysis, the average wave significance influences ship in the operational area.

Significant RMS heave amplitude rise depends on the submerged draft of the hull illustrated in Figure 7. Contrast response showing on draft changes at maximum waves on 90 degrees angle increases when the ship draft depicted by the yellow area turning red in the beam seas trajectory. The indicates a high probability of RMS under these red colors.

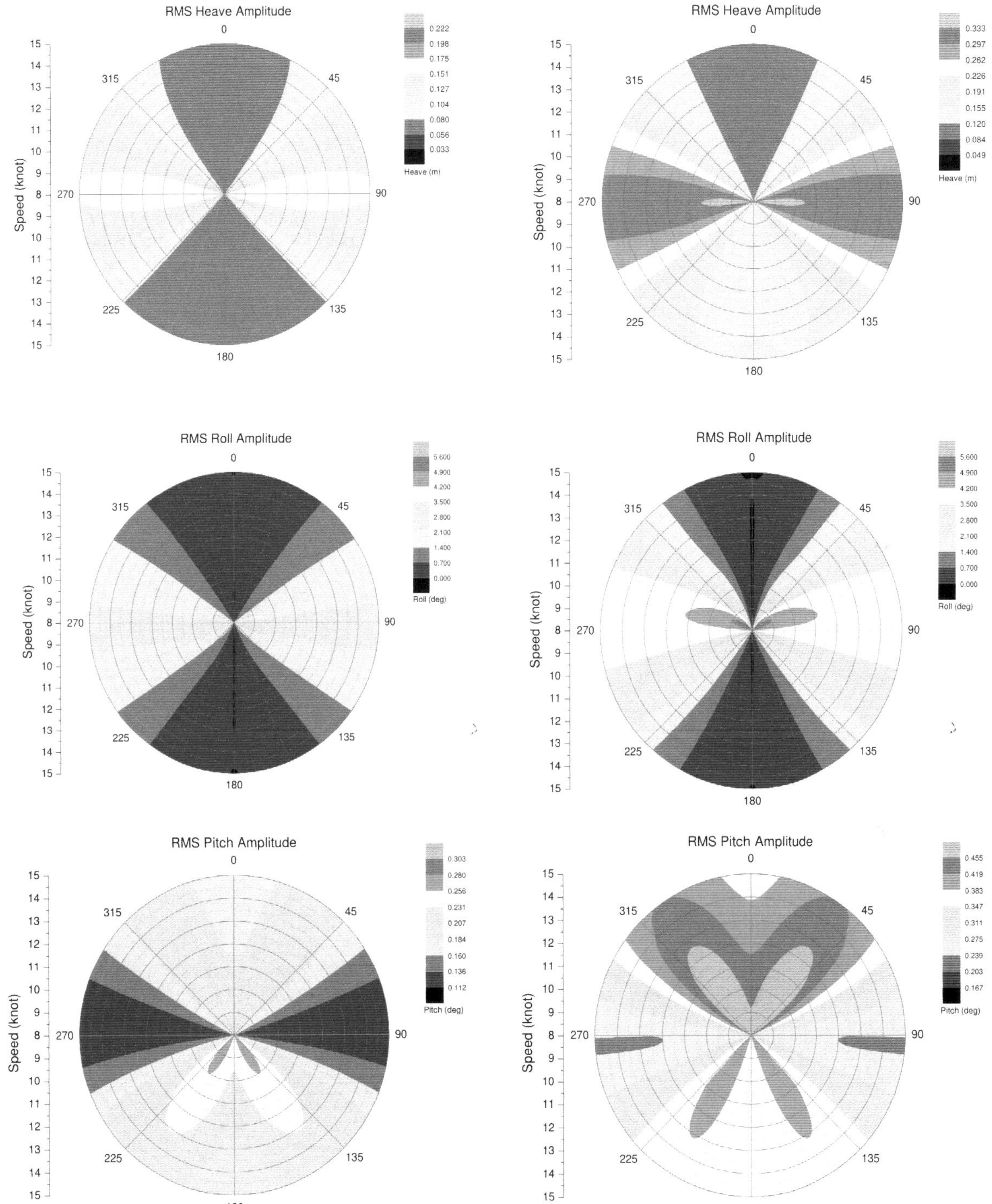

(a) (b)

Figure 7. Speed polar diagrams of heave, roll, and pitch in combination (a) draft 3.3 m and (b) draft 3.8 m with different draft and heading angles under sea states

Figure 6 also shows the RMS in the roll condition where the most roll condition occurs in the beam seas direction. The polar diagram shows the RMS condition of the Ro-Ro ship at a speed of 9 knots to 15 knots. This depiction is to illustrate the shape of the ship's motion that occurs. A similar incident occurred where an increase in response happened when the ship's draft changed. Furthermore, the polar graph on the maximum RMS pitch occurs in the direction of the following sea at 3.3 m conditions and increases sharply when 3.8 meters in the head seas direction.

Figures 6(a) and (b) display the polar diagrams of the heaving influences with speeds also wave heading angles at different drafts. The heaving increasing for wave heading in beam seas angles, the maximum of RMS occurs in 90° at the 3.8m draft.

The maximum response of heave and roll motions occurs at an angle of 90°, whereas highest pitch motion occurs at an angle of 135°. The ship also experiences high rollmotion value where the RMS averaged 2.17° for a 3.3m draft and 4° for a 3.8mdraft. The vessel can experience an almost high roll. Therefore, it is necessary to be careful of operations at a wave angle of 90°.

Among others, the criteria assessed by NORDFORSK include roll RMS, heave vertical acceleration, and bow acceleration. The results of the seakeeping analysis are presented in Tables4 to 6.

The analysis results related to the seakeeping criteria are given successively in Tables 4, 5, and 6, showing that all seakeeping criteria from [14] are all met both on ships with modifications of adding bilge with radius 0.75 meters or ships with modifications to the hull's bow.Each in both 3.3 m and 3.8 m models satisfy the comfort criteria from the NORDFORSK analysis. Still, it should be noted that model 1 in 3.8 m RMS heave and pitch conditions has a massive contribution to disrupt passengers, which occurs in sickness.Whereas in model 2 and model 3, after modification, a continuous decrease occurs in assessing the acceptance criteria at the comfort level so that the model is more comfortable to use in operation.

Table 4. Seakeeping Criteria, RMS of Roll (deg.)

Hull Modification	Draught (m)	Speed (knots)	Heading Angle μ			Criteria	Status
			180°	135°	90°		
Add bilgeradius of0.75 m	3.3	10	0.00	1.27	2.17	≤ 6 deg	✓
		13	0.00	1.09	2.17		✓
		15	0.00	1.05	2.17		✓
	3.8	10	0.00	1.96	4.00		✓
		13	0.00	1.76	4.00		✓
		15	0.00	1.64	4.00		✓
Add bilge radius of0.75 m plusbow modification	3.3	10	0.00	2.25	4.33		✓
		13	0.00	2.06	4.33		✓
		15	0.00	1.95	4.33		✓
	3.8	10	0.00	1.96	4.00		✓
		13	0.00	1.76	4.00		✓
		15	0.00	1.64	4.00		✓

Table 5. Seakeeping Criteria, RMS of vertical acceleration at FP (m/s^2)

Hull Modification	Draught (m)	Speed (knots)	Heading Angle μ			Criteria	Status
			180°	135°	90°		
Add bilge radius of 0.75 m	3.3	10	0.37	0.38	0.25	≤ 0.275 g =2.698	✓
		13	0.36	0.37	0.24		✓
		15	0.35	0.36	0.24		✓
	3.8	10	1.04	0.83	0.44		✓
		13	1.22	0.96	0.43		✓
		15	1.35	1.05	0.42		✓
Add bilge radius of0.75 m plus bow modification	3.3	10	1.04	0.84	0.43		✓
		13	1.22	0.97	0.42		✓
		15	1.35	1.05	0.42		✓
	3.8	10	1.03	0.84	0.44		✓
		13	1.21	0.97	0.43		✓
		15	1.34	1.05	0.42		✓

Table 6. Seakeeping Criteria, RMS of vertical acceleration at deck (m/s^2)

Hull Modification	Draught (m)	Speed (knot)	Heading Angle μ			Criteria	Status
			180°	135°	90°		
Add bilgeradius of0.75 m	3.3	10	0.09	0.13	0.23		✓
		13	0.11	0.15	0.22		✓
		15	0.13	0.16	0.22		✓
	3.8	10	0.52	0.40	0.36		✓
		13	0.53	0.41	0.35		✓
		15	0.54	0.41	0.34		✓
Add bilge radius of0.75 m plus bow modification	3.3	10	0.09	0.13	0.23	≤ 0.150 g =1.472	✓
		13	0.11	0.15	0.22		✓
		15	0.13	0.16	0.22		✓
	3.8	10	0.09	0.13	0.24		✓
		13	0.11	0.14	0.23		✓
		15	0.13	0.16	0.22		✓

5 RESISTANCE AND SEAKEEPING CORRELATION

Table 3 and Figure 4 showed that the effective speed of the Ro-Ro vessel if the size of engine is set in term of energy efficiency. In this case, the maximum speed(of Model 3) using engine with sizes 2x3200 HP is 11.5 knots. If the speed is increase into 13 knots and 15 knots then the sizes of engine will be 2 x 4688 HP and 2 x 8760 HP, respectively. Consequently, the cost of fuel consumption rises and hence the emission or pollution to the atmosphere.

In terms of seakeeping, observations were made at incoming angles of 180°, 135°, and 90°, and at draughts of 3.3 m and 3.8 m to seek the effect of increasing speed on ship performance and comfortability. According to Tables 2 to 4, the existing model, Model 2 and Model 3 fulfill the seakeeping criteria provided by NORDFORSK. However, it is apparent that the ship motions (rolling and vertical acceleration) are more intensive as the speed increases. This further may cause uncomfortability to crews and passengers. Therefore, all criteria must be accepted by NORDFORSK in order to assure motions are safe and safety for passengers.

6. CONCLUSIONS

Investigation of the resistance and seakeeping characteristics on a Ro-Ro vessel has been carried out using commercial design software (Maxsurf^{TM}) and CFD method quite successfully. The design software provided earlier estimation of resistance and seakeeping characteristics and the CFD technique has made those clearer through its advanced visualization capability.

Details of the conclusions include the effects of hull modification such as introducing the use of bilge radius and modification of front part of the vessel to reduce local wake and hence make the flow more streamline. Analysis showed that the optimum service speed is about 11.5 knots with the draught of 3.8m and the size of main engine approx. 2x3200 HP and the seakeeping performance of the boat is acceptable according to NORDFORSK. When the speed is increased to 13 and 15 knots the size of main engine upsurges incredibly more than twice of the speed of 11.5 knots. The seakeeping quality is still acceptable but motion is getting worse.

7. REFERENCES

1 T. Cepowski, "Determination of optimum hull form for passenger car ferry with regard to its sea-keeping qualities and additional resistance in waves," *POLISH MARITIME RESEARCH* , vol. 15, no. DOI: 10.2478/v10012-007-0057-2, pp. 3-11, 2008.

2 A. F. Molland, S. R. Turnock and D. A. Hudson, Ship resistance and propulsion: Practical estimation of ship propulsive power, United States of America: Cambridge University Press, 2017.

3 W. Mutmainnah, L. P. Bowo, A. Nurwahyudy, F. A. Prasetyo and M. Furusho, "Causative Factor Analysis of Passenger Ship Accident (Fire/Explosion) in Indonesia," in *Maritime Safety International Conference*, Indonesia, 2020.

4 K. Rawson and E. Tupper, Basic Ship Theory Volume 1 5th Edition Hydrostatics and Strength, London: Butterworth-Heinemann, 2001.

5 S. A. Harvald, Resistance and Propulsion of Ships, Canada: John Wiley & Sons, Inc, 1993.

6 Y. Cengel and J. Cimbala, Fluid Mechanics: Fundamentals and Applications., New York: The McGraw-Hill Company, 2006.

7 R. Bhattacharya, Dynamics of Marine Vehicle, Annapolis: Marryland: U.S. Naval Academy, 1978.

8 ITTC, ITTC Recommended Procedures and Guidelines: Seakeeping Experiments, S.1., International Towing Tank Conference, 2014.

9 I. Utama, M. Aliffrananda, A. Sulisetyono, D. Setyawan, S. I.K., D. Utama, W. Aryawan and Y. Hermawan, "Performance of Rescue Boat Operation when Operated in Waves," in *IOP Conf. Ser.: Mater. Sci. Eng.*, Surabaya, 2020.

10 Y.-H. Lin and C.-W. Lin, "Numerical Simulation of Seakeeping Performance on the Preliminary Design of a Semi-Planing Craft," *Journal of Marine Science and Engineering* , vol. VII, no. 199, pp. 1-24, 2019.

11 ITTC, ITTC - Recomended Procedures and Guidelines, Numerical Simulation of Capsize Behavior of Damaged Ships in Irregular Beam Seas, International Towing Tank Conference, 2014.

12 S. Denis and W. J. J. Pierson, On the Motions of Ship in Confused Seas, United Stated: The Society of Naval Architects and Marine Engineers (SNAME), 1953.

13 A. B. o. S. (ABS), Spectral-Based Fatigue Analysis for Vessel, America: Houston: ABS 2016, 2016.

14 NORDFORSK, Assessment of Ship Performance in a Seaway: The Nordic Co-operative Project: "Seakeeping Performance of Ships, Nordforsk, 1987.

15 A. Scamardella and V. Piscopo, "Passenger ship seakeeping optimization by the Overall Motion Sickness Incidence," *Ocean Engineering,* vol. 76, pp. 86-97, 2014.

16 G. A. P. Poundra, I. K. A. P. Utama, D. Hardianto and B. Suwasono, "Optimizing Trimaran Yacht Hull Configuration Based on Resistance and Seakeeping Criteria," in *Procedia Engineering*, Surabaya, 2017.

9. AUTHORS BIOGRAPHY

Rizky Chandra Ariesta is an Academic Assistant at the Ship Construction Research Group, Department of Naval Architecture, Faculty of Marine Technology, Sepuluh Nopember Institute of Technology. He received her S.T. (B.Eng) and M.T. (M.Eng) from a scholarship program in this faculty. His current research interests are computational mechanics, fatigue fracture, material science and engineering, sandwich structure, FRP Technolgy, ship stability and performance correlation.

M. Hafiz Nurwahyu Aliffrananda, S.T., M.T. is a young researcher in ship hydrodynamics. He frequently has the opportunity to be a lecturer assistant, and also has several experiences in projects and research at Sepuluh Nopember Institute of Technology, Department of Naval Architecture. He is interested in research that advances the use of computational fluid dynamics (CFD), particularly in ship dynamics, turbulent flow, and drag reduction techniques.

Soegeng Riyadi, Ph.D. student at Department of Naval Architecture Institut Teknologi Sepuluh Nopember (ITS). He has experience designing ships in several shipyards. He is a member Royal Institution of Naval Architects (RINA). Previous experience includes reducing ship resistance with foil at the back of the ship and investigating the operational mode of the fleet to achieve optimum fuel consumption with artificial Intelligence.

Professor I Ketut Aria Pria Utama, B.Eng, M.Sc, Ph.D, CEng, FRINA, FAIPI, is a Professor of Ship Hydrodynamics in the Department of Naval Architecture, Sepuluh Nopember Institute of Technology. He is a fellow and the Vice President of the Asian Region of the Royal Institution of Naval Architects (RINA). He is also a fellow of the Indonesian Academy of Sciences (AIPI). His research interests include ship resistance and propulsion, experimental and computational fluid dynamics, marine renewable energy, and maritime education.

SUPPORTING STRUCTURES FOR OCEAN THERMAL ENERGY CONVERSION (OTEC) SYSTEM: A SYSTEMATIC LITERATURE REVIEW

M Ari , Politeknik Perkapalan Negeri Surabaya, Indonesia.
Y S Hadiwidodo, Lab. of Ocean Structural, Materials and Production, Dep. of Ocean Engineering, Fac. of Marine Technology, Institut Teknologi Sepuluh Nopember, Surabaya, Indonesia.
Mukhtasor, Laboratory of Environmental and Ocean Energy, Dep. of Ocean Engineering, Fac. of Marine Technology, Institut Teknologi Sepuluh Nopember, Surabaya, Indonesia.

SUMMARY

Ocean thermal is one of renewable energy sourcesin which the seawater temperature difference is about 20°C. Several studies have been carried out related to the conversion systems, environmental, economics, including optimum siting of the conversion systems. Among others, supporting structures for such the systems play a great role in contributing to the feasibility of the project. However, it seems there has been no common concensus or guideline to which supporting structures fit best to a particular case under consideration. Further works are required to fill this gap so that a better understanding on the supporting structures may be presented. The purpose of this work is to conduct a systematic literature review of the available studies that attempted to deal with Ocean Thermal Energy Conversion (OTEC) structures. The type of the structures, the type of loads, and the most frequent critical structural parts are reviewed and presented in this paper.

NOMENCLATURE

OTEC	Ocean Thermal Energy Conversion
SLR	Systematic Literature Review
CWP	Cold Water Pipe
TWh	Terawatt-hours

1. INTRODUCTION

Marine energy technologies could give a lot of beneficial for islands and coastal areas. Seawater as an abundant resource can be used for the generation of electricity. Ocean Thermal Energy Conversion (OTEC) is one of marine energy technology which producing energy by takes advantage of the temperature difference between the sea surface and the deep sea or ocean to drive a thermodynamic heat engine that can generate electricity. The best regions for OTEC are usually found within 20° of the equator [1]. In addition, freshwater can be produced as a by product without consume much energy [2]. Small islands or small cities located in the coastal region require small OTEC plants that can generate between 1 and 10MW and produce between 1700 and 35,000 cubic meters of desalinated water per day, enough for communities of 4500 to 100,000 inhabitants. [3]. OTEC can also be used in hydrogen production. [4].
The technology is believed to have no effect on the thermal structure of the ocean, in that the total estimated available resource for OTEC is thought to be as high as 30 TW (terawatt) [5]. As an example, Indonesia is one of promising country for OTEC Development. Projections for ocean and tidal energy as renewable energies in Indonesia are estimated at 17.9 GW [6]. Rough estimation of economic potential of closed-cycle OTEC in Indonesia done in 2021 by Langer [7]. The research shows that the economic potential spreads over four provinces in Indonesiawithin an aggregated range of 0–2 GW with an annual electricity production of 0–16 TWh.
There are many papers discussing the system design of OTEC, such as concept design [8],design performance optimation [9], [10]and performance evaluation concept [11]. Some other papers focuse on economic optimazitation of OTEC, such as plant sitting and economic potential [7],[12], and economic assessment[13] , [14].
However, there are limited number of paper discussing the supporting structure of OTEC wich also play a great role in contributing to the feasibility of the project. Zang [15] review for the applied mechanical problems in ocean thermal energy conversion. But the papers mainly focuses on two aspects: steam turbine and the cold water pipe (CWP). The important role of OTEC CWP to the success of OTEC installation and operation is undeniable by the fact that numerous OTEC projects failed and then abandoned due to CWP problems such as Rio de Janeiro OTEC project in 1935 and even India OTEC project in 2003 [16].

Nevertheless, other structure problem should also being identified and analyzed. Therefore, this paper will focus on supporting structure of OTEC. The systematic literature review will be conducted to the available studies that attempted to deal with OTEC structures. A systematic literature review is a logical, linear process, where each part is informed by that preceding it. While the traditional literature review may give subjective prominence to the most interesting or influential studies, the systematic review is an objective and standardised exercise[17].

This systematic literature review (SLR) was conducted to answer the following review questions:

(1) What are type of OTEC supporting structure have been proposed by researchers?

(2) What are the load/stress being exposed to OTEC supporting structure?
(3) Which part is considered most critical to contributein the feasibility of the project?

2. METHODS

A protocol for the systematic review is developed to ensure that the study is transparent and rigorously carried out. The protocol should cover determining aim and research questions; literature searching and screening methods ; defining inclusion and exclusion criteria for the articles to be used as part of the review; data extraction strategies and narrative synthesis of qualitative data[18].

2.1 SEARCH STRATEGY AND SELECTION CRITERIA

The relevant keywords were used as search string to find related literature in Scopus database. "Ocean Thermal Energy conversion" was included as a major keyword. Various keywords related to structure were included to limit the specific boundary in which each study was conducted. Finally, the search was conducted after limiting the language to English. Therefore the search criteria are as follows :

TITLE-ABS-KEY(oceanANDthermalANDenergyANDconversion)ANDTITLE-ABS-KEY(struct*ORconstruct*ORmechanic*ORloadORstress)AND(LIMIT-TO(LANGUAGE,"English"))

There are 459 initial articles that had been identified using all keyword combinations from all sources in the search protocol.

2.2 LITERATURE SCREENING PROCESS

Screening process was done in several step. Figure 1 showed the overall SLR protocol and number of document result.Considering the latest development of OTEC to be reviewed in this paper, only literatures for the years 2011-2021 are included. There are 164 document result for these criteria. Among these data 23 are excluded due to duplication or incomplete information.

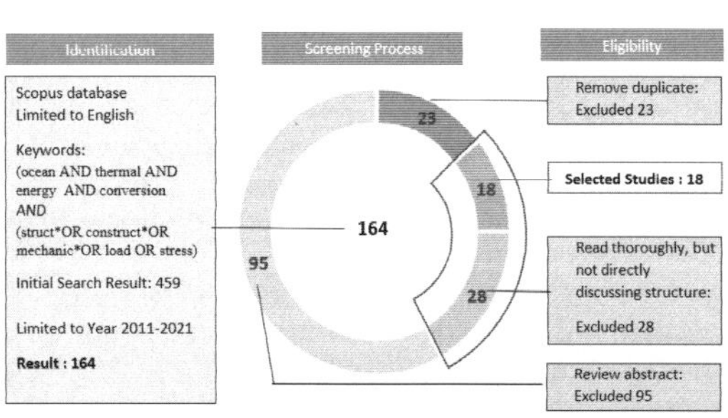

Figure 1 : SLR Protocol and result

After reviewing the keywords and abstracts of each paper, 95 additional articles that not directly related to the supporting structure of OTEC were excluded. This Leaves46 studies to be analysed in-depth. After read thoroughly the remaining studies, only 18article are selected to fulfil the research goals of this SLR

3. RESULT

3.1 TYPE OF OTEC SUPPORTING STRUCTURES

Basically, OTEC installations are subdivided into land based (onshore), fixed platform in the sea (offshore), or as floating constructs[3].

3.1 (a) Onshore

Shore-based OTEC are usually located in a tropicalcoastal region with the steep seabed. Since it was constructed on the shore, the influence of climate on the plants is less significant than that on the offshore floating plants, which contributes to the reduction of the maintenance costs. The construction of the shore-based plants is easy to implement, some challenges still existduring the operation. Earthquakes or eruptions would cause landslide and turbidity currents, which may damage the CWP laid on the seabed(Zhang et al., 2018).Furthermore, OTEC located near the coastline are highly vulnerable to erosion due to wave, current and thermal abrasion [19] . The onshore plant also suffers from the drawbacks of long intake pipes for pumping seawater toland, high initial construction cost and large amounts of energy to operate the pumps. It costs three times as much per unit power output as the sea-based plant due to the expense of the cold water pipe [20].

3.1 (b) Fix platform/Offshore

Fix platform OTEC power plants can be installed around 100 m deep which is closer to the cold water. In addition, the turbulent waves generated near the coast can be avoided. The problems include stress due to strong marine currents and large waves and the complicated delivery of the product. These platforms require massive pillars to ensure their stability, and the distribution of the energy produced requires submarine cables to reach the distribution power lines[3].

As the severe weather (typhoon and tsunami) often occurs in tropical regions, the capability to operate stably in the enormous wind and wave should be taken into account during the design of the OTEC plants. However, the famous Lockheed spar-type OTEC platform is one of successful examples (Zhang et al., 2018).

3.1 (c) Floating

The type of floating structure for OTEC are moored barge and grazing barge where spar, semi-submersible and shipshape are the most feasible to OTEC application[21]. Shipshape can be built from tanker FPSO which converted and modified to OTEC operation[22], [23]. As the same reason of offshore platform, medium – high power OTEC built from ship is closer to the cold water region and just use straight pipe to reach them. However in deep sea operations, waves, currents and winds will be a problem for the mooring system and energy distribution[3] .

3.2 LOAD/STRESS AFFECTED OTEC SUPPORTING STRUCTURE

The load acting on the OTEC structure depends on the location where the otec facility is located. In the offshore plant, several loads that affect the OTEC structure are hydrodynamic, seismic and wind. While the load on the onshore base is mostly seismic. Figure 2 illustrates how the OTEC facility reacts to the load.

3.2 (a) Hydrodynamic force

Ocean Thermal Energy Conversion (OTEC) is a renewable energy which generate electric based on the difference of between cold deep and warm ocean water temperature. Warm water with a temperature may be obtained instantly at ocean level, but cold water must be gotten through pipelines from the deep ocean. The hydrodynamic behavior of large-scale cold water pipes is an important consideration in system design and analysis. This load is generated by the waves and currents that cause the movement of the platform. eventually, it will provoke a couple connection between the platform or vessel and the cold water pipe [24].
Seawater currents generate movement of cold water pipe in the horizontal direction which causes the pipe to be subjected to bending stress. Bending loading fluctuates with different wavelengths and angles each time. In addition, seawater currents lead vortex-induced vibration [25].

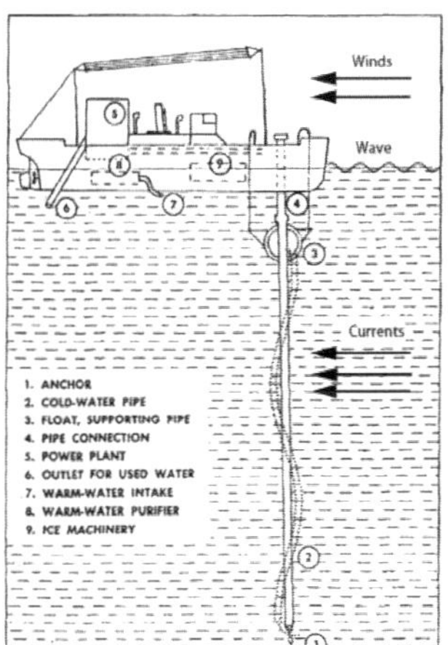

Fig. 2. The schematic diagram for cold water pipe (CWP) [15]

Hydrodynamic force may come fromthe warm surface-water intake pipes distributed around an OTEC plant. it can generate adequate momentum to motion of the platform[26].

The hydrodynamic force can be internal and external to the pipe. In addition to internal, external hydrodynamic forces can cause external viscous forces which in turn cause vortex shedding effects on the pipe [15].

3.2 (b) Seismic

Most countries whose territorial waters have the potential as a source of energy through OTEC are in the ring of fire. Earthquakes and eruptions occur frequently in the ocean environments, which would cause serious geologic disasters like turbidity currents. Turbidity current is the one kind of gravity flow containing primary sedimentary structures, which has a significant influence on the OTEC system, especially the land-based system [15]. Generally, turbidity currents can develop under extreme sea conditions, causing damage to the OTEC system's pipeline.In addition causing mechanical damage to OTEC structure, seismic in ocean can change temperature difference of sea water [27] . This may lead to a twenty-degree temperature difference needed for OTEC system not being achieved.

3.2 (c) Motion

Offshore constructions, such as the floating OTEC device, are subjected to winds, currents, and waves, which cause the platform to move. When extreme sea conditions, such as hurricanes and typhoons, occur, it is conceivable that the ocean platform will capsize if sea loads are not adequately addressed or operated. Although, under normal circumstances, the heave motion is restricted to 10% of the ocean depth. This limiting element for the OTEC device is the assurance of obtaining the calculated seakeeping performance of an actual example [15].

3.3 CRITICAL PART OF OTEC SUPPORTING STRUCTURE AN THEIR SCENARIOS

3.3 (a) Cold water pipe (CWP)

Cold water pipe is The most challenging component attached to the OTEC system. At the beginning of this paper is mention about the failure of CWP in Rio de Jainero and India. The size of CWP which large and long need more attention. CWP on the stationary platform will experience stresses due to current and wave loads and the weight of the pipe itself[25].Vibration and self-induced vibration of CWP will be the main issue that can damage the pipe[21], [28].

Since CWP is one of the important structures of OTEC, several scenarios have been tested using simulations and optimizations. Method joint of CWP to OTEC platform had been done by [16] who simulated top joint fixed, flexible and pinned, Fig. 3. And the bottom joint with and without clump weight.[29] provide a thruster on the payload located at the bottom end that can control and direct the flexible riser to reduce vibration due to external disturbance the pipe/riser.

The material use for CWP has been proposed. [25] use fiberglass-filled HDPE composite structure as cold water pipes on floating flatform of OTEC to obtain load characteristics that occurs dueto current and wave loads. [16] investigate among the material (steel, aluminium, FRP) which one is the most suitable for CWP. [30] developing the advanced fiber composite CWP, sandwich core rings for Lockheed Martin's (LM). They investigate cost production, material and fabrication process.

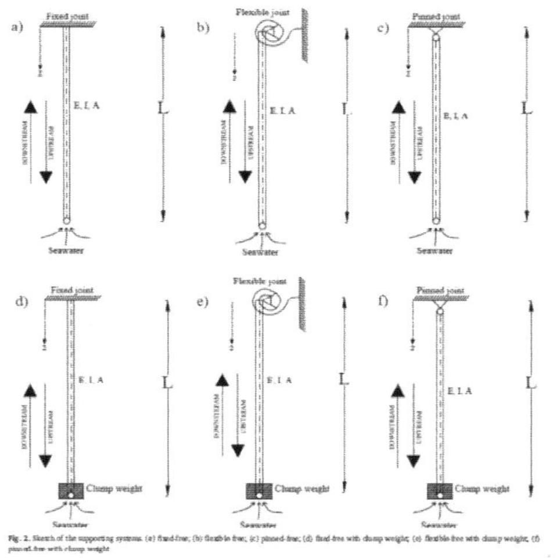

Fig 3. Sketch of the supporting systems. (a) fixed-free; (b) flexible-free; (c) pinned-free; (d) fixed-free with clump weight; (e) flexible-free with clump weight; (f) pinned-free with clump weight.

3.3 (b) Heat Exchanger (HE)
Although it is not a load-bearing structure, the heat exchanger is an important component. As a significant part of the OTEC system cycle, heat exchangers are often subject to corrosion. There are several studies on corrosion in heat exchangers. [31] examine the corrosion-resistant property of Electrodeposited Oxy-Nitriding Steel (EONS) plate as a substitute material for titanium. In his search for cheap materials for heat exchangers, [32]found pitting corrosion on aluminum materials.

3.3 (c) Mooring system
KeepingOTECfloating platform in its position under various motions is important. For this purpose a mooring system is needed. [33] assess the mooring performance for semi-submersible when subjected to the 'High Wave' and 'High Current environmental loads. Fig 4. descript mooring arrangement for this evaluation.
It has been proposed that the warm surface-water intake pipes distributed around an OTEC plant can generate adequate momentum to globally position a platform to overcome the second-order drift forces, thereby eliminating the need for additional power for thrusters or for mooring lines. It is evident that if the intake rate of the flow is high, there will be interaction among the locally created steady flow due to the intake, the incoming wave, and the ensuing platform motions[26]

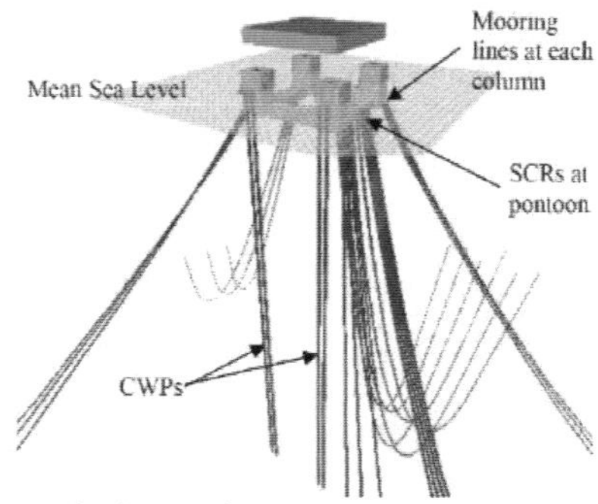

Fig 4. Mooring system analyzed[33]

4. DISCUSSIONS

The onshore OTEC structure is uncomplicated, but the potential for structural disturbances remains, particularly seismic and abrasion. Abrasion that occurs on the beach wall (turbidity) can cause siltation of the foundation of the CWP. The same thing has been expressed by [15]

In offshore otec structures, CWP is the most discussed component, with several scenarios with boundary conditions, such as fixed, pinned, and flexural joints with varied stiffness value and weights on the bottom. The main problem in CWP is the emergence of vibration [16]. Efforts to reduce vibration have been carried out, including by providing a thruster on the payload located at the lower end [29], a clump-weight is attached at the bottom end of the riser [28].). Further, [34] develop an effective vibration control for the OTEC system by the adaptive vibration control method to reduce the vibration of axially translating beam.
A mooring system is also needed for floating OTEC structures so that the influence of motion on the platform can be minimized. Likewise with the vibration on the CWP. Material for OTEC pipe also investigated to find out the most suitable. [16] investigate steel, alumminium, dan FRP. The next attempt is [30] who develop sandwich composite as CWP material.

The analysis that has been done has helped a lot in choosing the right OTEC structure, at least in terms of technical and economic feasibility. However, it is also necessary to develop a risk-based design analysis which considers structure, materials, and environmental issues.

There are an intergovernmental collaboration between countries resulted in the setting up of the Ocean Energy Systems Technology Collaboration Programme (OES) in partnership with International Electrotechnical Commission (IEC) to develop standards and protocols for the management of marine energy conversion systems, with emphasis on the conversion of wave, tidal and OTEC technologies into electricity. This are managed by the IEC TC-114 which is a technical committee (TC) which aim at developing a unified international standard in view of providing the guidelines for suitable site selection, design bases and productivity estimations for OTEC plants based on objective standards [20] It seems that the IEC would mostly concern on the electrical subject. In this SLR it is founded that it is still limited number of studies, specializing in structural analysis. Therefore, analytical recommendations areneeded in light of the need for guidelines in assessing supporting structures.

This literature review can lead to a point of departure and define the significance and originality of the expected contributions of future in-depth research to suggest developing OTEC structure.

5. CONCLUSIONS

This paper introduces type of OTEC supporting structure have been proposed by researcher. Three types of OTEC have been explored: onshore, offshore, and floating. Each form has been discussed in terms of its pros and downsides.Wave, current, and wind are the most dominant load in the ocean which generate hydrodynamic force to the OTEC platform. The cold water pipe, heat exchanger, and mooring system have been identified as the most significant components that need to be considered.For the purpose of resolving vibration difficulties, many boundary condition model scenarios for CWP as the OTEC critical structure have been examined.

6. REFERENCES

1 P. Breeze, "Marine Power Generation Technologies," Power Generation Technologies, pp. 323–349, Jan. 2019, doi: 10.1016/B978-0-08-102631-1.00014-6.

2 M. A. T. Chik et al., "Design optimization of power generation and desalination application in Malaysia utilizing ocean thermal energy," Jurnal Teknologi, vol. 77, no. 1, pp. 177–185, Nov. 2015, doi: 10.11113/JT.V77.4144.

3 J. Herrera, S. Sierra, and A. Ibeas, "Ocean Thermal Energy Conversion and Other Uses of Deep Sea Water: A Review," Journal of Marine Science and Engineering 2021, Vol. 9, Page 356, vol. 9, no. 4, p. 356, Mar. 2021, doi: 10.3390/JMSE9040356.

4 I. Dincer and M. F. Ezzat, "3.4 Renewable Energy Production," Comprehensive Energy Systems, vol. 3–5, pp. 126–207, Jan. 2018, doi: 10.1016/B978-0-12-809597-3.00310-2.

5 I. Dincer, M. A. Rosen, and F. Khalid, "3.8 Ocean (Marine) Energy Production," Comprehensive Energy Systems, vol. 3–5, pp. 335–379, Jan. 2018, doi: 10.1016/B978-0-12-809597-3.00316-3.

6 A. Development Bank, "Indonesia Energy Sector Assessment, Strategy, and Road Map – Update," 2020.

7 J. Langer, A. A. Cahyaningwidi, C. Chalkiadakis, J. Quist, O. Hoes, and K. Blok, "Plant siting and economic potential of ocean thermal energy conversion in Indonesia a novel GIS-based methodology," Energy, vol. 224, p. 120121, Jun. 2021, doi: 10.1016/J.ENERGY.2021.120121.

8 E. Jeon, G. Heo, I. Kim, H. Kim, and H. Jung, "Conceptual design for combined ocean thermal energy conversion using computational fluid dynamics and heat balance analysis," International Journal of Energy Research, vol. 44, no. 9, pp. 7477–7494, Jul. 2020.

9 Z. Wu, H. Feng, L. Chen, and Y. Ge, "Performance optimization of a condenser in ocean thermal energy conversion (OTEC) system based on constructal theory and a multi-objective genetic algorithm," Entropy, vol. 22, no. 6, Jun. 2020.

10 A. Giostri, A. Romei, and M. Binotti, "Off-design performance of closed OTEC cycles for power generation," Renewable Energy, vol. 170, pp. 1353–1366, Jun. 2021.

11 T. Yasunaga, K. Fontaine, and Y. Ikegami, "Performance Evaluation Concept for Ocean Thermal Energy Conversion toward Standardization and Intelligent Design," Energies 2021, Vol. 14, Page 2336, vol. 14, no. 8, p. 2336, Apr. 2021, doi: 10.3390/EN14082336.

12 L. Seungtaek, L. Hosaeng, M. Junghyun, and K. Hyeonju, "Simulation Data of Regional Economic Analysis of OTEC for Applicable Area," Processes 2020, Vol. 8, Page 1107, vol. 8, no. 9, p. 1107, Sep. 2020, doi: 10.3390/PR8091107.

13 L. Talluri, G. Manfrida, and L. Ciappi, "Exergo-economic assessment of OTEC power generation", doi: 10.1051/e3sconf/202123801015.

14 S. Rahmawati, A. I. Muna, and W. Wardhana, "Economic study on the construction of a 50 MW Ocean Thermal Energy Conversion (OTEC) facility in Banten Province, Indonesia," IOP Conference Series: Earth and Environmental Science, vol. 649, no. 1, p. 012015, Feb. 2021, doi: 10.1088/1755-1315/649/1/012015.

15 W. Zhang, Y. Li, X. Wu, and S. Guo, "Review of the applied mechanical problems in ocean thermal

energy conversion," Renewable and Sustainable Energy Reviews, vol. 93. 2018. doi: 10.1016/j.rser.2018.05.048.

16 R. Adiputra and T. Utsunomiya, "Stability based approach to design cold-water pipe (CWP) for ocean thermal energy conversion (OTEC)," Applied Ocean Research, vol. 92, 2019, doi: 10.1016/j.apor.2019.101921.

17 E. Purssell and N. McCrae, "How to Perform a Systematic Literature Review," How to Perform a Systematic Literature Review, 2020, doi: 10.1007/978-3-030-49672-2.

18 D. Ridley, "The Literature Review: A Step-by-Step GUide for Students 2nd Edition," SAGE Publication, 2012.

19 A. O. Sinitsyn, E. Guegan, N. Shabanova, O. Kokin, and S. Ogorodov, "Fifty four years of coastal erosion and hydrometeorological parameters in the Varandey region, Barents Sea," Coastal Engineering, vol. 157, p. 103610, Apr. 2020, doi: 10.1016/J.COASTALENG.2019.103610.

20 J. R. S. Doorga, O. Gooroochurn, B. A. Motah, V. Ramchandur, and S. Sunassee, "A novel modelling approach to the identification of optimum sites for the placement of ocean thermal energy conversion (OTEC) power plant: application to the tropical island climate of Mauritius," International Journal of Energy and Environmental Engineering, vol. 9, no. 4, pp. 363–382, Dec. 2018, doi: 10.1007/S40095-018-0278-4.

21 S. Xiang, P. Cao, R. Erwin, and S. Kibbee, "Otec cold water pipe global dynamic design for ship-Shaped vessels," in Proceedings of the International Conference on Offshore Mechanics and Arctic Engineering - OMAE, 2013, vol. 8. doi: 10.1115/OMAE2013-10927.

22 R. Adiputra, T. Utsunomiya, J. Koto, T. Yasunaga, and Y. Ikegami, "Preliminary design of a 100 MW-net ocean thermal energy conversion (OTEC) power plant study case: Mentawai island, Indonesia," Journal of Marine Science and Technology (Japan), vol. 25, no. 1, pp. 48–68, 2020, doi: 10.1007/s00773-019-00630-7.

23 Y. Song, "A study of OTEC application on deep-sea FPSOs," Journal of Marine Science and Technology 2018 24:2, vol. 24, no. 2, pp. 466–478, Jun. 2018, doi: 10.1007/S00773-018-0567-X.

24 S. Shi, J. Halkyard, N. Kurup, and L. Jiang, "Coupled Analysis Approach in OTEC System Design," Proceedings of the International Conference on Offshore Mechanics and Arctic Engineering - OMAE, vol. 4, pp. 795–802, Aug. 2013, doi: 10.1115/OMAE2012-83978.

25 M. Dalil, B. Wirjosentono, J. Koto, and A. Ginting, "Load characteristics of cold water pipe based on fibreglass-filled HDPE composites," in AIP Conference Proceedings, 2021, vol. 2342. doi: 10.1063/5.0046393.

26 B. Padmanabhan and R. C. Ertekin, "Interaction of Waves With a Steady Intake/Discharge Flow Emanating From a 3D Body," Journal of Offshore Mechanics and Arctic Engineering, vol. 133, no. 4, Nov. 2011, doi: 10.1115/1.4003057.

27 R. Fajaryanti, H. M. Manik, and C. Purwanto, "Application of multichannel seismic reflection method to measure temperature in Sulawesi Sea," IOP Conference Series: Earth and Environmental Science, vol. 176, no. 1, p. 012044, Jul. 2018, doi: 10.1088/1755-1315/176/1/012044.

28 R. Adiputra and T. Utsunomiya, "Linear vs non-linear analysis on self-induced vibration of OTEC cold water pipe due to internal flow," Applied Ocean Research, vol. 110, 2021, doi: 10.1016/j.apor.2021.102610.

29 W. He, S. Zhang, and S. S. Ge, "Boundary control of a flexible riser with the application to marine installation," IEEE Transactions on Industrial Electronics, vol. 60, no. 12, pp. 5802–5810, 2013, doi: 10.1109/TIE.2013.2238873.

30 A. Miller, T. Rosario, and M. Ascari, "Selection and validation of a minimum-cost Cold Water Pipe material, configuration, and fabrication method for Ocean Thermal Energy Conversion (OTEC) systems," International SAMPE Technical Conference, 2012.

31 H.-J. Kim, H.-S. Lee, Y.-K. Jung, and W.-Y. Oh, "Performance and applicability of new material Heat exchanger to substitute titanium," ASME 2016 10th International Conference on Energy Sustainability, ES 2016, collocated with the ASME 2016 Power Conference and the ASME 2016 14th International Conference on Fuel Cell Science, Engineering and Technology 1,59542, 2016.

32 M. P. Eldred et al., "Heat exchanger development for ocean thermal energy conversion," OCEANS'11 - MTS/IEEE Kona, Program Book, 2011, doi: 10.23919/OCEANS.2011.6107175.

33 K. Yen and X. Jing, "The Feasibility Study of OTEC Integration on a Semi-Submersible O&G

Production Platform," Offshore Technology Conference Asia 2018, OTCA 2018, Mar. 2018, doi: 10.4043/28434-MS.

34 X. He, W. He, Y. Liu, Y. Wang, G. Li, and Y. Wang, "Robust Adaptive Control of an Offshore Ocean Thermal Energy Conversion System," IEEE Transactions on Systems, Man, and Cybernetics: Systems, vol. 50, no. 12, pp. 5285–5295, Dec. 2020, doi: 10.1109/TSMC.2018.2870999.

7. AUTHORS BIOGRAPHY

M. Ari is student of doctoral in ocean engineering. He is a lecturer at Politeknik Perkapalan Negeri Surabaya.

Y S Hadiwidodo is a lecturer at Dep. of Ocean Engineering, Fac. of Marine Technology, Institut Teknologi Sepuluh Nopember, Surabaya. He is also work at Lab. of Ocean Structural, Materials and Production

Mukhtasor is a Professor at Dep. of Ocean Engineering, Fac. of Marine Technology, Institut Teknologi Sepuluh Nopember, Surabaya. He is also work at Laboratory of Environmental and Ocean Energy

FLOW MANAGEMENT INNOVATION TO SHIFT TOWARD SUSTAINABLE SOCIETY

H Doi and **T Shinoda,** Kyushu University, Japan

SUMMARY

With action on environmental issues becoming increasingly important, the maritime industry is facing a new paradigm shift for zero emission technology. Environmental issues should be urgently solved; however, this new type of paradigm shift means that the maritime industry will face a dilemma between the current growth-based economy and the imminent sustainable economy. In this paper, we apply the framework of the theory of constraints when considering the dilemma; the issues are then elucidated, and some holistic improvement solutions are considered.

1. INTRODUCTION

Environmental issues have become increasingly important in a global society, and they must be addressed immediately, which will require lifestyle changes. The current global society should shift from a growing economic society to a sustainable developed society.

Particularly, to realize low energy consumption in seaborne trade, specifically in ship production, the shipbuilding industry must urgently recognize a new paradigm for zero emission technology. This will press shipbuilding and maritime industries to alter management policies on the basis of Sustainable Development Goals proposed by the United Nations.

In the current market, which is based on promising long-term growth, a company's major competitive objective is to achieve a low-cost, productive method for increasing production while using the same resources and maintaining fixed costs. Hence, a productive system is essential for obtaining and maintaining price competitiveness. However, in a future market based on a new paradigm, a company's major competitive objective becomes less clear.

Different technologies are required to efficiently solve pending environmental issues.Lead time (LT) is defined as the time required from the beginning of the work to the delivery of the product. Shortening the LT is an efficient way of tackling environmental issues. One way of shortening the LT is to delay the beginning of new projects until the preceding project is completed.

Concerning LT, there are linkages and fluctuations in production systems, and they sometimes interrupt production progress. To address this issue, the theory of constraints (TOC) is applied to improve overall optimization by focusing on the constraint conditions of production systems [1] [2].

In this paper, an inherent dilemma, which belongs to the maritime industry including shipbuilders, is elucidated by mainly using the frameworks oftheTOCthinking and the TOC dilemma cloud analysis[3]We then search for solutions to the dilemma by taking the following steps:

1) Recognize magnitude of undesirable environmental effects
2)Review the environmental issues affected by economic activity and the second law of thermodynamics
3) Clarify the dilemma induced by a paradigm shift to a sustainable society
4) Investigate necessary seeds for a sustainable society
5) Consider holistic system improvement for a sustainable society
6) Refer to design stage improvements of shipbuilding by the TOC
7) Create new insights into holistic improvement by the TOC way for sustainable development society

2. REPRESENTATIVEFRAMEWORK IN TOC

For finding solutions to a complicated problem, we mainly apply two useful TOC frameworks: TOC thinking process and TOC dilemma clouds[3]

2.1 TOC TINKING PROCESS

Fig. 1 shows a typical structure of the TOC thinking process, which describes the cause and effect of a problem by using linkage between four boxes; the lowest box for cause, the highest box for effect, the left box for assumption, and the right box for reason. Filling the boxes executes an interpretation of the problem; when we make an action with cause under an assumption, the result appears with effect along with the expressed reason. Fig. 1 shows a simplified example of implication in the TOC thinking process; however, when lighting a candle, the assumption should be scrutinized. After consideration, different assumptions are observed; for example, no normal air or other gases are present. It means that a one-sided assumption arises, and we can correct the cause and effect. When we meet a different cause and effect, to improve upon the problem, we must reconsider the assumption and the way it connects.

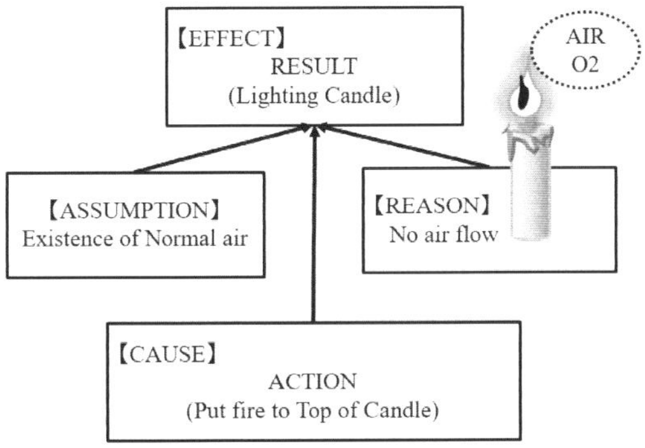

Figure 1 Typical structure of TOC thinking process (Describing lighting candle)

2.2 TOCCLOUDS

Fig. 2 shows a typical structure of TOC clouds that describe a dilemma using five linked boxes; the leftmost box "A" represents a common target, boxes "B" and "C" represent the need, uppermost box "D" represents the action to realize the need "B," and the lower box "D′" for the action to realize the need "C." When all the boxes have been filled and a conflict relation between actions "D" and "D′" appears, a dilemma is sometimes observed.

Fig. 2 also shows a simplified example of implication in TOC clouds. A way of considering the upper relationship A–B–D is explained. When we seek to realize common target A, it is necessary to realize the need "B" because of connected reasons A–B. When we seek to realize the need "B," it is necessary to conduct a proper action "D" because of the connection B–D.

A corresponding way of thinking occurs on the lower relationships A–C–D′. When we seek to realize common target A, it is necessary to realize the need "C" because of connected reasons A–C. When we seek to realize the need "C," it is necessary to conduct a proper action "D′" because of connected reasons B–D′.

Since both conflict actions are drawn by two logical thinking ways to realize common target, it is difficult to solve the dilemma. We must then compromise by determining one way of general thinking. TOC clouds take another way of thinking. It is necessary that we deeply consider the dilemma to induce undesirable effects in an existent system; solving the dilemma is an important process to improve the system.

Clouds shall be solved by breaking linkage of cloud structure. Solution shall be obtained by making following questions and replying the question;

Question (1),
"Why can't action D′ satisfy the need B?" serves to elucidate the reason for unlinking B and D. If we can link B and D′, the dilemma structure is broken and disappears. The answer to the question is the reason for unlinking B and D, and the solution invalidates the

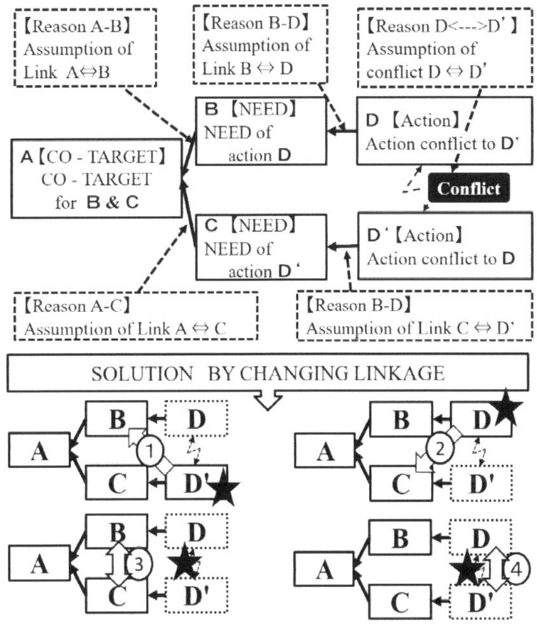

Figure 2 Sturucture and solution method of TOC clouds

reason for unlinking. The question, "Tell how to do?" draws out a concrete method.

Question (2)
"Why can't action D satisfy the need C?" seeks a solution by linking C and D′.

Question (3)
"Why can't B & C be satisfied simultaneously" seeks a solution by linking B and C.

Question (4)
"Why does D & D' conflict B?" seeks a solution by linking D and D'.

3. MAGNITUDE OF UNDESIRABLE EFFECT OF ENVIRONMENTAL ISSUES

Environmental issues have become a recognized matter following a series of abnormal global disasters, year after year According to TEEB[4], during the last half a century alone, biodiversity on earth (i.e., the bounty of the earth) has significantly decreased. Every aspect regarding biodiversity has rapidly decreased. Some examples of the biodiversity crisis are as follows: 40% of forest area on the earth has been destroyed during the past 300 years; 35% of mangrove forests have been destroyed in 20 years; after 1900, 50% of wetlands have been destroyed; and 30% of coral has been damaged because of fishery,pollution, disease, and coral reef bleaching. Additionally, the recent species extinction rate is 1000 times greater than the extinction rate covering the natural history of the earth, which shows that species extinction is rapidly proceeding. Fig. 3shows a futuretrend of the increasing amount of discharged CO_2 in the world resulting from fossil fuel consumption. From a scientific and technical perspective, the discharged CO_2 induces a serious global warming risk;an estimated

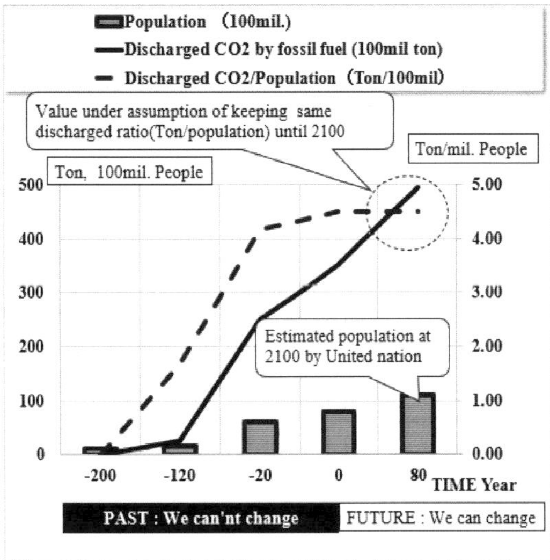

Figure 3 Trends and estimated discharged CO2 at 2100

amount of CO2 discharged by 2100 relates to an assumed population, with the discharge measured by tons per population. In other words, population growth impacts energy growth. There are countermeasures to decrease the risks of global warming; for example, low energy consumption. Zero emission technology is the highest priority issue for a sustainable developing society.

4. CONSIDERABLE FACTOR ON ENVIRONMENTAL ISSUE

4.1 THE SECOND LAW OF THERMODYNAMICS

In his work, The Economics of Sustainable Development, Herman E. Daly emphasized the importance of introducing the second law of thermodynamics into economics [5], where it becomes the prior condition of everything. It means that the current global society has to realize a paradigm shift from a growing economic society to a sustainable economic society.

4.2 MAJOR RISKS FOR THE CORRENT GLOBAL SOCIETY

In his work, Factfulness, Hans Rosling elucidated the importance of how to understand data.[6]. He showed five major risks:

1) Pandemic of infectious diseases,
2) Financial crisis,
3) Third World War,
4) Global warming, and
5) Extreme poverty.

In this paper, we apply TOC thinking to these five major risks. Fig. 4 shows the links between the risks, and intermediate effects are added to the linkage of risks. A negative loop appears in the risks showing the effect of the negative loop will gradually worsen. To avoid a negative loop, it is necessary to maintain a positive economy to avoid a financial crisis, which means maintaining energy consumption for the activity of the economy. Conflict among the five risks has been previously neglected because fossil fuel energy

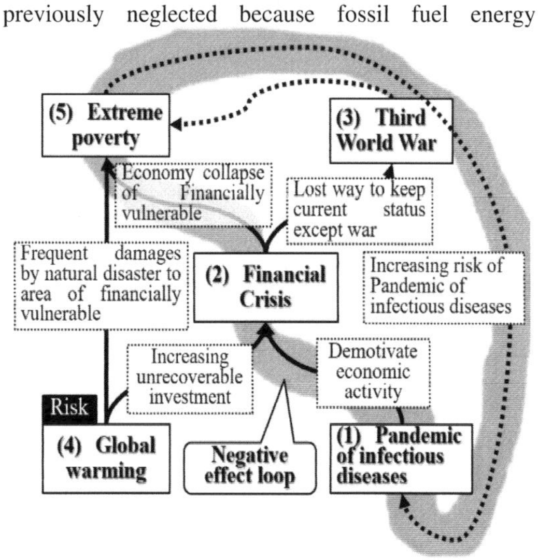

Figure 4 Relationship of the five major risks

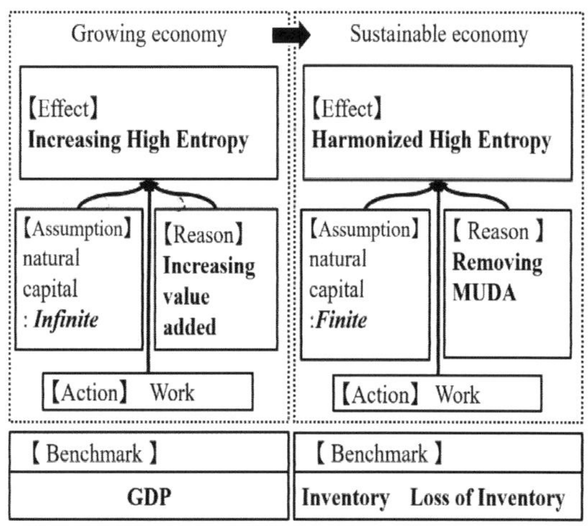

Figure 5 Growing economy and Sustainable economy by TOC thinking process

Consumption was small enough for the capacity of the earth. Now, however, the ratio is far from negligible. Fig. 5 shows the current growing economy and coming sustainable economy through the TOC thinking process. As for current growing economy, the assumption is that an infinite supply of resources form natural capital is always prepared for any market needs. When we work under this assumption, the added value of production is globally distributed, increasing high entropy. The reason is that distribution of production is driven by growing economies, meaning that added value for a growing economy istheonlymarket acceptance. The GDPbenchmark is adopted in the world. As for the sustainable developing economy,the assumption is that a finite supply of resources from natural capital.The expected effect is harmonized entropy by reducing MUDA. The benchmark will be Inventory.

4.3 MUDA IN A SUSTAINABLE ECONOMY

We call waste MUDA in Japanese [7], and we think MUDA of resource leads to a sustainable economy. It is important to remove MUDA in the sustainable development of an economy. In his work of Law of Paradox [8], Katsuhiro Nishinari introduced MUDA and introduced three assumptions: 1) mission, 2) period, and 3) point of view. As for sustainable economy, three assumptions are as follows;

1) mission : constructing sustainable developing society
2) period : eternal from now
3) point of view : resident on the earth

Under these assumptions, when we search MUDA, we consider its reason for each assumption. The reason for the mission assumption is "destroy eco-service," the reason for the period assumption is "discharged excessive-high entropy," and the reason for the point of view assumption is "make action to disharmony of the earth." In a sustainable development economy, reducing MUDA is to suppress these three reasons. To reduce MUDA, it is necessary to execute action to erase reasons as follows;

1) Not destroy Eco-service
2) Not discharge excessive high entropy
3) Not make action to disharmony of the earth

The seeds to realize above actions are;

1) Nature first (SDGs)
2) Less production & consumption Shifting to Zero emission
3) Promote policy of cooperating first

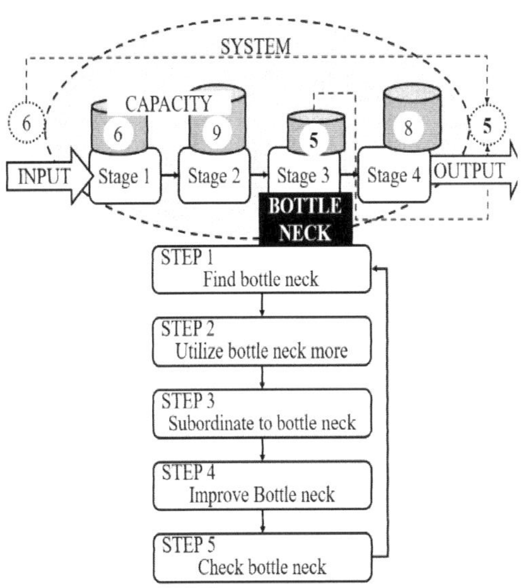

Figure6 Theory of Constraint and TOC improvement cycle

5. TOTAL OPTIMIZATION TOACCELERATE SYSTEM PERFORMING

To realize a sustainable economy, we have to produce zero emission products and services at the earliest opportunity. Delays may increase the environmental footprint. It is important to realize system improvements to deliver new products faster. Fig. 6 shows the Theory of Constraint, which means that system productivity is governed by a bottleneck capacity. Injection speed higher than the bottleneck shall increase inventory in the system. LT of the system is governed by the amount of inventory. Injection control to minimize inventory in the system is important to avoid MUDA in a sustainable economy.

6. HOLISTIC IMPROVEMENT ON THE DESIGN STAGE OF SHIPBUILDING

6.1 MOTIVATION TOC IMPLEMETATION

In the shipbuilding design department, the short-term problem of frequent drawing delays was encountered. Furthermore, coming projects brought the serious problem of a financial penalty, 10,000 USD per delayed drawing. This made the design department leaders fight to meet the due dates. For the design department, problems with delayed drawing were not new, but significant improvements could not be made. To avoid massive financial penalties, the leader of the design department decided to try the TOC method [9].

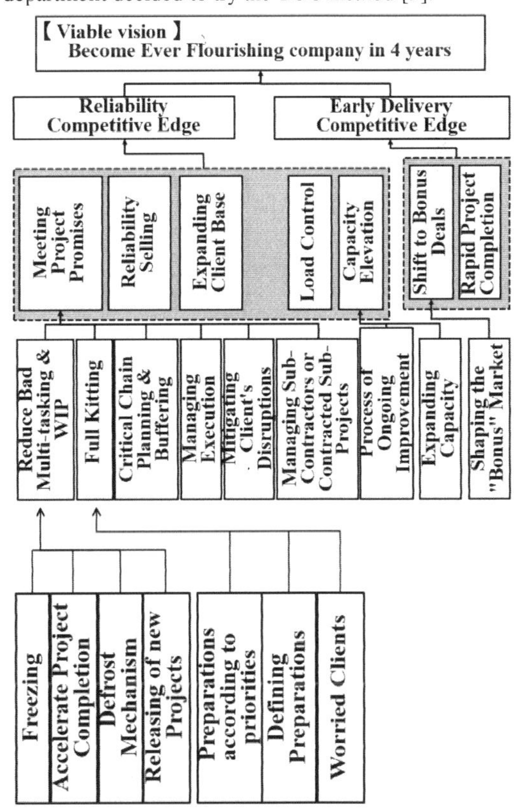

Figure 7 Strategy and Tactics tree for a project by TOC

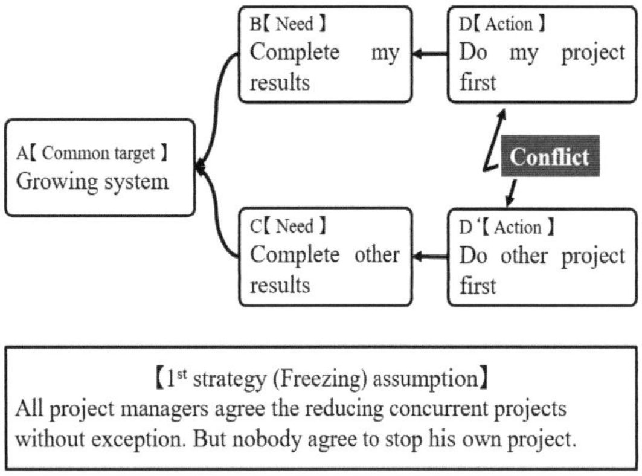

Figure 8 Dilemma by assemption of the first strategy (Freezing)

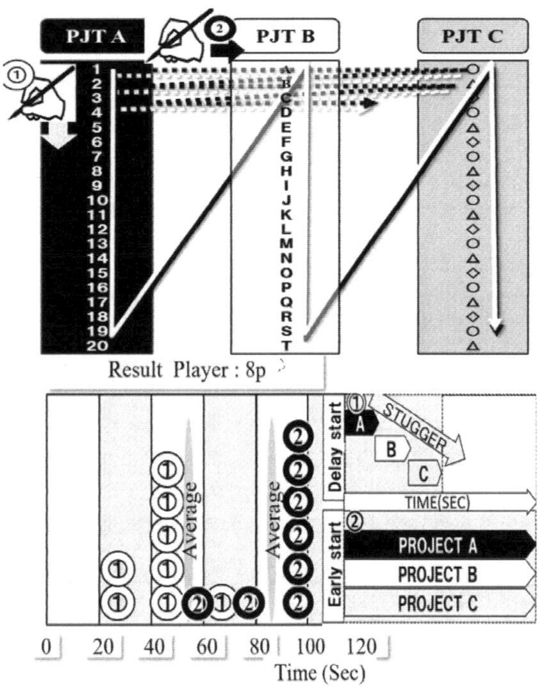

Figure 9 Results of Multitask game

6.2 STRATEGY AND TACTICS TREE VIA TOC

TOC prepared Strategy and Tactics tree (S and T tree) is an assumption tree used to realize a viable vision. Fig. 7 shows the S and T tree for the type of project. In the TOC, strategy means the desirable condition (what to be), and tactics means reasonable action (how to do). A viable vision is generally too high to be a desirable condition in a general approach. The TOC investigated a viable visionand then considered its assumption to realize the primary objective. The assumptions were the intermediate targets before realizing the top target. In the case of the design department, the top vision was meeting due dates. For a sustainable society, this viable vision is important and to be discussed by a diverse community.

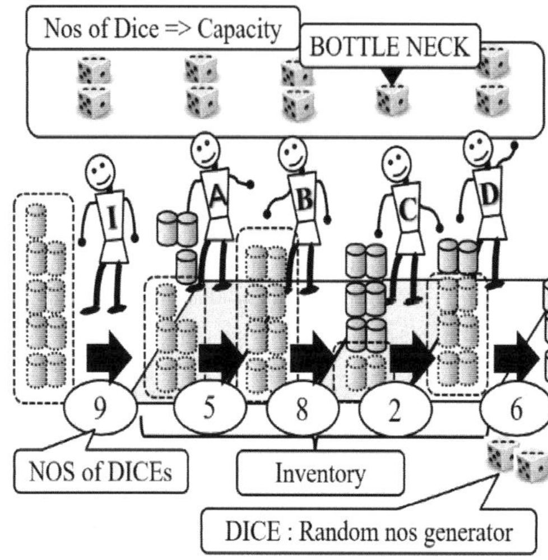

Figure 10 TOC dice game

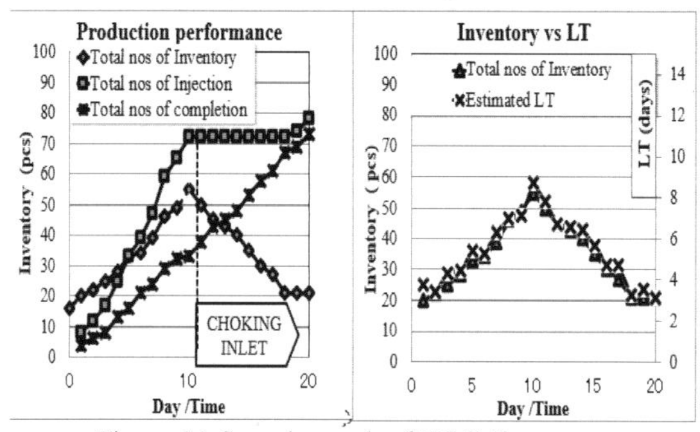

Figure 11 Sample result of TOC dice game showing Inlet control effect

6.3 DESIRABLE EFFECT BY INJECTION CONTROL

The first strategy of the S and T tree is freezing. To improve meeting a due date is to realize shortening LT. Fig. 8 shows the dilemma of assumption of the first strategy (freezing), which shows the difficulty in delaying the start. Generally, most managers encourage their staff to complete a job quickly and most staff begin as early as possible. Many initial tasks make the system suffer in a multitask condition. Bad multitasking means inducing longer LT.

Fig. 9 shows a Multitask game, which interactively compares starting early (concurrent processing) and starting late. It is understood that a delayed start is better and necessary for the system, but the actual execution is difficult in the real world.

Fig.10 shows an outline of the TOC dice game[10]. Settling the stage from the injection (I) to A–D, each stage adds value into inventory received from the previous stage and sends it to the next stage. Manufacturing results normally fluctuate daily, and the dice are random generators to express this fluctuation. The capacity of the stage is determined by the quantity of dice: one for stage C and two for all other stages, which sets stage C as a bottleneck in the system.

Members then enjoy a simulation of the system production for a month (20 days). Fig. 11 shows the inlet control effect on total inventory in the system and LT through the TOC dice game. In 10 days with all work stages full, the amount of injection into the system is almost double the amount passing the bottleneck, and inventory in the system rapidly increases. Members discussed how to reduce increased inventory and decided to choke the inlet until the inventory decreased in the initial quantity. Then, members operated the system for the last 10 days. They succeeded in reducing the inventory without affecting delivery. The estimated LT was obtained by the quantity of inventory divided by capacity. Choking the inlet was effective in controlling the quantity of inventory and LT in the system.

6.4 FORTY(40%) PERCENT LT REDUCTION IN THE DESIG STAGE IN A FEW MONTHS

To begin freezing, project priority, or deciding what projects will be suspended, was determined. Following this, the top management declared the freezing to all staff in the system. The official announcement was a big job for the top management of the system, as the responsibility for any delay rests on the top management. In Fig. 12, a new trend of publishing drawings was shared with concerned managers. After freezing commenced, the next box was started. They decided full kit for the suspended project. Well-done preparation led to a positive start to the project. This full kit was the condition to restart the suspended project. Then, the timing of the restart was discussed with mutual understanding of the point of timing. The suspended project was restarted by the announcement of the top management of the system. Fig. 13 shows the trend of shortening LT for projects. The TOC implementation was started at beginning of October 2012, and the declaration of freezing was announced on October 19, 2012. Inventory in the system was reduced. In the middle of December 2012, the standard LT over 8 months was decreased to about 4.3 months. After half a year, a minimum LT of 3 months was reached. Inlet control was quite effective in reducing the inventory and LT of the system.

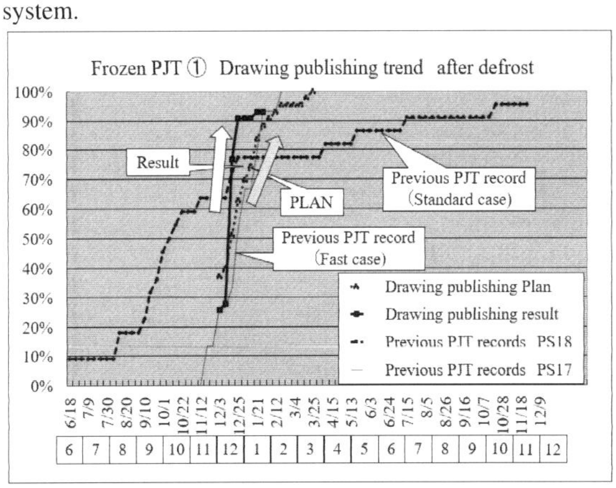

Figure 12 Increasing drawing publishing velocity by freezing

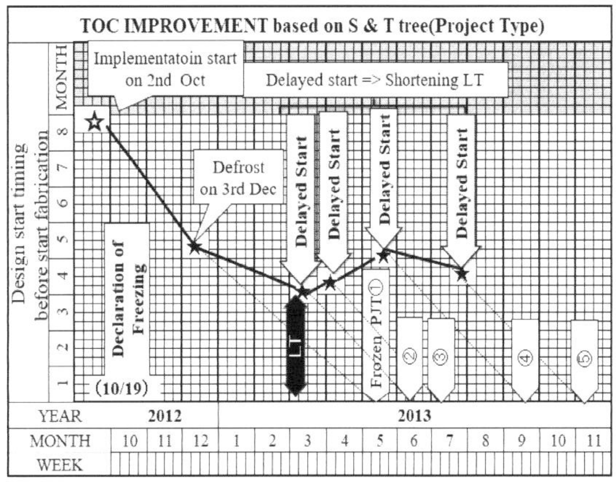

Figure 13 Reducing project LT by TOC way in short term

6.5 EXPECTED EFFECT OF CONTROLLING INVENTORY IN SYSTEM FOR A SUSTAINABLE SOCIETY

Most systems are expected to be high productivity. From an environmental viewpoint, value added is obtained with some material and manufacturing energy, shown in equation (1):

$$Inv_{i+1} = Inv_i + E_i (\text{or } HE) \quad (1)$$

where
Inv: Inventory in society
E : Energy
HE: High entropy

Equation (1) means the existence of a large amount of inventory has discharged high entropy. Of course, inventory needs some cost. A larger amount of inventory is not good for the environment or for financial aspects. Chocking the inlet of the system, shown in Fig. 14, is an effective method for reducing LT. The process

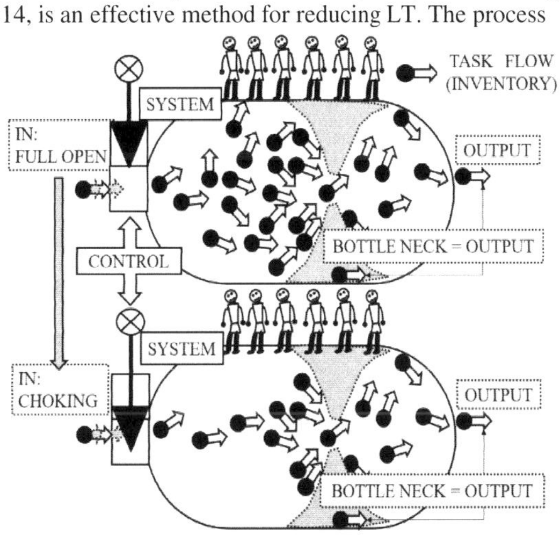

Figure 14 Effect in the system by choking inlet

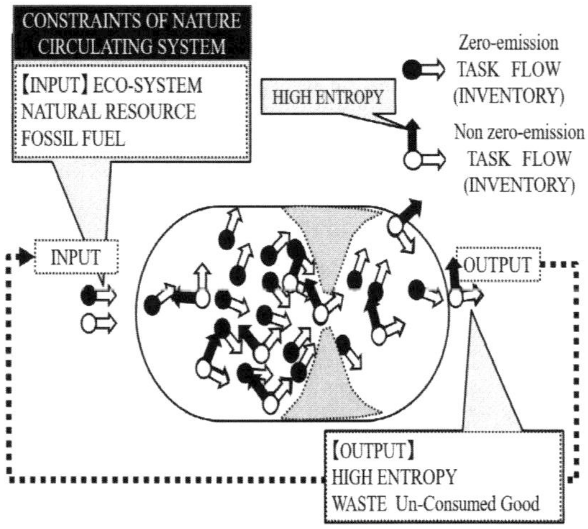

Figure 15 Nature circulating system diagram

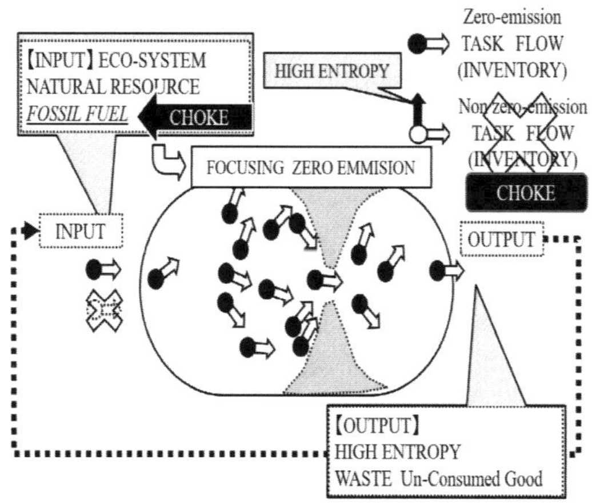

Figure 16 Input control on Nature circulating system diagram for sustainable developing society

of adding value into inventory is accompanied by discharging high entropy; reducing inventory contributes to environmental improvement. Fig.14shows the importance of inlet control for environmental issues.

Equation (2) is obtained by integrating equation (1):

$$\text{P-sold} + \text{P-unsold} = \text{Inv} + \text{HE} \quad (2)$$

where
P-sold : Product consumed by market
P-unsold: Product after the expiration date, which must be processed as waste disposal

Equation (3) is obtained by differentiating equation (1) in time:

$$\text{Pe} = \Delta\text{Inv} + \text{HEv} \quad (3)$$

where
Pe: Productivity of system
= total amount of product divided by time.
ΔInv: Inventory derivative
Depends on the market need
HEv: Discharging velocity of High entropy

As per the environmental viewpoint, productivity should be limited to prevent excessive discharging of high entropy. However, for the increasing global population, steady improvement is needed. The inventory derivative must be decreased to reduce high entropy. Excessive product and inventory are induced by input. If the input is controlled to realize a steady production amount, excessive product or inventory can be avoided while promoting zero emission technology. In a sustainable developing society, we can observe the developing status by checking the amount of input, which is simple and reliable. Fig. 15 shows a diagram of nature in a circulating system. In a natural circulating system, there are consuming systems and natural capital injection systems. In a current nature circulating system, injecting natural capital (eco-service, natural resource, and fossil fuel) will flow in a zero-emission process and a nonzero emission process. In a current growing economy, the nonzero emission process increases and expands the environmental risk. In a sustainable developing society, an improvement on injection of natural capital shall be done as shown in Fig. 16. For eco-service to be improved, recycling of natural resources should be promoted, choking fossil fuel. Improvement and control of inlet can be effective for system improvement.

7. CONCLUSION

Global society is facing a global warming risk, which is one of the five major risks shown by Hans Rosling. The environmental issue has a conflict between increasing necessary energy to support the increasing population and decreasing necessary consumption of fossil fuel energy, which must be solved for future generations. Zero emission technology is a priority solution for the risk of global warming. Reviewing the TOC holistic improvement results, the following conclusions are obtained:

Choking the inlet is highly effectivein reducing inventory and LT without losing outputs.

(1) Choking the inlet brings shorter LT and lower entropy in the system, with good financial effects.
(2) Reducing fossil fuel into the natural circulating system as input control is effective in improving a natural circulating system (high entropy reduces, the same output is kept, and zero emission technology advances).

8. REFERENCES

1. GOLDRATT, E.M. The Goal : A Process of Ongoing Improvement, Japanese edition, *Diamond Co.,ltd,* pp. 552, 2001.

2. GOLDRATT, E.M... Standing on The Shoulders of Giants: Production Concepts Versus Production Applications. The Hitachi Tool Engineering example. Gestão&Produção, 16(3), 333-343. *https://dx.doi.org/10.1590/S0104-530X2009000300002*、2009.

3. GOLDRATT, E.M. . The Goal 2 : It's Not Luck, Japanese edition, *Diamond Co.,ltd,* pp.400, 2002.

4. BRINK, T. P.The Economics of Ecosystems and Biodiversity in National and International Policy Making (TEEB - The Economics of Ecosystems and Biodiversity),*Routeledge.* pp.528,2012.

5. DALY, E,HERMAN. The economics of sustainable development, Japanese edition, *MisuzuShobCo.,ltd*, pp317,2005

6. ROSLING, H..*Factfulness*, Japanese edition, *Nikkei BP Co.,ltd,* pp336,2019.

7. OHNO, T. (1988). Toyota Production System: beyond large-scale production. *Oregon: Productivity Press*,1988.

8. NISHINARI, K.Law of Paradox,*ShinchoshaCo.,ltd,* pp.222,2017.

9. GOLDRATT, E.M. Critical Chain, Japanese edition, *Diamond Co.,ltd*, pp.384、2003.

10. COX, J., BERGLAND S. and JACOB, D. Velocity:Combining Lean, Six Sigma, andthe Theory of Constraints to Achieve Breakthrough Performance,*Free Press,USA,* , pp310,2009.

9. AUTHORS BIOGRAPHY

Hirofumi Doi is an Audit and Supervisory Board Member of Mitsui E&S shipbuilding Co.,ltd, Japan. He is also a doctor's course student of Kyushu University, Japan.He received B.A. in naval architecture from Hiroshima University. His main research field is design of equipment engineering in naval architecture and ship production system

Takeshi Shinoda is a professor of Kyushu University, Japan. He received D.Eng., M.A. and B.A. in naval architecture from Kyushu University. His main research field is equipment engineering in naval architecture, ship production system, safety engineering, sea transportation including containerized transportation and its facility, and environmental engineering concerning ship.

CONCEPT DESIGN OF FERROCEMENT PATROL VESSEL SUPPORTING INDONESIAN COAST GUARD PATROL OPERATION IN NATUNA SEA

P A S Sudiro, I N P Apriyanto, and **J Jupriyanto**, Republic of Indonesia Defense University, IDN

SUMMARY

Natuna Sea is one of the most priority maritime operation region because of its threats such IUUF is hing and Sovereignty Violation. Indonesian Coast Guard (Bakamla RI) is the agency which has role to enforce maritime law on Indonesian Sea Territories. But on the otherside, Indonesian Coast Guard has main problem at its patrol fleet. It's caused by several reasons such lack of patrol ship, lack of operational fund, and lack of raw materials industry capability. Ferrocement is a sustainableshipbuilding material that has potencies to solve the problems. It's easy to applicate, strong enough, andlow cost. The research aims to formulate the Concept Design of Ferrocement Patrol Vessel supportingIndonesian Coast Guard's patrol operation. The research uses Design For SixSigma (DFSS) method which consist of Define, Measure, Analyze, Design, and Verify (DMADV). It results the concept design of the ferrocement hull based on the requirements.

NOMENCLATURE

[Symbol]	[Definition] [(Unit)]
LOA	Length Over All (m)
B	Beam (m)
H	Height (m)
T	Draft (m)
GT	Gross Ton (tonnes)
Ø	Radius (mm)

1. INTRODUCTION

Natuna Sea is one of the furthermost regions in Indonesia. As a part of Riau Islands (KEPRI), it's rich with fishes and natural gas reserves [1]. But its riches are comparable with the threats which are happened frequently. Former Indonesian Coast Guard (Bakamla RI) Commandant, Vice Admiral (ret.) Achmad Taufiqoerrochman classified it into oneofthemostprioritymaritimeoperationregion [2].Most of thethreatsareIUUFishingandSovereigntyViolation.During the increase of China's assertiveness, a part of northern Natuna Sea which is crossed by nine dash lines is being disputed.

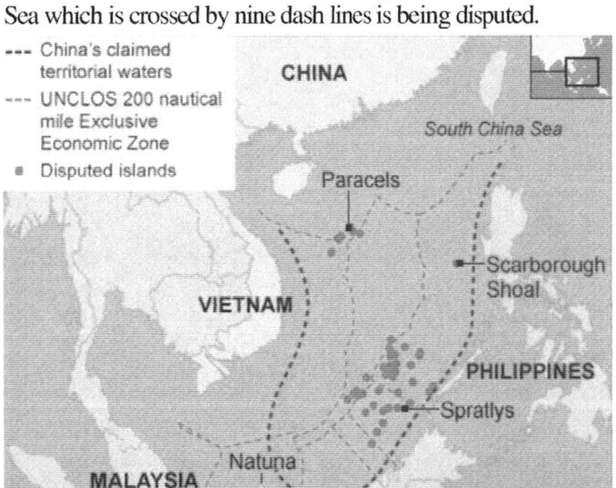

Figure 1: A part of Natuna Sea which is being disputed[1]

To enforce maritime law on Indonesian SeaTerritories include Exclusive Economic Zone (EEZ), the Indonesian Government established Indonesian Coast Guard (Bakamla RI). Indonesian Coast Guard (Bakamla RI) hasduties to hold security and safety patrol operation in Indonesian Territory and Indonesian Exclusive Economic Zone (EEZ) [3]. But on the otherside, Indonesian Coast Guard has main problem at its patrol fleet. It's caused by several reasons such lackof patrol ship, lack of operational budget, and lack of raw material industry capability. President Director of PT Krakatau Steel (Persero) Tbk, Silmy Karim said that most of the steel for the domestic infrastructures (include shipbuilding industries) have been supplied abroad that causingthe increase of production cost [4]. So farIndonesian Coast Guard (Bakamla RI) has only fourPatrol Vessel which are capable for patrol operation on Natuna Sea [5]. Meanwhile they can't be placed on Natuna Sea at the same time. Moreover, the operational fundis lower than the ship procurement cost[6].

Ferrocement is a composite structural material comprising thin sections consisting of cement mortar reinforced by a number of closely spaced layers of steel wire mesh [7]. It is a sustainableshipbuilding material that has potencies to solve the problems. It's easy to applicate, strong enough, andlow cost [8].

2. METHODOLOGY

The research methodology uses Design For SixSigma (DFSS). DFSS is derived from Six Sigma method which is applied at new product design [9].The methodology consists of Define the requirements, Measure the requirements, Analyze the potencies, Designthe concept, and Verify that the concept is solution of the problems [9]. The sequence of the method is shown by the flow chart below:

Figure 2: The Flow Chart of DFSS Method

3. RESULT AND DISCUSSION

The purpose of the research is to formulate the concept design of the ferrocement hull supportingIndonesian Coast Guard's patrol operation on Natuna Sea. Firstly, the requirement of the Patrol Vessel which be operated in Natuna Sea must be defined.

3.1 DEFINE

The requirement of the vessel has strong relation with the characteristic of sea where it operated. There are two subdivisions, firstly is environment characteristic and secondly is threat characteristic.

Environment characteristic is about wave's size. Wave's size determines how safe the vessel during its operation. It contains length and height of the wave. The formulas below aim to determine how safe the vessel during its operation [10]:

$$\text{Danger Wave Height} \geq \text{LOA} \times 30\%$$
$$\text{Danger Wave Length} \geq \text{Wave Height} \times 7$$

The wave height that reaches more than 30% of LOA will have enough potential energy to knock the vessel down [10]. Former Indonesian Coast Guard (Bakamla RI)Commandant, Vice Admiral (ret.) Dr. Desi Albert Mamahit, M.Sc. said that the wave height of the Natuna Sea can reach more than 5 m. Another source said that at the same time, the wave length can reach 200 m [11]. If the number is filled into the formulas with assumption that 200 m is Danger Wave Length, the results are:

$$200 \text{ m} = \text{Wave Height} \times 7$$
$$\text{Wave Height} = 200 \text{ m}/7$$
$$= 28{,}57 \text{ m}$$

28,57 m is assumed as Danger Wave Height. So that:

$$28{,}57 \text{ m} = \text{LOA} \times 30\%$$
$$\text{LOA} = 28{,}57 \text{ m}/30\%$$
$$\sim 95 \text{ m}$$

According to the result, the vessel will totally safe during its operation in Natuna Sea if its LOA is more than 95 m. Regularly Indonesian Coast Guard (Bakamla RI) operatesKapal Negara (KN) TanjungDatu that has dimensions below[12][13]:

LOA = 110 m
B = 15,5 m
H = 6,9 m
T = 4 m

Figure 3: Kapal Negara (KN) TanjungDatu [13]

There are two common threats on Natuna Sea: IUUFishingandSovereigntyViolation. IUU Fishing mostly done by Vietnamese fisherman. In just May 2021there are 24 Vietnamese Fishing Ships inside Indonesian EEZ [14]. It can be more threatening that Vietnam operates not only common Fishing Ships but also Maritime Militia Vessel sized about 400 GT [15]. Meanwhile Sovereignty Violation is done by China Coast Guard Vessel during their claim about nine dash lines region which crosses Indonesian EEZ on Natuna Sea. Both Vietnam Maritime Militia Vessel and China Coast Guard Vessel have similar hull construction. The vessels are designed for ramming. It can be seen at slab-like flat portions along the hull [16]. On the contrary, Indonesian Coast Guard Patrol Vessel hasn't adopted its construction. It will be dangerous when it makes contact with them on Natuna Sea.

Figure 4: China Coast Guard Vessel Hull [17] (above) and Indonesian Coast Guard Vessel Hull [13] (below)

3.2 MEASURE

There are basically two requirements of Indonesian Coast Guard Patrol Vessel that be operated in Natuna Sea:

3.2 (a) Size

The Vessel must have LOA more than 95 m for safety reason because of Natuna Sea's environment. The present Patrol Vessel that regularly be operated there, Kapal Negara (KN) Tanjung Datu has LOA 110 m. There is a ferrocement ship features that was built during

World War II. Its complete features are shown below [18]:

Type	: B7A1
LOA	: 111.6 m
B	: 16.5 m
H	: 10.7 m
T	: 8 m
Thickness	: 0.108 m

This table below shows comparison features between Kapal Negara (KN) TanjungDatu and B7A1 Type Ferrocement Ship:

Table 1: Comparison Features Between KN TanjungDatu and B7A1

Dimensions	Type of Vessel	
	TanjungDatu [12][13]	**B7A1** [18]
LOA	110 m	111,6 m
B	15,5 m	16,5 m
H	6,9 m	10,7 m
T	4 m	8 m

Because of the similarity, the feature of B7A1 can be acceptedasthe concept design of ferrocement patrol vessel. As for the oversized of the draught (T) will be used for sonar apparatus. The concept design of the patrol vessel will be planned as multi role vessel whose has some abilities beside as a patrol vessel. One of the abilities is research vessel.

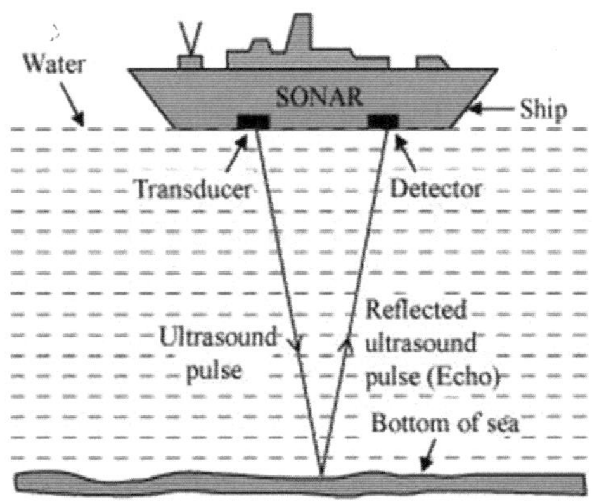

Figure 5: Ship's Sonar [19]

3.2 (b) Form

According to the threats, the Patrol Vessel on Natuna Sea must haveslab-like flat portions along the hulls for ramming. Indonesian Coast Guard Patrol Vessel hasn't adopted the construction. In fact, the main form of Ferrocement Ships that had ever built shows that the construction is suitable to the material.

Figure 6: Capella, a Ferrocement Ship that had ever built [20]

Capella is a Ferrocement Ship that was built by Germany during World War II. It was launched in 1943 and had been operated until 1988 [20]. The form had adopted the construction earlierto prevent collision with ice blocks on Baltic Sea. It proved that the construction is available to be applied on Ferrocement Ship, so that Ferrocement Ship will be safe for ramming.

3.3 ANALYZE

Ferrocement is a composite structural material comprising thin sections consisting of cement mortar reinforced by a number of closely spaced layers of steel wire mesh [7]. As a composite Ferrocement has two constituent components, cement mortar as matrix and wire mesh as reinforced [21]. Cement mortar is actually a particulate composite which contains cement as matrix, water as solvent, and sand as reinforced [22].

Figure7: Preparation of Ferrocement [23]

There is a comparison study shown that Ferrocement has some advantages over the other shipbuilding materials at three parameters: Material Cost, Fire Resistance, and Repairability [8]. This research will discuss deeply about three parameters: Mass, Strength, and Material Cost.

3.3 (a) Mass

Mass is common argument about Ferrocement, that like other cementitious materials it is viewed as heavy material[8]. In B7A1 case, the density of the material can be counted according of its further features [18]:

Displacement : 10,940 tons
Deadweight to Displacement Ratio : 0.53
Mortar Volume : 2,247.8 m^3

So that:

$$\text{Deadweight} = 0.53 \times 10{,}940 \text{ tons} = 5{,}798.2 \text{ tons}$$

$$\text{Deadweight} = \text{Displacement} - \text{Mass}$$
$$\text{Mass} = \text{Displacement} - \text{Deadweight} = 10{,}940 \text{ tons} - 5{,}798.2 \text{ tons} = 5{,}141.8 \text{ tons} = 5{,}141{,}800 \text{ kg}$$

From this result, the density of the material can be counted:

$$\text{Density} = \text{Mass}/\text{Volume} = 5{,}141{,}800 \text{ kg}/2{,}247.8 \text{ m}^3 = 2{,}287.48 \text{ kg}/\text{m}^3$$

The density number is evidently lower than Aluminium (2,700 kg / m^3) that known as one of the lightest shipbuilding materials.

3.3 (b) Strength

The section of B7A1's body plan is shown below:

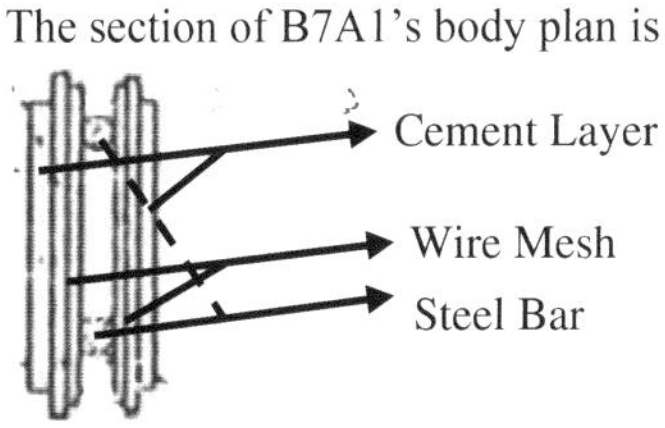

Figure 8: The Body Plan of Type B7A1 [18]

There is the research used the same material with the same form into Bending Test [24]. Its result was a 5 m × 0,15 m × 0,02 m sample can hold until 2,532 N force [24]. The following figure shows assumption that the ramming area as size as 1 m × 1 m × 0.108 m:

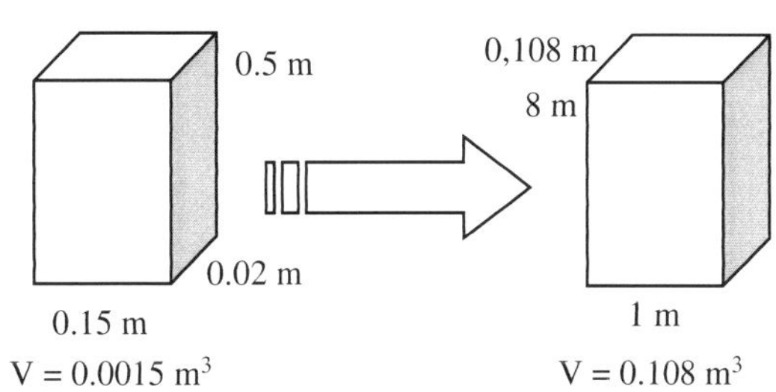

Figure 9: Comparison between test specimen (left) and assumed ramming section (right)

From the comparison, the volume ratio between assumed ramming section and test specimen is:

$$\text{Volume Ratio} = 0.864 \text{ m}^3/0.0015 \text{ m}^3 = 576$$

The maximum strength that can be held by the section can be assumed as:

$$\text{Strength (assumed)} = 576 \times 2{,}532 \text{ N} = 1{,}458{,}432 \text{ N}$$

If the vessel is rammed by another vessel with 4,000 tonnes (4,000,000 kg) and acceleration 0.5 knot/s (0.257 m/s^2) the ramming force is:

$$\text{Ramming Force} = 4{,}000{,}000 \text{ kg} \times 0.257 \text{ m}/\text{s}^2 = 1{,}028{,}000 \text{ N}$$

The assumed strength is obviously higher than the maximum ramming force.

3.3 (c) Material Cost

The bottom surface area can be counted as:

$$\text{Bottom Surface Area} = \text{Mortar Volume}/\text{Thickness} = 2{,}247.8 \text{ m}^3/0.108 \text{ m} = 20{,}812.96 \text{ m}^2$$

The materials which were used to build Type B7A1 [18] are:

Table 2: List of The Materials Price

Materials	**Composition**	**Amount**	**Price**
Mortar	Black Sand : Cement : Water 1,4 : 1 : 0.5	2,247.8 m^3	detailed below
	Cement	775.1 m^3	Rp2,083,081,250 [25]
	Black Sand	1,085.14 m^3	Rp200,750,900 [26]
Reinforced Steel	ø 6 mm	1,381,824 kg	Rp12,403,023,250 [27]
Wire Mesh	M 12 (Type)	20,812.96 m^2	Rp3,125,607,081 [28]
TOTAL			Rp17,812,462,480

As comparison, the procurement of KN Tanjung Datu needed Rp208,267,000,000 [29].

3.4 DESIGN

From the former step, the hull design of Indonesian Patrol Vessel can be concepted like the following figure:

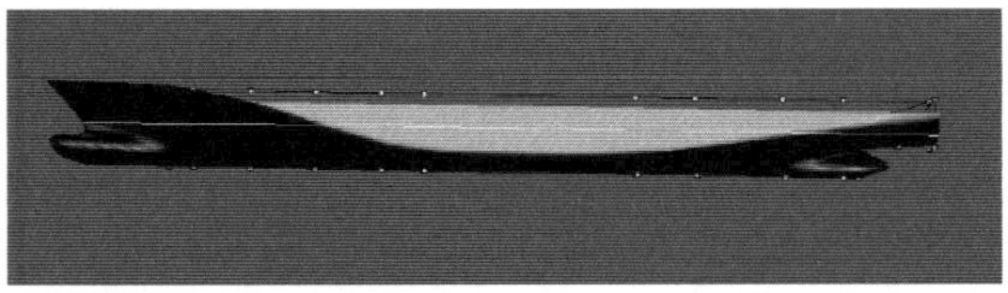

Figure 10: Concept Design of Indonesian Patrol Vessel

By optimizing with Maxsurf Software, the dimension of the concept design can be described as:

LOA	: 110.25 m
B	: 14.52 m
H	: 10.6 m
T	: 5.04 m
Displacement	: 4,711 tonnes

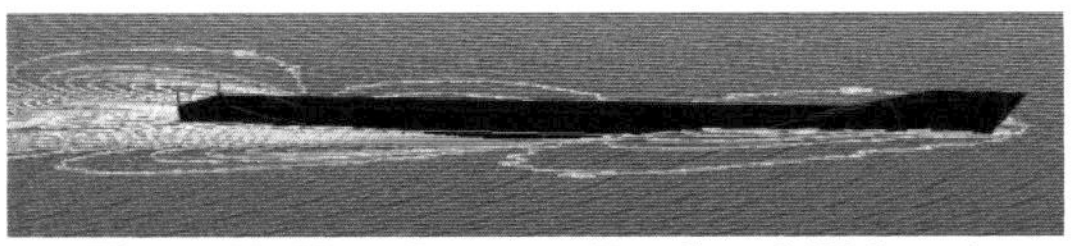

Figure 11: Wave Pattern at Top Speed (23 knots)

The figure above shows the wave pattern at top speed (23 knots). The ship makes transverse wave length at maximum one third of LOA. So that the ship is safe to be operated [30].

3.5 VERIFY

Table 3: Verification Result of The Concept Design to Indonesian Coast Guard's Requirement

Parameters	Result	Description
LOA	110.25 m	qualify the requirement
Strength	1,458,432 N (assumed)	have enough strength to hold 4,000 tonnes vessel
Cost	Rp17,812,462,480 (hull)	lower than the procurement cost of former Patrol Vessel

From the result, the concept design can be claimed as the qualified design of Indonesian Coast Guard Patrol Vessel, especially based on three main discussed parameters: Size, Strength, and Cost. This concept design has been conformable for safety requirement to be operated in Natuna Sea. The strength, although it is just an assumption, can show that the concept design will strong enough to dampen ramming with another vessel. From the calculation of hull building cost, the procurement cost can be predicted much lower than former Patrol Vessel. In other words, the vessel will available to be equipped with additional features.

Figure 12: Comparison between The Concept Design (above) and KN Tanjung Datu hull form (below) [13]

The hull form has already had slab-like flat portions to make ability for ramming. The bow has been also equipped with Bulbous Bow to break the wave in order to reduce pitch movement.

4. CONCLUSIONS

There are two subdivisions of the requirement, firstly is environment characteristic and secondly is threat characteristic. Environment characteristic is about wave's size to determine how safe the vessel during its operation. Common threats in Natuna Sea are IUUFishingandSovereigntyViolation. There are two requirements of Indonesian Coast Guard Patrol Vessel that be operated in Natuna Sea: the LOA more than 95 m for safety and theslab-like flat portions along the hull for ramming. Considering the requirement and the result of the research, Ferrocement is suitable to be used as ship building material for the future Indonesian Coast Guard Patrol Vessel that will be operated in Natuna Sea.

5. ACKNOWLEDGEMENTS

The research is presented for developing performance of Indonesian Coast Guard (Bakamla RI) especially during patrol operation in Natuna Sea. Then, the research is presented for increase Maritime Security in order to realize Indonesia as World Maritime Fulcrum. The research is mostly an approach, to introduce ferrocement as sustainable future ship building material. Therefore, the further researchesare still be needed.

6. REFERENCES

1. VASWANI, K., 'The sleepy island Indonesia is guarding from China', *Retrieved from https://www.bbc.com/news/world-asia-29655874*, 20th October 2014.

2. TAUFIQOERROCHMAN, A., 'PENDEKATAN OPERASIONAL DOM', *Translations of KONSEP OPERASI MARITIM INDONESIA*, 2018.

3. Article No. 61, Law No. 32 of 2014.

4. YUNIANTO, T. K. 'Prospek Industri Galangan Kapal untuk Mendongkrak Konsumsi Baja RI'. *Retrieved from https://katadata.co.id/happyfajrian/berita/5f33b61a3dd93/prospek-industri-galangan-kapal-untuk-mendongkrak-konsumsi-baja-ri.* 12th August 2020.

5. HASAN, S. 'Slide 12', *PENGUATAN KELEMBAGAAN SISTEM KEAMANAN LAUT INDONESIA*, *Retrieved from https://kkp.go.id/an-component/media/upload-gambar-pendukung/DitJaskel/publikasi-materi-2/sistem-kelautan/PPT%20Penguatan%20Kelembagaan%20Sistem%20Keamanan%20Laut%20Indonesia.pdf?fbclid=IwAR2aE96LkGk3ab9t805X8dcUnL5aM9CfqkxWklf0elyKgqN-Fah2klNF3L4,* March 2021.

6. *Ibid.*, 'Slide 11'.

7. AMERICANCONCRETEINSTITUTE (ACI), 'Ferrocement', *Retrieved from https://www.concrete.org/topicsinconcrete/topicdetail/Ferrocement?search=Ferrocement,* n.d.

8. WHANG, B., 'CONCLUSIONS AND RECOMMENDATIONS', *COMPARISON STUDY OF ALUMINIUM, FERRO-CEMENT, AND FIBER-REINFORCED PLASTIC FOR SMALL CRAFT IN KOREA*, 1972.

9. FRANCISCO, M. *et al.*, 'Design for six sigma integrated product development reference model through systematic review', *Retrieved from* https://www.emerald.com/insight/2040-4166.htm, 2020.

10. TREDUP, S., 'Dangerous waves and your boat', *Retrieved from https://www.oceannavigator.com/dangerous-waves-and-your-boat/*, 31st March 2011.

11. PURWANTO, B. R. *et al.*, 'ANALISIS PENGARUH GELOMBANG LAUT TERHADAP PENGGUNAAN BIAYA DAN JENISKAPAL PATROLI UNTUK OPERASI PENGAMANANLAUT NATUNA UTARA', *Retrieved from http://dx.doi.org/10.24895/MIG.2020.22-2.1187121*, 28th October 2020.

12. BROWN, N. *et al.*, 'Indonesia>Auxiliaries – Coast guard', *Jane's Fighting Ships 2018-2019*, 2018.

13. ADJIE, H., 'KN TanjungDatu 1101: Dimensi Laksana Frigat, Berstatus Kapal Patroli Penjaga Pantai', *Retrieved from https://www.indomiliter.com/kn-tanjung-datu-1101-dimensi-laksana-frigat-berstatus-kapal-patroli-penjaga-pantai/*, 27th November 2021.

14. INDONESIA OCEAN JUSTICE INITIATIVE, 'IUU Fishing Di LautNatuna Utara – Mei 2021'. *Retrieved from https://oceanjusticeinitiative.org/2021/06/14/iuu-fishing-di-laut-natuna-utara-mei-2021/,* 14th June 2021.

15. ZIYADI, A. 'Kapal TK1482, Kapal Nelayan Dengan Standar Militer Milik Vietnam'. *Retrieved from https://militermeter.com/kapal-tk1482-kapal-nelayan-dengan-standar-militer-milik-vietnam/,* 8th August 2021.

16. AXE, D., 'Huge New Chinese Ships Are Made For Ramming', *Retrieved from https://www.forbes.com/sites/davidaxe/2021/01/15/huge-new-chinese-ships-are-made-for-ramming/?sh=1191d22f1c19,* 15th January 2021.

17. CHINA DAILY, 'Ocean patrol ship launched from Wuhan shipyard', *Retrieved from http://www.chinadaily.com.cn/a/202102/08/WS6020fae1a31024ad0baa81b4_3.html,* 2nd August 2021.

18. SHIP STRUCTURE COMMITTEE, 'U.S. Concrete Ships Built During World War II', *SURVEY OF EXPERIENCE USING REINFORCED CONCRETE IN FLOATING MARINE STRUCTURES*, 1984.

19. TZANETOS, A. *et al.*, 'A new metaheuristic method for optimization: Sonar Inspired Optimization', *Retrieved from http://dx.doi.org/10.1007/978-3-319-65172-9_35*, August 2017.

20. Anonym, 'Betonschiff Capella', *Retrieved from https://www.schiffsspotter.de/Museums_und_Traditionsschiffe/Betonschiff_Capella,* n.d.

21. DANTES, K. R. et al., 'PENDAHULUAN', *Translation of Composites Manufacturing and Testing,* 2017.

22. *Ibid.,* 'JENIS-JENIS KOMPOSIT'.

23. SHAHEEN, Y. B. et al., 'DEVELOPMENT OF LIGHT WEIGHT FERROCEMENT COMPOSITE PLATES', *Retrieved from https://www.researchgate.net/publication/338697760,* 20th January 2020.

24. RISMAWAN, et al., 'ANALISA KEKUATAN LENTUR BAHAN FERROCEMENT BERPENGUAT KAWAT ANYAM SEBAGAI BAHAN DASAR *MODULAR FLOATING PONTOON*', *Retrieved from https://ejournal3.undip.ac.id/index.php/naval/article/view/7121/6888,* October 2014.

25. RUANGARSITEK, 'Harga Semen Per Sak', *Retrieved from https://ruangarsitek.id/harga-semen-per-sak/,* 17th August 2021.

26. ARDININGTYAS, I., 'Harga Pasir Bangunan Per Kubik Terbaru 2021', *Retrieved from https://udhargabangunan.com/harga-pasir.html,* 19th August 2021.

27. Anonym, 'Harga Wire Mesh Semua Ukuran Terbaru September 2021', *Retrieved from https://hargaper.com/harga-besi-wiremesh.html,* n.d.

28. JAYASTEEL, 'Harga Wiremesh Per Meter persegi terbaru', *Retrieved from https://www.jayasteel.com/2017/12/harga-wiremesh-per-meter-persegi-terbaru.html?m=1,* n.d.

29. ASHAB, M. B., 'KN TanjungDatu 1101, Kapal Patroli Terbesar Berbiaya Rp208,2 M', *Retrieved from https://nasional.sindonews.com/berita/1274695/14/kn-tanjung-datu-1101-kapal-patroli-terbesar-berbiaya-rp2082-m,* 18th January 2018.

30. US NAVAL ACADEMY (USNA), 'RESISTANCE AND POWERING OF SHIPS', *Retrieved from https://www.usna.edu/NAOE/_files/documents/Courses/EN400/02.07%20Chapter%207.pdf,* n.d.

7. AUTHORS BIOGRAPHY

Pratondo Ario Seno Sudiro holds the current position of student at Republic of Indonesia Defense University (RIDU). He is responsible for researches about Ship Building Material, especially about Ferrocement. His previous experience includes Warfare Innovation Continuum Workshop 2021 'Hybrid Force 2045' by Naval Postgraduate School, National Maritime Seminar 2021 by SESKOAL, Focus Group Discussion by Faculty of Remote Sensing RIDU, and Online Workshop '3D Ship Modelling' by Mechanical Engineering UI.

I Nengah Putra Apriyanto holds the current position of lecture at Faculty of Defense Technology, RIDU. He has just been promoted as Head of Bureau of Academic, Student, and Cooperation, RIDU. He is responsible for researches about Electronic Warfare, Naval Operation, and System Engineering. His previous experience includes lecture at SESKOAL, STTAL, and Naval Academy. He also wrote some books about Maritime Strategy, Electronic Warfare, and National Defense. Before be on duty at RIDU, he was Deputy Commander of STTAL.

Jupriyanto Jupriyanto holds the current position of lecture at Faculty of Defense Technology, RIDU. He is responsible for Consultant of BAPPENAS and researches about Indonesia Defense Industry. His previous experience includes Consultant of BALITBANG KEMHAN, Technological Capability Development Expert (Early warning and maritime security enforcement technology system) for Grand Design Badan KeamananLaut (Bakamla RI), Technology Analysis Expert for Local Content and Offset program in Indonesia Ministry of Defense, and System and structure engineering expert for UAVs technology development in Defense Industry.

EFFECT OF RAMP DOOR MATERIAL ON THE WIRE ROPE FATIGUE LIFE AND WINCH POWER

I P S Asmara, F Hamzah, T Karyono, M T Wahyudi, and **S F Kumaratih,** Politeknik Perkapalan Negeri Surabaya, Indonesia

SUMMARY

The object of this research was the passenger deck ramp door of the ferry ro-ro ship. This research compared the material of the installed ramp door that uses ASTM A36 steel and the alternative design that uses ST52 steel. The differentchemical composition made the yield strength ofST52 steel was higher than ASTM A36 steel. Accordingly, the alternative design hadthinner plate than the existing design. The weight of the existing constructionwas 9.34 tons, while the alternative structure was 8.21 tons. Based on the construction weight, the diameter for the wire rope was 9 mm, with a 20% increase in the estimated lifetime from 10 years for the installed ramp door and 12 years for the alternative design. The power efficiency of the electro-hydraulic winch was increased by 15.2% from 33.91 kW for the installed ramp door and 28.76 kW for the new design.

NOMENCLATURE

C	Carbon
Mn	Manganese
P	Phosphorus
S	Sulphur
Si	Silicon
Cu	Copper
Mo	Molybdenum
Cr	Chromium
V	Vanadium
Nb	Niobium
Ni	Nickel

1. INTRODUCTION

The ASTM A36 steel is the most widely used for construction. One of the shipyards in East Java considers applying a higher-grade material for the structural component of the vessel to decrease the light deadweight. Recently, this shipyard uses ASTM A36 steel in most of the ship's parts. The shipyard has a contract with the supplier of material for a RORO-ship construction. The shipyard has experience in shipbuilding and using similar material for a long time, so there has never been a further analysis between the strength of the material and the costs incurred. Therefore, in this study, we are comparing the use of ASTM A36 steel and other construction materials.

The object of this research is the passenger deck ramp door of the 1500 GT ferry RORO-ship. This ramp door is built using ASTM A36 steel material. In this research, the comparison material that we are using is ST52 steel. Both of the materials are carbon steel with a C content of ≤0.25% for ASTM A36 steel [1] and 0.25% for ST52 steel [2]. The increase of C content of the material indicates the increasing yield strength and tensile strength [3-4]. This also applies to the amount of Mn content, both ultimate tensile strength and yield strength are increased with the increment of Mn content [5]. The amount of Mn content of the material is 0.85% for ASTM A36 steel [1] and 0.90% for ST52 steel [2]. It indicates that the yield strength value of ST52 steel is greater than ASTM A36 steel, so it is possible to make the alternative design for the ramp door using ST52 steel material with a thinner plate size so that the construction weight can be reduced.

The weight of the ramp door construction likely affects the diameter of the wire rope used for the drive system of the ramp door. The wire rope used on ramp door construction with less weight will have a higher life estimation and lower winch power requirement.

2. METHODS

In this research, we need to calculate the total weight of the ramp door construction. **Figures** 1 and 2 are the top view of the 2D construction drawing of the existing passenger deck ramp door.

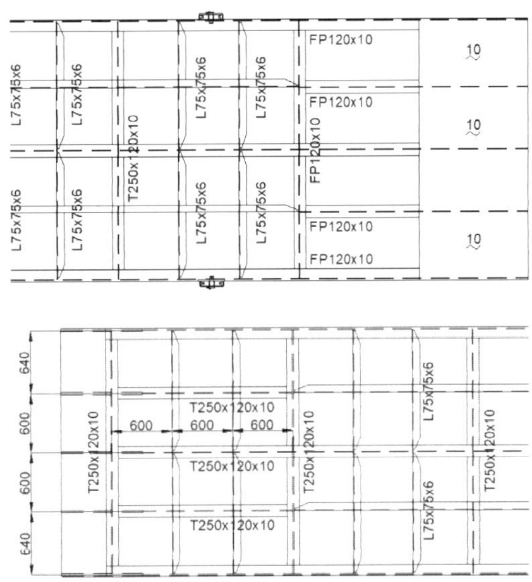

Figure 1: The top view of 2D construction drawing of the front part (above) and hinge (below)

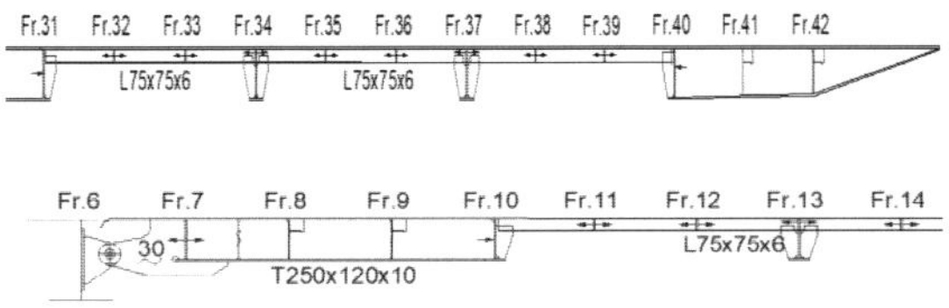

Figure 2: The side view of the 2D construction drawing of the front part (above) and hinge (below)

The material used for the existing ramp door was ASTM A36. **Table** 1 shows the chemical composition of the material.

Table 1: Chemical composition of ASTM A36 steel

Chemical Composition	Amount of Content (%)
Carbon *(C)*	0.25 max.
Manganese *(Mn)*	0.85
Phosphorus *(P)*	0.030
Sulphur *(S)*	0.030
Silicon *(Si)*	0.400 max.

Source:Mechanical Investigations of ASTM A36 Welded Steels with Stainless Steel Cladding

This study developed a new design of ramp door construction using ST52 steel material with the chemical composition shown in **Table** 2.

Table 2: Chemical composition of ST52 steel

Chemical Composition	Amount of Content (%)
Carbon *(C)*	0.25
Manganese *(Mn)*	0.90
Phosphorus *(P)*	0.035
Sulphur *(S)*	0.040
Silicon *(Si)*	0.38
Copper *(Cu)*	0.20
Molybdenum *(Mo)*	0.06
Chromium *(Cr)*	0.15
Vanadium *(V)*	0.008
Niobium *(Nb)*	0.008
Nickel*(Ni)*	0.20

Source:Microstructure and Mechanical Properties of Friction-stir Welded ST52 Steel Joints

Firstly, we were developing a 3D model of the ramp door construction drawing using Autodesk Fusion 360 according to the 2D construction shown in **Figures** 1 and 2. The second stage was to export the 3D model of the ramp door to Ansys Workbench with the IGES format. The loads of this ramp door are SUV cars that go to the passenger deck with a total weight per car was 2.5 tons. Since the vehicle is a moving load, the analysis considered the impact load factor [6] using **Equation** 1.

$$K = 1 + 20/(50+L) \tag{1}$$

Where:

K = Impact load factor

L = Construction length (m)

The value of K is 1.276. So, the load received by the ramp door is 3.190 tons. Figure 3 shows normal force on the tire of an SUV car is 0.798 tons.

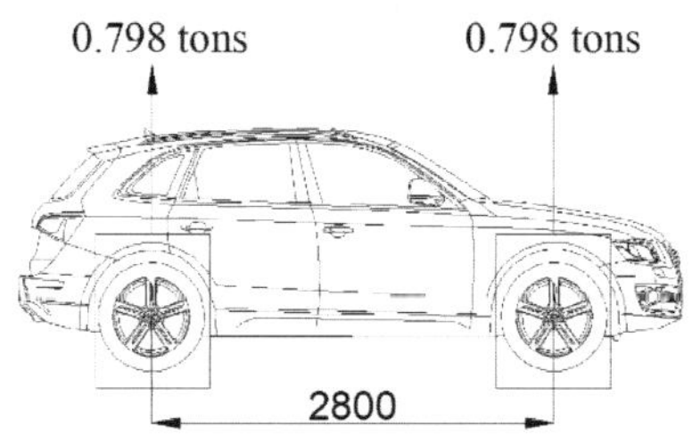

Figure 3: Load distribution on car tires

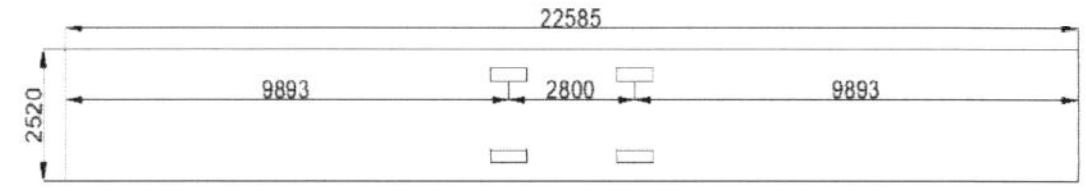

Figure 4: Load location on case 1

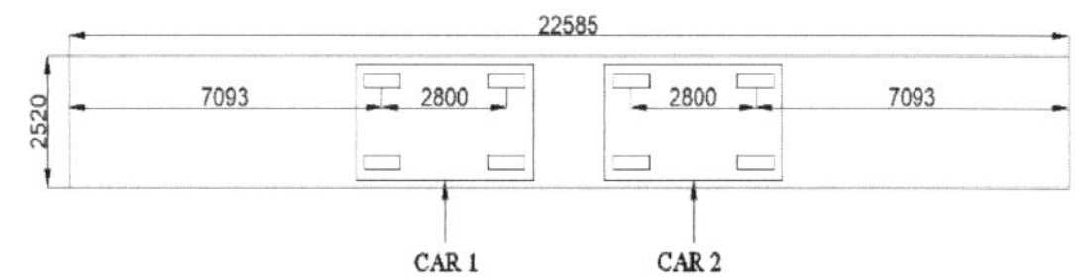

Figure 5: Load location on case 2

In this research, there will be two loading cases. Case 1 is a loading scenario for one car, and case 2 for two cars load. Figures 4 and 5 are the position of the load from the top view of the ramp door.The stress analysis considers the simple support condition. The boundary conditions are the hinge condition at the deck connection and roll condition at the other end part of the ramp door.

Based on the weight of the ramp door, we use Rudenko calculation [7] to determine the pulley design and wire rope selection of the winch system. **Equation** 2 shows the breaking load calculation of the wire rope.

$$P = S\,.\,K \tag{2}$$

Where:

P = breaking load of wire rope (tons)

S = load that carried by wire rope (tons)

K = safety factor

The lifetime of wire rope depends on the design of the pulley system. The design variables of the pulley system are rope diameter, rope material, pulley diameter, the number of bends, as well as working force, and the level of working conditions. Equation 3 shows the estimation of wire rope lifetime affected by the variables.

$$N = \frac{z}{\alpha\,.\,Z_2\,.\,\beta\,.\,\varphi}$$

Where:

N = lifetime estimation of wire rope

z	= wearingbends of wire rope
α	= cycle per month
Z_2	= constant dependingon pulley system
β	= number of bends on pulley system
φ	= constant (2.5)

The number of wearing bends depends on the working force, type of wire, and the ratio of rope-pulley diameters. All variables in the existing and new design of the winch system are the same, except the working force due to the less weight of the new design of the ramp door.

The winch uses an electro-hydraulic motor for its drive system. It controls the movement of the wire rope passing thru the pulley system. The power of the electro-hydraulic winch depends on the tensile force of the wire rope known as rated pull [8] in the kN unit and the lifting speed of the ramp door.**Equation** 4 shows that the required power for the winch depends on torsion and drum revolution.

$$P = 2 \pi n T \qquad (4)$$

Where:

P	= power of the hydraulic motor (kW)
T	= torque of hydraulic motor (kN.m)
n	= rotation of motor (rpm)

3. RESULT AND DISCUSSION

3.1 MAIN DIMENSION OF THE RAMP DOOR

Table 3 shows the dimension of the ramp door. The span was 22.585 m, the breadth was 2.52 m, and the height was 0.258 m using a fabricated 250 mm T beam, 75 mm L stiffener, and 8 mm plate. **Table** 4 provides the summary of structural weight.

Table 3:Main dimension of the ramp door

Dimension	Value	Unit
Length	22585	mm
Breadth	2520	mm
Height	258	mm
Slope	12	degree

Table 4: Weight of the ramp door

Material	Quantity	Density	Weight (kg)
Plate 8 mm	8 sheets	7.85	4521.60
Plate 10 mm	6 sheets	7.85	4239.00
Profile L	13.6 bars	7.85	576.50
Total Weight			9337.10

3.2 THE 3D MODEL OF THE RAMP DOOR

The 3D modelling of the ramp door as shown in **Figure** 3 was done based on the 2D construction and then exported to IGES format to be analysed on Ansys Workbench.

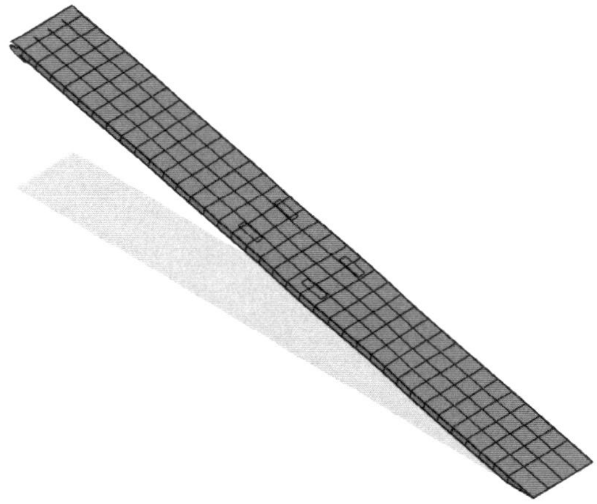

Figure 6: 3D model of ramp door

3.3 ENGINEERING DATA

The material used on the existing ramp door construction was steel ASTM A36. Table 4 shows the mechanical properties of the material. The properties were the input of engineering data for the stress and deformation analysis using Ansys Workbench.

Table 5:Mechanical properties of steel ASTM A36

Mechanical Properties	Value	Unit
Tensile yield strength	250	MPa
Ultimate strength	400-550	MPa
Elongation	23	%

Source:Mechanical Investigations of ASTM A36 Welded Steels with Stainless Steel Cladding

3.4 MESHING

The meshing process is dividing the components into smaller elements that connected by nodes. It aims to produce the accurate analysis results. The ideal meshing size for ramp door was 76 mm.

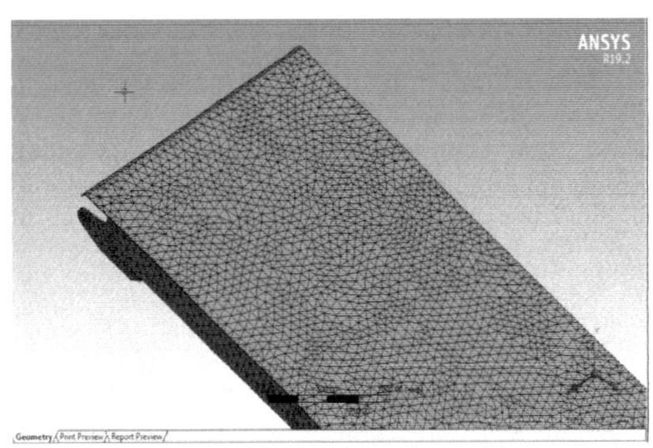

Figure 7: Meshing of ramp door model

3.5 LOAD AND SUPPORT

The model received four single forces shown in **Figure** 7. The single load was 7.83kN. The boundary conditions of the ramp door structure were hinge support at the higher deck connection and roll support at the other edge of the ramp door attaching to the lower deck of the vessel. **Figure** 7 presents the position of load and support for case 1 and **Figure** 8 for case 2.

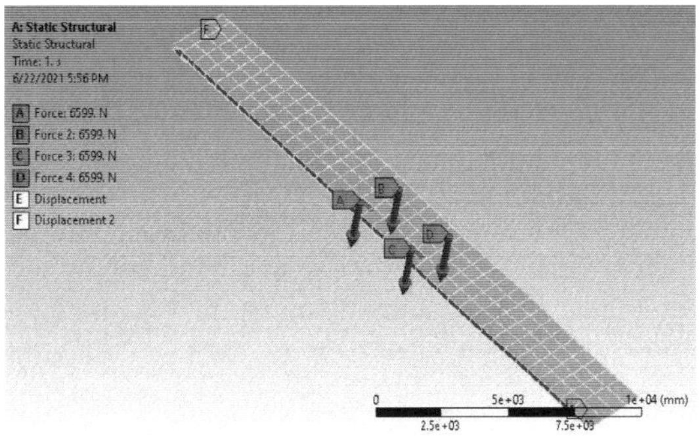

Figure 8: Load and support for case 1

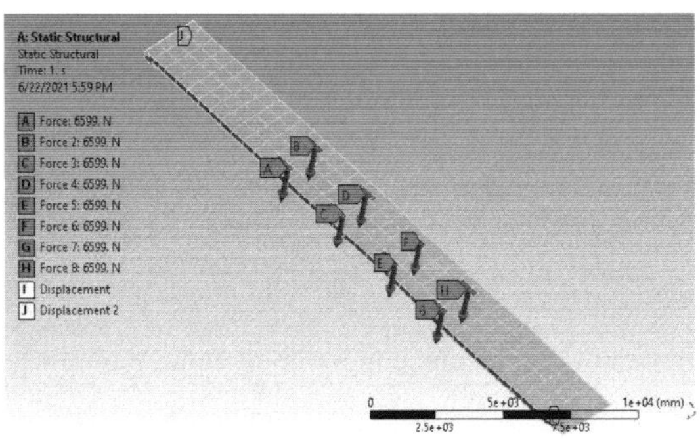

Figure 9: Load and support for case 2

3.6 STRESS AND DEFORMATION ANALYSIS

Figures 10 and 11 show the von-Misses stress of the ramp door structure for cases 1 and 2, respectively. **Table** 6 summarizes the maximum stress. The stress in case 2 was relatively higher than in case 1 by considering the yield strength of the respective material. The safety factor of case 2 was less than case 1 to optimize the efficiency of winch power.

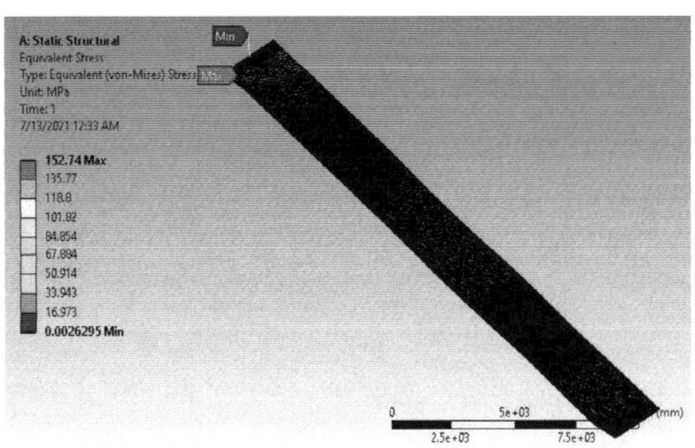

Figure 10:The von-Misses stress of case 1

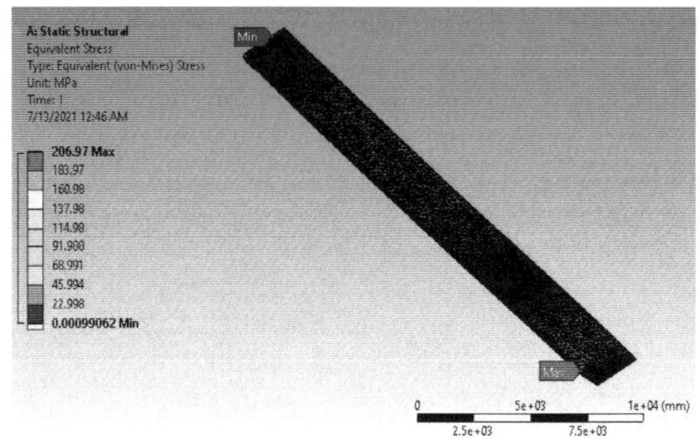

Figure 11:The von-Misses stress of case 2

Table 6: Maximum stress of the ramp door

Load Case	Max. Stress (MPa)	Safety Factor
Case 1	152.74	1.637
Case 2	206.97	1.208

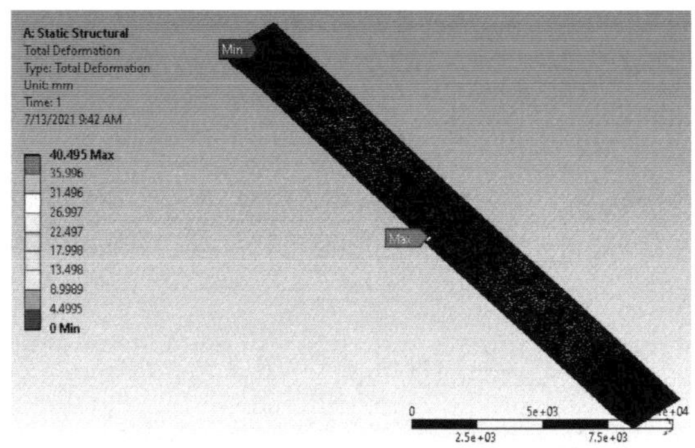

Figure 12:Deformation of Case 1

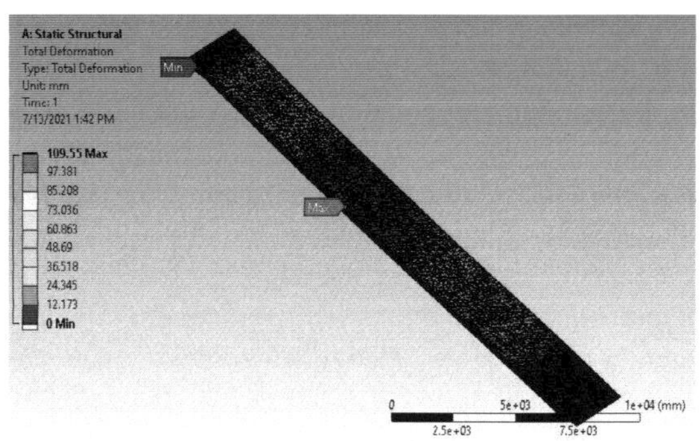

Figure 13:Deformation of Case 1

Table 7: Maximum deformation of the ramp door

Load Case	Max. Deformation (mm)
Case 1	40.50
Case 2	109.55

Figures 12 and 13 represent the deformation of case 1 and case 2, respectively. The location of maximum deformation was at the middle part of the ramp door. It was as same as the location of the maximum stress. The maximum deformation for case 2 washigher than the twice of deformation for case 1 as seen in **Table** 7.

3.7 CALCULATION VALIDATION

This study compared the stress and deformation analysis with the empirical equation of the bending moment theory using **Equation**5 [9].Table 8 shows the error percentage between manual calculation and software analysis results.

$$\sigma = \frac{M \cdot y}{I}$$

Where:

M = maximum moment (N.mm)

y = center weight of cross-section area

I = moment of inertia (mm^4)

Table 8: Error percentage between 2 calculations

Load Case	Error Percentage
Case 1	0.204%
Case 2	1.553%

3.8 ALTERNATIVE DESIGN

The alternative design of the ramp door were using steel ST52 material that has a higher value of yield strength than steel ASTM A36 of the existing ramp door. Data in Table 8 is the mechanical properties of steel ST52.

Table 9:Mechanical properties of steel ST52

Mechanical Properties	Value	Unit
Tensile yield strength	337	MPa
Ultimate strength	526	MPa
Elongation	35%	

Source: Microstructure and Mechanical Properties of Friction-stir Welded ST52 Steel Joints

The alternative design has the same main dimension as the installed ramp door, and here are the dimensions of ramp door stiffeners:

1. Profile T: 250 x 120 x 10 mm
2. Profile L: 75 x 75 x 6 mm
3. Deck plate : 22585 x 2520 x 6 mm

The difference from the existing ramp door was the thickness of the deck plate, the deck plate of the ones was 2 mm less than the installed ramp door. **Table** 10 shows the weight of the alternative ramp door.

Table 10: Weight of the ramp door

Material	Requirement	Density	Weight (kg)
Plate 6 mm	8 sheets	7.85	3391.2
Plate 10 mm	6 sheets	7.85	4239
Profile L	13.6 sticks	7.85	576.504
	Total Weight		8206.704

Figures 14 and 15 show the maximum von Misses stress of case 1 and case 2. **Figures** 16 and 17 are the output of deformation analysis for case 1 and case 2. Tables 11 and 12 are the summary of the stress and deformation analysis.

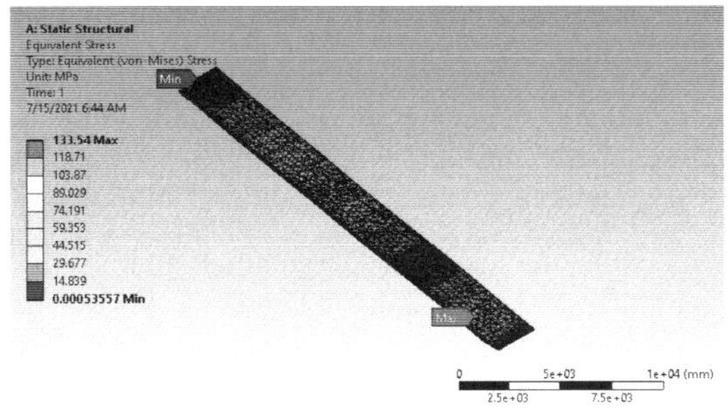

Figure 14: Ramp door stress (Case 1)

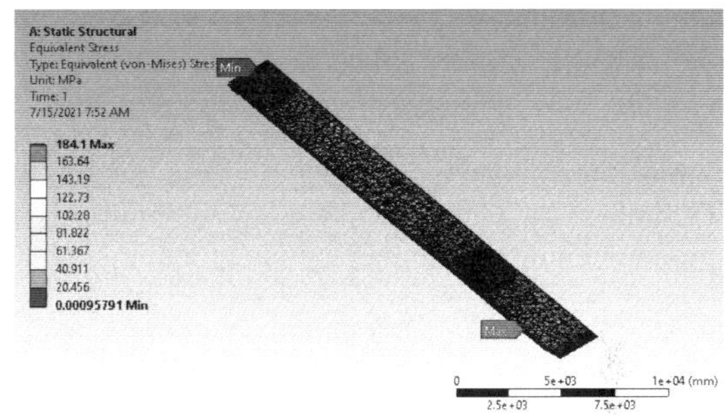

Figure 15: Ramp door stress (Case 2)

Table 11: Maximum stress of the ramp door

Load Case	Max. Stress (MPa)	Safety Factor
Case 1	133.54	2.583
Case 2	184.1	1.874

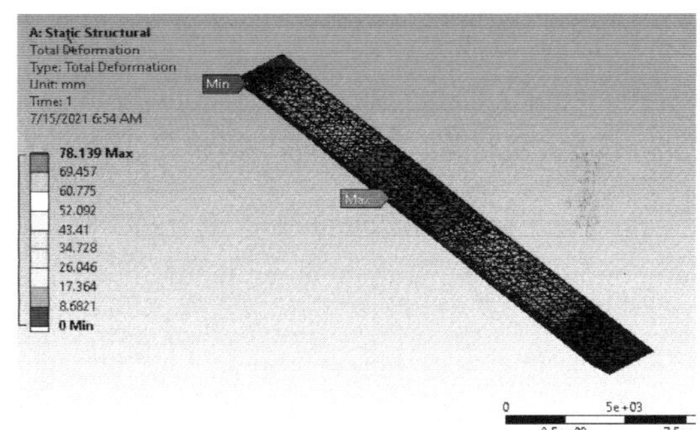

Figure 16: Ramp door deformation (Case 1)

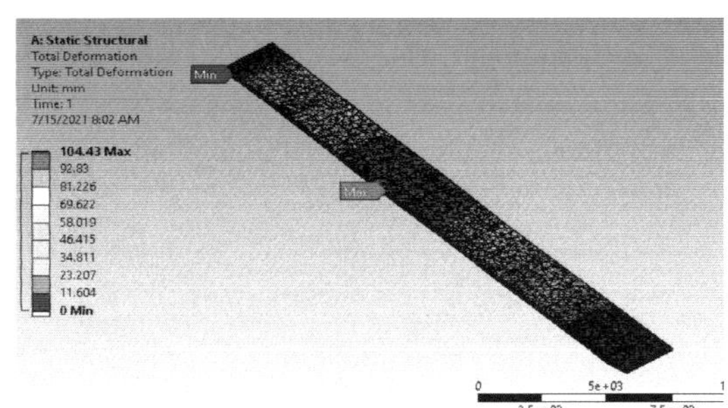

Figure 17: Ramp door deformation (Case 2)

Table 12: Maximum deformation of the ramp door

Load Case	Max. Deformation (mm)
Case 1	78.139
Case 2	104.43

3.9 TOTAL COST OF RAMP DOOR MATERIAL

According to the material list of the existing ramp door and alternative design, we could calculate the cost of ramp door material. **Table** 13 is the table of material costs for the installed ramp door. **Table** 14 shows material costs for the alternative ramp door. The tables show that the material costs were not significantly different. The material cost for the alternative design was only 2% higher than the existing one.

Table 13: Material cost of existing ramp door

Material	Requirement	Cost /unit	Cost
Plate 8 mm	8 sheets	Rp 7.350.000,-	Rp 58.800.000,-
Plate 10 mm	6 sheets	Rp 12.775.000,-	Rp 76.650.000,-
Profile L	13.6 sticks	Rp 604.000,-	Rp 8.456.000,-
	Total Cost		Rp 136.054.000,-

Table 14: Material cost of alternative ramp door

Material	Requirement	Cost /unit	Cost
Plate 6 mm	8 sheets	Rp 5.681.000,-	Rp 45.448.000,-
Plate 10 mm	6 sheets	Rp 14.130.000,-	Rp 84.780.000,-
Profile L	13.6 sticks	Rp 628.000,-	Rp 8.540.000,-
	Total Cost		Rp 138.768.000,-

3.10 WIRE ROPE AND WINCH POWER

The 6 x 37 IWRC wire rope was used for the ramp door according to the wire rope characteristic that is usually applied for crane, hoist, and lifting. We selected the 6 x 37 IWRC wire rope with a diameter of 9 mm that has a breaking load of 5.62 tons and breaking stress of 180 kg/mm2 for both ramp doors. **Table** 15 is the table of lifetime estimation of the wire rope for the existing and the new ramp door. The required winch power for the new design was 15.2% lower than the power required by the existing winch as seen in **Table** 16.

Table 15:Lifetime estimation of wire rope

Ramp door	Working Load	Lifetime Estimation
Existing	5.154 tons	10 years
New design	4.530 tons	12 years

Table 16: Calculation result for the winch requirement

Ramp door	Tensile Force of Winch	Winch Power
Existing	8.418 kN	33.912 kW
New design	7.406 kN	28.763 kW

4. CONCLUSION

The stress on case 1 of the existing ramp doors is accepted but not for case 2 because the safety factor is less than 1.67 by the yields strength of the material. The alternative design has a safety factor of more than 1.67 for both cases. The lifetime of wire rope is 10 years for the installed ramp door and 12 years for the alternative design. The material cost of the new design is 2% higher than the existing one but it requires 15.2% lower than the winch power of the existing one.

5. REFERENCES

1 Preedawipat, P., et al. 'Mechanical Investigations of ASTM A36 Welded Steels with Stainless Steel Cladding'. *MDPI Open Access Journal,* 2020.

2 Küçükömeroğlu, T., et al. 'Microstructure and Mechanical Properties of Friction-stir Welded ST52 Steel Joints'. *International Journal of Minerals, Metallurgy and Materials,* 2018.

3 Concepcióna, V L., Lorusso, H N., Svoboda, H G., 'Effect of carbon content on microstructure and mechanical properties of dual phase steels'. *International Congress of Science and Technology of Metallurgy and Materials, SAM – CONAMET,* 2013.

4 Calika, A., Duzgunb, A., C, Osman Sahin., Ucard, N., 'Effect of Carbon Content on the Mechanical Properties of MediumCarbon Steels'. *De Gruyter,* 2014.

5 Zhihao, Z., Yi, M., Jianzhong, C., 'Effect of Mn on Microstructures and Mechanical Properties of Al-Mg-Si-Cu-Cr-V Alloy'. *Connecting Repositories(CORE),* 2012.

6 Roehman, F., 'AnalisaPerencananJembatan Kali WulanDesaBungoKecamatanWedungKabupaten DemakuntukBangunanAtas'. *Engineering Journal UNISFAT,* 2009.

7 Rudenko, N., 'Mesin Pemindah Bahan'. *Erlangga,* 1992.

8 Prambudi, B., 'Analisa Teknik danEkonomiPerancangan Konstruksi Ramp door padaKapal Ro-Ro Barge'. *Repository ITS,* 2018.

9 K H, Sunggono., 'Teknik – Sipil'. *NOVA,* 1984.

10 Kumaratih, S F., 'Analisis Fatigue Life Struktur Ramp door Passenger Deck danTali Baja Kapal

Ferry Ro-Ro 1500 GT'. *Final Project PPNS*, 2021.

6. AUTHORS BIOGRAPHY

I P S Asmara holds the current position of lecturer at Politeknik Perkapalan Negeri Surabaya (PPNS). He is responsible for teaching the subject of ship strength. His previous teaching experience includes the subject of mechanics of material, finite element method and material handling equipment.

F Hamzah has been a lecturer at Marine Engineering Department of PPNS since 1987. He is responsible for teaching the subject of technical drawing and ship construction.

T Karyono is responsible for delivering the subject of ship structure at Shipbuilding Engineering Department of PPNS since 1995.

M T Wahyudi is also a lecturer at Shipbuilding Department of PPNS. He is responsible for teaching the subject of destructive and non-destructive material test

S F Kumaratih is a student at Design and Manufacture Engineering Study Program of PPNS. She has been learning about applied mechanics, design product, material handling and crane, theory and application of CAD/CAM, and production management.

ANALYSIS OF MIGRATION PROCESS FROM OHSAS 18001:2007 TOWARDS ISO 45001:2018 AT OHS SERVICE COMPANY AS SUPPORTING THE MARITIME COMPANIES

D R Widiana, S K Sari, and **I L Kusminah**, Politeknik Perkapalan Negeri Surabaya,Indonesia

SUMMARY

In 2018, the international standard of theOHSManagement Systemwas switched from OHSAS 18001:2007 to ISO 45001:2018. As one of the OHS service companies that havebeen implementing the OHSAS standard, it's necessary to migrate to ISO 45001to support its existence as a supporting industry for maritime companies. A study was conducted to determine the strategies that companies could use in the migration process.Gap Analysis and SWOT Analysis methodsare used to determine the migration strategy. This method combines several studies to formintegratedresearch with amore objectivemethodsstarting from measuring the level of company readiness to migrate using the Gap Analysis method, then proceeding with determining the migration strategy using the SWOT method, and ends with the determination of recommendations for improvement.Based on the results of the gap analysis, it's known that the company's readiness to migrate is 87.45%with 14 findings of improvement. A migration strategy was prepared using a SWOT analysis which results in 7 migration steps, one of the steps is to improve the findings.Finally, the company could migrate and carryout the ISO 45001:2018 standard.

1. INTRODUCTION

Being the most important optionof sea transportation, ships are still the main product of most maritime companies. In addition to being a product of a company, ships are also transportation to support companies in the maritime sector for shipping activities. Therefore, maintaining and ensuring the condition of ships is an obligation for maritime companies so they can operate safely. To ensure that every component of the ship is in a safe condition, periodic testing and inspection of the equipment must be carried out. These tests and inspection activities can be carried out by the company internally or using the assistance of external parties such as an OHS Service Company.

Assupport for maritime companies, one of the OHS Service Companies in Indonesia offers a variety of inspection and testing services regarding the maritime world, ranging from material testing (Positive Material Identification, Hardness Test, Wall Thickness Measurement, etc.), welding results testing (Radiography Test, Magnetic Test, Penetrant Test, etc.), Lifting Equipment Inspection, Electrical Equipment Inspection, Fire Protection Inspection, and several other inspections. In addition to focusing on providing services to other companies, to create a safe workplace, OHS Service Companies also implement an OHS Management System. OHSAS 18001:2007 is the standardreferencechosen by OHS Service Companies to implement OHS Management System to be able to compete in the international world.

The Occupational Health and Safety Assessment Series (OHSAS) 18001: 2007 is part of the OHS management system that was developed by the British Standard Institute (BSI) in 1999.Along with the times, ISO-based international standards such as ISO 9001 and ISO 14001 have developed until the emersion of high–levelstructures and several requirements that are more relevant to the current situation.Finally, with the agreement of various countries that are members of ISO agreed to issue an OHS management system standard ISO 45001: 2018. With the implementation of ISO 45001 in March 2018, companies with OHSAS 18001 standard must migrate to the ISO 45001 standard within 3years after the approval of theISO 45001 standard.

According to the International Accreditation Forum Mandatory Document (2018) in BSN (2019) regarding the requirements for migration of OHSAS 18001:2007 to ISO 45001:2018. The level of change required depends on the maturity and effectiveness of the existing management system and organizational structures also the practices.To carry out the migration process, a company needs a strategy so that the migration process has a clear goal.

Based on that background, research was conducted on the migration of OHSAS 18001: 2007 to ISO 45001: 2018. Before the migration process, a Gap Analysis was carried out between the OHSAS standard and the ISO standard to determine the company's readiness and become a consideration in determining the strategy to use inthe migration to ISO 45001: 2018. The tool that will be used in the Gap Analysis method is a checklist. The checklist will make it easier for researchers to identify what requirements have been implemented and which are not implemented or owned by the company, so that recommendations for improvement can be immediatelygiven.

The migration process to ISO 45001 is expected to prevent work accidents. Based on the results of research conducted by Kleinova and Szaryszova (2014), the ISO standard is better at preventing workincidents and accidents than the OHSAS 18001 standard. The SWOT method that will be used to determine the migration strategy in this study isthe extend of research methods that have been conducted by Pesce (2018), where this method is used in the process of evaluating the effectiveness of ISO 14001.In addition to developing and updating methods from previous research, this research also combines several studies to form integrated research with more objectivemethods, starting from measuring the level of company readiness to migrate using the Gap Analysis method, then proceeding with determining the migration strategy using the SWOT method, and ends with the determination of recommendations for improvement based on the results of the Gap Analysis.Finally, it is expected that the company will be able to meet all the requirements of ISO 45001 and be able to migrate in time.

2. LITERATURE REVIEW

2.1 OHSAS 18001 AND ISO 45001

OHSAS 18001 is anOHS management standard that was first developed by the British Standards Institution (BSI) in 1999, which was refined in 2007. OHSAS 18001 is similar to ISO performance standards.The part numbering system, requirements for documentation, training, leadership involvement, written in a formal language,and form of structure is similar to ISO standards.As the case with ISO standards, OHSAS 18001 is also based on ideas of worker involvement, safety culture, and continuous improvement.
According to Dentch (2018) in BSN (2019), it is known that there are several differences from the OHSAS 18001 standard with ISO 45001, including:

1 ISO 45001 adopts a High-LevelStructure consists of 10 clauses that aim to facilitate integration with other management systems.
2 In the High-Level Structure there is a good leadership role. ISO 45001 demands the integration of OHS aspects into the company's management system to be able to encourage Top Management to have a strong leadership role towards OHS Management Systems. While in OHSAS 18001,OHS responsibility is delegated to the OHS manager.
3 In ISO 45001, organizations are also required to identify and control other risks and opportunities related to the organization's sustainability. While OHSAS 18001 only considers risks.
4 ISO 45001 emphasizes and encourages the participation and contribution of workers in developing theOHS Management System. Meanwhile, OHSAS 18001 does not clearly explain the participation of workers in theOHS Management System.
5 ISO 45001 requires suppliers of goods and services to meet OHS requirementsbecause they are one of thefactors that affectthe performance of OHS organizations. While OHSAS 18001 did not specifically mention the claim.
6 ISO 45001 is dynamic in all clauses and uses a process approach. Meanwhile, OHSAS 18001 is based on procedures so it is not dynamic.

2.2 MIGRATION PROCESS

ISO 45001 is an international standard on occupational safety and health management systems that replaces the OHSAS 18001 standard. According to the International Accreditation Forum Mandatory Document (2018) in BSN (2019) regarding the requirements for migration of OHSAS 18001:2007 to ISO 45001:2018, the level ofthe changes requireddependson the maturity and effectiveness of the existing management system and organizational structures and practices.Therefore, an impact analysis or gap assessment is highly recommended to identify real-time and resource impacts. For organizations that havebeen implementing OHSAS 18001:2007 and wantto migrate to ISO 45001:2018 it is recommended to follow these steps:

1 Get a copy of ISO 45001:2018.
2 Identification of differences in OHS Management Systems that need to be addressed to meet new requirements.
3 Develop an implementation plan.
4 Ensure the suitability of any required competencies and create awareness for all parties who influence the effectiveness of OHS Management Systems.
5 Upgrade existing OHS Management Systems to meet new requirements and provide verification for its effectiveness, and
6 It is necessary to discuss with the certification institution for arrangements migration if there is.

2.3 GAP ANALYSIS

According to the IT Infrastructure Library (ITIL) in Yahya.R, et al (2018), gap analysis is an activity that compares two types of data and identifies the differences.Typically, gap analysis is used to compare two types of requirements.Gap analysis is often used on one type of topic or category, making gap analysis efficient in identifying which sectors or aspects need improvement.
Gap analysis is effective because the checklists are structured and appropriate to the topic.The checklist includes all requirements and it is hierarchically created in its review.The checklist will include frequently asked questions and provide an overview of the topic or category being evaluated.The questions in the checklist are comprehensive and detailed.The steps used to make a gap analysis are as follows:

2.3 (a) Score Determination

The scoring system used in the Gap Analysis is shown in Table 2. 1 :

Table 2.1 Gap Analysis Score

Score	Definition
1	If the organization or company does not understand what is needed and does not do it
2	If the organization or company understands the importance of the activity but does not do it
3	If the organization or company has the document but has not implemented it, or if the organization or company has applied it but does not record it
4	If the organization or company is involved in an activity

Score	Definition
	but it is not consistent
5	If the organization or company is running well (activities are carried out consistently)

(Yahya, 2018)

According to Yahya (2018) in Table 2.1, the Gap Analysis Score was divided into 5 categories, for organizations that do not understand the needs, organizations that already understand but does not implement, organizations that already have documents but have not implemented or have implemented but have not documented, already implemented but not yet consistent, and the last one has been running consistently. The greater the score obtained by the organization, the better the implementation of a policy.

2.3 (b) Checklist Assessment

The checklist assessment was carried out by the respondents followingthe current organizational conditions.The selected respondents have adequate competence.The assessment is carried out using a scoring system according to Table 2. 1.

2.3 (c) Gap Assessment

Gap Assessment aims to find out how big the gap is in the company.The percentage value is obtained by adding up the scores of each variable and divided by the maximum value in that variable.The smaller the gap, the better it is.To measure readiness, a percentage value is obtained following the company's readiness in its implementation as shown in Table 2.2 :

Table 2.2 Gap Analysis Range

Percentage	Description
75% - 100%	Organization ready to complete ISO 45001
50% - 74%	Organizations still have to clean up for the preparation of ISO 45001
1% - 49%	OHS Management systems Organization urges improvement because it is far from ISO 45001 requirements

(Yahya, 2018)

According to Yahya (2018) in Table 2.2, the Gap Analysis Range is divided into 3 categories, for organizations that are ready to complete ISO 45001, organizations that are still preparing for ISO 45001, and organizations that are still far from the requirements of ISO 45001.

2.4 SWOT ANALYSIS

SWOT analysis is a strategic planning method used to evaluate the strengths, weaknesses, opportunities, and threats in a project or business speculation.This process involves determining the specific objectives of the speculated business or project and identifying the internal and external factors that support and which do not achieve these goals.The following is a diagram of the SWOT analysis in Figure 2.1:

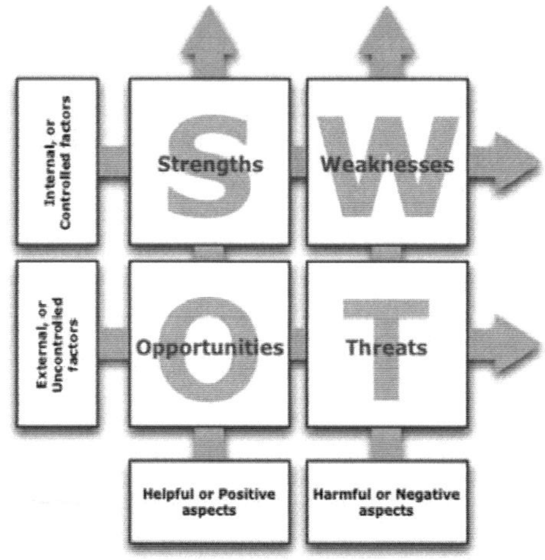

Figure 2.1 SWOT Analysis
(source:Sarsby, 2012)

Based on the SWOT diagram in Figure 2.1, there are 4 components in the SWOT analysis consisting of:

1. Strengths; internal factors that support the company in achieving its goals. Supporting factors can be in the form of resources, expertise, or other advantages that may be obtained due to financial resources, image, advantages in the market, and good relations between buyers and suppliers.
2. Weaknesses; internal factors that hold the company from achieving its goals. Inhibiting factors can be in the form of incomplete facilities, lack of financial resources, ability to manage, marketing expertise, and company image.
3. Opportunity; external factors that support the company from achieving its goals. External factors that support the achievement of goals can be in the form of policy changes, changes in competition, changes in technology and developments in supplier, and buyer relationships.
4. Threats; external factors that hold the company in achieving its goals. External factors that hold the company can be in the form of the entry of new competitors, slow market growth, increased bargaining power of major suppliers and buyers, changes in technology, and new policies.

3. RESEARCH METHODOLOGY

In solving the problems found in this research, a research method is needed. The method used in this research is the gap analysis method. This method was done by comparing several references in a checklist. The reference standards used in the preparation of this checklist are ISO 45001: 2018, OHSAS 18001: 2007, and implementation in companies. This checklist is compiled in the form of a table with a gap comparison model, where the gap in question is a comparison between ISO and OHSAS standard requirements with the implementation carried out by the company. The preparation of this gap analysis checklist resulted in 201 points of requirements written in interrogative sentences. These requirement points are grouped according to 7 clauses of requirements in the ISO standard, where each clause consists of several sub-clauses. In addition to containing requirements points, several other things are included in this checklistthat is document requirements, availability of company documents, results of fulfillmentof requirements, implementation assessment, and findings/observations during data collection.

To perform this analysis, it is necessary to have OHS management system documents made by the company, some other data obtained from interviews with several employees of the company concerned underthe gap analysis checklist, as well as direct observations at the company. The results obtained from data collected are then given a scored assessment according to the gap analysis score and gap analysis range. The data obtained is used to evaluate the gaps and findings that exist in the requirements of ISO 45001: 2018. The existing findings will be corrected according to the time available before the external audit of the migration process.After the gap was analysed, to determine the migration strategy, a SWOT analysis was carried out based on the results of the gap analysis.

4. DATA PROCESSING

4.1 GAP ANALYSIS

Based on data processing that was carried out using a gap analysis checklist, the results of the company's readiness to migrate were 87.45% with 12,55% gap score.The percentage level of readiness for migration to ISO 45001 is smaller than the level of implementation of OHSAS 18001 in companies which is at a value of 92.20%. This shows that there are differences from the requirements of the ISO and OHSAS standards, where ISO requirements are more complex than OHSAS requirements according to the literature review in sub-chapter 2.1. The percentage of migration readiness level is taken from the average readiness value of each clause, where the readiness value of each clause is also obtained from the average readiness value of each sub-clause. Then the percentage of assessment in each sub-clause is obtained from the comparison of the total value of readiness with the total maximum value in each of the sub-clauses. The percentage of migration readiness level in each sub-clause of ISO 45001 isshown in Table 4.1 :

Table 4.1 Migration Readiness Percentage

Sub - Clauses	Migration Readiness Percentage
4.1	100,00%
4.2	100,00%
4.3	91,67%
.4.4	80,00%
5.1	87,30%

Table 4.1 Migration Readiness Percentage

Sub - Clauses	Migration Readiness Percentage
5.2	89,47%
5.3	90,48%
5.4	80,00%
6.1	89,21%
6.2	86,57%
7.1	100,00%
7.2	92,00%
7.3	86,67%
7.4	86,67%
7.5	79,63%
8.1	85,07%
8.2	89,74%
9.1	86,89%
9.2	90,32%
9.3	83,33%
10.1	80,00%
10.2	81,25%
10.3	82,61%

Based on the data in Table 4.1, it can be seen that from 23 sub–clauses of ISO 45001 requirements consisting of sub-clauses 4.1 to sub-clause 10.3. There are 20 sub-clauses that have findings/observations and 3 other sub-clauses have fulfilled the 100% requirement.The findings found from the results of the gap analysis checklist are summarized in Table 4.2 :

Table 4.2 Gap Analysis Findings

No	Findings / Observation
1	Not consistent in meeting the requirements of ISO 45001
2	Not all information is documented
3	Occupational Disease Aspects have not been included in the OHS Policy
4	Risk assessment identified each job, should be identified for each potential hazard
5	Hazard identification does not consider leadership & cultural factors in the organization
6	HIRADC is not updated regularly
7	The company has not planned a way to evaluate the effectiveness of actions to address risks
8	The target of the Safety Program is not routinely implemented
9	The company does not identify and control documented information of external origin
10	There is no Assembly Point
11	The employee fatigue level evaluation form has not been adjusted in the scope of the company
12	There are findings of work environment measurement results (lighting is less than minimum value)
13	The company has not determined the frequency and method for evaluating compliance
14	There is no Occupational Disease reporting scheme

Based on the data presented in Table 4.2, it can be seen that from 12,55% gap score of the company's readiness to migrate there are 14 types of findings that cause imperfections in the level of readiness for migration to ISO 45001, starting from the company's consistency in meeting the requirements in the absence of an occupational disease reporting scheme.

4.2 SWOT ANALYSIS

ASWOT analysis consists of four components in the analysis consisting of strength, weakness, opportunity, and threat. This component of the SWOT analysis is obtained from the results of the previous gap analysis checklist and also observations at the company which will become input data in the SWOT matrix.The following is a SWOT matrix that will produce a strategy in the migration process to ISO 45001 which is shown in Table 4.3 :

Table 4.3 Migration Process SWOT Matrix

	S ***Strength***	**W** ***Weakness***
	1. Top management is fully committed 2. The company has implemented OHSAS 18001 standard 3. There is a specific person responsible for the migration process	1. There are 14 types of findings that do not meet the requirements of ISO 45001 2. Not all employees understand themigration process
	S – O Strategy	**W – O Strategy**
O ***Opportunity*** 1. The external audit process is carried out online 2. The external auditimplementation schedule is postponed	1. Company leaders are committed to the migration of the OHSAS 18001 standard to ISO 45001 (S1,O1,O2) 2. Document all ISO 45001 requirements is in soft-file form(S1, S2, S3, O1) 3. Prepare and ensure ISO 45001 requirements for the external audit process to the maximum extent possible(S1, S2, S3, O2)	1. Fix findings to meet ISO 45001 requirements during external audit preparation time(W1, O2) 2. Provide socialization/briefing to all employees regarding ongoing activities (especially representatives employees involved in external audits)(W2, O1)
		W – T Strategy
T ***Threat*** The migration process is limited in time by the external audit fromthe certification institute	**S – T Strategy** Split the task for the preparation of ISO 45001 requirements so that it can be completed on time(S1, S2, S3, T1)	1. Split the task for the preparation of ISO 45001 requirements so that it can be completed on time(W1, T1) 2. Maximizing employee participation so that targets can be met andthe employees have a better understandingof the activities(W2, T1)

Based on Table 4.3, there are four our types of strategies can be used for the migration process to ISO 45001, namely the S – O, S – T, W – O, and W – T strategies. This strategy was obtained by taking advantage of strengths and opportunities (S and O) to overcome or reduce weaknesses and threats (W – T).

4.3 MIGRATION STRATEGY

The steps/company migration strategy based on the results of the SWOT analysis is as follows:

1 Company leaders are committed to the migration of the OHSAS 18001 standard to ISO 45001
2 Split the task for the preparation of requirements of ISO 45001
3 Maximizing employee participation so that the targets can be met and employees more understand the migration process
4 Improve findings to meet ISO 45001 requirements
5 Documenting all ISO 45001 requirements in the soft-file form
6 Prepare and ensure all ISO 45001 requirements are available during the external audit process
7 Provide socialization/briefing to all employees regarding ongoing activities (especially representatives thatinvolved in the external audit process)

4.4 RECOMMENDATIONS

One of the steps in the ISO 45001 migration strategy is to improve the findings of the ISO requirements from the gap analysis result. In Table 4.2 it is known that there are 14 findings that require improvement. Therefore, before migrating, it is necessary to make improvements to the existing findingsfollowingsteps of the ISO 45001 migration strategy. Recommendations for improvements to the findings/observations that can be made in the company are shown in Table 4.4:

Table 4.4 Recommendations

No	Findings / Observations	Recommendations
1	Not consistent in meeting the requirements of ISO 45001	Improving the implementation of OHS culture in the company through: ➢ Periodic safety awareness socialization ➢ Safety campaign ➢ Periodic rewards for employees with the best performance ➢ Safety induction video screening in the company lobby
2	Not all information is documented	Improve the application of information documentation through: ➢ Policy for making reports on each activity ➢ Creation of an online documentation system
3	Occupational Disease Aspects have not been included in the OHS Policy	Revise the OHS Policy by adding aspects of Occupational Diseases
4	Risk assessment identified each job, should be identified for each potential hazard	Revise the hazard identification plan from the identification of each job to the identification of each potential hazard

Table 4.4 Recommendations

No	Findings / Observations	Recommendations
5	Hazard identification does not consider leadership &cultural factors in the organization	Revise the OHS hazard identification manual by considering leadership and cultural factors in the company
6	HIRADC is not updated	Updating HIRADC

No	Findings / Observations	Recommendations
	regularly	
7	The company has not planned a way to evaluate the effectiveness of actions to address risks	Plan ways to evaluate the effectiveness of actions to address risks
8	The target of the Safety Program is not routinely implemented	Making reminders for every safety program that will be implemented
9	The company does not identify and control documented information of external origin	Identify and control documented information of external origin
10	No Assembly Point	Designing assembly point
11	The employee fatigue level evaluation form has not been adjusted with the scope of the company	Revise the employee fatigue level evaluation form according to the scope of the company
12	There are findings of work environment measurement results (thelighting is less than minimum value)	Perform lighting repairs in the workplace by changing lamps to meet lighting minimum limit A work environment measurement report (lighting) is also made after inspection both at the time of the initial measurement to the measurement report after repair In addition to the measurement report, a follow-up report is also made as evidence that repairs have been made in the form of replacing lamps
13	The company has not determined the frequency and method for evaluating compliance	Arranging methods to evaluate compliance
14	There is no Occupational Disease reporting scheme	Creating an Occupational Disease reporting scheme

Based on Table 4.4 which discusses recommendations for improvement of findings in the company, it is known that of the 14 findings, there are 10 findings with recommendations that can be implemented directly through improvement design or document creation, while for the other 4 findings the recommendations that can be given are limited to suggestions in the future due to the need for periodic reviews for improvement.

5. CONCLUSION

Based on the results of data processing and analysis that has been carried out, it can be concluded that the level of company readiness to migrate to ISO 45001 is 87.45% with 14 findings of improvement.With this level of readiness, companies can migrate to ISO 45001 with a 7-step migrationstrategy, one of them is to improve existing findings.Recommendations were given consistingof direct improvements to 10 findings of requirements and suggestions for other findings that require continuous monitoring. Finally, the company could migrate and implementthe ISO 45001:2018 standardand could maintain its existence as an OHS service company that supports maritime companies that prioritize safety through the implementation of the OHS Management System.

6. REFERENCES

1. Department of Occupational Safety and Health, 'Guidelines for Hazard Identification, Risk Assessment and Risk Control (HIRARC)', *Ministry of Human Resources: Malaysia*, 2008.

2. ISO 45001, 'Occupational Health and Safety Management Systems – Requirements with Guidance for use', *ISO copyright office*, 2018.

3. Kleinova, R and Szaryszova, P., 'The New Health and Safety Standard ISO 45001:2016 and its Planned Changes', *International Journal of Interdisciplinarity in Theory and Practice ITPB – NR : 3,* 2014

4. OHSAS 18001:2007, 'Occupational Health and Safety Management Systems – Requirements', *OHSAS Project Group*, 2007.

5. OHSAS 18002:2008, 'Occupational Health and Safety Management Systems – Guidelines for the Implementation of OHSAS 18001 : 2007', *OHSAS Project Group*, 2008.

6. Pesce, M., et al, 'SWOT Analysis of the Application of International Standard ISO 14001 in the Chinese Context. A Case Study of Guangdong Province', *Sustainability*, 2018

7. Sarsby, A, 'A Useful Guide to SWOT Analysis', *PansophixOnline: Nottingham*, 2012.

8. Yahya, R, et.al, 'Analysis of OHSAS 18001:2007 standard renewal towards ISO 45001:2018 at PT. Power Plant Indonesia by Using Gap Analysis method', *EDP Sciences*, 2018.

7. AUTHORS BIOGRAPHY

Dika Rahayu Widiana, S.ST., MT., Ph.D, holds the current position of lecture at Politeknik Perkapalan Negeri Surabaya (Shipbuilding Institute of Polytechnic Surabaya). She is responsible for analysing the data. Her previous experience includes being a lecture at the occupational health and safety engineering department.

Silvy Kurnia Sari, holds the current position of student at Politeknik Perkapalan Negeri Surabaya (Shipbuilding Institute of Polytechnic Surabaya). She is responsible for taking the data. Her previous experience includes attending on the job training at one of occupational health and safety service company.

Imah Luluk Kusminah, ST., MT, holds the current position of lecture at Politeknik Perkapalan Negeri Surabaya (Shipbuilding Institute of Polytechnic Surabaya). She is responsible for writing the manuscript. Her previous experience includes being a lecture at the occupational health and safety engineering department.

EVALUATION FOR BALLAST TANK DRAINAGE BY HYDRAULICS EXPERIMENT

T Watanabe, Department of Naval Architecture and Ocean, Graduate School of Engineering, Kyushu University, Japan
T Shinoda, Department of Marine Systems Engineering, Kyushu University, Japan
G Liu, Department of Urban and Environmental Engineering, Graduate School of Engineering, Kyushu University, Japan
H Obata, Outfitting Design Department, Namura Shipbuilding Co., Ltd., Japan
T Nakamori, Hull Design Department, Namura Shipbuilding Co., Ltd., Japan
K Kuroki, Hull Design Department, Hull Design Section, Namura Shipbuilding Co., Ltd., Japan

SUMMARY

The quality of the ballast water drainage system during the cargo handling process of ships is an important issue for those involved in marine transportation such as cargo owners and shipping companies as it is significantly related to their revenue shipping. An excess charge is incurred if moored beyond the scheduled time due to the loading time, and the deadweight tonnage superfluity residual water reduces the deadweight tonnage. To minimize the loss of their transportation, this paper addressed two problems: shortening the ballast drainage time and minimizing the amount of residual water in the ballast tank. The objective of this study was to construct an optimal ballast drainage system to solve these problems. In this paper, hydraulic experiments modelling the drainage of the ballast tank were conducted to evaluate the drainage system's design policy and reveal the optimum design condition of a ballast tank.

1. INTRODUCTION

Vessels, especially those carrying specific cargo, such as crude oil tankers, may sail without cargo. In such a situation, the ballast water system maintains the draft by taking in seawater as a weight, lowering the centre of gravity and increasing stability. As a result, the system provides sufficient seaworthiness for vessels in trim, stability, strength, and propulsion efficiency even when empty.

Currently, most of the research on ballast water focuses on ballast water filtration systems to prevent environmental pollution, and there are few studies on the drainage efficiency of ballast water. Improving the drainage efficiency of ballast water shortens the port's loading time and reduces the loss of deadweight. Since these effects bring significant benefits to ship owners and shipping companies, they contribute to increasing the added value of ships.

Structural frames such as side girders, longitudinal frames, and transverse frames with drain holes and lightening holes separate the ballast tank, and this complicated structure hinders drainage. Therefore, in the hull equipment design, various measures have been taken to improve drainages, such as increasing the drain hole area in the ballast tank and changing the position of the pump bell mouth and the pumping method. However, this problem is a trade-off between hull strength and construction cost, and there is currently no optimal design method.

This study aims to propose the optimum design method of the ballast tank by hydraulic model test and numerical fluid calculation and to develop an experimentally verified calculation model to evaluate the drainage efficiency.In this paper, hydraulic model tests were conducted to verify the effects of design conditions such as pump flow rate, bellmouth position, and drain hole size on flooding efficiency. A miniature model of a ballast tank with a scale of 1:10 was created and tested on a ballast tank of a typical bulk carrier. This experiment is the most elemental content of this study to determine the basic design principles for the ballast tank design.

2. SCALED EXPERIMENT MODEL

Figure 1 shows the schematic diagram of the 4×4ballast tank model. Define the direction fromthe bow to the stern as the positive direction of the X-axis and the direction fromthe centre line to the port side as the positive direction of the Y-axis. In this paper, the area enclosed by the transverse and longitudinal frames in the double bottom framing system is called a compartment, and the area composed of multiple adjacent compartments divided by longitudinal is called a large compartment. Therefore, the arrangement of the compartments can be expressed as M x N (M is the amount in the X direction, N is the amount in the Y direction). Moreover, the location of a compartment can be expressed as (m-n) (m is the position in the X direction, n is the position in the Y direction).

As shown inFigure 2. the hydraulic experimental equipment in this study consistsof a ballast tank model, a ballast pump system, and an ultrasonic sensor.

The central part of each frame is removable and has a 5mm deep mortise to be replaced with a different-sized drain hole. The design of the 1/10 scale model is according to the ballast tank of a typical bulk carrier.

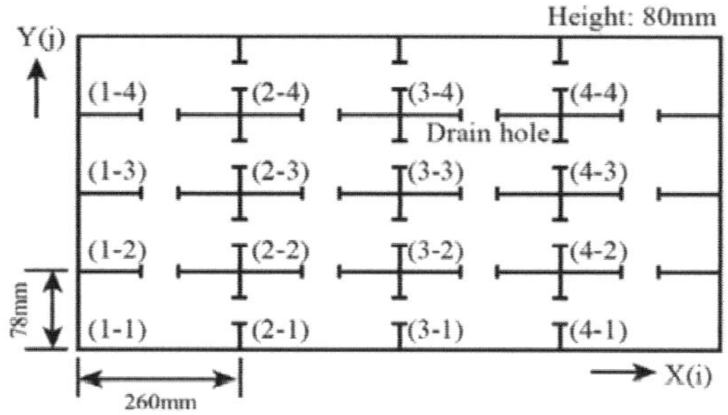

Figure 1: Schematic diagram of ballast tank model

The model is made of acrylic plates, and methylene dichloride is used to bond them. This chemical substance melts the surface of the acrylic plate and solidifies it again to bond it, and it can eliminate leaks. The inner frame is 2 mm thick, and the outer and bottom plates are 5 mm or 10 mm to reduce deflection. Auxiliary structural materials

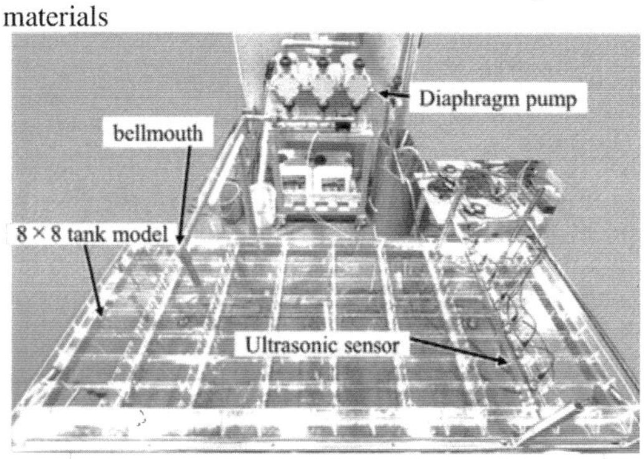

Figure 2: Experiment devices

that do not affect the result of the experiment are also used to suppress deformation. The size of a compartment is 260 mm x 77 mm (length in X direction and length in the Y direction),the height of the longitudinal frame is 30 mm, and the height of the transverse is 80 mm.It should be noted that the "1:10 scale "refers to the scaling in geometric length. The length and width are 1/10 of the full-scale ballast tank, and the height of the longitudinal frame is also 1/10 of the full size. There are two patterns of experimental models, a 4x4 compartment ballast tank model and an 8x8 compartment ballast tank model. Some longitudinal frames of the 8x8 ballast tank model can be changed in height by inserting a plate, allowing experiments with fewer compartments than 8x8 compartments. Since the transverse frame is usually a closed structure in a double bottom, only a suitable height is selected. The design of the experimental ballast tank model does not introduce the consideration of non-dimensional parameters, but this could be an extension in future research. In addition, to facilitate model making and numerical simulation, the design omits ancillary structures such as stiffeners and panels of longitudinal frames and simplifies circular holes to square.The influence of these structural changes will also be explored in future research.

The ballast water tank referredtoin this study has 10×18 compartments and three side girders, but in this paper, this structure was simplified to examine each of the factors that are expected to affect drainage efficiency. Therefore, we conducted experiments using a 4x4 tank and clarified how changes in the bell mouth installation compartment, pump flow rate, and drain hole height and width affect drainage.Adjusters are joined to every six legs of the model abutment, which allows trim adjustment. Since the acrylic model is deformed by self-weight and external factors such as moisture, it is tough to eliminate the influence of trim in the experiment. However, with a general analogue spirit level and a precision level (LEVELNIC DL-m1) the trim is adjusted in all compartments to ±1mm/m.

A diaphragm pump (TACMINA, FXD-FXW-8) is used as a ballast pump system composed of three reciprocating diaphragm pumps with different phases to reduce the pump pulsation so that the suction can be stable. Also, for the convenience of model making and numerical simulation, bell-mouth is simplified as a long straight pipe with an outer diameter of 32 mm and an inner diameter of 25 mm, made of vinyl chloride. It is inserted from the top of the tank and placed with the lower edge at the height of 4 mm from the bottom with a small spacer which has less effect on drainage.

Ultrasonic sensors (KEYENCE, FW-02) are used to measure the change over time in the water level, and distance measurement is based on the principle of ultrasonic reflection. The measuring range is 50~200 mm, and the response speed is 250 milliseconds.

3. DETAIL OF HYDRAULICS EXPERIMENT

The change of capacity of the pump can experiment with the pumping rate to adjust the flow rate by manipulating an electric inverter of the experimental model scale pump. In this paper, the pumping rate is set in 5 steps from 3.3L/min. to 16.7L/min. at the same intervals, which corresponds to from 200ton/h. to 1,000 ton/h. of a full-scale tank. In the experiments, freshwater (tap water) was used because the influence of viscosity is considered lower than that of seawater or saltwater. The ultrasonic sensor's output data is a voltage signal rather than a direct distance and has a linear relationship with the distance. Therefore, the experimental water level can be obtained according to the following method.

First, inject water into the ballast tank model to reach the initial water level w_i and let it stand. The fluctuation of the water surface can be observed by the display screen of the ultrasonic sensor. When the data becomes stable, the standing of water can be completed. Before switching on the pump, turn on the data logger (time t_0) to record initial data for 5 seconds or more. Then start the pump (time t_1) and observe the change of the water level in the tank. When the water level drops to almost stable (time

t_2),record the final water level w_f, then turn off the pump and wait about 30 seconds before stopping the data logger (time t_3) to store enough voltage data corresponding to the final water level. Finally, the data is imported into the computer, and the water level movement curve of the experiment is obtained according to formula (1).

$$w(t) = \frac{w_f - w_i}{\overline{V_f} - \overline{V_i}} \cdot \left\{V(t) - \overline{V_f}\right\} + w_f \qquad (1)$$

where ($\overline{V_i}$) is the initial average voltage between time t_0 ~ t_1 and ($\overline{V_f}$)is the final average voltage between time t_2~t_3.

The integrated discharge volume is calculated from the water level data of each compartment by the following formula (2).

$$D(t) = \sum \left\{w_{xy,i} - w_{xy}(t)\right\} \cdot A \qquad (2)$$

A is the area of the compartment, and $w_{xy,i}$ is the initial water level of (x-y).

To show the reproducibility of the experiments, Figure 3 compares the change over time in the water level of one experiment with the change over time in the average water level of 10 experiments in the same experiment condition.Eight sensors are utilized to measure the water level.Since there are 16 compartments in the 4x4 tank model experiment, water level data is obtained from two experiments in which the sensors are placed at x = 1,2 and x = 3,4. Here, since the data presented in Chapter 4 is the sum of the data of two experiments, it is necessary to discuss whether the same phenomenon occurs in multiple experiments under the same conditions.In Figure3, the experimental conditions are 6×1 tank model, bell mouth location (1-1), pump flow rate 3.3 L/min., initial water level 30mm, drain hole size W.9mm x H.6mm. The value acquired by the sensor slightly fluctuates because it contains noise and the wave motion on the water surface, but since there is no significant difference compared to the 10-time's average, the reproducibility for each experiment was guaranteed.

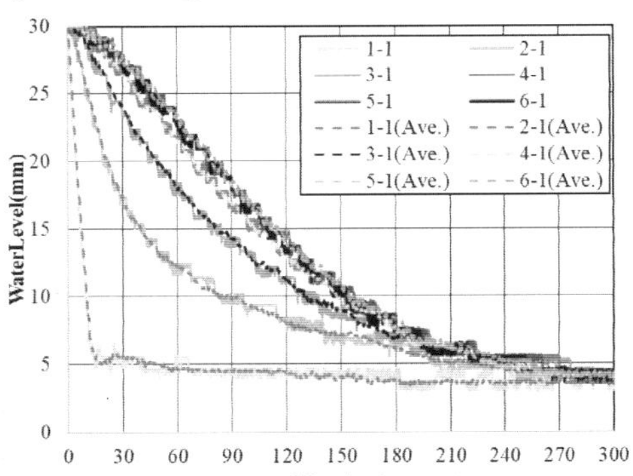

Figure 3: Comparison of changes in water level over time of 1experiment and an average of 10experoments

It should be noted that the use of this measurement method has drawbacks. Ultrasonic sensors can only measure thewater level change of a small area in the model, not a larger area. If the water surface has a large curvature, the water level characteristics of the area cannot be expressed. Moreover, since it is only a point of the water level, the measurement results will fluctuate.

4. RESULTS OF HYDRAULIC MODEL TEST

This chapter describes the results and considerations of three hydraulic model tests.The size of the drain hole expresses the width and height as W. × H. For example, a width of 24 mm and a height of 6 mm is expressed as W.24 mm x H.6 mm.

4.1 BELLMOUTH LOCATION

The standard location of the bellmouth installation compartment is (2-2). This experiment investigates the drainage difference in the four patterns in which the bellmouthlocationischanged, as shown below; (2-1), moved from (2-2) to the centre line side in the ship width direction, (1-2), moved to the bow side in the ship length direction, and (1-1) that combines them. In this experiment, the drain hole size is W.24mm x H.9mm, the pump flow rate is 16.7L/min., which is equivalent to 1000ton/h. in the actual ship scale, and the initial water level is 50 mm for all patterns.Figure 4 shows the total discharge volume change over time.

It is clear that the drainage efficiency is highest when locating the bell mouth in the centre of the tank, but in this case, the construction of piping for drainage is also complicated. Therefore, it is best to install it at the end of the tank in terms of design and construction.

From the graph, the movement in the width direction of the ship dramatically contributes to the deterioration of drainage efficiency compared with the movement of the bell mouth in the length direction. It is less efficient to install it on the centreline side, especially (1-1) significantly reduces the discharge volume.

Figure 4: Integrating discharge with different bellmouth locations
(Pumping rate: 16.7L/min.,
Initial water level: 50mm,
Drain hole size: W.24mm×H.9mm)

4.2 PUMPING FLOW RATE

This experiment examines the effect of pump flow rate on drainage efficiency. The following five patterns of experiments were compared. The pump flow rates are 3.3L/min., 6.7L/min., 10.0L/min., 13.3L/min., and 16.7L/min. which correspond to200ton/h., 400ton/h., 600ton/h., 800ton/h., and 1000ton/h. on the actual ship scale, respectively.In this experiment, the drain hole size is W.9 mm x H.18 mm, bellmouth compartment is (2-2), and the initial water level is 50 mm for all patterns.

Figure 5 shows the total discharge volume change over time.Comparing the water levels at 60 seconds after the pump starts, where the total discharge volumeis increasing linearly, the pump flow rate and discharge volume show a linear relationship in 3.3L/min., 6.7L/min., and 10.0L/min., whereas they are less than a linear relationship in 13.3L/min. and 16.7L/min.. In particular, when the pump flow rates are 13.3L/min. and 16.7L/min., there is no significant difference in discharge volume. Therefore, even if the pump flow rate is increased further, the effect is negligible.Assuming that the pump flow rate and the ballast pump price are in a proportional relationship, the most suitable flow rateis considered 10.0L/min. to 13.3L/min..

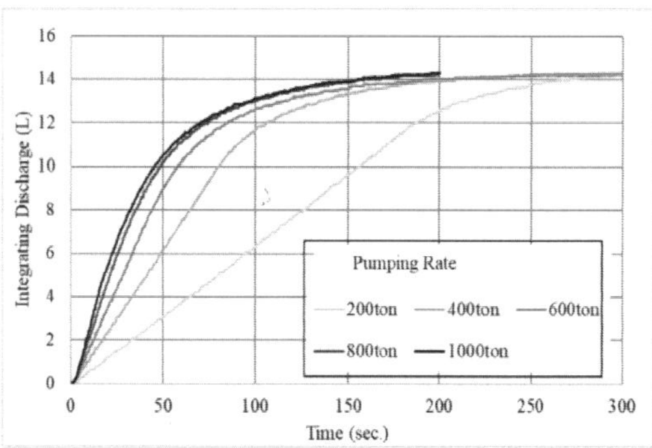

Figure 5: Integrating discharge with different pumping rates (Bellmouth location: (2-2), Initial water level: 50mm, Drain hole size: W.9mm×H.18mm)

4.3 DRAIN HOLE SIZE

This experiment compares the difference in the discharge volume due to the size of the drain hole.In thisexperiment, the pump flow rate is 16.7 L/min, bellmouth compartment is (2-2), and the initial water level was 50 mm for all patterns.

Figure 6 shows the total discharge volume change over time.Figure 6 (a) shows the time course of discharge volume when the height of the drain hole opening is changed, assuming that the opening width is 9 mm, which is a constant value. From this comparison, the discharge volumes of W.9 mm × H.18 mm and W.9 mm × H.24 mm are almost identical. This result indicates that increasing the height of the drain hole has little effect on improving drainage efficiency.

Figure 6 (b) compares the changes in discharge volume overtime when the width of the drain hole opening is changed, with the drain hole height set to a constant value of 9 mm. As the width of the opening increases, the drainage volume increases. In this case, W.24 mm x H.9 mm has the most effective drainage, and a significant difference can be seen compared to W.18 mm x H.9 mm. Even if the drain holes have the same area, a hole with a large width is more advantageous in structural strength than a hole with a large height. Therefore, a hole with a low height and a large width is rational.

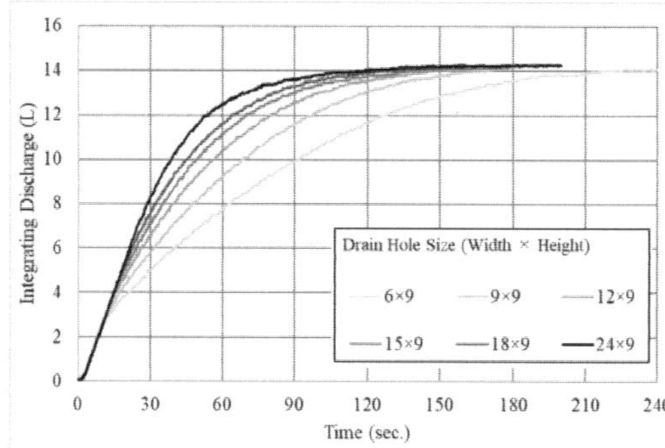

(a) In the case of drain hole height constant 9mm

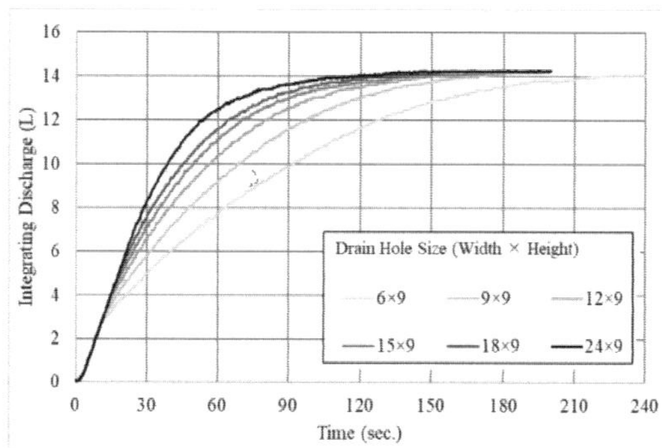

(b) In the case of drain hole width constant 9mm

Figure 6: Integrating discharge with different drain hole sizes (Bellmouth location: (2-2), Pumping rate: 16.7L/min., Initial water level: 50mm)

5. CONCLUSIONS

This paper explained in detail the hydraulic experiment equipment and the data processing method firstly.

Hydraulic model tests modelled the ballast tank and drainage process to develop a drainage system shortening drainage time and minimising residual ballast water. From this result, the following findings were obtained as the design policy of the ballast tank.

1) Locating the bell mouth on the centre side in the ship width direction of the ballast tank dramatically

improves drainage efficiency than positioning it to the centre side in the ship length direction.

2) There is a limit to improving drainage efficiency by increasing the pump flow rate. Even if only the pump capacity is increased, it lowers the water level in only the bell mouth compartment and increases the water level difference from the surrounding compartments, which does not improve the overall drainage.

3) Increasing the width of the drain hole is more advantageous for improving efficiency than increasing the height. Considering the structural strength, a shape with a lower height and a larger width is advantageous than a drain hole with the same area.

In this paper, the circle drain hole was changed to a square. However, it has not investigated the effect in detail. In the future, it will be necessary to examine the effect of changing shape and the number of holes while maintaining the same area.

6. REFERENCES

1. FUKUCHI, N., SHINODA, T., 'A Practical Method for Analysing the Indeterminable Problems in the Design of Ship Equipment (Part 1) - The System Collecting Oil and Water Through the Tank Bottom Structure', *J. Soc. Nav. Arch. Japan,* 1989.

2. EMORI, I., SAITO, K., SEKIMOTO, K., 'Model Test Theory and Application', *Gihodo booksJapan*, 1973.

7. AUTHORS BIOGRAPHY

T Watanabe is a doctoral student in the Department of Naval Architecture and Ocean Engineering, Kyushu University. In this research, he is mainly responsible for the experiment of ballast water drainage.

T SHINODA is a professor at Kyushu University, Japan. He received D.Eng., M.A. and B.A. in naval architecture from Kyushu University. His main research field is equipment engineering in naval architecture, ship production system, safety engineering, sea transportation including containerized transportation and its facility, and environmental engineering concerning ship.

G LIU is a doctoral student in the Department of Urban and Environmental Engineering, Kyushu University. In this research, he is mainly responsible for the numerical simulation of ballast tank models on OpenFOAM, and the programming of the simplified mathematical model.

H Obata holds the current position of deputy general manager of the outfitting design department at Namura Shipbuilding Co., Ltd. He has outfitting design experience includes drain course arrangement of ship.

T Nakamori, D.Eng. holds the current position of manager of the hull design section at Namura Shipbuilding Co., Ltd. He has structural design experience of the ship includes strength evaluations of the drain hole.

K Kurokiis responsible for hull structural design at Namura Shipbuilding Co., Ltd. He has 12 years of structural design experience of the ship includes strength evaluations of the drain hole.

FIRE RISK ASSESSMENT IN UPSTREAM OIL AND GAS COMPANY TO PREVENT CATASTROPHIC EFFECTS IN OFFSHORE FACILITIES

A M Zulpikar, D R Widiana, Y Praharsi, A T Nugraha, and **M R Dhani,** Politeknik Perkapalan Negeri Surabaya, Indonesia

SUMMARY

Operations in upstream oil and gas companies had a high risk of fire. The huge fire that occurred at the liquified petroleum gas (LPG) facility can cause the LPG production process to stop for approximately five months. In this case, the offshore facilities must de-pressurize and shut down the production wells to cut off the supply of hydrocarbons to the onshore facilities. This study aims to assess the fire risk in the hydrocarbon containment: crude oil and LPG storage tanks. The method used was Task Risk Assessment (TRA) which adjusted to the company. The results in crude oil storage tanks obtained 3 high-risk potential hazards were hydrocarbon gas release, hydrocarbon liquid release from the bleeder vent, and crude oil spills. In LPG sphere storage tanks, there are 2 high-risk potential hazards were gas clouds from pipes and LPG spills. The recommendation is carry out the control based on the hierarchy.

1. INTRODUCTION

Operations in upstream oil and gas have some activities were at offshore compressing and processing oil and gas, produced water treatment, daily oil sampling, tank inspection, tank cleaning, pigging operation, diesel bunkering, drilling activity, well intervention activity, and pipeline hydrocarbon transferring from offshore to onshore. Where offshore activities will stop when onshore encounters problems, some onshore activities that are at risk of causing fires are Crude Oil Loading, Crude Oil Tank inspection, Crude Oil Tank cleaning, Piping inspection, daily sampling, grass cutting routine, LPG Transfer, civil project work, painting activities, electrical equipment maintenance. These activities are in Crude Oil Tank and LPG Storage Tank facilities. The object of this research is an area that has a high level of fire risk. These areas are 3 Crude Oil Storage Tanks (A/B/C) and 4 LPG Sphere Storage Tanks (A/B/C/D). The object was chosen because it is the area with the most hydrocarbons, this is based on the company's Technical Datasheet, P&ID, and PFD.

In the process of the work being carried out, the company has performed hazard identification and risk assessment, to reduce the impact of existing risks. However, the potential fire hazards have not been fully identified, so they still have a large potential risk. In a period of 5 years (2016-2020), 4 fire incidents resulted in equipment damage, production processes being disrupted, and the company experiencing financial losses [7]. The biggest fire in the company occurred on March 7, 2018, in the De-Propanizer column area which caused the LPG production process to be stopped for 5 months.

Therefore, it is necessary to understand all the potential fire hazards in the workplace, know the procedures that apply in the workplace, and know the causes of fires and fire risk mitigation. Fire risk assessment (FRA) is a process for characterizing the risks associated with fires given to fire scenarios, their likelihood of occurrence, and their potential impacts. FRA can use various worksheets, such as HIRADC worksheet[3].But, this method only shows the fire hazard of a work activity, there is no detailed description of why the hazard can occur. Another FRA study, the fire hazard controls were not described in detail. The control is only carried out administratively and using personal protective equipment [1]. FRA must be carried out with depth identification. How come the fire hazard can occurs and how to control it, should be explain [6]. In this study, FRA was carried out using the company's task risk assessment worksheet (TRA). TRA is a method by identifying the potential that arises during the activity and gives elaboration about the action to control these hazards and prevent work accidents. TRA has the advantage of being able to describe potential hazards in more detail than other methods [2].

The TRA method has the advantage of being able to provide an explanation of how the fire hazard can occur, and how to control the hazard. However, there are still few studies that identify fire hazards using the TRA method. Therefore, in this study, FRA is carried out using the TRA method to prevent fires and minimize their impact.

2. METHODOLOGY

The first step of this research is data collection, both primary and secondary data. Primary data is obtained from interviews and brainstorming with the operators in crude oil storage tanks and LPG storage tanks. Secondary data collected is the standard operating procedure, layout, and company's risk matrix. The next step is to determine the activities to be identified by compiling the sequence of work and the dangers of each activity to determine the factors that cause fire accidents.

2.1 FIRE RISK ASSESSMENT

Fire risk assessment (FRA) is a process for characterizing the risks associated with fires given to fire scenarios, their likelihood of occurrence, and their potential impacts [1]. FRA process in this study used a quantitative method. Quantitative risk assessment usually involves two basic components, identification of scenarios and their likelihood, and quantification of the consequences of these scenarios [4]. FRA in this study was carried out by using the Task Risk Assessment (TRA) worksheet. This research begins with the identification of potential fire hazards, then an analysis of the likelihood, severity, and risk values is carried out according to the company's risk matrix. From the results of the risk value, if the value is high, control must be carried out so that the risk value can go down as low as possible. The control is carried out using the principle of the hierarchy of control, namely elimination, substitution, engineering control, administrative control and the last is the use of appropriate personal protective equipment [5].

2.2 RISK MATRIX

The risk assessment carried in this study used a risk matrix that has been applied to the company. The following is an image of the risk matrix used:

SEVERITY					
5 Catastrophic	5 Medium	10 Medium	15 High	20 High	25 High
4 Major	4 Medium	8 Medium	12 Medium	16 High	20 High
3 Moderate	3 Low	6 Medium	9 Medium	12 Medium	15 High
2 Minor	2 Low	4 Medium	6 Medium	8 Medium	10 Medium
1 Insignificant	1 Low	2 Low	3 Low	4 Medium	5 Medium
	1 Rare	2 Unlikely	3 Possible	4 Likely	5 Almost Certain
	LIKELIHOOD				

Figure 1 Risk Matrix

Based on Figure 1, it can be seen that there are 3 levels of risk; low, medium, and high. If the level of risk is high, it is considered unacceptable, and it needs control. Risks defined as a medium are considered tolerable. Even if this risk is within the tolerable range, efforts should still be made to reduce it to an ALARP level. This is the principle of ALARP and should be demonstrated. As Low As Reasonably Practicable (ALARP) principle is that the residual risk shall be as low as reasonably practicable. The term "reasonably practicable" means that risk is reduced to risk levels such that further reduction measures would be so disproportionate to the probability and consequences of occurrence that it would be objectively unreasonable to implement them. Reasonably practicable involves weighing a risk against the trouble, time, and money needed to control it. Thus, ALARP describes the level to which management expects to see workplace risks controlled. The acceptable risk criterion is described as LOW, this is the area in the Risk Matrix that is colored green. Any risk that falls below the tolerable lower limit of the area is widely accepted. Note that if a risk is acceptable it does not necessarily prevent improvement from starting if it is economical, easily identifiable, and practicable. [7]

2.3 LIKELIHOOD

The likelihood of the hazard being realized is categorized by reviewing and choosing a descriptor. The choice is based on the information available to the team and based on the team's knowledge and experience of the industry and company data [7]

2.4 SEVERITY

The severity of the hazard being realized is categorized by reviewing and choosing a descriptor. The team reviews and chooses the severity which best represents the seriousness of the possible consequences should the incident occur. All four types of severity must be considered (People, Environment, Asset Damage, and Reputation). Each of the severity categorized is represented on the Risk Matrix by a description (e.g. Catastrophic). The descriptor chosen must represent the severest category agreed upon by the team. For example, the Environment severity was slight but the Asset/Production severity was major then major is the overall severity rating which should be used [7]

3. RESULTS AND DISCUSSION

From the identification and assessment of fire risk in crude oil storage tanks, the results obtained are as follows:

Table 1 Risk Table at Crude Oil Tank

High Risk	Moderate Risk	Low Risk
CRUDE OIL STORAGE TANK		
- Hydrocarbon Gas release (Reid Vapor Pressure) - Hydrocarbon liquid release from the bleeder vent - Crude oil spills	- Pipe corrosion - Soil subsidence - Grass cutting machine with gasoline fuel - Use of portable generators (power for civil electrical equipment and tools)	- Ineffective injection of corrosion inhibitors - Incompetent people conducting pressure safety valve certification - Emissions from diesel engines (vehicle entry) - Entanglement in electrical cables and pipes

In crude oil storage tank there are 11 potential fire hazards. These potential hazards are classified into 3 potential hazards with a high level of risk, 4 potential hazards with a medium level of risk and 4 potential hazards with a low level of risk.

Three potential hazards with a high level of risk in this crude oil storage tank come from the same activity, it is crude oil loading. Crude oil loading is oil transfer from tank to the consumer vessel through a pipeline, so that's why this activity has a high risk of fire potential [8]

Table 2 Risk Table at LPG Storage Tank

High Risk	Moderate Risk	Low Risk
LPG STORAGE TANK		
- Gas clouds from pipes - LPG spills/leaks	- High-pressure gasses - Grass cutting machine with gasoline fuel - Soil subsidence	- Electrical equipment short circuit - Incompetent people conducting control and safety valves maintenance - Emissions from diesel engines (vehicle entry) - Project equipment damaged

In LPG storage tank there are 9 potential fire hazards. These potential hazards are classified into 2 potential hazards with a high level of risk, 3 potential hazards with a medium level of risk and 4 potential hazards with a low level of risk

The high-risk of potential hazard should be control to reduce the risk level [7]

3.1 CRUDE OIL TANK

Fire Potential High Risk are: Hydrocarbon Gas release (Reid Vapor Pressure), Hydrocarbon liquid release from the bleeder vent, Crude oil spills.

Control Measure:
Elimination: N/A

Substitution: N/A

Engineering control:
- Install Gas Detector and alarm (sound and visual)
- RVP (Reid Vapor Pressure) control at the upstream of Crude Oil Tank
- Passive fire protection at Crude Oil Tank, Valves, and pipelines
- Heat Sensing for activated alarm and fire deluge system
- Cathodic Protection for Tank and transfer pipeline
- Provide roof leg boot to control the emission from crude oil tank
- Provide early streamer lightning protection
- Provide Grounding and Bonding system
- Emergency Shut Down Valve at Crude Oil Tank system
- Provide bundwall and drain line to close system
- Oil Spill Contingency Plan : Oil Spill Response team, OSRE availability, trained persons, etc.
- Oil Spill Response Equipment (OSRE): Solid Boom, Sorbent Boom, Sorbent Pad, Transfer Pump, Power Pack, Dispersant Skimmer, Oil recovery tank, etc.

Administrative control:
- Preventive Maintenance for Detector, Alarm, Emergency Response Equipment, and all related equipment
- Risk Based Inspection for Crude Oil Tank: NDT and DT
- Routine Corrosion Inspection for Crude Oil Tank and pipeline
- Routine Visual Check for Crude Oil Tank and pipeline
- Emergency Response Plan Procedure
- Emergency Response Team (certified Advanced Fire Fighting, Management of Major Emergency, First Aider, Oil Spill Response, and Rescue)
- Provide Safety Sign for hydrocarbon hazard, escape route, and Muster Point
- Provide CCTV

PPE:
- Mandatory: Safety Helmet, Safety Glasses, Standard Fire Retardant coverall, hand gloves, and safety shoes
- Special: Breathing AirApparatus, Fire Fighting Suit

3.2 LPG STORAGE TANK

Fire Potential High Risk are: Gas clouds from pipes, and LPG spills/leaks

Control Measure:
Elimination: N/A

Substitution: N/A

Engineering control:
- Install Gas Detector and alarm (sound and visual)
- Passive fire protection at LPG Storage Tank, Valves, and pipelines
- Heat Sensing for activated alarm and fire deluge system
- LPG Recovery pond
- Cathodic Protection for Tank and transfer pipeline
- Provide early streamer lightning protection
- Provide Grounding and Bonding system
- Emergency Shut Down Valve at LPG Storage Tank system
- Pressure Safety Valve at LPG Storage Tank

Administrative control:
- Preventive Maintenance for Detector, Alarm, Emergency Response Equipment, and all related equipment

- Risk Based Inspection for LPG Storage Tank: NDT and DT
- Routine Corrosion Inspection for LPG Storage Tank and pipeline
- Routine Visual Check for LPG Storage Tank and pipeline
- Emergency Response Plan Procedure
- Emergency Response Team (certified Advanced Fire Fighting, Management of Major Emergency, First Aider, Oil Spill Response, and Rescue)
- Provide Safety Sign for hydrocarbon hazard, escape route, and Muster Point
- Provide CCTV

PPE:
- Mandatory: Safety Helmet, Safety Glasses, Standard Fire Retardant coverall, hand gloves, and safety shoes
- Special: Breathing Air Apparatus, Fire Fighting Suit

4. CONCLUSION

From the fire risk assessment that has been carried out, there are five potential fire hazards that have a high level of risk. They are Hydrocarbon Gas release (Reid Vapor Pressure), Hydrocarbon liquid release from the bleeder vent, Crude oil spills, Gas clouds from pipes, and LPG spills/leaks. This potential hazard must be controlled to reduce the level of risk. The controls provided are sorted according to the control hierarchy; elimination, substitution, engineering control, administrative control and the use of appropriate personal protective equipment. For further analysis, a special study is needed to assess the readiness of facilities and emergency response teams in the company.

5. ACKNOWLEDGEMENTS

We would like to thank all those who have helped in the process of preparing this paper.

6. REFERENCES

1. Aprillia, SA., 'Fire Risk Assessment Analysis And Design Of Active Fire Protection In The Workshop Area (Case Study: Fabrications and Shipbuilding Service Company)', Diploma thesis, Shipbuilding Institute of Polytechnic Surabaya, 2017.

2. Dedi F., 'Study of Hazard Identification and Risk Assessment Using TRA (Task Risk Assessment) Method in Machine Maintenance Workshop Unit 1 as an Effort to Prevent Work Accidents at PT. PJB UBJ O and M Tanjung Awar – Awar', 2017.

3. Lathif L., 'Design of Automatic Fire Extinguishing System on Panel Using Carbon Dioxide Total Flooding Based on Fire Risk Assessment Identification (Case Study: Steel Smelting Company)', Shipbuilding Institute of Polytechnic Surabaya, 2020.

4. National Fire Protection Association, 'NFPA 551 Guide for the Evaluation of Fire Risk Assessment', 2019.

5. Soehatman, R. 'Fire Management, Concept and Implementation', 2020.

6. Shu-jiao Tonga, Zong-zhi Wub, Ru-jun Wanga, Hao Wu, 'Fire Risk Study of Long-distance Oil and Gas Pipeline Based on QRA', 2016

7. Upstream Oil and Gas Company, 'HSE Management', 2020.

8. Yi Zhou, Xiaogang Zhao, Jianyu Zhao, Du Chen, 'Research on Fire and Explosion Accidents of Oil Depots', 2016.

7. AUTHORS BIOGRAPHY

Asep M Zulpikaris a master student of Safety and Risk Department at Shipbuilding Institute of Polytechnic Surabaya (Politeknik Perkapalan Negeri Surabaya). He is responsible for taking the data. He is experience in oil and gas safety expert more than 10 years.

Dika Rahayu Widiana,Ph.D, holds the current position of lecture at Politeknik Perkapalan Negeri Surabaya (Shipbuilding Institute of Polytechnic Surabaya). She is responsible for analysing the data. Her previous experience includes being a lecture at occupational health and safety engineering department.

Yugowati Praharsi, Ph.D, holds the current position of lecture at Politeknik Perkapalan Negeri Surabaya (Shipbuilding Institute of Polytechnic Surabaya). She is responsible for analysing the data. Her previous experience includes being a lecture at business management department.

Anggara Trisna, ST., MT, holds the current position of lecture at Politeknik Perkapalan Negeri Surabaya (Shipbuilding Institute of Polytechnic Surabaya). He is responsible for writing the manuscript. His previous experience includes being a lecture at automation engineering department.

Mey Rohma Dhani, S.ST., MT., holds the current position of lecture at Politeknik Perkapalan Negeri Surabaya (Shipbuilding Institute of Polytechnic Surabaya). She is responsible for writing the manuscript. Her previous experience includes being a lecture at occupational health and safety engineering department.

THE DESIGN AND SIMULATION OF THE LINEAR ENGINE GENERATOR USED FOR ELECTRIC PROPULSION SYSTEM OF THE MARINE TRANSPORTATION AND OFFSHORE GAS/OIL PLATFORMS

M Li, U Ngwaka, D Wu, F Zhang and **G Chen,** Department of Mechanical Engineering, School of Engineering, University of Birmingham, UK

SUMMARY

Maritime shipping and offshore gas/oil platforms account for a large amount of greenhouse gas emissions. It is beneficial to use the linear engine generator (LEG) fuelled by low or zero-carbon fuel as the electric propulsion system for marine transportation and offshore gas/oil platforms to achieve net-zero carbon emission. The LEG could replace the diesel engine and gas turbine for marine applications, with its advantages of simple mechanical structure and adaptability to various fuels, such as hydrogen, ammonia, methane, biofuels, and various energy sources like nuclear and solar energy. This article introduces the ammonia/hydrogen-fuelled LEG prototype for electric propulsion system for marine transportation. The LEG parametric sensitivity analyses are done for performance evaluation through simulation modelling. Furthermore, a detailed design of 1.8 kW electrical output LEG is presented.

NOMENCLATURE

v	Velocity (m s^{-1})
BDC	Bottom dead centre
CNG	Compressed natural gas
EPS	Electrical propulsion system
FFKM	Perfluoroelastomer
FPE	Free-piston engine
GHG	Greenhouse gas
HFO	Heavy fuel oil
IC	Internal combustion
IMO	International Maritime Organization
LEG	Linear engine generator
LG	Linear generator
LNG	Liquefied natural gas
PM	Permanent magnet
SI	Spark ignition
TDC	Top dead centre

1. INTRODUCTION

Based on the studies of Greenhouse Gas (GHG) published by the International Maritime Organization (IMO), the share of shipping sector CO_2 global emissions increased from 2.76% in 2012 to 2.89% in 2018 [1], and it is expected to increase to 12% ~18% by 2050 [2]. Besides, the shipping industry accounts for asmuch as 15% NO_X emissions and 8% SO_2 emission [3]. Except for maritime shipping, gas turbines or diesel engines are also used to generate power in offshore gas/oil platforms [4]. As for the diesel engine and turbine engine, the main fuels at present are fossil fuels, such as conventional heavy fuel oil (HFO). Liquefied Natural Gas (LNG) could replace more polluting fuels like HFO to achieve emissions reductions, specifically CO_2, NO_X, SO_X, and particulate matter, but it could not help thoroughly deal with the emission challenges [5]. Moreover, incorporating the waste heat recovery system in these traditional engines adds the complexity of the whole system and slightly decreases the power output [6, 7].

The electrical propulsion system (EPS) has the advantages of pollutant emission reduction over classical mechanical systems [8, 9], reducing noise and vibration [8], offering more freedom in the design of the ship and the layout of the main engines[10]. Therefore, to meet improved energy efficiency and low emissions requirements, the EPS is a promising choice for ships and offshore platforms. Compared to the diesel-driven and turbine-driven or steam-driven electrical propulsion systems [11], the linear engine generator (LEG) could be a better choice for its less complicated structural mechanism, variable compression ratio, the possibility of multi-fuel adaptation, and higher part-load efficiency [12, 13].

The free-piston engine (FPE) technology was invented in the 1920s [14], and with the development of electric motor technology, it was coupled to the linear generator to work as a linear engine generator (LEG). Several prototypes of the LEG have been investigated. In 1998, Sandia National Laboratories developed and tested the internal combustion (IC) LEG prototype. The prototype was powered by various fuels, such as natural gas, hydrogen, propane, and n-heptane [15]. In 1999, an opposed-piston LEG prototype was developed at West Virginia University, with a power output of about 316 W [16]. An open-cycle external combustion LEG was designed in 2014 [17] and developed in 2018 [18] and in 2020 the semi-closed-cycle LEG was developed [19] at Newcastle University.

Research in LEGs fuelled by hydrogen and other low-carbon fuels is receiving more attention because of the current global drive to reduce emissions. In 2009, compressed natural gas (CNG) and hydrogen were used to study LEG's performance. Test results show that, compared with CNG, the heat release rate of hydrogen fuel LEG increased, and fuel consumption decreased [20]. However, the hydrogen-fuelled LEG requires precise ignition, injection, and the air-fuel ratio [21]. In 2019, a numerical investigation on hydrogen-fuelled LEG was conducted. The analysis shows that the hydrogen-air mixture has poor homogeneity due to the slow movement of the piston before ignition. However, the mixture's homogeneity was optimized at increased stroke [22]. Notwithstanding, the hydrogen storage onboard is a major concern. Meanwhile, ammonia, a carbon-free hydrogen carrier, is regarded as a promising fuel. In 2013, the testing of a 4-stroke two-cylinder spark-ignition engine (SI engine) fuelled by ammonia hydrogenation was conducted and the results indicated that adding hydrogen to the air-ammonia mixture is necessary to improve ignition and increase ignition combustion speed. The ammonia/hydrogen ratio mainly depends on the load and less on the engine speed [23]. In 2020, the ammonia/hydrogen premixed combustion was investigated and the results show that when the equivalent ratio is about 1.10 ~ 1.20, there is improved performance of the stability of the ammonia flame with reduced NO_X emissions. Additionally, the high-pressure and high-temperature environment helps to significantly reduce NO_X emissions from ammonia/hydrogen combustion, which indicates the potential of low NOx LEG fuelled by ammonia/hydrogen [24].

An ammonia/hydrogen dual-fuel LEG prototype is being developed for EPS based on the studies above. The LEG under investigation is of the Joule Cycle external combustion type proposed by Wu and Roskilly [17]. It could separate the combustion process from the piston chambers and have various heat resources compared to the former LEG prototypes. The working principles[25], the preliminary experiment [26], and the combustion characteristics of premixed ammonia/ hydrogen in a linear Joule engine generator are investigated[24, 27]. Based on the simulation, the LEG has an electrical power output of about 1.8 kW and an engine efficiency of about 35%. A new configuration of the Joule Cycle LEG is presented, and the technical challenges and the simulation results will be discussed in this paper.

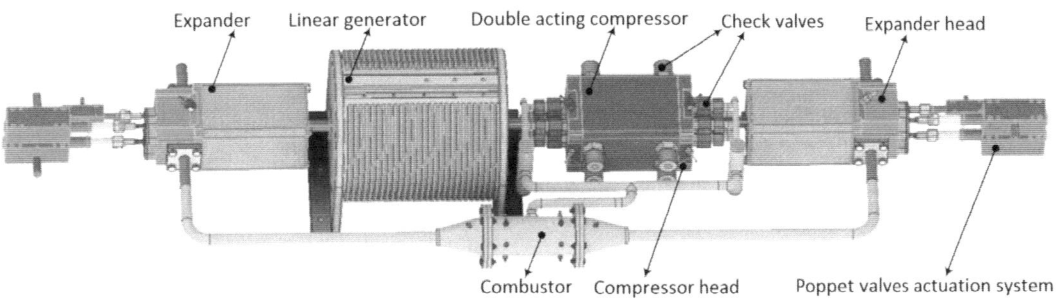

Fig. 1 Configuration of the LEG

2. WORKING PRINCIPLEANDTHE CONFIGURATION

The Joule Cycle LEG uses an external heat source such as the combustor fuelled by ammonia/hydrogen or the heat exchanger. Moreover, this piston engine generator consists of a double-acting compressor, two expanders, a linear generator (LG) with some accessories. The configuration of the LEG is described in Fig. 1 and the flow track of the working fluid through the whole system is shown in Fig. 2. The suction process happens in the double-acting compressor, and the ambient air could be compressed to about 15.24 Bar, which is an isentropic process in an ideal state. Then the exhaust check valves, will open for a while to let the high-pressure gas flow into the combustor. After being heated here, the high-pressure, high-temperature working fluid will flow to the expanders. Pressure loss exists through the combustor resulted from the fluctuation of the compressed working fluid and the flow resistance in the pipes. The intake valves of the expanders open for a while, when the high-pressure, high-temperature working fluid will push the expander piston to move from the top dead centre (TDC) to the bottom dead centre (BDC), which is another ideal isentropic process. During this period, some of the work extracted by the piston/cylinder is used to drive the compressor and the linear generator.

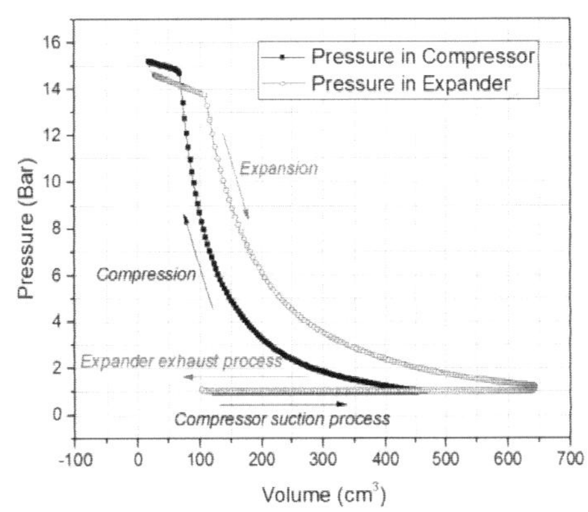

Fig. 2 Pressure VS. Volume

3. LEG DESIGN IN DETAIL

3.1 LINEAR GENERATOR

The Simcenter Magnet and the Simcenter Amesim were used to build and study the LG model and the whole LEG system during the simulation. The design of the LEG parts follows the principle combining the simulation results and the engineering achievability.

The linear generator (LG) is chosen from the following

two types, the long translator LG in Fig. 3 (a) and the short translator LG in Fig. 3 (b). They are both tubular permanent magnet (PM) motors, but the geometry sizes of the stators and the translators are different. Based on the simulation, the comparison of the magnetic forces of the two types LGs is shown in Fig. 3 (c). The magnetic forces are simulated while the exact velocity of $v=5\cdot sin(100/3\cdot\pi)$m/s drives the translators. As a result, the RMS (Root Mean Square) of the short translator LG is about 579.13 N, while the RMS of the long translator LG is about 450.99 N, which reflects the short one has better performance. Besides, the simulation results show that the lighter translator mass is more beneficial to the overall LEG performance. Therefore, the shorter translator LG is chosen.

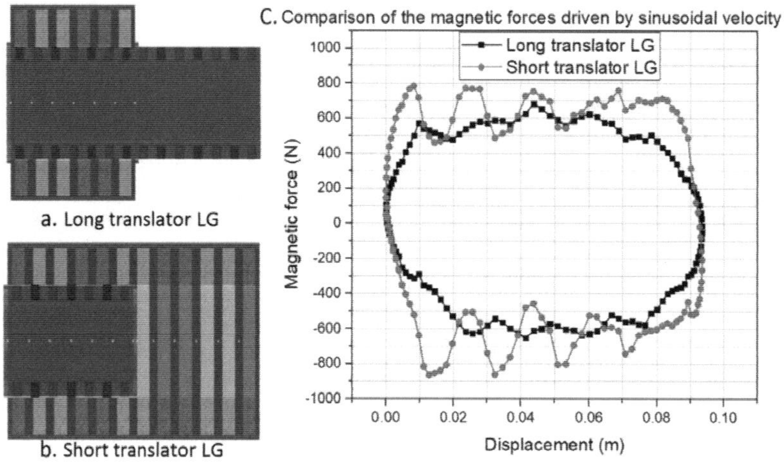

Fig. 3 Comparison of the configurations and magnetic forces of the two types of linear generators

The short translator LG is a 3-phase, 6-slot/7-pole tubular PM motor. It is equipped with modular stator windings in which the coils of each phase are arranged next to each other. The power output of the short translator LG can be seen in Fig. 4 (a), where the power outputs of the three phases are described in three different lines respectively and the total power output is the sum of the instantaneous values of three phases. Due to the different distributions of the coils and magnets, the number and the values of each power output peak are different. The power output of phase B has the most significant peak value of about 2.2 kW, the peak value of power output A has the smallest one 1.52 W, and the peak value of total power output is about 3.25 kW. As for the design of the linear generator, one of the challenges is to ensure the axiality tolerance of the translator and stator. Based on the assemble experience, there are large attractive and repulsive forces between permanent magnets and coils. In this case, the chrome-plated support shaft and bracketsare separated to facilitate adjustment of the shaft position and ensure axiality tolerance, and the diagram of the modified support method can be seen in Fig. 4 (b).

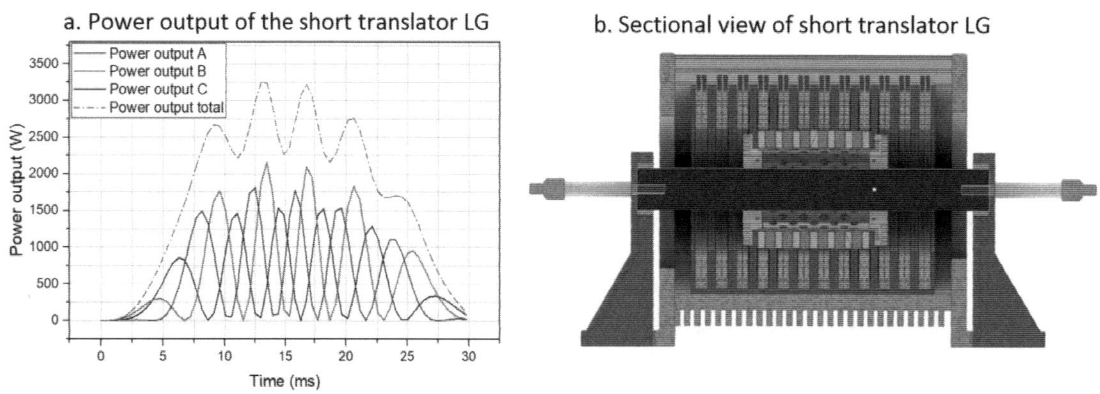

Fig. 4 Power output of the short translator LG and thesectional view of the short translator LG

3.2 ENGINE DESIGN

Thegeometry sizes of the linear engine should be determined based on the performance of the LG.

Investigation on the compressor and expanders' optimal bore diameters and their effect on system efficiency are shown in Fig. 5 (a). The compressor and expander's optimal match will benefit from achieving the best engine efficiency. The optimal bore diameter of the expander (Bexp) will increase along with the increase of the bore diameter of the compressor (Bcomp). However, the engine efficiency decreased to some extent. When the Bcomp is 60 mm, the optimal Bexp is about 78 mm to get the engine efficiency of about 37.5%. If the Bcomp changed to 80 mm, the optimal Bexp increased to about 95 mm while the engine efficiency decreased to about 33.5%. In this case, to obtain an acceptable engine performance and avoid the engine sizes being too small to arrange the valves, the bore diameter of the compressor is set to be 70 mm, and the bore diameters of the expanders are selected as 82 mm.

The check valves are placed on the cylinder heads as well as the cylinder surfaces. The sizes of the orifice area will affect the flow of the working fluid so that the orifices should be managed properly within the limited surfaces. The effect of the orifice area on the engine efficiency is shown in Fig. 5 (b), where the engine efficiency increases with the increase of the orifice area. Besides, when the orifice area is below 600 mm^2, the engine efficiency is more sensitive to the orifice area of the expander valves than the compressor valves, and the exhaust valve orifice area than the intake valves orifice areas. When the expander exhaust valve orifice is below 200 mm^2, the engine efficiency will decrease dramatically. When the Orifice area is above 600 mm^2, its effect on the engine efficiency is negligible.

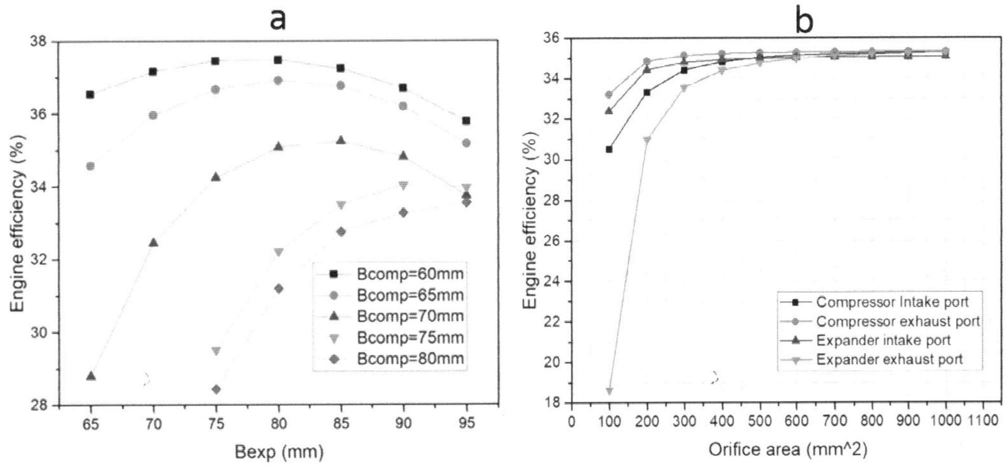

Fig. 5 Effects of the chamber bore diameters (a) and the valves orifice areas (b) on engine efficiency

The double-acting compressor will be beneficial to decrease air leakage and increase the power density. Due to the limited space, the intake check valves are installed on the cylinder surface while the exhaust valves are set on the cylinder heads on both ends. Both intake and exhaust valves are check valves, whose configuration is shown in Fig. 6 (a). Five exhaust check valves are located in the compressor head shown below in Fig. 6 (b), while four intake check valves are installed on the surface of the cylinder shown in Fig. 1. The intake checkvalves' position is corresponding to the position where the air start to be sucked into the compressor. Moreover, considering the layout and seal of the rod connecting the pistons, the compressor head is divided into two parts (the larger green head and the smaller red one), which will help install and fix the rod seal.

The total orifice area is set to be about 500 mm^2 on four intake check valves used. The exhaust valve orifice area should be larger than 150 mm^2, for there is a positive relationship between the LEG performance and the orifice area within a certain range. In this case, the flow channel's diameter of the exhaust check valves is set to be D= 6.5 mm, while the number of the exhaust check valve is five. Besides, the cracking pressure is set to be less than 0.1 Bar, which will benefit the engine performance too. Regarding the high temperature in the compressor, Perfluoroelastomer (FFKM) is used as the O-ring to ensure there will be no leakage.

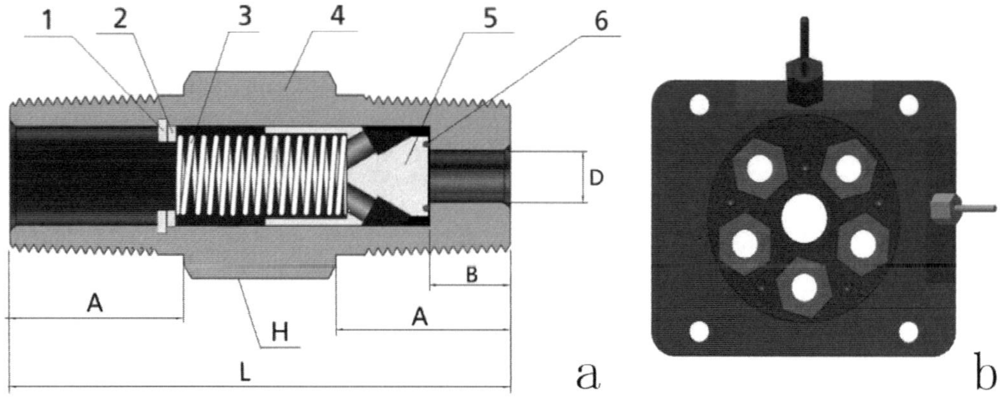

Fig. 6 (a) Sectional view of check valve and (b) Layout of the exhaust check valves on the compressor head

There are two expander chambers, and the expander heads are set to be on both sides ends. The two intake valves and one exhaust valve can be seen in Fig. 7, where the blue valves are the intake valves and the red one is the exhaust valve. In the expander head design, valve seats are used to protect the expander head and avoid leakage. Besides, valve guides made of bronze are used to support and guide the valves, so that good lubrication and relatively small position changes will be achieved. Considering the pressure in the chamber of the expander head could reach about 11.2 Bar, rod seals are needed to ensure there is no leakage. The poppet valves' actuation system is designed specially, unlike traditional IC engines. To open the intake ports, the intake poppet valves need to be pulled to offset the air force, resulting from the pressure difference between the high-pressure, high-temperature intake working fluid and the exhaust air remained in the expander chamber. A similar mechanism is used to control the exhaust poppet valve.

Fig. 7 Poppet valves of the Expander chambers

The expander piston design is another focus during the designing of the LEG, which is supposed to ensure the pistons could work well with small friction and avoid leakage. In the floating linear engine, the friction coefficient of the oil control ring is about 0.04 to 0.14 [28]. For the automotive piston ring with different coating, the friction coefficients of the ring/cylinder liner pairs range from 0.13 to 0.25 [29]. Therefore, the piston rings in the LEG should be among these ranges. Adopting the piston ring design in [30], the final design of the expander piston in the LEG could be seen in Fig. 8, where 4 graphite piston rings are used to help avoid leakage, and keyway structures are used to strengthen the connection of different parts of the piston as well as avoid the leakage of the high-pressure working fluid. The graphite used here is a self-lubricating material with a friction coefficient of about 0.1 [31] and max suitable temperature is about 980°C. Therefore, it should be a good choice for the LEG prototype.

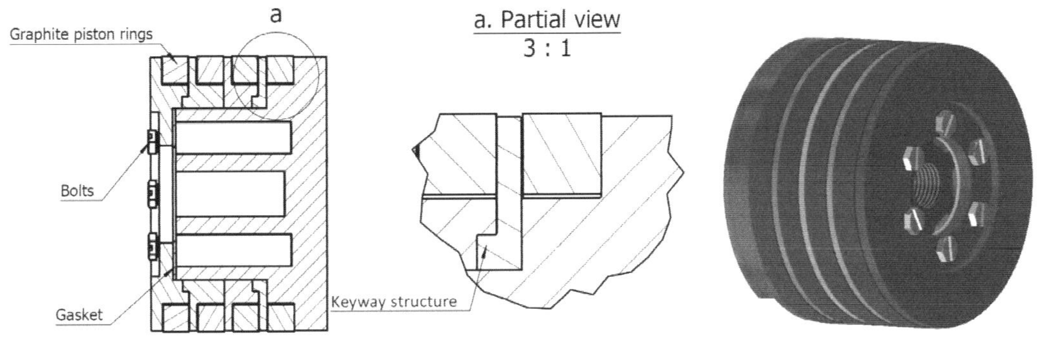

Fig. 8 Schematic view of expander piston

The main parameter settings and the simulation results of the LEG model built by Simcenter Amesim can be seen in Table. 1. The mean mechanical power output is about 2158.85 W and the heat from the combustor/heat exchanger is about 6160 W, in this case, there is anengine efficiency of about 35.05%. Considering the LG efficiency, the whole system's efficiency is about 30.59%.

Table. 1 Specifications and performance

Parameters	Unit	Value
Compressor bore	mm	70
Expander bore	mm	82
Stroke	mm	120
Mover mass	kg	16.576
Compression ratio	[-]	15.24
Peak temperature	°C	860.62
Engine efficiency	%	35.05
Heat from the combustor	W	6160
Mean mechanical power output	W	2158.85
Generator efficiency	%	87.28
Mean electric power output	W	1884.26
Peak piston speed	m/s	4.9
Frequency	Hz	15

4. CONCLUSION

To achieve the net carbon emission and meet the developing need of the new ships and offshore gas/oil platforms, the LEG-powered electrical propulsion system would have a promising future. Advanced simulation tools are used to guide and integrate LEG design from the system level to the component level.

However, during the study, it is sometimes paradoxical between the needs of the simulation results and the possibility of achieving the ideal state. In this case, compromises need to be adopted to ensure there is acceptable performance. Smaller engine geometry sizes and larger orifice areas will lead to better engine efficiency. However, it would be challenging to manage the position of the valves when the bore diameters are too small. Therefore, the bore diameter of the compressor is chosen as 70 mm, and the expander bore diameter is 82 mm. Besides, the orifice areas are determined to ensure enough space to install the valves and the engine efficiency will not be affected. The double-acting cylinders are beneficial for the higher power density. However, considering the layout of the valves and the shaft, two expander cylinders are used to ensure there is enough space for the poppet valves and the shaft. The mass of the whole moving parts could also play an important part in the engine performance, and smaller LG translator and double-acting compressor piston are used to decrease the total mass. However, because there are two expander chambers, there will be two expander pistons. Besides, to avoid the high-pressure working fluid leakage, a unique piston graphite piston ring is installed.

In conclusion, there are various details of the LEG parts that have been considered to achieve an improved LEG performance. As a result, the Lab-scale LEG prototype has been developed to achieve a stable state and a projected performance. Further studies focus on the control system and the methods to increase engine efficiency. As for the LEG application in the electric propulsion system, the linear alternator and the other parts need to be sized based on actual needs.

5. ACKNOWLEDGEMENTS

The authors thanks for the finance support from EPSRC (Engineering and Physical SciencesResearch Council, United Kingdom) on the project, Powering Carbon-free Autonomous Shipping:Ammonia/Hydrogen dual-fuelled Linear Engine-Generator (EP/S00193X/1).

6. REFERENCE

1. IMO, "Fourth IMO GHG Study 2020," International maritime organization, London, 2021.

2. IMO, "Second IMO GHG Study 2009," International maritime organization,London, 2010.

3. E. J. Williams, B. M. Lerner, P. C. Murphy, S. C. Herndon, "Emissions of NO_x,SO_2, CO, and HCHO from commercial marineshipping during Texas Air Quality Study (TexAQS) 2006" JOURNAL OF GEOPHYSICAL RESEARCH, VOL. 114, D21306, 2009.

4. J. E. Barrera, E. Bazzo, and E. Kami, "Exergy analysis and energy improvement of a Brazilian floating oil platform using Organic Rankine Cycles," Energy, vol. 88, pp. 67–79, 2015.

5. C. Nuchturee, T. Li, and H. Xia, "Energy efficiency of integrated electric propulsion for ships – A review," Renew. Sustain. Energy Rev., vol. 134, p. 110145, 2020.

6. E. Thibaut and B. Leforgeais, "Selection of power from shore for an offshore Oil and Gas development," in PCIC Europe, 2015, pp. 1–7.

7. A. R. Årdal, "Challenges with integration and operation of offshore oil & gas platforms connected to an offshore wind power plant," in 2014 Petroleum and Chemical Industry Conference Europe, 2014.

8. L. Hengrong, "Design on IPNCS of electrical propulsion ship based on real-time Ethernet," Energy Procedia, vol. 16, pp. 1707–1713, 2012.

9. K. Fagerholt, "Maritime routing and speed optimization with emission control areas," Transp. Res. Part C Emerg. Technol., vol. 52, pp. 57–73, 2015.

10. A. Ait Allal, Toward an evaluation of marine fuels for a clean and efficient autonomous ship propulsion energy, vol. 13. Materials Today: Proceedings, 2019.

11. J. Song, Y. Song, and C. –. Gu, "Thermodynamic analysis and performance optimization of an Organic Rankine Cycle (ORC) waste heat recovery system for marine diesel engines," Energy, vol. 82, pp. 976–985, 2015.

12. B. Jia, D. Wu, A. Smallbone, U. C. Ngwaka, and A. P. Roskilly, "Dynamic and thermodynamic characteristics of a linear Joule engine generator with different operating conditions," Energy Convers. Manag., vol. 173, 2018, doi: 10.1016/j.enconman.2018.07.098.

13. O. Noren and R. Erwin, "The future of the free-piston engine in commercial vehicles," SAE, 1958.

14. P. R. Pateras, "Motor-compressor apparatus," U.S. Patent 1,657,641, 1928.

15. P. Van Blarigan, N. Paradiso, and S. Goldsborough, "Homogeneous charge compression ignition with a free piston: A new approach to ideal otto cycle performance," SAE Tech. Pap., no. x, 1998, doi: 10.4271/982484.

16. P. Famouri, "Design and testing of a novel linear alternator and engine system for remote electrical power generation," IEEE Power Eng. Soc., 1999.

17. D. Wu and A. P. Roskilly, "Design and parametric analysis of Linear Joule-cycle Engine with out-of-cylinder combustion," in Energy Procedia, 2014, doi: 10.1016/j.egypro.2014.11.1034.

18. B. Jia, D. Wu, A. Smallbone, C. Lawrence, and A. P. Roskilly, "Design, modelling and validation of a linear Joule Engine generator designed for renewable energy sources," Energy Convers. Manag., vol. 165, pp. 25–34, Jun. 2018, doi: 10.1016/J.ENCONMAN.2018.03.050.

19. U. Ngwaka et al., "Parametric Analysis of a Semi-Closed-Loop Linear Joule Engine Generator using Argon and Oxy-Hydrogen Combustion," Energy, vol. 217, p. 119357, Nov. 2021, doi: https://doi.org/10.1016/j.energy.2020.119357.

20. Y. Woo, Y. Lee, and Y. Lee, "The performance characteristics of a hydrogen-fuelled free piston internal combustion engine and linear generator system," Int. J. Low-Carbon Technol., vol. 4, no. 1, pp. 36–41, 2009, doi: 10.1093/ijlct/ctp003.

21. U. Ngwaka et al., "Evaluation of performance characteristics of a novel hydrogen-fuelled free-piston engine generator," Int. J. Hydrogen Energy, vol. 46, no. 66, pp. 33314–33324, 2021, doi: 10.1016/j.ijhydene.2020.02.072.

22. C. Yuan, Y. Jing, C. Liu, and Y. He, "Effect of variable stroke on fuel combustion and heat release of a free piston linear hydrogen engine," Int. J. Hydrogen Energy, vol. 44, no. ue 36, pp. 20416–20425, 2019.

23. S. Frigo and R. Gentili, "Analysis of the behaviour of a 4-stroke Si engine fuelled with ammonia and hydrogen," Int. J. Hydrogen Energy, vol. 38, no. 3, pp. 1607–1615, Feb. 2013, doi: 10.1016/j.ijhydene.2012.10.114.

24. F. Zhang, G. Chen, D. Wu, T. Li, Z. Zhang, and N. Wang, "Characteristics of Ammonia/Hydrogen Premixed Combustion in a Novel Linear Engine Generator," Proceedings, vol. 58, no. 1, p. 2, Sep. 2020, doi: 10.3390/wef-06925.

25. D. Wu, A. S. Jalal, and N. Baker, "A Coupled Model of the Linear Joule Engine with Embedded Tubular Permanent Magnet Linear Alternator," in Energy Procedia, 2017, doi: 10.1016/j.egypro.2017.03.571.

26. D. Wu, "A preliminary experimental study on a lab-scale Linear Joule Engine prototype," Energy Procedia, vol. 158, pp. 2244–2249, 2019.

27. F. Zhang, G. Chen, D. Wu, T. Li, Z. Zhang, and N. Wang, "Characterising premixed ammonia and hydrogen combustion for a novel Linear Joule Engine Generator," Int. J. Hydrogen Energy, vol. 46, no. 44, pp. 23075–23090, 2021, doi: 10.1016/j.ijhydene.2021.04.110.

28. Z. Westerfield, "A Study of the Friction of Oil Control Rings Using the Floating Liner Engine," SAE Int. J. Engines, p. 9, 2016.

29. S. Wan, "Comparison of the scuffing behaviour and wear resistance of candidate engineered coatings for automotive piston rings," Tribol. Int., vol. 106, pp. 10–22, 2017.

30. U. Ngwaka, B. Jia, C. Lawrence, D. Wu, A. Smallbone, and A. P. Roskilly, "The characteristics of a Linear Joule Engine Generator operating on a dry friction principle," Appl. Energy, vol. 237, 2019, doi: 10.1016/j.apenergy.2018.12.081.

31. M. B. Peterson and R. L. Johnson, "Friction studies of graphite and mixtures of graphite with several metallic oxides and salts at temperatures to 1000 F," Natl. Advis. Commitee Aeronotics, 1956.

7. AUTHORS BIOGRAPHY

Conflicts of Interest: The authors declare no conflict of interest.

Dawei Wu is a Senior Lecturer in Mechanical Engineering and an EPSRC (Engineering and Physical Sciences Research Council) Fellow. Dawei is also a part of the Vehicle Technology Research Centre.Dawei has published over 50 peer reviewed papers and book chapters in the field of Mechanical and Marine Engineering, mainly on power generation systems, decarbonisation, cryogenic energy applications, novel linear engine generators and zero carbon propulsions. He has received major grants from EPSRC, British Council, and European Commission. Dawei is the leader of the team and responsible for the whole project.

Ugo Ngwaka is a research associate at the University of Birmingham, and he graduated from Newcastle University with a Ph.D. degree. Ugo has published several papers related to the linear engine generator and now he is responsible for the design and experiment of the linear generator.

Gen Chen used to be the research associate at the University of Birmingham, he graduated from Xi'an Jiaotong University with a Ph.D. degree, and he is currently involved in the industry.

Mingqiang Li is a Ph.D. student at the University of Birmingham. Mingqiang has published two papers and at present, he focuses on the simulation and design of the linear engine generator.

Fangyu Zhangis a Ph.D. student at the University of Birmingham and she has published several papers related to the simulation of combustion and design of the combustor.

COMPARATIVE STUDY OF SHIP RESISTANCE ON MONOHULL AND CATAMARAN MODEL OF 15 GT FISHING BOAT

H Siswanti; M Musta'in and **A R Farisa,** Politeknik Negeri Madura, Indonesia

SUMMARY

On small boats, hull design that provides a smaller value of ship resistance is usually preferred. Therefore, it is necessary to consider the selection of the hull shape provides optimum resistance.The used of multihull known as alternative to reduce ship resistance.This study investigated comparison of the resistance of fishing vessels ±15 GT.The analysis will be carried out on 4 ship hull models, U-shaped and V-shaped monohull, streamlined and asymmetrical hull catamaran.Assuming the main dimension is constant, the differences in displacement, WSA, frictional resistance and total resistance were analyzed.Furthermore, a comparative analysis was carried out at several speed variations.Based on the study, it is known that the catamaran hull model provides less resistance than the monohull. The asymmetric catamaran provides the smallest ship resistance, but the reduction of ship resistance is accompanied by significant reduction in displacement. While the U-shaped monohull model shows the opposite

NOMENCLATURE

Loa	Length over All (m)
Lpp	Length perpendicular (m)
B	width (m)
H	Height (m)
T	Draught (m)
Vs	Speed (knot)
WSA	Wetted Surface Area (m^2)
Rf	Frictional Resistance (KN)
Rt	Total Resistance (KN)

1. INTRODUCTION

Resistance is the important factor in ship design, in connection with the development of the ship efficiency. One of the things that affect the ship resistance is the shape of the hull. The shape of the immersed hull will affect the fluid flow characteristics around the ship, hence causing the resistance to increase or decrease. If the resistance increases, the power of engine and also the fuel oil consumption increase respectivelly. On small boats a hull design that provides smaller resistance is usually preferred. Therefore, in designing the ship, it is necessary to consider the selection of the shape of the hull that provides the optimal resistance value. So with a small main engine power, the ship can achieve designed service speed. The shape of the hull such as catamaran and trimaran is known to have an effect on reducing ship resistance, when compared to monohull [1].

On passenger ships the use of the catamaran hull has been known as an alternative to reduce ship resistance [2]. In recent years, the use of the catamaran hull for fishing vessels has been widely studied. Because the use of a catamaran also provides an advantage in terms of ship stability [3,4]. The use of a catamaran hull also reduce the ship's resistance and the effect is to reduce fuel consumption [5].

In this study, a comparison of the resistance value of fishing vessels ±15 GT with a length of 13 meters will be analysed. The analysis will be carried out on 4 ship hull models, namely U-shaped monohull, V-shaped monohull, streamlined hull catamaran and asymmetrical hull catamaran. Assuming the same length (L), width (B) and draft (T), the differences in displacement, wetted surface area, frictional resistance and total resistance at a service speed of 16 knots were analysed. Furthermore, a comparative analysis of the ship resistance of each model was carried out at several speed variations (50-150% Vs).

2. METHODS

1. Designed 4 ship hull models, namely U-shaped monohull, V-shaped monohull, streamlined hull catamaran and asymmetrical hull catamaran. The design assuming the same length (L), width (B) and draft (T)
2. Calculation of wetted surface area, volume displacement and displacement, and then analysed by comparing each models.
3. Calculate, modelled and analysed of ships resistance each models
4. comparative analysis of the ship resistance of each model was carried out at several speed variations (50-150% Vs)

3. RESULT AND DISCUSSION

3.1 MODELLING

The catamaran fishing vessel worked in this research have main dimension as follows:

Loa	: 13 m
Lpp	: 12.52 m
B	: 7 m
H	: 2 m
T	: 1,2 m
Vs	: 16 Kn

Assuming the same length (L), width (B), Height (H) and draft (T), 4 models of the hull were made, namely U-shaped monohull, V-shaped monohull, streamlined hull catamaran and asymmetrical hull catamaran. The four hull models are presented in Figure 1, Figure 2, Figure 3 and Figure 4.

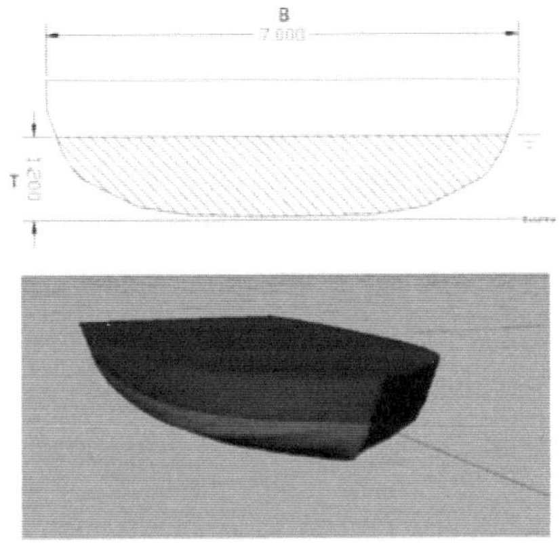

Figure 1: Model A, U-shaped monohull

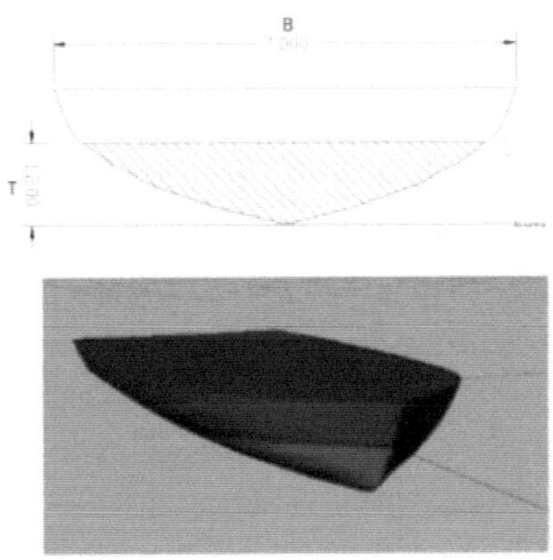

Figure 2: Model B, V-shaped monohull

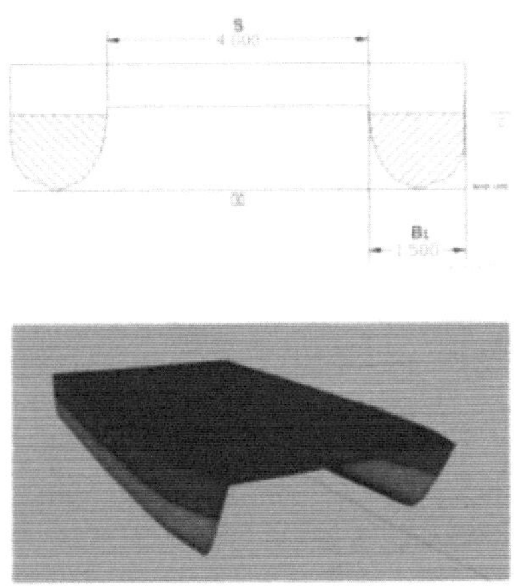

Figure 3: Model C, Streamlined hull catamaran

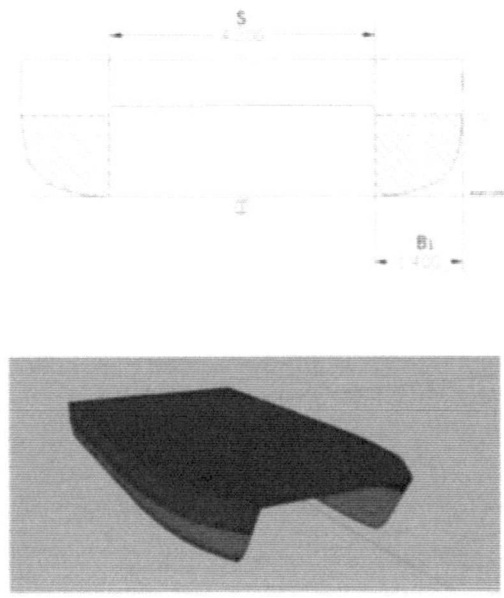

Figure 4: Model D, Asymmetrical hull catamaran

3.2 SHIP DISPLACEMENT COMPARISON

After the four ship models are designed, then the next step is to determine the displacement volume, displacement and Wetted Surface Area (WSA). The values for the four models are shown in table 1.

Table 1: Comparison each Model

Model	**WSA [m^2]**	**Volume *Displacement* [m^3]**	***Displacement* [ton]**
A (Monohull U)	85.905	53.634	55.23
B (Monohull V)	77.302	43.235	44.32
C (Streamline Catamaran)	74.142	31.881	32.68
D (Asymmetric Catamaran)	73.138	30.481	31.24

The WSA and displacement are used as comparison charts for each ship model to find out the differences between each model. The comparison graph is shown in Figure 5 and Figure 6.

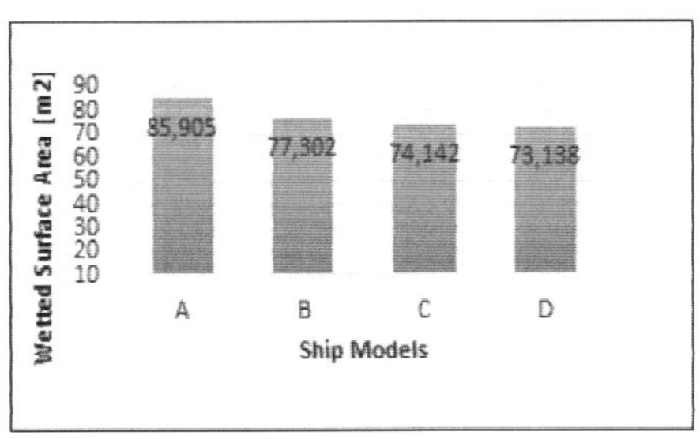

Figure 5: WSA comparison each model

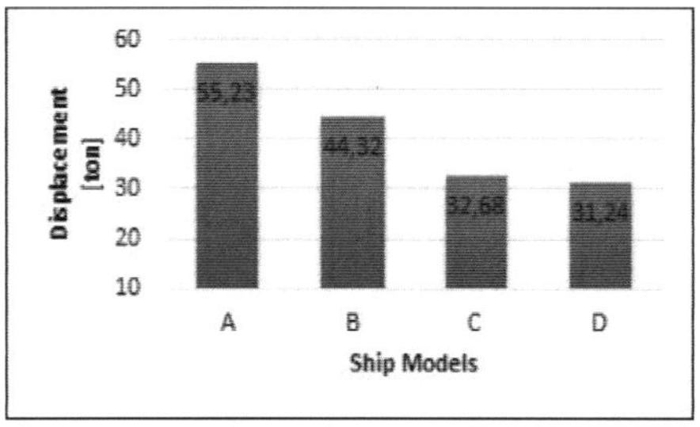

Figure 6: Displacement comparison each model

Changes in the shape of the immersed hull affect to changes of WSA and displacement in each model.

The wetted surface area of the catamaran is decrease about 13,6-14,8 % compared to the monohull. Shape changes of the immersed hull and the decrease in wetted surface influence the decrease in displacement. This is also confirmed by theory and other studies showing similar results [5,6].

The next step is calculate the ship's resistance at full speed (16 Knots) of each model ship. The ship resistance calculation using the Fung method. Besides the total resistance, the frictional resistance in each model is also compared. The results are shown in table 2 and the frictional resistance graphic is shown at figure 7.

Table 2:Friction & Total Resistance

Model	Friction Resistance (Rf) [kN]	Total Resistance (Rt) [kN]
A (Monohull U)	41.039	51.8
B (Monohull V)	40.015	50.273
C (Streamline Catamaran)	20.307	26.531
D (Asymmetric Catamaran)	19.612	25.668

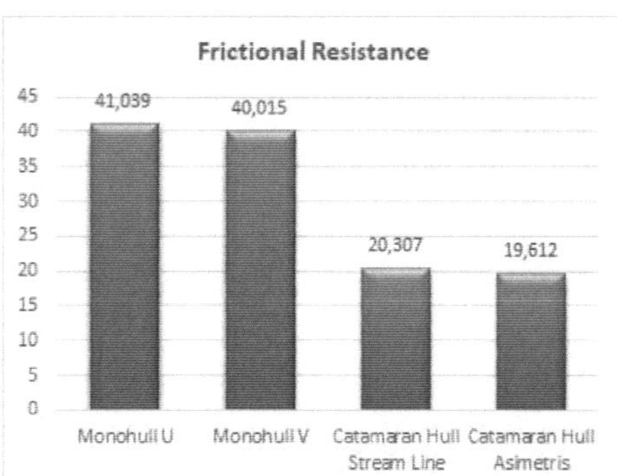

Figure 7: frictional resistance graphic of each model

This frictional resistance is influenced by the speed of the vessel, the viscosity of the fluid and the surface area immersed in the water. Therefore, with a decrease in WSA, the frictional resistance of the catamaran is also smaller than that of the monohull. The frictional resistance of the U-shaped monohull (model A) is the largest compared to other models, because this model has the largest WSA. otherwise, the frictional resistance of asymmetric catamaran (model D) shows the smallest results, because it has the smallest WSA. Meanwhile total resistance is affected by ship displacement. Therefore, the total resistance value of the catamaran model is also smaller than the monohull model, proportionally by the displacement. The results of this study show consistency with the theory [6].

Furthermore, a comparative analysis of the ship resistance of each model was carried out at several speed variations (50-150% Vs). The result of ship resistance of each model at several speed variations shown in figure 8.

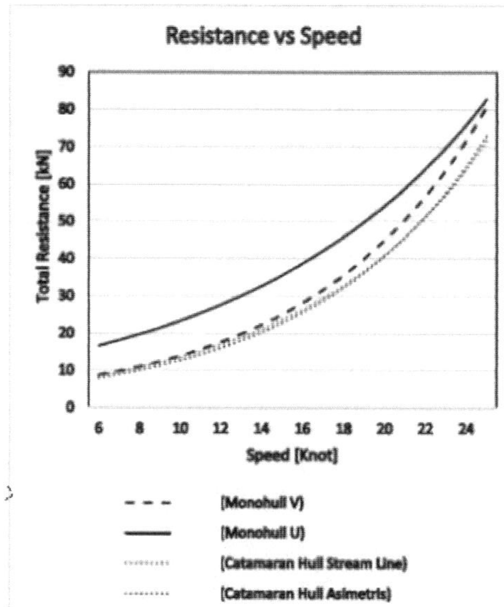

Figure 8: ship resistance of each model at several speed variations

The graph in Figure 8 shows that in all variations of speed the catamaran and V shaped monohull model resistance are lower than the U-shaped monohull model. In another study it was explained that the hard chine or V bottom shape has a smaller ship resistance when compared to the U round bottom or round flat bottom shape [7].

However, at low speeds the V monohull model and the catamaran have almost the same resistance. As the speed increases, the deviation of the ship's resistance between the catamaran and monohull models is getting further. This shows that at low, medium, and high speeds the catamaran hull model provides less resistance than the monohull. The asymmetric catamaran hull model provides the smallest ship resistance.This result confirm by theory and other research.

4. CONCLUSIONS

Based on the analysis, it is known that catamaran hull model provides less resistance than the monohull. The asymmetric catamaran hull model provides the smallest ship resistance value, but the reduction in the ship resistance value is also accompanied by a significant reduction in displacement compared to other ship models. While the U-shaped monohull model shows the opposite.

5. REFERENCES

1. Samuel, Iqbal. M, Utama, I.K.A.P "An Investigation Into The Resistance Components Of Converting A Traditional Monohull Fishing Vessel Into Catamaran Form", International Journal of Technology ISSN2086-9614, 2015

2. Kos, S., Brčić. D, Frančić, V, "Comparative Analysis Of Conventional And Swath Passenger Catamaran', 2009.

3. Samuel, Kim. D.J., Iqbal, M., Bahatmaka, A., Prabowo, A.R., "Modification of Traditional Catamaran to Reduce Total Resistance: Configuration of Centerbulb", Maritime Transportation and Harvesting of Sea Resources" , Taylor & Francis Group, London, ISBN 978-0-8153-799-5, 2018.

4. Putranto, T., Aryawan, W.D., Kurniawati, A.D., Setyawan, D., Pribadi, S. R. "Resistance and Stability Analysis for Catamaran Fishing Vessel with Solar Cell in Calm Water", MATEC Web of Conferences 159, 01059, 2018.

5. Tupan. J, et al "A Comparison of Monohull, Catamaran, Trimaran Vessels Based on Operational Review of Fuel Use", International Journal of Engineering Research & Technology (IJERT) Vol. 9 Issue 12, 2020.

6. Harvald, AA. SV., 'Tahanan dan Propulsi Kapal", Airlangga University Press, 1992

7. Y. Novita, "Studi Tentang Bentuk Kasko Kapal Ikan di Beberapa Daerah di Indonesia," *Jurnal Ilmu kelautan dan Perikanan,* pp. 16(4): 240-249., 2006.

6. AUTHORS BIOGRAPHY

Heni Siswanti, S.T., M.T. is lecturer in the Department of Shipbuilding Engineering, Politeknik Negeri Madura where she is responsible for the courses of Principal of Naval Architecture and Ship Structure. She is also PhD student at the Department of Naval Architecture, Sepuluh Nopember Institute of Technology. Her research interests include ship construction, advanced ship materials, and ship design.

M. Musta'in, S.T., M.T. is lecturer in the Department of Shipbuilding Engineering, Politeknik Negeri Madura. He is responsible for the courses of ship resistance and propulsion. His research interests include are ship resistance and marine machinery system.

Ainun Rina Farisa is student in the Department of Shipbuilding Engineering, Politeknik Negeri Madura. Under the supervision of Heni Siswanti and M. Musta'in She conducted the computer modelling of some small ship design and resistance.

EXPERIMENTAL STUDY OF THE STRENGTH OF COMPOSITE MATERIAL MADE OF CARBON FIBRES WITH A MIXTURE OF KEVLAR FIBRES AND PINEAPPLE FIBRES AS AN ADVANCE MATERIAL FOR MINE HUNTER HULL MATERIAL

Cahya Kusuma, Sekolah Tinggi Teknologi Angkatan Laut, Indonesia and Royal Institution of Naval Architecture, UK

SUMMARY

The advantages of sandwich material compared to metal materials are its lighter weight and low magnetism, making it suitable for use as minehunting ship hull material. In this study, the authors will make a carbon fiber sandwich with a mixture of Kevlar fiber and pineapple sisal fiber compared to GRP (Glass Reinforced Polyester). This research uses the technique of making Vacuum Assessment Resin Infusion (VARI). Strength test with tensile test, impact test and hardness test using ASTM standard. Based on the test results, carbon composite with a mixture of pineapple fiber has a maximum stress of 70,8008 N/mm2 which can be used as a hull material with a maximum length of 24m. While the carbon sandwich with kevlar has a maximum stress of 382,408N/mm^2, GRP has a maximum stress of 127.24N/mm^2. Carbon fiber material with kevlar can be used for boats longer than 24m.

1. INTRODUCTION

One of the main tasks of mining vessels is Mine Counter Measurement (MCM). One thing that is very important to note is the characteristics of the cause of the mine explosion which in this case is triggered by changes in the magnetic field in the surrounding area[1][2]. So minesweepers and sappers must have a low magnetic hull so as not to trigger a mine explosion. Therefore, by taking this into account, a vessel with material properties that has a low or zero magnetization value is needed to support the Minecountermeasurement operation of this type of vessel.
Taking this into account, one of the materials that has a very low or zero magnetization value and is reliable and strong is a composite material[3][4][5]. In this study, the authors innovate to replace kevlar fibre with pineapple sisal fibre to be used as an innovation in the manufacture of composites. The use of pineapple sisal fibre is because Indonesia is a tropical country that is rich in biological natural resources, especially fibre so that the three composite materials will be easy to find and inexpensive[5][6]. This research uses the technique of making vacuum assessment resin infusion (VARI)[7][8][9].

2. METHODOLOGY RESEARCH

In this study, the tensile strength, impact, and hardness of the sandwich composite material will be tested with carbon fibre material with a mixture of kevlar fibre, and sisal pineapple fibre. Tensile testing will be validated by tensile testing of Glass Reinforced Polyester (GRP).

2.1 KEVLAR FIBRES

One sheet of kevlar material can be seen in Figure 1

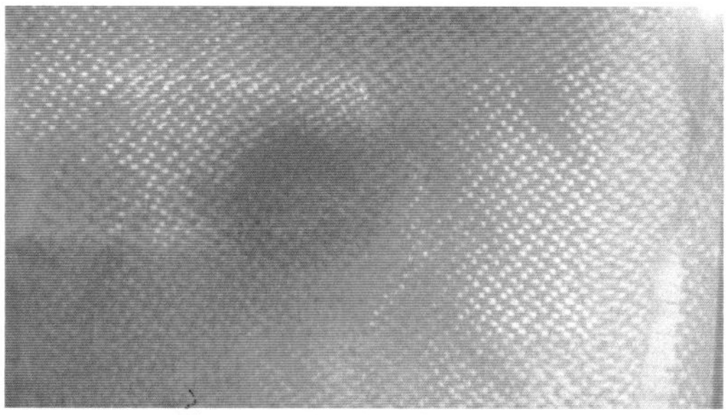

Figure 1. Kevlar Fibre material

2.2. SISAL PINEAPPLE

The sisal pineapple fibre is washed first to remove dirt and then the fibre is taken with a predetermined fibre length, generally 30 cm, then dried in the sun for about 3 hours or until dry and then soaked in NaOH solution with a concentration of 5% solution for about 3 hours. 2 hours later the fibre was washed again with clean water with the aim of removing the NaOH solution adhering to the fibre and then dried again for 3 hours in the sun so that the fibre could be used for composite materials used as reinforcement.. One sheet of pineapple fibrematerial can be seen in Figure 2

Figure 2. Pineapple fibre material

2.2.SPECIMEN MANUFACTURING

The manufacture of specimens in this study consisted of 2 specimens, namely carbon composite specimens with Kevlar and sisal pineapple specimens. Layer composition as in the table 1 and 2.Specification of Kevlar specimen consists of 4 layers of carbon fibre, 4 layers of Kevlar fibre and 4 layers of carbon fibre as shown in table 1.

Table 1. Specification of Kevlar specimen

No	Fibres Content	Weight (gram/m)
1	Carbon Fibre	300
2	Carbon Fibre	300
3	Carbon Fibre	300
4	Carbon Fibre	300
5	Kevlar Fibre	175
6	Kevlar Fibre	175
7	Kevlar Fibre	175
8	Kevlar Fibre	175
9	Carbon Fibre	300
10	Carbon Fibre	300
11	Carbon Fibre	300
12	Carbon Fibre	300

Specification of Sisal pineapple consists of 4 layers of carbon fibre, 4 layers of Sisal pineapple fibre and 4 layers of carbon fibre as shown in table 2.

Table 2. Specification of sisal pineapple specimens

No	Fibres Content	Weight (gram/m)
1	Carbon Fibre	300
2	Carbon Fibre	300
3	Carbon Fibre	300
4	Carbon Fibre	300
5	Sisal pineapple Fibre	175
6	Sisal pineapple Fibre	175
7	Sisal pineapple Fibre	175
8	Sisal pineapple Fibre	175
9	Carbon Fibre	300
10	Carbon Fibre	300
11	Carbon Fibre	300
12	Carbon Fibre	300

2.3. VACUUM ASSESSMENT RESIN INFUSION

Vacuum Assessment Resin Infusion :

a) Clean the workbench, make sure it is level and grease it with magic gloss wax.
b) Glue the mesh cloth using spray adhesive then smooth it out.
c) Install layer by layer 4 layers of carbon fibre, 4 layers of pineapple fibre, 4 layers of carbon fibre
d) Glue sealant tape around the specimen so that the plastic becomes vacuum
e) Place on the left and right sides of the suction and pressure hoses that have been lined with mesh cloth and peel ply cloth.
f) Mix 5 kg of refractory resin with 100 ml of Mepox A catalyst, with a mixture ratio of 10% of the resin volume.
g) Furthermore, the resin insertion process is approximately 10-15 minutes. The resin insertion time depends on the size of the specimen being made.
h) Wait for the specimen to cool for approximately 6-8 hours. This cooling process depends on the area and thickness of the specimen being made.
i) After the specimen sheet is completed, the sheet is cut according to the shape of the tensile, impact, and hardness test specimens.

The process of making specimens with the VARI method can be seen in Figure 3.

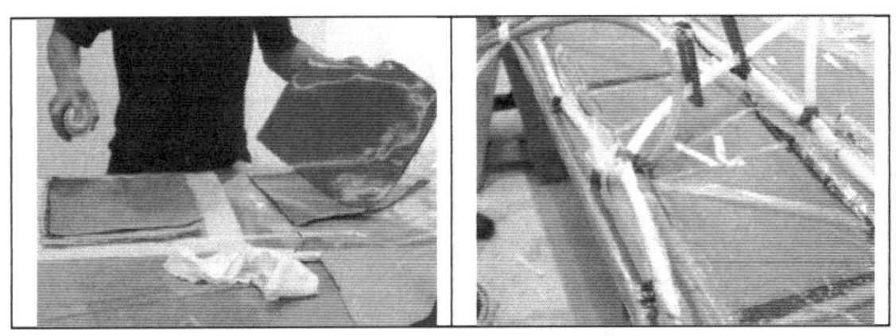

Figure 3. VARI Processing

3. RESULT AND DISCUSSION

Hardness Test (ASTM E384), and Impact Test (ASTM D – 256)

3.1 TENSILE STRENGTH

3.1 (a) Tensile Strength Kevlar Fibre with carbon at table 1.

No	Max_Stress N/mm^2	Max_Strain %
1	385,21	9.9
2	379.32	9.8
3	381.5234	10
	380.4217	9.9

The average tensile strength of Kevlar Fibre with carbon is 380N/mm^{2}elongation 9.9%.

3.1 (b) Tensile Strength Pineapple Fibre with carbon

The results of the Tensile Strength Pineapple Fibre with carbon test in the table 2.

Table 2. Tensile Strength Pineapple Fibre with carbon

No	Max Force	Max_Disp	Max_Stress	Max_Strain
	Kn	mm	N/mm^2	%
1	4.53125	9.671	70.8008	29.3061
2	3.96875	10.9365	62.0117	33.1409
3	3.93	8.02075	61.5234	24.3053

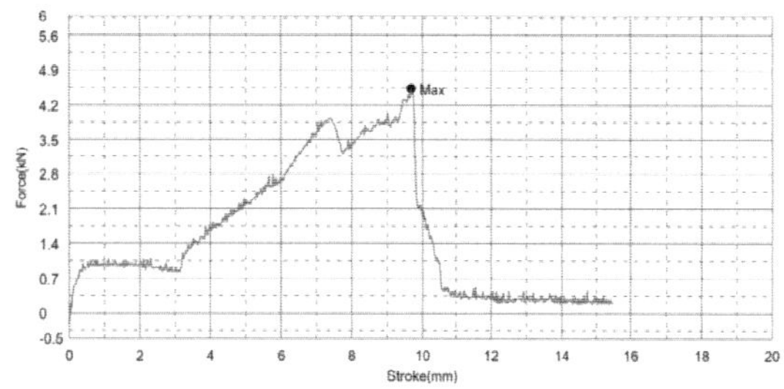

Figure 4. Pineapple Fibre Composite Tensile Test -1

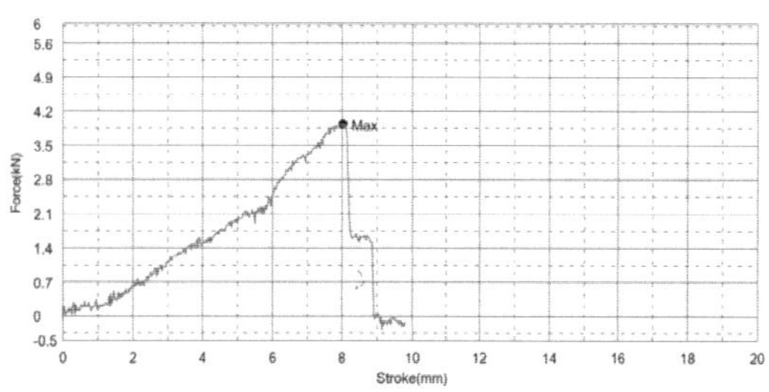

Figure 5. Pineapple Fibre Composite Tensile Test -2

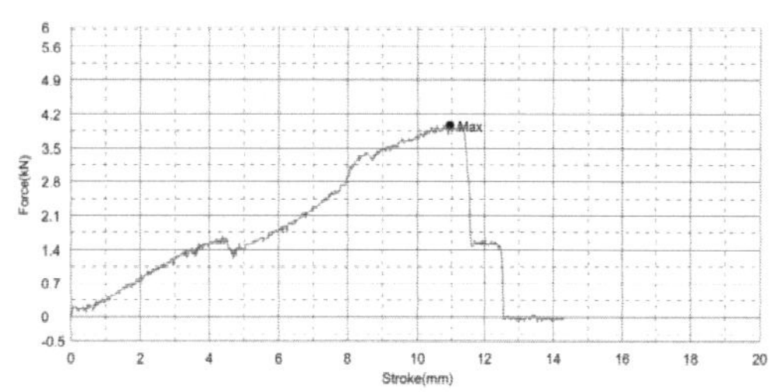

Figure 6. Pineapple Fibre Composite Tensile Test -3

Carbon composite with a mixture of pineapple fibre has a tensile strength value of 3.96875 kN, a maximum displacement of 10.9365 mm, a maximum stress of 70.8008 N/mm2, and a maximum strain of 33.14%.

3.1 (c) Tensile Strength Glass Reinforced Plastic

GRPspecimen in Figure 7.The tests carried out are Tensile Test BKI Rules for Fibreglass Reinforced Plastics Ships – 2016[10]

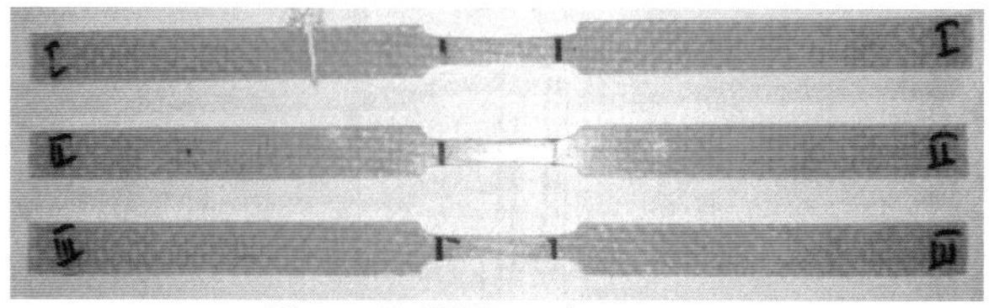
Figure 7. GRPspecimen

The results of the tensile strength of GRP in the table 3.

Table 3. Tensile strength of GRP

Experiment	Sample			Mean
	1	2	3	
Max Force(N)	10281.3	10118.8	9662.5	10020.9
Break Load(N)	9275	9106.25	9250	9210.42
Tensile Strength(N/mm^2)	122.648	123.239	120.38	122.088
Elongation(%)	14.38	15.78	14.936	15.032

3.2. HARDNESS TESTER

Hardness test in this study using the SHORE D and MOSH tools. Test results using the SHORE D tool in the table 5.

Table 5. Hardness tester used Shore D

SHORE D	Specimen			Mean
	1	2	3	
Carbon Kevlar	89	90	89	89.3
Carbon Pineapple	90	87	88	88.3

Test results using the MOSH tool in the table 6.

Table 6. Hardness tester used MOSH

MOSH	Specimen			Mean
	1	2	3	
Carbon Kevlar	7	6	6.5	6.5
Carbon Pineapple	4	5	5	4.6

In the hardness test, the kevlar fibre composite has the highest strength, namely 89.3 for the D-SHOR method and 6.5 for the MOSH method.

3.3. IMPACT TEST

Impact test results of Pineapple fibre with Carbon in Table 7.

Table 7. Impact test results of Pineapple fibre with Carbon

Angle β	Angle α	Result (J/mm^2)
41	70	0.023
28	70	0.031
21	70	0.034

Impact test results of kevlarfibre with Carbon in Table 8.

Table 8. Impact test results of Kevlarfibre with Carbon

Angle β	Angle α	Result (J/mm^2)
10	70	0.0373
8	70	0.0377
7	70	0.0378

From the impact test data, it was found that Kevlar carbon fibre has the highest impact value with a value of 0.0378 J/mm2 with an average of 0.0376 J/mm2 in each test sample.

4. CONCLUSION

Based on the tensile strength data, carbon composite with a mixture of pineapple fibre has the highest strength value with a tensile force value of 3.96875 kN, a maximum displacement of 10,9365 mm, a maximum stress of 70,8008 N/mm2, and a maximum strain of 33.14% and the Indonesian classification bureau for ships with a length of 24m are required to have strength[11][12] with a tensile strength of 85Mpa.
Base on Indonesian classification bureau for special shipsRules For Fibreglass Reinforced Plastic ShipsTensile strength : 98 N/mm^2[10].
The material is more appropriate for ships with a length of less than 24 m for the hull, while for ships in general it is suitable for the interior of the ship.
Based on the average tensile strength value of Kevlar Fibre with carbon and validation with GRP as well as the rules of the Indonesian Classification Bureau, it is concluded that carbon fibre with Kevlar can be used as material for mine boats longer than 24m.

5. ACKNOWLEDGEMENTS

The author would like to thank you for the help or encouragement from colleagues, especially the Sekolah Tinggi Teknologi Angkatan Laut (STTAL), Surabaya, Indonesia.

5. REFERENCES

1 A. S. Erickson and L. J. Goldstein, *Chinese Mine Warfare*, no. 3.

2 Ø. Midtgaard and M. Nakjem, "Unmanned Systems for Stand-off Underwater Mine Hunting," *Underw. Def. Technol.*, no. 2027, pp. 1–7, 2016.

3 T. Surdia and S. Saito, "Pengetahuan Bahan Teknik," 1985.

4 N. H. Sari, A. Taufan, and I. Yudhyadi, "Ketahanan Bending Komposit Hybrid Serat Batang Kelapa/Serat Gelas Dengan Matrik Urea Formaldehyde," *J. Energi Dan Manufaktur*, vol. 5, no. 1, pp. 91–97, 2012.

[5 F. Rubino, A. Nisticò, F. Tucci, and P. Carlone, "Marine application of fibre reinforced composites: A review," *J. Mar. Sci. Eng.*, vol. 8, no. 1, 2020.

6 S. M.M, *Composite Materials Handbook,*. New York.: McGraw-Hill Book., 1984.

7 2014:141 Notoatmodjo *et al.*, "No 主観的健康感を中心とした在宅高齢者における健康関連指標に関する共分散構造分析Title," *Proc. 8th Bienn. Conf. Int. Acad. Commer. Consum. Law*, vol. 1, no. hal 140, p. 43, 2014.

8 M. R. Ricciardi *et al.*, "A new cost-saving vacuum infusion process for fibre-reinforced composites: Pulsed infusion," *J. Compos. Mater.*, vol. 48, no. 11, pp. 1365–1373, 2014.

9 S. Shevtsov, I. Zhilyaev, S. H. Chang, J. K. Wu, J. P. Huang, and N. Snezhina, "Experimental and numerical study of vacuum resin infusion for thin-walled composite parts," *Appl. Sci.*, vol. 10, no. 4, 2020.

10 Biro Klasifikasi Indonesia, "Rules For The Classification And Construction Part 3. Special Ships Volume V Rules For Fibreglass Reinforced Plastic Ships" 2016 Edition

11 Biro Klasifikasi Indonesia, "RULES FOR SMALL VESSEL UP TO 24 M Consolidated Edition 2021," vol. VII, 2021.

12 V. V. BKI, "Volume vii rules for small vessel up to 24 m 2013 edition biro klasifikasi indonesia," vol. VII, 2013.

9. AUTHORS BIOGRAPHY

Cahya Kusuma, Department of Mechanical Engineering, Sekolah Tinggi Teknologi Angkatan Laut (STTAL) Surabaya, Indonesia. Active as a member of RINA since 2020

THE EFFECT OF ADDITIONAL INTERCEPTORS ON SHIP RESISTANCE AND LIFT FORCE CASE STUDY PATROL SHIP 40 METER

Cahya Kusuma, Sekolah Tinggi Teknologi Angkatan Laut, Indonesia and Royal Institution of Naval Architecture, UK
Wawan Kusdiana, Sekolah Tinggi Teknologi Angkatan Laut, Indonesia
Sutrisno, Sekolah Tinggi Teknologi Angkatan Laut, Indonesia
Riko Fendiawan, Sekolah Tinggi Teknologi Angkatan Laut, Indonesia
Agoes Santoso, Marine Engineering Department Institut Teknologi Sepuluh Nopember, Surabaya, Indonesia, and Royal Institution of Naval Architecture, UK

SUMMARY

This paper examines the effect of adding Interceptor to ship resistance and ship lift force. The research was carried out by making a model of the Patrol 40 trim by stern without an Interceptor and with the addition of an interceptor with Maxsurf software, then tested with Fine Marine (Numeca) software. The test results showed a decrease in resistance after the addition of the Interceptor at a speed of 12 knots decreased by 16,196 kN or 42.9%, at a speed of 18 knots decreased by 86.356 kN or 63.9% and at a speed of 24 knots decreased by 122,699 kN or 62,92%. The lift force generated at a speed of 12 knots 1,134,977 kN or the stern rises 0.2 cm, at a speed of 18 knots 1,135,598 kN or the stern rises 20.1 cm and at a speed of 24 knots 1,132,403 kN or the stern rises 32.5 cm.

1. INTRODUCTION

The initial design of the 40 meter patrol boat did not use an Interceptor. The addition of the Interceptor is considered because ships that experience Trim by Stern or ships tend to sink more at the stern. With the addition of this Interceptor, it will analyze the changes in resistance and lift generated by the addition of the Interceptor and whether the Interceptor is in accordance with the resistance and lift force to support the ship's trim. This research was conducted by making a ship model without an Interceptor and by adding an interceptor with Maxsurf software, then testing it with Fine Marine (Numeca) software.

The interceptor is a metal plate mounted vertically a few centimeters below the transom. Its working principle is to deflect the fluid flow that passes through the keel of the ship, thereby causing a lifting force at the stern of the ship.

Figure 1. Interceptor on patrol boat 40m

The height of the interceptor installed will affect the efficiency of the ship's lift force against the ship's speed according to the speed of the ship to be used[1][2][3][4]. The resistance angle created by the interceptor reduces unwanted bend angles due to fluid flow that is bent by the interceptor at high speed and the speed can be increased up to 4-5 thereby saving fuel up to 14%[5].Effectiveness on hydrodynamic performance of high-speed planig crafts on monohulls and catamarans shows a remarkable reduction in resistance of up to 15% for monohull models and a reduction in resistance of up to 12% for catamaran models[2].The addition of an interceptor also affects the trim[4] and maneuverability of the ship[6]. By reducing the resistance, the ship's speed can be increased[7][8][9][10]

2. RESULT AND DISCUSSION

2.1 PATROL BOAT 40M MODEL

Specification of 40 meter patrol boat :

Length (LOA)	: 46.23 Meters
Length Water line (LWL)	: 42.02 Meters
Max Breadth (B. Max)	: 7.90 Meters
Waterline Breadth (B.WL)	: 6.90 Meters
Draft	: 2.20 meters
Maximum Speed	: 24.5knots

The first step is to draw a model using maxsurf software according to the ship's line plan data, then render it into a solid model in Figure 2.

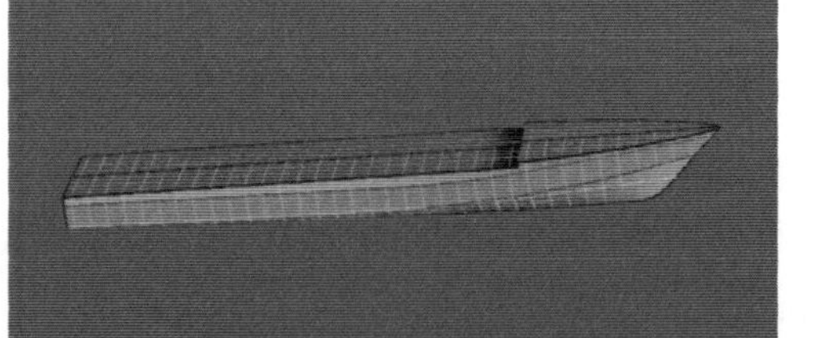

Figure 2. Rendering model 40 meter patrol boat.

The second step is to use the Autodesk Inventor software, the ship model from maxsurf, which has already been made, an interceptor model on the stern transom of the ship model in Figure 3.

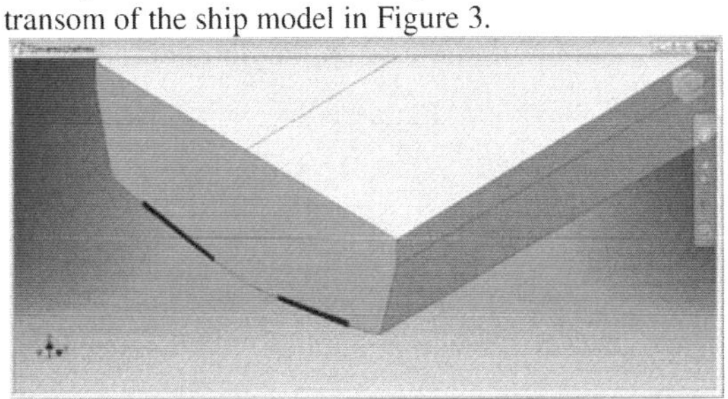

Figure 3. Interceptor model

2.1 TESTING THE MODELWITHOUT INTERCEPTOR

Testing the model without an interceptor using Fine Marine (Numeca) software at speeds of 5 knots, 12 knots, 18 knots and 24 knots.

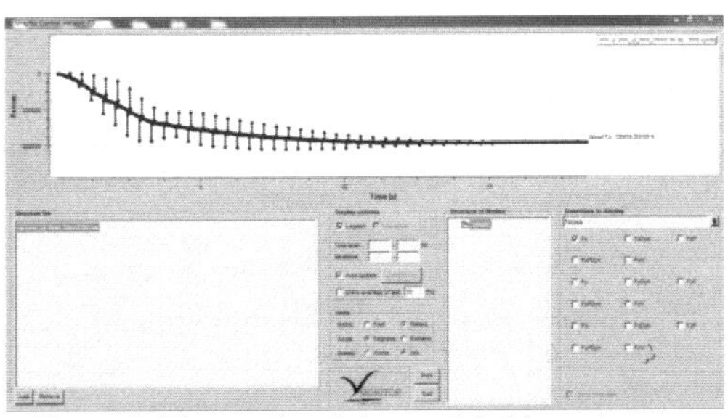

Figure 4. The results of the model ship resistance test without an Interceptor at a speed of 24 Knot.

Ship resistance without an interceptor can be seen in the table 1 and figure 5.

Table 1.Ship Resistance Without Interceptor

Without Interceptor	Velocity Ship(Knot) 5	12	18	24
Ship resistance	4956.77	37828.8	135726	195019
Trim by stern(o)	0.00581	0.08837	1.45264	2.71843
Trim by stern(m)	0.00174	0.02642	0.43428	0.81249

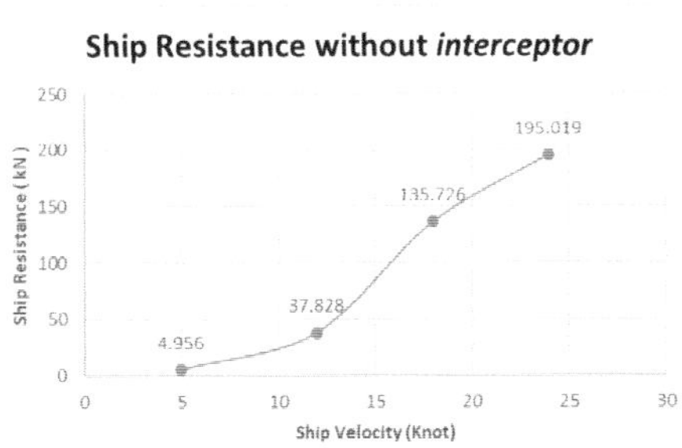

Figure 5. Ship resistance without Interceptor.

2.2. TESTING THE MODELWITH INTERCEPTOR

In this test, the amount of resistance generated by the interceptor is calculated and the magnitude of the lift force caused by the interceptor is calculated. The interceptor height variable to be tested with fine marine software (Numeca) is from a maximum height of 30 mm, 25 mm, 20 mm, and 15 mm, while the speed variable used is 5 knots manoeuvring speed, 12 knots economical speed, 18 knots cruising speed. Knots and full speed 24 Knots.

2.2 (a) Interceptor Testing With A Variable Height Of 30 mm At 4 Variations Of Speed (5 Knots, 12 Knots, 18 Knots And 24 Knots)

The test was carried out at a blade height of 30 mm Interceptor at a speed of 5 knots to a full speed of 24 knots.One of the values of resistance and lift force of a variable ship is 30mm at a speed of 24 knots in Figure 6.

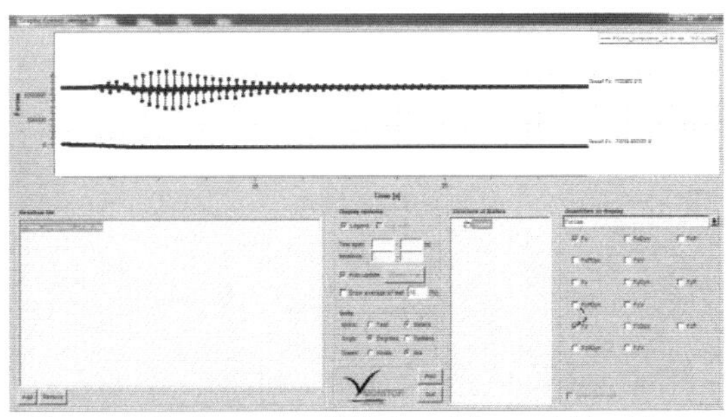

Figure 6. Resistance and lift force of a variable ship is 30mm at a speed of 24 knots

Table 2. Ship Resistance With Interceptor 30 mm At 4 Variations Of Speed (5 Knots, 12 Knots, 18 Knots And 24 Knots)

Interceptor 30mm	Velocity Ship(Knot) 5	12	18	24
Ship resistance(N)	3259.99	21632.01	49368.68	72319.90
Angle(α)	0.003144	0.069338	0.675523	1.088750
Lift Force(N)	1133431	1134977	1135598	1132403

2.2 (b) Interceptor Testing With A Variable Height Of 25 mm At 4 Variations Of Speed (5 Knots, 12 Knots, 18 Knots And 24 Knots)

Table 3. Ship Resistance With Interceptor 25 mm

Interceptor 25mm	Velocity Ship(Knot) 5	12	18	24
Ship resistance(N)	3473.67	21998.31	51267.78	75058.82
Angle(α)	0.001591	0.073848	0.779029	1.267924

Lift Force(N)	1133673	1134570	1135374	1132496

2.2 (c) Interceptor Testing With A Variable Height Of 20 mm At 4 Variations Of Speed (5 Knots, 12 Knots, 18 Knots And 24 Knots)

Table 4. Ship Resistance With Interceptor 20 mm

Interceptor 20mm	Velocity Ship(Knot)			
	5	12	18	24
Ship resistance(N)	3464.49	22036.77	50005.39	74063.89
Angle(α)	0.000940	0.072500	0.723187	1.210530
Lift Force(N)	1133691	1134928	1135072	1131991

2.2 (d) Interceptor Testing With A Variable Height Of 15 mm At 4 Variations Of Speed (5 Knots, 12 Knots, 18 Knots And 24 Knots)

Table 5. Ship Resistance With Interceptor 15 mm

Interceptor 15mm	Velocity Ship(Knot)			
	5	12	18	24
Ship resistance(N)	2088.26	21892.25	47696.82	74498.00
Angle(α)	0.001279	0.091400	0.692723	1.201721
Lift Force(N)	1133789	1134847	1135542	1132311

From the data on the Interceptor height variable in tables 2 to 5, the lift force that affects the trim by stern height of the ship is calculated, as a result of the lift force generated by the interceptor. The results of all test variables are made into one table of test calculation results in the figure:

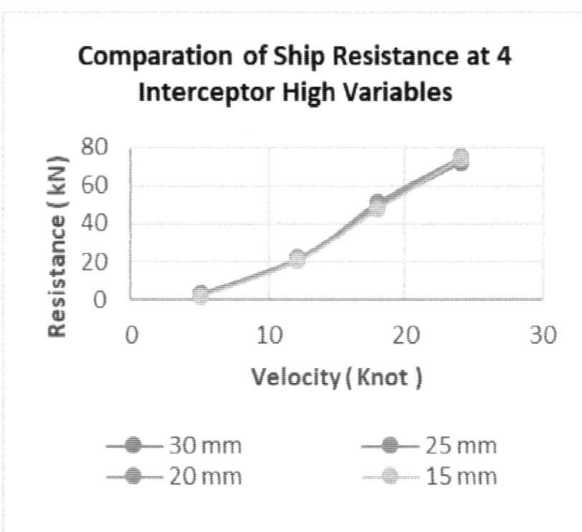

Figure 7. Comparation of Ship Resistance at 4 interceptor high variables

Comparison of lifts using an interceptor on 4 speed variables at figure

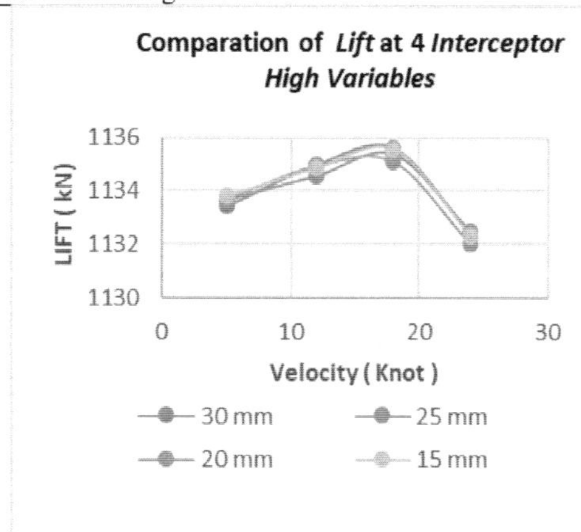

Figure 8. Comparation of lift at 4 interceptor high variables

Comparison of ship resistance and lift with a 30mm high interceptor at a variable speed of 5 knots to 24 knots in the table

Table6.Comparison of ship resistance and lift with a 30mm high interceptor

Velocity	Ship Resistance (kN)		Lift due The
	With Interceptor	Without Interceptor	Interceptor (cm)
5	4.96	3.259	0.094
12	37.828	21.632	2.07
18	135.726	49.368	20.19
24	195.019	72.319	32.55

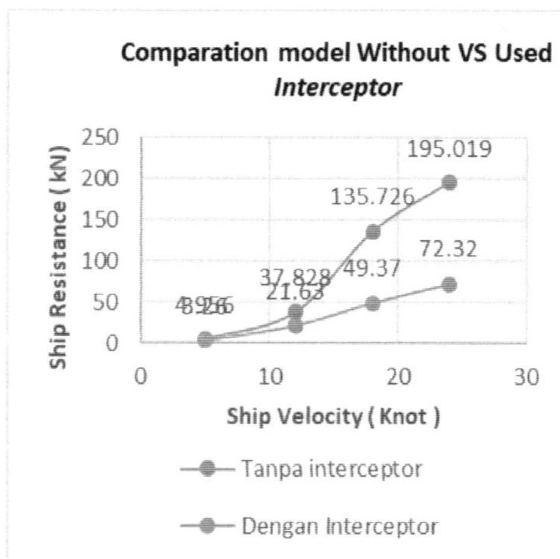

Figure 9. Comparation Model Without VS Used Interceptor

From the comparison curve picture above, it can be seen that the change in resistance after the addition of the 30mm Interceptor, namely the speed of 5 knots of resistance decreased by 34.2%, the speed of 12 knots of resistance decreased by 42.9%, the speed of 18 knots of resistance decreased to 63.9% and the speed of 24 knots of resistance decreased. to 62.92%. From the curve above, it is known that the installation of the interceptor resulted in a large decrease in resistance at each speed tested.

3. CONCLUSION

From the analysis carried out on the addition of the Interceptor on the 40 meter patrol boat, it can be concluded:
a.Changes in resistance that occur at each speed tested decreased by 75.08 kN on average from the resistance without an interceptor.
b.The value of the lift force generated at the stern of the ship after the addition of the interceptor at each test speed is an average of 1,134,326 kN
c.In the test variables, all the values of resistance and lift force meet the criteria needed by the ship because the difference in resistance and lift force is very small.The installation of the interceptor is proven to increase the ship's lift force which increases the trim position of the ship and in the end can reduce the total resistance of the ship.
For further research, it is necessary to make a ship model whose boundary area is only on the part where the interceptor is installed. In order for the calculation to focus more on changes in resistance, lift force and other phenomena due to the interceptor so that it is more detailed

4. ACKNOWLEDGEMENTS

The author would like to thank you for the help or encouragement from colleagues, especially the Sekolah Tinggi Teknologi Angkatan Laut (STTAL), Surabaya, Indonesia.

5. REFERENCES

1 A. G. Avci and B. Barlas, "An experimental investigation of interceptors for a high speed hull," *Int. J. Nav. Archit. Ocean Eng.*, vol. 11, no. 1, pp. 256–273, 2019.

2 M. H. Karimi, M. S. Seif, and M. Abbaspoor, "An experimental study of interceptor's effectiveness on hydrodynamic performance of high-speed planing crafts," *Polish Marit. Res.*, vol. 20, no. 2, pp. 21–29, 2013.

3 A. G. Avci and B. Barlas, "The usage of interceptors in high speed crafts," no. December, 2016.

4 H. Ghassemi, M. Mansouri, and S. Zaferanlouei, "Interceptor hydrodynamic analysis for handling trim control problems in the high-speed crafts," *Proc. Inst. Mech. Eng. Part C J. Mech. Eng. Sci.*, vol. 225, no. 11, pp. 2597–2618, 2011.

5 A. G. Avci, B. Barlas, and A. I. Ölçer, "An Investigation of Fuel Efficiency in High Speed Vessels by Using Interceptors," no. January, pp. 127–141, 2018.

6 M. A. Hafiz, A. Sulisetyono, and A. Baharuddin, "Study of the Interceptors Effect on Ship Manoeuvrability Using the Open Free Running Test Method," *IOP Conf. Ser. Mater. Sci. Eng.*, vol. 1052, no. 1, p. 012040, 2021.

7 S. Resistance, "FundamentaJs of Ship Resistance and," no. July, 2003.

8 O. M. Faltinsen, *Hydrodynamics of high-speed marine vehicles*, vol. 9780521845. 2006.

9 W. Frederick and N. Y. J. Wiley, "The resistance and propulsion of ships / by William F . Durand .," pp. 1859–1958, 1958.

10 A. F. Molland, S. R. Turnock, and D. A. Hudson, *Ship resistance and propulsion: Practical estimation of ship propulsive power*, vol. 9780521760. 2011.

9. AUTHORS BIOGRAPHY

Cahya Kusuma Department of Mechanical Engineering, Sekolah Tinggi Teknologi Angkatan Laut(STTAL)Surabaya, Indonesia. Active as a member of RINA since 2020

Wawan Kusdiana,Department of Mechanical Engineering, Sekolah Tinggi Teknologi Angkatan Laut(STTAL)Surabaya, Indonesia .

Sutrisno,Department of Mechanical Engineering, Sekolah Tinggi Teknologi Angkatan Laut(STTAL)Surabaya, Indonesia .

Riko Fendiawan, Department of Mechanical Engineering, Sekolah Tinggi Teknologi Angkatan Laut(STTAL), Surabaya, Indonesia

Agoes Santoso,Marine Engineering Department Institut Teknologi sepuluh Nopember, Surabaya, Indonesia, Indonesia and Royal Institution of Naval Architecture, UK

RE-DESIGN SOLID BOX AND AIR BAG CONFIGURATIONS ON UNSINKABLE SMALL PASSENGER BOAT

W R Hetharia, **E R de Fretes, J Tupan**and **E J Legwurnitua** , Facuty of Engineering - Pattimura University Ambon, Indonesia

SUMMARY

Previous works were conducted by the authors concerning the performance of unsinkable small passenger boat. The boat was designed and constructed in model scale and full-scale ship. The model was tested in tank and full-scale ship was tested at open sea to verify the design parameters. The research is continued in order to improve the boat performance. A re-design process was executed in order to modify the boat,air bags and centralised air supply system. The modified unsinkable boat was tested with the full passengers at sea for floating and sinking purposes. The required volume of solid boxes and air bags on board and outside are1.12 m^3, 0.563 m^3 and 0.213 m^3 respectively. The result of test at sea showed that the modified boat was floating, unsinkable at static condition even with full water in board and keep running with flooding conditions. It is found that the boat is stable due to its positive value of GMt and GZ. The recommendation was made for future application for the stakeholders.

NOMENCLATURE

[Symbol]	[Definition] [(Unit)]
Δ	Weight displacement (tonnes)
B	Beam (m)
BMt	Radius Metacentric (m)
C_B	Block coefficient
DWT	Deadweight (tonnes)
GM_T	Initial metacentre (m)
GZ	Righting arm (m)
H	Height (m)
KB	Vertical centre of buoyancy (m)
L	Length (m)
L_{OA}	Length overall (m)
L_{WL}	Length of waterline (m)
LWT	Lightweight (tonnes)
T	Draft (m)
V	Speed (knot)
W	Total boat weight (tonnes)
ρ	Specific weight (1.025 ton/m^3, SW)

1. INTRODUCTION

Problem concerning the safety of small passenger speed boat was continued from the previous study by the authors. This study was initiated by collecting the data concerning the accidents of sinking and capsizing. A new concept was developed by introducing solid boxes and air bags. The concept of solid boxes was developed by the authors from the previous work of tuna long boat fishing boat [1, 2]. Similar concept was introduced into small passenger speed boat. However, initial computation proved that the solid boxes would not fully apply due to the spaces required for the passengers. In this case expended air bags were applied. In the normal condition the air bags are in folding condition and were stored in their places inside and outside the boat (under boat dampers). Those air bags will be expanded by supplying the air from central air compressors. The study was continued with theoretical computation of boat's parameters. A model-scale boat was developed and tested on the tank to validate the computation. The results of test proved that the model was afloat when fill with water and return to its initial position after being capsized [3, 4]. In addition, a prototype full-scale boat was constructed and tested at the sea to verify the theoretical computation and results of model test. The sea trial confirm that the boat was afloat [5]. This paper describes the recent study of application of central air bag system to the air bags placed on board, side and top of the boat roof. With the existing configuration, the sea trial proved that the boat still afloat in static and running condition while full of water inside.

2. LITERATURE REVIEW

2.1 SAFETY ISSUE OF SMALL BOATS

The role of fast passenger transport in the archipelago regions contributes to development of the regions, particularly in Eastern Indonesian regions. The need of fast transport with small payloads makes this kind of boat is favourable to be operated in this region. In fact, some marine accidents occur due to harsh sea condition and boat configuration which cause the loss of life and boat. In this case, most of the boats are sink or capsize at sea [6, 7]. In addition, a study was conducted concerning the effect of boat dimension of stability parameter [8]. However, the stability parameter could not extend more due to limitation of small boat size. A survey carried out by our research team in this last three months concluded that most of those small passenger boats were not provided with safety equipment. Regular safety equipment like life jacket, fife buoy, compass and marine radio are not available on board. The response from boat operators are: the journey is in short distance and the destination port may be viewed. Even so, there are a lot of sea accident happen every time. There should be a serious intention from related institution concerning this safety issue.

2.2 THE CONCEPT OF SHIP FLOATING

A ship can be float when there is a balance between the total weight and weight displacement. In this case, both two parameter should be equal in quantity [9, 10,and 11]. This relationship is showed as follows:

$$\Delta = W \quad (1)$$

where: Δ = Weight displacement
W = Total weight

The total weight of a ship consists of two weight components which are lightweight (LWT) and deadweight (DWT) [12, 13, 14, 15]:

$$W = LWT + DWT \quad (2)$$

The LWT component includes the boat weight, engine and equipment and DWT component consists of passengers and boat operators, luggage and fuel.

The weight displacement of boat (Δ) is obtained from:

$$\Delta = C_B \times L \times B \times T \times \rho \quad (3)$$

where: L = boat length B = boat beam
T = boat draft C_B = block coefficient
ρ = specific weight (1.025 ton/m^3, sea water)

To achieve boat balance at the draft (T), then the total boat weight (W)equals to weight displacement (Δ).
The incoming water will reach the boat when the draft (T) is greater than ship height (H). In this case, we have the relation of:

$$W_{total} > \Delta = C_{Bh} \times L \times B \times H \times \rho \quad (4)$$

where:
C_{Bh} = block coefficient at ship height.

Since the LWT and part of DWT components are constant then the additional weight is the incoming water on board. The innovation of unsinkable small passenger boat is explained further in the references [3, 4 and 5]. The general concept is described as follows (Figure 1).

The total volume inside the boat is consist of empty space, structures and solid boxes. Our computation showed that when the empty volume filled with water then the boat will sink. To reduce the empty space then sufficient air bags should inserted in the boat. This condition will end-up with the reducing of incoming water. In addition, the existence of passengers on board will reduce the incoming water. The application of air bags should be limited in order to provide the space for the passengers.

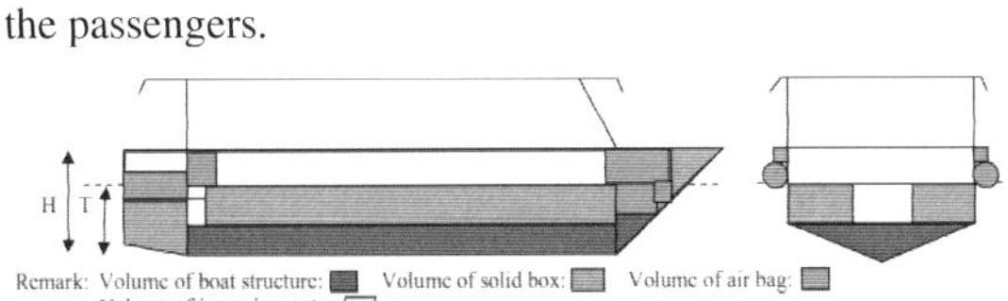

Figure 1: Composition of solid boxand air bag on board

When the boat is full of water and passengers, it will loss the stability due to the low metacentric height and the effect of free water surface on board. In this case, additional air bags should be provided outside the boat. This configuration will add the buoyancy and provide righting moment to increase the stability parameter.

3. METHODOLOGY

3.1 BOAT DATA FROM PREVIOUS WORKS

The boat data in this study is collected from the previous works [3, 4 and 5]. The data is presented in this following Table 1.

Table 1. The Dimensions of Full-Scale Boat

No	Boat parameters	Value	Unit
1	Length Overall, L_{OA}	6.61	m
2	Length of Waterline, L_{WL}	6.35	m
3	Beam, B	1.40	m
4	Draft, T	0.40	m
5	Height, H	0.65	m
6	Speed, V	13	knot

Other boat data are described as follows:
Passenger: 14 persons Operator: 1 person
Hull Material: FRP Autonomy: up to 30 n.m
Luggage: @ 20 kg Fuel: 40 l
Prime mover: 40 HP (outboard)
Equipment: anchor, fuel tank, air bags, air supply systems

The lines plan was developed by using Maxsurf software. Hydrostatics and stability parameters were computed by this naval software. The lines plan and general arrangement of the boat are presented in Figures 2 and 3.

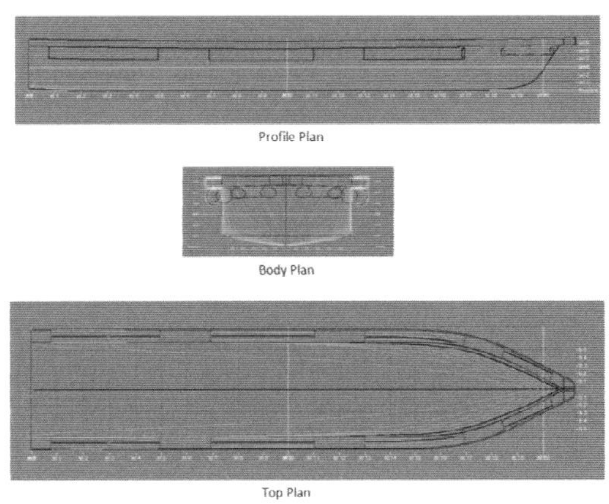

Figure 2: The lines plan

3.2 AIR SUPPLY SYSTEM

In this study, the boat is equipped with central air supply system. The system consists of 2 units of DC 12V air compressor powered by accumulator 12 Volt. Those components are hanged under the roof ceiling at aft part of the boat. The air compressors are connected to the air bags by air hose. Some valves are attached in order to control air supply by the system

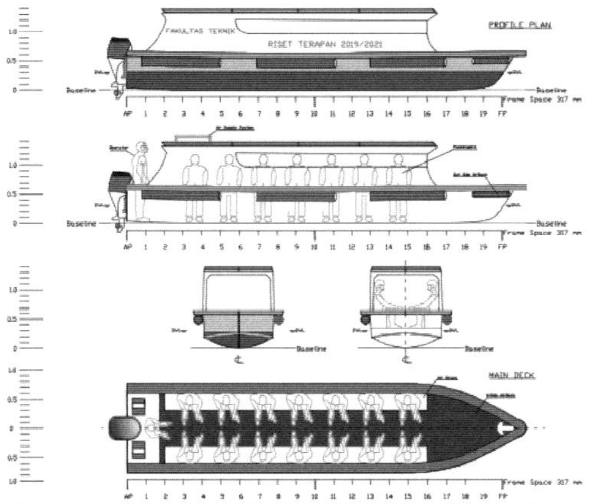

Figure 3: General arrangement

3.3 FULL-SCALE TEST PROCEDURE

The boat was tested in the real sea condition. In the test, the total 14 passengers and 1 operator are on board. The loads test were made for: empty boat, passengers and operator, passengers, operator and full water, steady test and running test.

4. RESULTS AND DISCUSSIONS

The documentation of the test are presented at the following Figures 4 and 5. The results of sea test are presented in Table 2. In this study, there are 4 scenario

Figure 4. Static test with load variation

Figure 5.Running test (full passengers + incoming water)

Table 2. Results of sea trial

No	Boat parameters	Value	Unit
A	**Empty Condition**		
1	Draft at midship	0.243	m
2	Draft at AP	0.255	m
3	Freeboard	0.377	m
4	KB	0.149	m
5	BMt	0.582	m
6	Weight displacement	1.125	tons
7	Weight displacement to ship height	4.142	tons
8	Total boat weight (empty)	1,120	tons
	Resume: The boat is afloat		
B	**Passengers and Operator**		
1	Draft at midship	0.395	m
2	Draft at AP	0.451	m
3	Freeboard	0.225	m
4	KB	0.235	m
5	BMt	0.446	m
6	Weight displacement	2.176	tons
7	Weight displacement to ship height	4.142	tons
8	Total boat weight	2.170	tons
	Resume: The boat is afloat		
C	**Passengers, Operator + water**		
1	Draft at midship	0.583	m
2	Draft at AP	0.673	m
3	Freeboard	0.037	m
4	KB	0.345	m
5	BMt	0.302	m
6	Weight displacement	3.706	tons
7	Weight displacement to ship height	4.142	tons
8	Total boat weight (empty)	3.701	tons
	Resume: The boat is afloat		
D	**Passenger, operator, water, air bag**		
1	Draft at midship	0.515	m
2	Draft at AP	0.605	m
3	Freeboard	0.105	m
4	KB	0.307	m
5	BMt	0.417	m
6	Weight displacement	3,138	tons
7	Weight displacement to ship height	4.142	tons
8	Total boat weight (empty)	3,129	tons
	Resume: The boat is afloat		

Table 3. Stability parameters of the unsinkable boat

No	Parameters	Unit	Scenario			
			A	B	C	D
1	Metacentre GMt	M	0.347	0.140	0.165	0.195
2	Max GZ	M	0.213	0.059	0.006	0.027
3	Angle GZ max	deg.	43.6	24.6	8.2	11.8

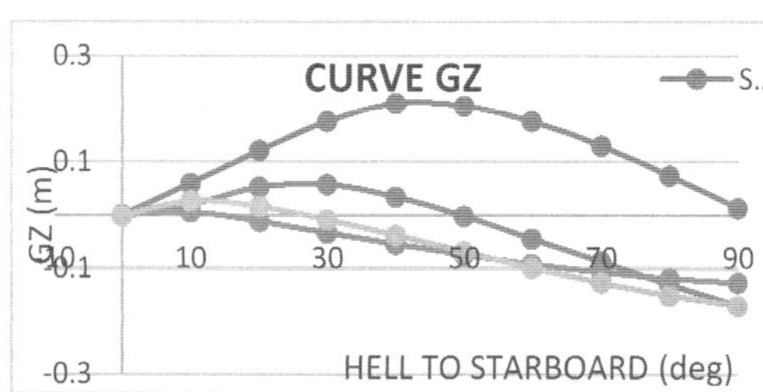

Figure 6. GZ curve of unsinkable boat

The results presented in Table 2 shows hydrostatics parameters while Table 3 and Figure 6 show stability parameters of the boat. There are four scenarios concerning hydrostatics and stability parameters carried out in this study i.e. empty boat (A), boat and passengers (B), boat, passengers and full water (C) and passengers, full water and inside air bags (D). In addition, the boat was tested on statics and running conditions in fully loads and water. It may be seen from Table 2 that all hydrostatics parameters of the boat are changing depending on the load on board. The critical condition exists at scenario C which the boat was not inserted by air bags inside. It is clearly seen that in all conditions the boat are floating (not sink) due to their total weights are less than weight displacement at ship height. Even the boat is not provided with inside air bags, it is still afloat. This condition was achieved because there are still solid boxes attached on board and additional air bags attached at outside of the boat. However, it is concluded that to be safe in boat operation then the inside air bags should be applied. As it is showed in the scenario D that there is a reserve good volume to reduce the incoming water, Therefore, the scenario D should be applied in critical condition. Therefore, to keep the boat afloat in the critical condition the following volume of solid boxes and air bags should be applied:

- Volume of solid boxes : 1.120 m^3
- Volume of inside air bags : 0.563 m^3
- Volume of outside air bags: 0.213 m^3

Where the amount of volume of solid boxes, inside air bags and outside air bags are computed and determined during the study.

From Table 3 and Figure 6, It is clearly seen that all scenarios have positive GMt and GZ. This means that at the range of hell angles from 0 to the positive area of GZ curve the boat still has righting moment to return it back at the right position when there is external forces acting to tilt the boat. The boat stability parameters will vanish at the GZ curve reach the zero value at certain hell angle. It can be seen that scenario C has high risk in capsizing followed by scenario D and B.

5. CONCLUSION AND RECCOMENDATION

5.1 CONCLUSION

Some points may be concluded from the results of study which are:

- ➢ The full-scale boat will float when the total boat weight is less than total weight displacement at ship height.
- ➢ The solid boxes with the volume of 1.120 m^3 and air bags with the volume of 0.563 m^3 installed on board contribute to reduce the incoming water.
- ➢ The outside air bags give an additional displacement weight as much as 0.213 m^3 to float the boat.
- ➢ Since the value of GMt and GZ are positive for all scenarios then the boat is still stable at initial condition. However, this situation will be crucial when the boat is filled with water at a small angle of inclination.

5.2 RECOMMENDATION

The result of study show that the boat is still afloat even in the extreme loads. Therefore, it is recommended that:

1 Add side air bags with larger diameter to prevent capsizing and additional buoyancy.
2 There should be a requirement that the boat should be provided with air boxes and air bags to avoid the boat from sinking and capsizing
3 The production scenario should be provided in order to serve demand from stakeholders

6. ACKNOWLEDGEMENTS

Special thank is given to The Ministry of Research, Technology and Higher Education, The Republic of Indonesia for the research grant of "Penelitian Terapan Unggulan Perguruan Tinggi" 2018 - 2021 to execute this research.

7. REFERENCES

1. HETHARIA, W. R, de LIMA, E, J, '*Design and Evaluating of Unsinkable Tuna Long Boat', 'Proceeding of the 6th Regional Conference on Marine Technology (MARTEC 2008)', University of Indonesia, Jakarta 26 – 27 August,* 2008

2. HETHARIA, W. R, METEKOHY, O., 'Design and Application of Unsinkable Tuna Longboat for Local Fishermen', *Occasional Papers No. 58 (March 2017), Kagoshima University, Japan,* 2017.

3. HETHARIA, W. R., GASPERSZ, F., LOUHENAPESSY, Y and FENINLAMBIR A., 'The Application of Solid Boxes on Small Passenger Speed-Boat', *IOP Conference Series; Earth and Environmental Science Vol. 333 Issue 1, IOP Publishing,* 2019.

4. HETHARIA, W. R., de FRETES, E. R., GASPERSZ F. and Louhenapessy, Y., '*A Model Scale of Unsinkable Small Passenger Speed Boat', Proceeding of International Conference ICSOT 25-26 November 2019, Kampus Undip Semarang,* 2019.

5 HETHARIA, W. R, de FRETES, E. R., GASPERSZ, F and LOUHENAPESSY Y, '*The Configuration of Solid Boxes and Air Bags for Full-Scale Unsinkable Small Passenger Boat', Proceeding of The 4th International Conference on Marine Technology SENTA 2019, 5 -6*

Desember 2019, Institut Teknology Surabaya, ITS, Surabaya, 2019.

6 ANONIMOUS, *'Data Kecelakaan Kapal Laut', Basarnas Ambon*, 2016,

7 HETHARIA, W. R., de FRETES, E. R., GASPERSZ, F and TIANOTAK A. J., *Kajian Tentang Beberapa Aktifitas Masyarakat Maluku di Sektor Kelautan', Prosiding Kongres Budaya Maluku - Ambon,* 2014

8 HETHARIA, W. R., GASPERSZ, F, FENINLAMBIR, A., *'Evaluasi Parameter Stabilitas Kapal Penumpang Kecil', Prosiding Seminar Nasional Fakultas Teknik Universitas Pattimura Ambon, 26 April 2018, ISSN : 2620-3995,*2019.

9 BARRAS, B. and DERRETT, D. R., *'Ship Stability for Masters and Mates', Butterworth – Heinemann, 6th edition, An Imprint of Elsevier Ltd., Linacre House, Jordan Hill, Oxford OX2 8DP*, 2006.

10 BIRAN, A. B., 'Ship Hysrostatics and Stability', *Technion – Faculty of Mechanical Engineering, 1st publication, Butterworth – Heinemann, An Imprint of Elsevier Ltd.*, Linacre House, Jordan Hill, Oxford OX2 8DP, 2003.

11 MOORE, C. S., *'Intact Stability – The Principles of Naval Architecture Series', The Society of Naval Architects, SNAME, Jersey City, New Jersey*, 2010.

12 *PARSONS, M. G., 2003, 'Parametric Design', Ship Design and Construction - Chapter 11, Volume. 2. SNAME Publication, Jersey City, NJ, USA,* 2003.

13 WATSON, D. G. M., *'Practical Ship Design', Elsevier Ocean Engineering Book Series, Volume I, ELSEVIER*, 1998.

14 LEWIS, E. V., *'Principles of Naval Architecture, Stability and Strength', SNAME Publication, Vol. I, Jersey City, New Jersey.* 1988

15 Van DOKKUM, K, *'Ship Knowledge - A Moderen Encyclopedia', DOKMAR, Enkhuizen, The Netherland,* 2003

8. AUTHORS BIOGRAPHY

Wolter Roberth Hetharia, FRINAmembership No 00209111, currentlyholds the position of lecturer at Faculty of Engineering, Department of Naval Architecture, Pattimura University. He is responsible for the courses of ship design and resistance. Several papers have been published concerning ship resistance and fishing boats. He also presented some papers at RINA conference and other international conferences. He and his team executed several research projects funded by Department of Higher Education, Research and Technology.

Eliza R de Fretes is the lecturer at The Faculty of Engineering, Department of Naval Architecture, Pattimura University. He is responsible for the course of ship hydrodynamic and stability. Several papers have been published concerning shiphydrodynamic and stability.

Jacobus Tupan, is the lecturer at The Faculty of Engineering, Department of Naval Architecture, Pattimura University. He is responsible for the course of ship economic and transportation.

ESTIMATION OF EXHAUST GAS EMISSIONS OF LONGLINEVESSELS 51-100 GT AT NIZAM ZACHMAN OCEANIC FISHING PORT

Vita R. Kurniawati, Choirin Alvia Syahrin, and **Yopi Novita,** IPB University, Indonesia

SUMMARY

The increase of exhaust emissions from longline vessels has long-term impacts. Accordingly, management measures are necessary, and therefore, information about the current status of exhaust emissions from longline vessels is required. This research was intended to estimate exhaust emissions of longline vessels based on fuel consumption. The result was presented in Fuel Use Intensity (FUI) per fish production calculated in kg and 1,000 Rupiah income. This research was conducted at Nizam Zachman Oceanic Fishing Port, focusing on longline vessels between 51-100 GT. The result shows that the average fuel consumption is 179 tons per vessel, contributing 1.328 kg of CO, 1.964 kg of NOx, 3.393 kg of SOx, and 54 kg of PM. Furthermore, inaccuracy issues found in statistical reports caused FUI of longline vessels to be remarkably higher than the world FUI for a similar group. This research concludes that the operation of longline vessels was fuel inefficient.

1. INTRODUCTION

The increase of the earth's temperature by 1.5°C was caused by increasing exhaust gas emissions from fishing vessels, contributing 3.3% of the total exhaust gas emissions produced by the world [1]. The major driver of the increase in exhaust gas emissions is excessive fishing vessel operations which induce various long-term impacts on the environment. However, data from tuna fisheries showed that tuna catches in Indonesia were decreased from 20,000 Mt in 1995 to 8,000 Mt in 2010 [2], indicating that excessive fishing efforts are not in line with productivity. This fact confirms that actions are needed to improve efficiency in fishing in respect of fuel consumption and emission production. In addition, these actions will support Indonesia's commitment to the Paris Agreement 2015 to overcome 26% greenhouse gas emissions independently.

The number of fishing vessel operations tend to increase from 2003 to 2012with an average growth rate of 1.69% [3]. The increase in operations also occurred in longline vessels. In 2010, the operation in Indonesia was 916 units, increased to 1,083 units in 2011, and increased again in 2012 as many as 1,179 units [2]. The operation of these vessels requires significantamount of fuel due to technological advancement on board, such as line throwers, line haulers, and freezers which require more energy than other types of fishing vessels. According to [4,5], fuel cost for longline vessels in the Indian Ocean is responsible for 60% of the total production cost. Given that emission has a linear correlation with fuel consumption, significant fuel use in longline vessels produces higher emissions than other fishing vessels [6].

Exhaust gas emissions from longline vessels need to be reduced by managing fuel efficiency [7]. Accordingly, information about the current status of exhaust gas emissions from longline vessels is required as the base line for management measures. However, little information is available regarding fuel emission and emission production, hence actions cannot be doneoptimally. Globally, some researches have been published discussing the significant fuel consumption on fishery [8, 9, 10], yet little information is found regarding Indonesian fishery as the second largest tuna producer. In Indonesia, previous research on exhaust gas emissions from fishing vessels has been conducted in Papuan waters [11] and the Indian Ocean [12]. However, thestudy doesnot provide a general description of the performance of longline vessels. Therefore, this research was intended to provide information at the general scope that can be used as a reference for controlling gas emissions through the efficiency of fuel use intensity, particularly for the longline vessel in Indonesia. Specifically, it aimed to calculate energy consumption, estimate exhaust emissions, and calculate fuel use intensity (FUI) per production unit of the catch from the operation of 51-100 GT longline vessels. The deliveries are expected to contribute in controlling excessive exhaust gas emissions from longline vessels.

Thisstudy is conducted by analysing fuel consumptionof longline vessels based atthe fishing port from which emission will be estimated. Considering data availability, Nizam Zachman Oceanic Fishing Port has been chosen to gather fishing data. The port is one of the home bases forIndonesian longline vessels [13]. At least 20% of the vessel population in the port is longline vessels [14,15] and about 70% of which is larger vessels between 51-100 GT [16]. Furthermore, information about exhaust gas emissionof longline vessels will be beneficialnot only for the government but also for Nizam Zachman Oceanic Fishing Port which is currently developing the eco-fishing port [17]. Researchers and academics can also use this information to participate in managing exhaust emissions from longline vessels.

2. METHODS

Vessel records, including fuel consumption and landed catches from 2018 to 2019, were collected from Nizam Zachman Oceanic Fishing Port. Data were collected from two main sources, statistic reports and vessel monitoring systems (VMS). Additional information was also gathered from enumerators and fishers.

Calculation of energy consumption, emission as well as FUI were carried out in the following steps:

1) The average fuel consumption per vessel per trip was calculated based on statistical data.
2) Subsequently, another fuel calculation was estimated from VMS record to verify the results using the followingapproaches:
   a) Fuel consumption for the main engine was calculated using the Fyson equation [18].
   FOT main engine = fc × t × HP engines....... (1)
   where,
   FOT = Fuel of Tank (kg),
   fc = Fuel consumption rate (0,19 kg/HP/hour),
   HP = Horse power (HP),
   t = Trip durations (hour).
   b) Fuel consumption of auxiliary engines wasestimated using this Ishida equation[19].
   FC auxiliary engines= 0,2 × O × L............. (2)
   where,
   FC = fuel consumption (kg),
   O = rated output (kg),
   L = load factor (100%).
   Assumption: total fuel consumption of auxiliary engines is 20% of the total fuel consumption of main engine [20].
   c) These results are then added to obtain the average fuel consumption per trip of longline vesselbased on VMS data. Compared to statistical reports, data gathered from VMS was found to be more accurate, and hence it was used for further calculation.
3) Subsequently, the amount of CO, NOx, SOx, and PM were estimated using Tier 1 formula[21].
   $ETrip = 10^{-3} \times \sum e \sum j \sum m \sum p\ (FCe,j,m,p \times EFe,j,m,p)$(3)
   where,
   ETrip = emission over a complete trip (tonnes),
   FC = fuel consumption (tonnes),
   EF =fuel consumptionbased on emission factor(kg/tonne),
   I = pollutant,
   e = engine category (main, auxiliary),
   j = engine type (slow-, medium-, and high-speed diesel, gas turbine and steam turbine).
   m =fuel type (bunker fuel oil, marine dieseloil/marine gas oil),
   p = the different phase of trip (cruise, hotelling, manoeuvring).

Emission factors used in this study are based on the following table.

Table 1Emission factor (kg/tonne)

Engine/ Fuel	CO	NO_x	SO_x	PM
MDO (Marine Diesel Oil)/Diesel fuel	7.4	10.9	19	0.3

Source: [21]

4) For further investigation, FUI was calculated as the representation of the fuel use relative to its productivity expressed in kg fish and Rp 1.000 revenue. Hence, this research uses two types of FUI as follows:
   a) FUI_{fish},
   $$FUI_{fish} = \frac{\sum FC}{\sum TC} \quad .... (4)$$
   b) FUI_{income},
   $$FUI_{income} = \frac{\sum FE}{\sum TR} \quad .... (5)$$
   where,
   FUI_{fish} = Fuel use intensity per kg fish
   FUI_{income} = Fuel use intensity per Rp 1.000 revenue
   FC = Fuel consumption (kg)
   TC = Total catch (kg)
   FE = Fuel cost (Rupiah)
   TR = Total revenue (calculated in Rp 1.000)
5) Similar to fuel data, catch data gathered from the statistical reportalso require further verification. Henceanother calculation was carried out to estimate the actual fish hold capacity. Since limited data were found, the calculation of fish hold capacity were approached using cubic number (CUNO), fish hold volume, and stowage factor formula [18, 22].
   a) Firstly, CUNO of the vessel is calculated using the following equation [18]:
   CUNO (m^3) *= LOA × B × D*.............................(6)
   where,
   LOA = length overall (m)
   B = breadth (m)
   D = depth (m)
   b) The CUNO result is then used to estimate fish hold capacity based on[22] work, which suggested the typical percentage of fish hold volume toward its CUNO.
   c) Based on projected fish hold volume, fish hold capacity was determined using this equation [18]:
   Hold capacity (tonnes) = SF × V......................(7)
   where,
   SF = storage factor (tonnes/m^3)
   V = fish hold volume (m^3)
   Assumption: longline vesselapplies a frozen storage system hence the stowage factor is 0.6 tonnes/m^3.

3. RESULT

3.1 LONGLINE VESSEL OPERATING SYSTEM

The longline vessels operation system at Nizam Zachman Oceanic Fishing Port generally consists of several stages, namely steaming, setting, drifting, and hauling.This fishing operation uses longline fishing gear with the target catch of large pelagic fish such as yellowfin tuna

(*Thunnus albacares*), skipjack tuna (*Katsuwonus pelamis*), and albacore (*Thunnus alalunga*).

The first stage of longline operations begins with steaming; it is a trip to the fishing ground. Commonly, longline vessels from Jakarta fishing port measuring more than 30 GT go to the Indian Ocean (high seas) for fishing,specifically infishing areas 572 and 573. This is following the Regulation of the Minister of Maritime Affairs and Fisheries of the Republic of Indonesia Number 59/PERMEN-KP/2020 concerning Fishing Track and Fishing Equipment in the Fisheries Management Area of Republic Indonesia (WPPNRI) and the high seas [23]. Vessel speed to fishing ground ranges from 4 – 6 knots, and it took up to 15 days to reach the fishing ground.

The next stage is setting. Setting is the process of attaching baitsto the hooks and deploying fishing gear from the stern area. This stage, on average, starts from 07.00 to 13.00 local time. The speed of the vesselat this stage ranges from 3-5 knots.

The drifting stage was conducted after the setting process was completed. The captain will turn off the main engine to save fuel, but the auxiliary engine was still turned onto support the freezer and deck equipment such asline haulers.

Hauling is the process of lifting fishing gear and collecting catches from hooks. Unlike setting, this process was conducted in the middle or bow area of the vessel. The longline vessel usually uses a line hauler to lift the mainline. Furthermore, the branch line ace is used to lift the branch line. Generally, this process starts at 19.00 to03.00 local time. The speed of the vessel at this stage ranges from 2-3 knots.

According to operation stages, it can be concluded that the main engine turns on for approximately 16 to 17 hours per day. The duration ofthe setting stage is approximately 5 to 7 hours, while for the hauling stage, it is 8 to 9 hours. The main engine is turned off during drifting for approximately 7 hours.

3.2 FUEL CONSUMPTION OF LONGLINE VESSEL OPERATIONS

Fuel consumption per trip varies depending on the distance and the duration of the engine operating time [7]. Furthermore, internal factors such as the condition and characteristics of the engine, engine performance, type of fuel, and the condition of the vessel's motion also affect the consumption rate [7]. External factors such as weather, wave height, wave period, and wind speed are also influential [24].

The average fuel consumption by longline vessels measuring 51-100 GT from 2018 to 2020 can be seen in Figure 1. In 2018 and 2019 fuel consumption has the same value at 21,500 kg even though the average trip duration in from 2018 to 2019 increased by 28.3%. The value indicates that the statistical data for that year is less accurate. This was suspected due to an error in recording or filling out the supplies form. In 2020 the port conducted a better data collection. Therefore, it can be observed that the fuel consumption data showed an increase and a variation. The data collection method was improved again in June 2020 by implementing afuel record based on fuel invoices.

Before June 2020, fuel consumption record is just a generalization with no valid fuel invoice. The estimation is roughly derived from fuel subsidy quota applied by the government, which was 21,500 kg per vessel [15]. Therefore, when regulation to attach valid invoice is applied,the fuel consumption rate increased by 3.84%.

Based on statistical data, the fuel consumption data for longline vessels is relatively small and impossible for vessel between 51 – 100 GT. Therefore, other data were gathered from VMS. However, data from VMS only consists of vessel speed and engine running time. Therefore, using equation (1) and (2), fuel consumption for each vessel was estimated and the result is presented in Figure 1.

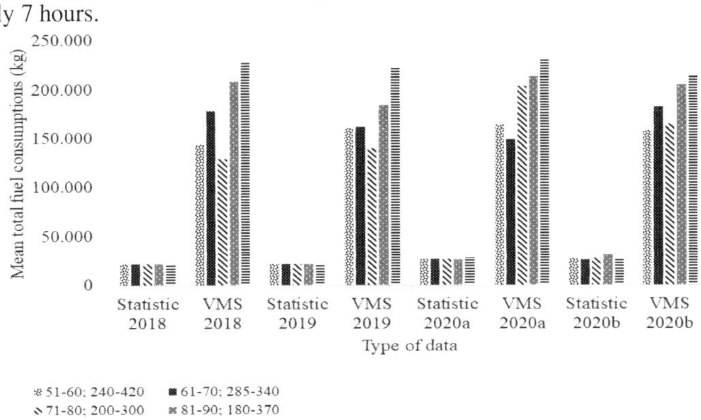

Figure 1. Average fuel consumption for longline vessels based on statistical data from Nizam Zachman Oceanic Fishing Port and VMS data for 2018-2020

The figure 1 shows a significant difference between fuel consumption depicted from statistical data and VMS calculation. In detail, the differences are presented in percentage discrepancy, as seen in Table 2. Based on the interviews, the port authority only recorded the amount of fuel one time when the vessel left the port (Port Clearance document).In fact, when at sea, fuel can be refilled 4 to 5 times, with a refilling capacity is around 15-25 tons of diesel fuel. Practically, refilling is required as the fishing ground is far away, fuel tank capacity is limited to 6-10 tons, and the trip may take more than three months. Due to significant discrepancies, the following calculation uses fuel consumption estimated from VMS data.

Table 2 Average fuel consumption per vessel per trip (kg) statistical data with invoices for 2020 and VMS

GT	Engine Power (HP)	Average Fuel Consumption per vessel per trip		
		Statistical Data (kg)	VMS Data (kg)	Deviation (%)
51-60	240-420	25,000	157.955	84
61-70	285-340	26,250	182.327	86
71-80	200-300	26,250	163.952	84
81-90	180-370	35,623	204.738	83
91-100	280-370	25,000	213.130	88
	Average	**27,625**	**184.420**	**85**

3.3 ESTIMATION OF EXHAUST EMISSIONS FROM LONGLINE VESSEL OPERATIONS

The type of fuel used by longline vessels at Nizam Zachman Oceanic Fishing Port is diesel fuel which has a density of 860 kg/m^3 [25]. This fuel type produces primary pollutants: CO, NOx, SOx, and Particulate Matter (PM) [21]. Using equation (3), emission from consumed fuel was calculated and the result is shown in Table 3. The table revealed that, on average, the vessel emitted more than 1,000 kg of CO, NOx, and SOx andabout 40-70 kg of PM. The 91-100 GT vessels produced the highest emissions, whilst the least is the vessel from the group of 71-80 GT.

Exhaust emissions have many negative impacts, both health and environmental impacts [26]. Exhaust emissions by vessels have been classified as carcinogenic by the WHO in 2012, which are harmful to human health [27]. In addition to health impacts, these gas emissions will negatively impact the environment, such as droughts, erratic rainfall seasons, rising sea levels, declining productivity of the agricultural sector, and increasing intensity of forest fires [28].

The International Maritime Organization (IMO) regulates global shipping safety, security, and environmental performance, including emissions production. The regulation was adopted in the International Convention for the Prevention of Pollution from Ships – the Marine Pollution Convention (MARPOL) in 1973 and then modified by Protocol 1978concerning limiting the harmful impacts fromvessel emissions on air quality [29]. Since May 19, 2005, the regulation has been effective and aims to reduce CO, NOx, SOx, and PM from engines [30].

Emissions from fishing vessel must be controlled to reduce greenhouse gas effect that induce global warming. This is in line with Indonesia's commitment to the 2015 Paris Agreement [31], which states that Indonesia will independently reduce its greenhouse gas emissions by 26%.

Table 3 Total exhaust emissions of 51-100 GT longline vessels at Nizam Zachman

GT Intervals	Average fuel consumption (Tonnage)	Exhaust gas emission			
		CO(kg)	NOx(kg)	SOx(kg)	PM(kg)
51-60	152	1.126	1.658	2.891	46
61-70	164	1.212	1.785	3.112	49
71-80	147	1.086	1.599	2.788	44
81-90	204	1.512	2.228	3.883	61
91-100	226	1.671	2.461	4.289	68
Mean		**1.328**	**1.946**	**3.393**	**54**

3.4 FUEL USE INTENSITY (FUI) OF LONGLINE VESSELS

Fuel Use Intensity (FUI) is the measurement for energy efficiency that compares fuel performance per specified

production unit. In this study, FUI is calculated per one kilogram of landed catches and per Rp 1,000 revenue.
According to Table 4, the FUI of longline vessel operation is 10 kg of fuel/kg of fish, equivalent to 13,248 litres of fuel/ton of fish. These results have a significant difference from the world FUI for tuna fisheries. Based on [32], the world FUI for longline vessels is 2.92 kg of fuel/kg of fish or equivalent to 3,400 litres of fuel/ton of fish. This significant difference is due to several factors.The main factor is the number of landed longlines catch reported totheport is not the total catch per trip. Based on the interviews with the captain and port officers, the recorded catch is only the final catch which is brought when the vessel is back to the port. Meanwhile, the actual catch is not reported to the port;hence it is not recorded the annual statistical data of Nizam Zachman Oceanic Fishing Port. This is allegedly due to the practice of transshipment.

In addition, the FUI per Rp 1,000 revenue is 2,107. This means to produce Rp 1,000 revenue, at least Rp 2,107 is required for fuel expenses. Given that the fuel cost is significantly higher than the revenue, this confirms that transshipment practice happened, resulting in unreported catch data.

Table 4 FUI longline vesselsin Nizam Zachman Oceanic Fishing Port

GT Intervals	Average of Catches (kg)	Average of Catches Cost (Rp) per Vessel	Average of Fuel Consumption (kg) per Vessel	Average of Fuel Consumption Cost (Rp)	Fuel Use Intensity			
					Kg Fuel/kg Fish	Intervals per vessel (kg)	Rp fuel/Rp fish	Interval per vessel (Rp)
51-60	20,858	362,635,738	152,149	590,233,877	7	2 - 99	1.628	363.204.000-1.355.961.600
61-70	14,890	258,947,188	163,797	635,417,857	11	2 - 33	2.454	445.530.240-1.098.230.400
71-80	18,986	364,451,438	146,713	569,145,746	8	2 - 88	1.562	363.450.240-1.503.048.960
81-90	27,947	488,620,613	204,360	792,775,232	7	3 - 145	1.622	422.547.840-1.516.838.400
91-100	15,825	268,019,190	225,752	875,760,828	14	3 - 175	3.268	482.630.400-1.576.592.640
Average	**19,701**	**348,534,833**	**178,554**	**692,666,708**	**10**	**3 – 175**	**2.107**	**363.204.000-1.576.592.640**

Transhipment is the transfer of fish catches by operating vessels to fish carriers or similar vessels. This activity occurred in the middle of the sea, therefore difficult to control by the port office. Based on the interview, the transfer occurred 4-5 times with different transfer capacities depending on the holding capacity of the carrier vessel. To confirm this statement, it is necessary to define the fish hold capacity of the longline vessel, which is calculated based on [22]. Using equations (6) and (7), the average fish hold 51-100 GT longline vessels ranging from 30 to 62 tons.
Meanwhile, based on Nizam Zachman Oceanic Fishing Port statistical data, the average catch only ranges from 15-17 tons. The comparison results explain that the average catch does not even fulfill half of the fish hold capacity and emphasize inaccuracy issues in catch data. According to [33] transshipment practice occurred because of tuna industry face difficulty in meeting the demands for fresh of catchdue to long trip duration, the retribution system is notappropriately implemented, and there is no legal mechanism for transshipment.

Figure 3 shows the calculation of emission in respect of FUI. The vessels that produced the most SOx gas emissions measured 61-70 GT, and the lowest were vessels measuring 91-100 GT. Overall, vessels measuring 91-100 GT produce the lowest average exhaust emissions. This is because the number of sample vessels operating is less than other vessel sizes. In addition, vessels of this size have less trip duration and lower engine power than other vessels. The dominant longline vessels operating 91-100 GT produce more exhaust gas emissions than other vessels. Vessels measuring 71-80 GT and 81-90 GT emit relatively similar exhaust emissions. This is because the difference in trip duration is not significant.

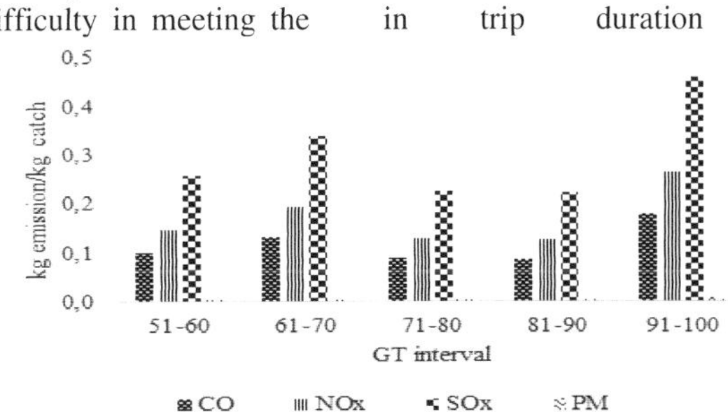

Figure 2 Exhaust gas emission (kg) per catches (kg)

4. CONCLUSIONS

1) Based on VMS data, the fuel consumption of longline vessel operations measuring 51-100 GT at Nizam Zachman Oceanic Fishing Port Jakarta is 178,554 kg. This value has exceeded the subsidy quota determined by the port, which is 21,500 kg per vessel.
2) Exhaust emissions by longline vessels are different for each type. The pollutants CO, NOx, SOx, and PM produced were 1,321 kg, 1,946 kg, 3,393 kg, and 54 kg, respectively.
3) The FUI value of longline vessels is 10 kg of fuel/kg of fish at the cost of one rupiah of fuel per 1,000 Rupiah income of 2,107. The average exhaust emissions per catch production value is 0.07 kg CO/kg fish, 0.10 kg NOx/kg fish, 0.18 kg SOx/kg fish, and 0.003 kg PM/kg fish.

5. ACKNOWLEDGEMENTS

The authors would like to thank to the Authority of Nizam Zachman Oceanic Fishing Port and the Directorate General of Surveillance for Marine and Fisheries Resources for assistance during this study.

6. REFERENCES

1. UNEP,'Emissions Gap Report 2019 Executive Summary',*Executive summary of United Nations Environment Programme*, 2019.

2. IOTC, 'Review of the Statistical Data and Fisheries Trends for Tropical Tunas', *Indian Ocean Tuna Commission,* 2013.

3. KKP, 'Analisis Data Pokok Kelautan dan Perikanan', *Kementerian Kelautan dan Perikanan, Republik Indonesia*, 2014.

4. ROCHMAN, F., JATMIKO I., FAHMI Z., 'Dinamika Industri Rawai Tuna di Pelabuhan Benoa', *Marine Fisheries*, 2014.

5. ISAMUDDIN,'Pembiayaan Usaha Perikanan Tangkap di Pelabuhan Perikanan Samudera (PPS) Nizam Zachman, Jakarta', Institut Pertanian Bogor, 2014.

6. DAVIE, S., MINTO, C., OFFICER, R., LORDAN, C., JACKSON, E.,'Modelling Fuel Consumption of Fishing Vessels for Predictive Use', *ICES Journal of Marine Science,* 2015.

7. UTAMI, T.K., PURININGSIH, F.S.,'Penghitungan Kadar Emisi Gas Buang di Pelabuhan Belawan', *Warta Penelitian Perhubungan,*2019.

8. MUIR J. Fuel and energy use in the fisheries sector: approaches, inventories and strategic implications. FAO Fish Aquac Circ. 2015;(C1080):I–94.

9. PARKER RWR, TYEDMERS PH. Fuel consumption of global fishing fleets: current understanding and knowledge gaps. Fish Fish [Internet]. 4 July 2014. 2015;16(4):684–96. Available from: http://dx.doi.org/10.1111/faf.12087

10. TYEDMERS PH, WATSON R, PAULY D. Fueling global fishing fleets. Ambio. 2005;34(8):635–8.

11. WARDHANA, E.M.,'Estimasi Jejak Karbon Kapal Perikanan Berdasarkan Moda Operasi - Study Case di Perairan Wilayah Papua', Institut Teknologi Sepuluh November, 2016.

12. SURYANTO, WUDIANTO,'Model Estimasi Konsumsi Bahan Bakar Kapal Ikan Huhate dan Rawai Tuna', *Jurnal Penelitian Perikanan Indonesia,* 2017.

13. IOTC,'Indonesia National Report to The Scientific Committee of The Indian Ocean Tuna Commission', *Indian Ocean Tuna Commission,* 2015.

14. KKP,'Profil Pengelolaan Perikanan Perikanan Samudera Pelabuhan Nizam Zachman Jakarta', *Kementerian Kelautan dan Perikanan, Republik Indonesia,* 2016.

15. PPSNZJ,'Laporan Statistik PPS Nizam Zachman Jakarta 2010', *Pelabuhan Perikanan Samudera Nizam Zachman Jakarta,* 2011.

16. PPSNZJ,'Buku Statistik Pelabuhan Perikanan Samudera Nizam Zachman Jakarta Tahun 2018',*Pelabuhan Perikanan Samudera Nizam Zachman Jakarta,* 2019.

17. DJPT-KKP,'Rencana Strategis (Renstra) Pelabuhan Perikanan Samudera Nizam Zachman Jakarta tahun 2020-2024', *Direktorat Jenderal Perikanan Tangkap, Kementerian Kelautan dan Perikanan, Republik Indonesia,* 2020.

18. FYSON, J.,'Design of Small Fishing Vessels', *Fishing News Book,*1985.

19. ISHIDA, T.,'Emission Estimate Methods of Air Pollution and Greenhouse Gases from Ships', *Journal of The Japan Institute of Marine Engineeing,* 2003.

20. KKP,'Peraturan Menteri Kelautan dan Perikanan Republik Indonesia No 13/PERMEN-HP/2015', *Kementerian Kelautan dan Perikanan, Republik Indonesia,* 2015.

21. TROZZI, C.,'Emission Estimate Methodology for Maritime Navigation', *US EPA 19th International Emissions Inventory Conference,* 2010.

22. LAFI, L., NOVITA, Y.,'Volume dan Kapasitas Muat Palka Kapal Tuna Longline Tipe Taiwan dan Bagan Siapi-Api di Pelabuhan Perikanan Nizam Zahman, Jakarta', *Buletin PSP,* 2008.

23. KKP,'Peraturan Menteri Kelautan dan Perikanan Republik Indonesia Nomor 59/PERMEN-KP/2020 tentang Jalur Penangkapan Ikan dan Alat Penangkapan Ikan di Wilayah Pengelolaan Perikanan Negara Republik Indonesia dan Laut Lepas', *Kementerian Kelautan dan Perikanan, Republik Indonesia,* 2020.

24. PRPIĆ-ORŠIĆ, J., VETTOR, R., SOARES, C.G., FALTINSEN, O.M.,'Influence of Ship Routes on Fuel Consumption and CO2 Emission', *Maritime Technology and Engineering – Guedes Soares & Santos (Eds),* 2015.

25. PERTAMINA,'Spesifikasi Solar/Biosolar', *Pertamina,* 2020.

26. VAN, T.C., RAMIREZ, J., RAINEY, T., RISTOVSKI, Z., BROWN, R.J.,'Global Impacts of Recent IMO Regulations on Marine Fuel Oil Refining Processes and Ship Emissions', *Transportation Research Part D,* 2019.

27. SILVERMAN, D.T., SAMANIC, C.M., LUBIN, J.H., BLAIR, A.E., STEWART, P.A., VERMEULEN, R., COBLE, J.B., ROTHMAN, N., SCHLEIFF, P.L., TRAVIS, W.D.,'The Diesel Exhaust in Miners Study: A Nested Case-Control Study of Lung Cancer and Diesel Exhaust', *Journal of the National Cancer Institute,* 2012.

28. WAHYUDI, J., 'Mitigasi Emisi Gas Rumah Kaca', *Jurnal Litbang,* 2016

29. IMO,'International Convention for the Prevention of Pollution from Ships MARPOL 73/78: the Regulations for the Prevention of Air Pollution from Ships (Annex VI)', *International Maritime Organization,* 1997.

30. IMO,'Marine Environment Protection Committee (MEPC), 70th Session, 24-28 October 2016', *International Maritime Organization,* 2016.

31. UNFCCC,'The Paris Agreement 2015', *United Nations Framework Convention on Climate Change,* 2016.

32. TYEDMERS, P.,'Fisheries and Energy Use', *Encyclopedia of Energy,* 2004.

33. HUDAYANA, A., UTAMI, T.N.,'Analisis Kebijakan Larangan Transhipment (Alih Muat) Tangkapan Perikanan Terhadap Kinerja Usaha Penangkapan Ikan Kapal Longline (Studi Kasus Pelabuhan Perikanan Samudera Nizam Zachman DKI Jakarta)', *Journal of Economic and Social of Fisheries and Marine,* 2017.

8. AUTHORS BIOGRAPHY

Choirin A. Syahrin is an undergraduate student in the Department of Fisheries Resources Utilization, Faculty of Fisheries and Marine Sciences, IPB University. She is concerned with sustainable fisheries by researching environmental impacts on ship operations for her thesis.

Vita R. Kurniawati is a lecturer in Department of Fisheries Resources Utilization, Faculty of Fisheries and Marine Sciences, IPB University. She is joining the Division of Fishing Vessel, Fishing Gear and Fishing Technology and be responsible for lecturing fishing vessel related modules. Her expertise is fishing vessel production and operation. Recently, she is interested in energy related research in fishing vessel operation.

Yopi Novita is a lecturer in Department of Fisheries Resources Utilization, Faculty of Fisheries and Marine Sciences, IPB University. She is joining the Division of Fishing Vessel, Fishing Gear and Fishing Technology and be responsible for lecturing fishing vessel related modules. Her expertise is fishing vessel design and she is interested in the ship dynamic and internal capacity related research, specifically for fishing vessel.

ELECTRICAL RESISTIVITY AS PREDICTION OF HIGH-VOLUME FLY ASH CONCRETE PERFORMANCE IN MARINE ENVIRONMENT

F D Dwi Putra, W D Pratiwi, I Santiasih, K D Wulandari, K Abdullah, and **N Arumsari**, Politeknik Perkapalan Negeri Surabaya, Indonesia

SUMMARY

Reinforced concrete for marine structures requires specific properties since it faces an aggressive environment. The interaction of concrete with seawater causes damage both physically and chemically. The primary damage to reinforced concrete in the marine environment is chloride-induced corrosion of reinforcement. Chloride ions diffuse into the concrete body until it reaches the reinforcement surface. If the chloride concentration reaches a critical value, then corrosion begins to attack. This paper studies the resistivity of high-volume fly ash (HVFA) concrete as measured by the four-probe (Wenner) method. This study tested HVFA concrete with 30-50% FA substitution compared to normal concrete. Fly ash characterization was carried out using XRD, XRF and SEM techniques. The resistivity of HVFA with 30% and 40% of FA replacementare higher than the normal concrete. This data indicates that HVFA concrete has better resistance to chloride-induced corrosion than normal concrete. In addition, HVFA has a longer service life than normal concrete in chloride environments such as seawater.

NOMENCLATURE

HVFA	High-volume fly ash
ρ	Electrical resistivity
a	probe-space
L	Cylinder height
d	Cylinder diameter
V	Voltage measured
I	current applied

1. INTRODUCTION

Reinforced concrete structure exposed to marine environment majorly undergo to chloride induced corrosion damage. Chloride ions migrate in the body of concrete reaching the metal surface, and at a certain chloride concentrationreferred to as the critical concentration, the corrosion attack begins to occur. Hence, durability of reinforced concrete may be determined by chloride penetration through concrete cover. Moreover, the diffusion coefficient of chloride within concrete is a parameter used to measure the service life of concrete [1], [2].

The method of AASHTO T259 (salt ponding) quantifies the diffusion coefficient by measuring chloride profiles at various concrete depths. This technique serves as appropriate data describing chloride ion movement and is not affected by conductor may present in concrete [3]. However, it needs about 90 days after curing, and even extended duration for higher quality concretes.

The standard test of AASTHO T227 (ASTM C1202), commonly known as the rapid chloride permeability test (RCPT) have adopted by many works. This method indirectly evaluates the permeability of concrete by measuring the amount of current that pass through a sample with specific dimension for 6 hours at a voltage of 60V DC. However, the current passes designate the movement of all ions in the pore solution, not just chloride. Hence, the presence of ions other than chloride affects the measurement result. In addition, any conducting materials, such as reinforcement and conductive fibre will bias the results [3]. Moreover, the value gain from RCPT did not seem to associate to diffusion coefficient for fly ash concretes [4].

Once the test's main objective is to control the product quality and develop a new product to a specific requirement, an alternating approach to predict chloride diffusion should be applied. The resistivity test of concrete is a unique alternative short-term method to predict chloride diffusion coefficient even though the test quantifies the chloride content indirectly. In addition, one of the main parameter determining the rate of active corrosion of the steel reinforcement is electrical resistivity of concrete [5].

Several factors affect the inherent electrical resistivity of concrete, namely pore size distribution and interconnection, conductivity of pore fluid, degree of saturation, and temperature[6]. Hence, the pore properties of the different concretes can be evaluated by resistivity measurement of concretes saturated with the same electrolyte at a constant temperature.

High-volume fly ash (HVFA) concrete has been studied and promoted as high-performance concrete. Yazici[7] found that fly ash replacement of 30-60% cement significantly decreases chloride penetration depth. Chalee [8] obtained that concrete with higher fly ash replacement and lower water to binder ratios produced better chloride resistance under the marine environment. Mehta [5] states that properly cured high-volume fly ash concrete products with low water content and appropriate aggregate grading are homogenous in microstructure and durable. In addition, the incorporation of fly ash in concrete production offers economic and ecological benefits since fly ash is a waste material.

Evaluation of HVFA concrete composed of fly ash produced from Indonesian power plants is fascinating. The main reasons are fly ash is abundant, and Indonesia's vast seas require various infrastructures to optimize the utilization of marine resources. This study analysed resistivity of HVFA concrete compared to normal concrete. Three samples with 30-50 fly ash replacement were evaluated based on the 28-day compressive strength and resistivity.

2. MATERIALS AND METHOD

2.1 Materials.
These materials were used for concrete mixtures: Ordinary Portland cement (OPC), known as Cement Type I according to ASTM C150, fly ash, crushed coarse aggregate, sand or fine aggregate, $CaCO_3$ powder, $Ca(OH)_2$ powder and demineralized water. Fly ash was obtained from Paiton Power Plant, Indonesia. Oxide constituent of fly ash was measured with X-Ray fluorescence (XRF), while X-ray diffraction of fly ash was conducted to determine the phases and the degree of amorphous.

Table 1. Some oxides and phases of fly ash

Oxide	CaO	SiO_2	Al_2O_3	Fe_2O_3	MgO	SO_3
%	17.79	37.94	14,29	15.45	7.12	1.39
Phase	Quartz	Pe	C_3A Cubic	C_3A Na C	C_4AF	hkl
%	6.8	3.7	1.1	2.4	2.1	73.3

Br : Brownmillerite (Si, Mg)
Pe : Periclase
C_3A Na C : C_3A Na Cubic
C_4AF : C_4AF (Colville)
An : Anhydrite
Ma : Magnetite
Mag : Maghemite
hkl_ : amorphous phase

As shown at Table 1, the sum of SiO_2, Al_2O_3, Fe_2O_3 were 67.68% then the FA was categorised as C class of ASTM 618. Furthermore, the fly ash constitutes of 26.7% crystalline and 73.3 % amorphous phase.

Morphology of fly ash particles was observed with the scanning electron microscope (SEM), as presented in Figure 1. It seems that majorly fly ash is composed of spherical particles with heterogeneous size.

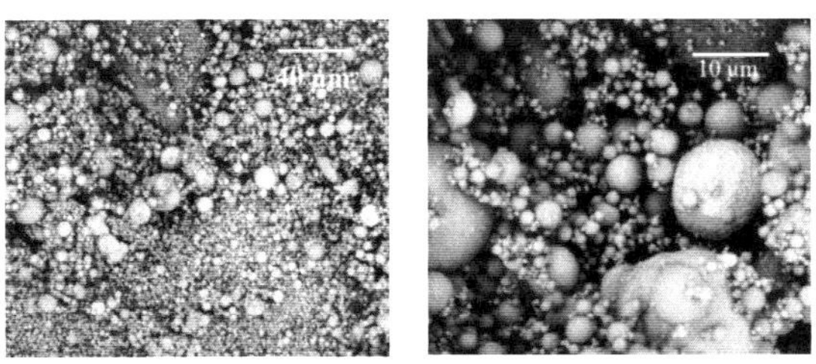

Figure 1. SEM image of fly ash examined with a magnification of 1000x (left side) and 5000x (right side)

Natural sand with a fineness modulus 2.77 and crushed coarse aggregate No.67 (nominal size 4.75 - 19 mm) determined by ASTM C33 were used in this study. The $CaCO_3$ powder, $Ca(OH)_2$ powder and demineralized water were obtained from local supplier.

2.2 Samples preparation

Four concrete samples were observed and denoted as NC, FA-30, FA-40, and FA-50. The two-digit number following "FA" indicates the degree of cement replacement by fly ash, in weight percent. For example, FA-30 indicates that 30% of the cement was replaced by fly ash. The complete mix proportion is shown in Table 2, include sand, coarse aggregate, water, calcium carbonate, and calcium hydroxide. Addition of calcium carbonate promotes cement hydration, as studied in previous work [9].

Table 2. Mix proportion of concrete in kg/m^3

Sample ID	OPC	FA	Sand	Agr	$CaCO_3$	$Ca(OH)_2$	W
NC	420	-	748.6	1221	-	-	210
FA-30	273	126	748.6	1221	12.6	8.4	210
FA-40	231	168	748.6	1221	12.6	8.4	210
FA-50	189	210	748.6	1221	12.6	8.4	210

Agr : crushed coarse aggregate; W : water

Cylindrical concrete specimens of 100 mm diameter and 200 mm height for compressive strength test and resistivity test were prepared refers to ASTM C192. The samples were mixed in a concrete mixer for 5 minutes and poured into the moulds. After 24 hours, specimens were demoulded and moist cured for 28 days. Compressive strength test was carried upon ASTM C39 at an age of 28 days. Surface resistivity has been measured with four probe resistivity test (Wenner array probe) method after immersed the cured samples in 5M NaCl solution for 28 days. In this technique, the four electrodesarelocated in a straight line and equally spacedas earlier study [10] as shown in Figure 2. The high concentration of electrolyte was used to minimize the effect of ion other than chloride on the electrical measurement, as stated in the previous work.

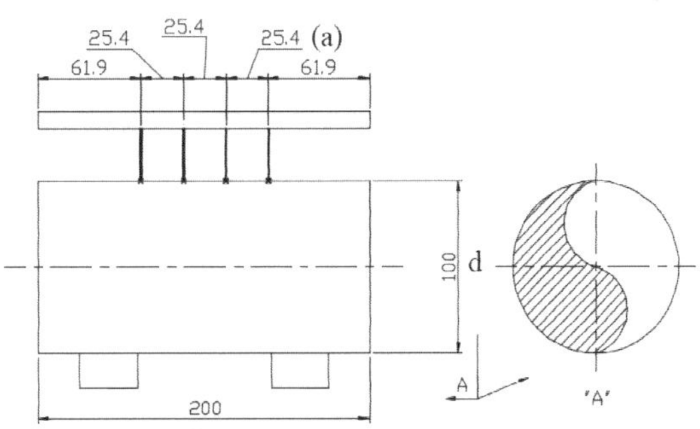

Figure 2. The dimension of specimen (mm) and the probe position of the resistivity test of Wenner array method.

The AC-current was applied through the outer probe, while the potential was measured between the inner probe [3]. For semi-infinite uniform body, The resistivity then calculated according to the formula [3][11][12] :

$$\rho = 2\pi a \, (V/I) \quad (1)$$

ρ : resistivity, ohm. cm
a : distance between probe, cm
V : measured potential, Volt
I : applied current, Ampere

Nevertheless, the probe is in contact with finite dimensions of bodies. The correction factor, K, was introduced by Morris [11]:

$$\rho = (\rho_{app}/K) \qquad (2)$$

where,

ρ = corrected resistivity
ρ_{app} = apparent resistivity (Value from measurement)
K = correction factor

The K factor depend on the specimen dimension and the probe space and the corresponding K value for this study is 1.41 [11].

3. RESULTS AND DISCUSSION

The compressive strength of concrete is presented in Table 3.

Table 3. Compressive strength (Fc) of 28-day concrete

No	Sample ID	Fc, MPa
1	NC	31
2	FA30	28
3	FA40	29
4	FA50	19

Table 3 indicate that replacement of cement with fly ash up to 40% is slightly decrease the compressive strength of 28-day concrete. In contrast, a marked reduction of concrete strength is observed for 50% fly ash replacement.

Table 4. Resistivity and apparent resistivity of concrete observed by Wenner Array method

No	Sample ID	Apparent resistivity (ohm.m)	Resistivity (ohm.m)
1	NC	32.907	23.338
2	FA30	53.708	38.091
3	FA40	43.642	30.952
4	FA50	22.236	15.770

The apparent resistivity of the concrete saturated with NaCl is shown in Table 4. The resistivity calculated by dividing the apparent resistivity with K. It can be seen from the table that the resistivity varies with the fly ash content of concrete.

Fig. 3 shows the compressive strength (MPa) and concrete resistivity (ohm.m). Substitution of cement with fly ash up to 40% resulted in a negligible effect on compressive strength but a markable increase in resistivity compared to normal concrete as the control sample. However, both resistivity and compressive strength decrease significantly when the level of fly ash replacement rises to 50%.

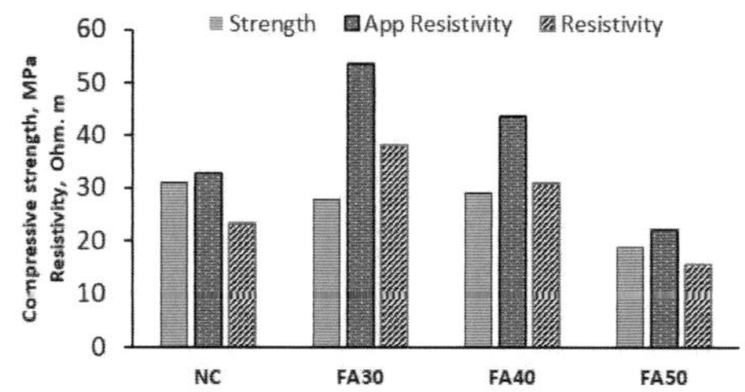

Figure 4. The compressive strength and resistivity of concrete

The resistivity of concrete increases due to the decrease in porosity. In addition, the decrease in porosity caused by the pore refinement of fly ash substation is reflected to improvement of the strength, too[13]. The resistivity of material indicates it ability to resist the passage of electrical current. In the sound concrete, the higher resistivity means the higher capacity to withstand the chloride ingress.

Diffusivity and conductivity of charged species *i* in concrete are associated with its partial conductivity, as expressed in Nernst-Einstein equation (2) [13]. This equation is applicable when concrete is assumed as a solid electrolyte.

$$D_i = \frac{RT\sigma_i}{Z_i^2 F^2 C_i} \qquad (2)$$

Where:

D_i: the diffusivity of species *i* (cm^2/s);
R : the gas constant (8.314 J/mol.K);
T : absolute temperature (K);
σ_i : partial conductivity of species *i* (S/cm);
Z_i: charge of species *i;*
F : Faraday's constant (96500 Coulomb/mol);
C_i: the concentration of species *i* (mol/cm^3).

The resistivity is the opposite of conductivity. Afterward, the conductivity of the sample can be calculated from its resistivity.

In general, diffusion is described by Fick's law:

$$\frac{\partial C_i}{\partial t} = D_i \frac{\partial^2 C_i}{\partial^2 x} \qquad (3)$$

The formula of Equation (3) is well-known as Fick's second law. The solution of equation [3] is:

$$C_i(x,t) = C_{i0}\left(1 - erf\frac{x}{2\sqrt{D_i t}}\right) \qquad (4)$$

where,

D_i : diffusivity of species *i*;

$C_i(x, t)$: concentration of species i at distance x and time t;

C_{io}: the concentration of species i at x = 0.

When Fick's law is used to describe the diffusion process, it has the following assumptions: (i)The fluxing species must not react with the matrix, and (ii)The matrix is homogenous in structure and composition. From equation [2], it can be seen that if the concentration profile is known and assuming the value of C_{io}, the diffusivity of ions can be calculated from the equation. The concentration profile usually measured by salt-ponding, but can be time-consuming[13]. Alternatively, the diffusivity calculated from Equation (2).

The threshold chloride contents required to initiate corrosion in reinforced concrete, expressed as $C(x,t)$in Eq. (4). The concentration of chloride at the concrete surface (C_{io}) is related to sea-water composition. By assuming threshold chloride concentration, then the service life of certain reinforced concrete structure with certain cover depth can be predicted. Conversely, the cover depth can be approximated when the service life is set.

The resistivity can be used as an electrical indicator of chloride penetration resistanceof concrete. Since the test can be simply conducted, the test should be applied as a quality control of chloride resistant [14]. Hereafter, concrete exposed to marine must be addressed to such a test.

This study found that the concrete with 30% and 40% of FA replacement shows the higher resistivity than the normal concrete, even though the strength of that sample is lower than the normal concrete. This fact indicates that fly ash contributes the chloride resistant of concrete. The fly ash contributes the improvement of concrete resistivity by physical dan chemical mechanism The chloride resistance of concrete is affected by the pore system and the chemical composition of the binder[15]. The small and disconnected pores usually provide high chloride resistant. This study is consistent with many previous studies that stated that incorporation of fly ash in concrete refined the microstructure of concrete [16]–[20][21]. The low compressive strength of FA-50 indicates that the fly ash amount exceeding the optimum value. The pore refinement supported by fly ash pozzolanic reaction is suppressed by pore formation from unreacted portion of fly ash.

In addition, the alumina content of fly ash provides the chloride binding capacity due to the formation of Friedel's salt. The amount of binding generally increases with the (available) alumina content. The FA-50 sample has the most alumina content, conversely it has the highest resistivity. Since the fly ash has lower activity than cement, then fly ash contribution in pore refinement is limited to certain amount. While fly ash amount exceeding the limit, it tends to generate the more unreacted portion and the more porous structure. Consequently, the improvement of chloride binding capacity of FA-50 was counteracted by the higher porosity created by unreacted part. Thus, the optimum amount of fly ash in this study is 40%.

4. CONCLUSION

1. Electrical resistivity measurement of concrete with 4 probe Wenner method provide the simple way for quality control of marine concrete.
2. Incorporation of fly ash to the concrete up to a certain limit amount provide the improvement of resistivity. Above the limit, the chemical contribution of fly ash is counteracted by reduction of physical properties.
3. The optimum concrete of this study is HVFA with 40% of fly ash replacement.

5. ACKNOWLEDGEMENTS

The authors gratefully acknowledge the financial support of the Shipbuilding Institute of Polytechnic Surabaya under the DIPA research grant 2021.

6. REFERENCES

1 M. Bitaraf and S. Mohammadi, "Analysis of chloride diffusion in concrete structures for prediction of initiation time of corrosion using a new meshless approach," *Constr. Build. Mater.*, vol. 22, no. 4, pp. 546–556, 2008, doi: 10.1016/j.conbuildmat.2006.11.005.

2 A. Farahani and H. Taghaddos, "Prediction of Service Life in Concrete Structures based on Diffusion Model in a Marine Environment Using Mesh Free, FEM and FDM Approaches," *J. Rehabil. Civ. Eng.*, vol. 8, no. 4, pp. 1–14, 2020, doi: 10.22075/JRCE.2020.19189.1380.

3 K. D. Stanish, R. D. Hooton, and M. D. . Thomas, "Testing the Chloride Penetration Resistance of Concrete : A Literature Review," 1997.

4 E. G. Moffatt, M. D. A. Thomas, and A. Fahim, "Performance of high-volume fly ash concrete in marine environment," *Cem. Concr. Res.*, vol. 102, no. June, pp. 127–135, 2017, doi: 10.1016/j.cemconres.2017.09.008.

5 Cement Concrete & Aggregates, "Chloride Resistance of Concrete," *Cem. Concr. Aggregates Aust.*, no. June, pp. 1–37, 2009.

6 H. Layssi, P. Ghods, A. R. Alizadeh, and M. Salehi, "Electrical Resistivity of Concrete: Concepts, applications, and measurement techniques," *Concr. Int.*, no. May, pp. 41–46, 2015.

7 H. Yazici, "The effect of silica fume and high-volume Class C fly ash on mechanical properties, chloride penetration and freeze-thaw resistance of self-compacting concrete," *Constr. Build. Mater.*, vol. 22, no. 4, pp. 456–462, 2008, doi: 10.1016/j.conbuildmat.2007.01.002.

8 W. Chalee, C. Jaturapitakkul, and P. Chindaprasirt, "Predicting the chloride penetration of fly ash concrete in seawater," *Mar. Struct.*, vol. 22, no. 3, pp. 341–353, 2009, doi: 10.1016/j.marstruc.2008.12.001.

9 W. D. Pratiwi, Triwulan, J. J. Ekaputri, and H. Fansuri, "Combination of precipitated-calcium carbonate substitution and dilute-alkali fly ash treatment in a very high-volume fly ash cement paste," *Constr. Build. Mater.*, vol. 234, p. 117273, 2020, doi: 10.1016/j.conbuildmat.2019.117273.

10 M. Thomas, T. Moffatt, H. Yi, and D. E. Smith, "Electrical Methods for Estimating the Chloride Resistance of Concrete," *ACI Fall 2013 Conv.*, no. 2821, 2013.

11 Apgar, "Practical Evaluation of Resistivity of Concrete in Test Cylinder using A Wenner Array Probe," *Cem. Concr. Res.*, vol. 66, no. December, pp. 37–39, 1996.

12 P. Ghosh and Q. Tran, "Correlation Between Bulk and Surface Resistivity of Concrete," *Int. J. Concr. Struct. Mater.*, vol. 9, no. 1, pp. 119–132, 2015, doi: 10.1007/s40069-014-0094-z.

13 X. Lu, "Application of the Nernst-Einstein equation to concrete," *Cem. Concr. Res.*, vol. 27, no. 2, pp. 293–302, 1997, doi: 10.1016/S0008-8846(96)00200-1.

14 A. A. Ramezanianpour, A. Pilvar, M. Mahdikhani, and F. Moodi, "Practical evaluation of relationship between concrete resistivity, water penetration, rapid chloride penetration and compressive strength," *Constr. Build. Mater.*, vol. 25, no. 5, pp. 2472–2479, 2011, doi: 10.1016/j.conbuildmat.2010.11.069.

15 M. Thomas, "The durability of concrete for marine construction," in *Marine Concrete Structures*, Elsevier Ltd, 2016, pp. 151–170.

16 M. Şahmaran, I. Ö. Yaman, and M. Tokyay, "Transport and mechanical properties of self consolidating concrete with high volume fly ash," *Cem. Concr. Compos.*, vol. 31, no. 2, pp. 99–106, 2009, doi: 10.1016/j.cemconcomp.2008.12.003.

17 W. Chalee, P. Ausapanit, and C. Jaturapitakkul, "Utilization of fly ash concrete in marine environment for long term design life analysis," *Mater. Des.*, vol. 31, no. 3, pp. 1242–1249, 2010, doi: 10.1016/j.matdes.2009.09.024.

18 D. Wang, X. Zhou, Y. Meng, and Z. Chen, "Durability of concrete containing fly ash and silica fume against combined freezing-thawing and sulfate attack," *Constr. Build. Mater.*, vol. 147, pp. 398–406, 2017, doi: 10.1016/j.conbuildmat.2017.04.172.

19 P. Dinakar, K. G. Babu, and M. Santhanam, "Durability properties of high volume fly ash self compacting concretes," *Cem. Concr. Compos.*, vol. 30, no. 10, pp. 880–886, 2008, doi: 10.1016/j.cemconcomp.2008.06.011.

20 P. Nath and P. Sarker, "Effect of fly ash on the durability properties of high strength concrete," *Procedia Eng.*, vol. 14, pp. 1149–1156, 2011, doi: 10.1016/j.proeng.2011.07.144.

21 T. Cheewaket, C. Jaturapitakkul, and W. Chalee, "Initial corrosion presented by chloride threshold penetration of concrete up to 10 year-results under marine site," *Constr. Build. Mater.*, vol. 37, pp. 693–698, 2012, doi: 10.1016/j.conbuildmat.2012.07.061.

22 M. D. A. Thomas, R. D. Hooton, A. Scott, and H. Zibara, "The effect of supplementary cementitious materials on chloride binding in hardened cement paste," *Cem. Concr. Res.*, vol. 42, no. 1, pp. 1–7, 2012, doi: 10.1016/j.cemconres.2011.01.001.

23 A. Costa and J. Appleton, "Chloride penetration into concrete in marine environment - Part I: Main parameters affecting chloride penetration," *Mater. Struct. Constr.*, vol. 32, no. 4, pp. 252–259, 1999, doi: 10.1007/bf02479594.

7. AUTHORS BIOGRAPHY

Febrian Dany Dwi Putra is a student of magister program of Safety and Risk Engineering in Politeknik Perkapalan Negeri Surabaya (Shipbuilding Institute of Polytechnic Surabaya)

Wiwik Dwi Pratiwi is a lecturer and researcher of materials of construction in Politeknik Perkapalan Negeri Surabaya (Shipbuilding Institute of Polytechnic Surabaya). In recent years, she focused on high volume fly ash cement and concrete. She is also an author of some conference papers and journals on the relevant topic.

Indri Santiasih is an Indonesian born, and based Senior Lecturer at Department of Safety Engineering Politeknik Perkapalan Negeri Surabaya (Shipbuilding Institute of Polytechnic Surabaya). Over the past sixteen years, she has been worked in PPNS as a lecturer and a scientist. Her research subject is toxicology, indoor air pollution especially particulate matter exposure and safety and health as well.

Kiki Dwi Wulandari holds the current position of lecturer at Department of Ship Building Engineering, Politeknik Perkapalan Negeri Surabaya (Shipbuilding Institute of Polytechnic Surabaya). Her previous experienceincludes innovation for construction materials and structure, green concrete materials, geopolymer, and self-healing concrete materials.

Kharis Abdullah holds the current position of lecturer at Department of Shipbuilding Engineering, Politeknik Perkapalan Negeri Surabaya (Shipbuilding Institute of Polytechnic Surabaya). His previous experience includes marine design, materials and structure.

Nurvita Arumsari holds the current position of Lecturer at Politeknik Perkapalan Negeri Surabaya (Shipbuilding Institute of Polytechnic Surabaya) in Marine Engineering department. She is responsible for teach, research and community service as tri dharma. She is interested in reliability and statistics modelling research. Her previous experience includes analyst in research centre for Regional Development and Community empowerment. Besides that, the author also often collaborates with other researchers or practitioner discussing about the field of energy prediction and reliability for maintenance.

DUAL UTILIZATION OF WASTE THERMAL AND CRYOGENIC ENERGY ON BOARD AN LNG FUELLED MARINE VESSEL USING ORC TECHNOLOGY

S Farrukh, D Wu, R Al-Dadah, Department of Mechanical Engineering, School of Engineering, University of Birmingham, B15 2TT, United Kingdom.

SUMMARY

In the last two decades, Liquified Natural Gas has emerged as the major fuel in marine vessels. Besides the energy obtained from combustion, LNG has 830 kJ/kg of cold energy as well, which is normally wasted during regasification. One way to utilize this energy is with the use of an Organic Rankine Cycle for two-fold recovery of the waste exhaust gas heat from the LNG engine and the LNG cryogenic energy. This paper investigates the utilization of LNG engine waste heat and cryogenic energy using a regenerative ORC combined with the Direct Expansion of LNG. A parametric sensitivity analysis was conducted, using different working fluids to maximize the first and second law efficiencies. The effect of the condensation and evaporation pressures on the system performance have been investigated. Different fire-retardants were mixed with the pure working fluids to measure the effect on system performance. The highest thermal and exergy efficiency was achieved by ethylene, 29.07 % and 35.19 %, respectively. Out of the fire-retardant mixtures, ethane/R23 (0.1/0.9) gave the best performance, achieving a thermal and exergy efficiency of 21.81 % and 28.18 %, respectively. It was found that fire retardants limit the performance of the system. However, their inclusion is a necessity, and further study is needed to investigate a wide range of potential fire-retardant compounds.

NOMENCLATURE

DE	Direct Expansion
n	Efficiency (%)
h	Enthalpy (kJ/kg)
s	Entropy (kJ/kg-K)
E_x	Exergy (kJ)
Q	Heat transfer (kJ)
LNG	Liquified Natural Gas
$\dot{m}$	Mass flow rate (kg/s)
NOx	Nitrogen oxides
ORC	Organic Rankine Cycle
SOx	Sulphur Oxides
T	Temperature (℃)
W	Work (kW)
WF	Working Fluid

1. INTRODUCTION

In the past two decades, the major world economies have shifted towards environmentally friendly fuels due to increasing global temperatures and greenhouse gas emissions. Liquefied low-emission fuels have replaced the traditional fossil fuels due to low emissions and high energy quality [1] [2]. The most popular fuel that has been commercialized is LNG, especially in the maritime sector, with the introduction of LNG fueled engines onboard sea vessels. These engines provide high efficiency and emit fewer particulates, such as NOx and SOx, as compared to conventional fuel engines and conform to the Tier III standards of IMO (International Maritime Organization) [3].

LNG is a cryogenic fuel and is stored at -162 ℃ at atmospheric pressure. It must be degasified to NG before it can be used for combustion in the engine. The regasification of LNG releases approximately 830 kJ/kg of cold energy, which is typically dissipated to the environment via seawater or ambient air [4]. Recently, this waste cryogenic energy has attracted the attention of many researchers, who have developed different methods of cryogenic energy utilization. One of the most researched applications of cryogenic energy utilization is power generation systems, with the first system being built in 1976 by Grienpentrog and Sacharendt [5]. They proposed a closed gas turbine cycle that utilized cryogenic energy to cool the working fluid temperature at the compressor inlet. However, the use of Brayton cycles onboard sea vessels is unfeasible, owing to the high temperature requirement. In this regard, Organic Rankine Cycles have emerged as a compact solutionsince they can utilize both the low and medium grade exhaust gas as the heat source and the ultra-low temperature LNG as the cold source. Many studies have been conducted on ORCs for dual waste heat and cryogenic energy utilization, while maintaining relatively high efficiency and simple cycle configuration. Ng et al. [6] designed a simple ORC system for cryogenic energy recovery onboard an LNG powered sea vessel. They achieved a maximum thermal efficiency of 21 %, using heptane as the working fluid. Tsougranis and Wu [7] studied two different ORC configurations onboard a sea vessel, the single-stage and the two-stage regenerative ORC. They found that the two-stage ORC almost always performs better than the single-stage ORC at temperatures above 250℃. Although ORC systems exhibit high performance on their own, they can only utilize the thermal exergy of the LNG, and the mechanical exergy is normally wasted. Thus, most systems integrate the Direct Expansion (DE) cycle with the ORC to utilize both the thermal and mechanical exergy of the LNG, which increases system performance. To prove that the ORC + DE combined system provided better efficiency than the standalone ORC system, Liu and Gao [8] performed a comparative study and found that, at the same operational parameters, the ORC + DE system performed better than the standalone ORC system. They achieved a thermal efficiency of 23.4 % for the combined system and 14.1 % for the standalone ORC system.

The selection of the working fluid is an important consideration in the system design process, as it can directly influence the system's performance. For LNG cryogenic energy recovery, the criterion established by different researchers includes cost, environmental impact, low freezing point, thermal stability at high temperatures and low specific heat [9] [10]. Based on this criterion, low molecular hydrocarbons have emerged as the first choice working fluids. A comparative study was performed, separately, by Sun et al. [11] and Liu et al. [4] on different hydrocarbons and refrigerants to measure the effect on the ORC system performance. The studies concluded that hydrocarbons perform better than refrigerants at the same operating conditions. This conclusion has been verified in other studies [12] [13]. Although hydrocarbons perform better than refrigerants, it must be acknowledged that hydrocarbons are highly flammable, and this risk is increased manifold onboard sea vessels with limited space and escaping opportunities. This makes the addition of fire retardants to the hydrocarbons essential in maritime applications. Kang et al. [14] performed a parametric analysis on zeotropic mixtures containing hydrocarbons and fire retardants. They found that fluorine-based hydrocarbons have the highest fire-retardant tendencies. They achieved the best system performance with R245fa/R600a (0.9/0.1) mixture at a heat source temperature of 110 °C. Song et al. [15] found that the mixture of cyclohexane/R141b (0.5/0.5) is non-flammable and produced 13 % more power than pure cyclohexane, at an evaporation temperature of 200 °C. However, limited data is available regarding fire retardant and hydrocarbon mixtures for cryogenic energy utilization.

This paper will investigate the effect of evaporation and condensation pressures on the performance of three ORC configurations, namely the simple ORC, ORC+DE and the regenerative ORC with DE for dual utilization of exhaust gas waste heat energy and LNG cryogenic energy. The pure working fluids used for the analysis are ethylene, ethane, and propane, which have previously been validated as excellent working fluids in low temperature applications [7]. Fire retardant compounds are added to these working fluids to measure the effect on system performance.

2. METHODOLOGY

2.1. SYSTEM DESCRIPTION

The proposed system configurations are shown in Fig. 1. The simple ORC consists of a pump, evaporator, expander, and condenser. The exhaust gas temperature has been limited to 150 °C, because of the new technique called

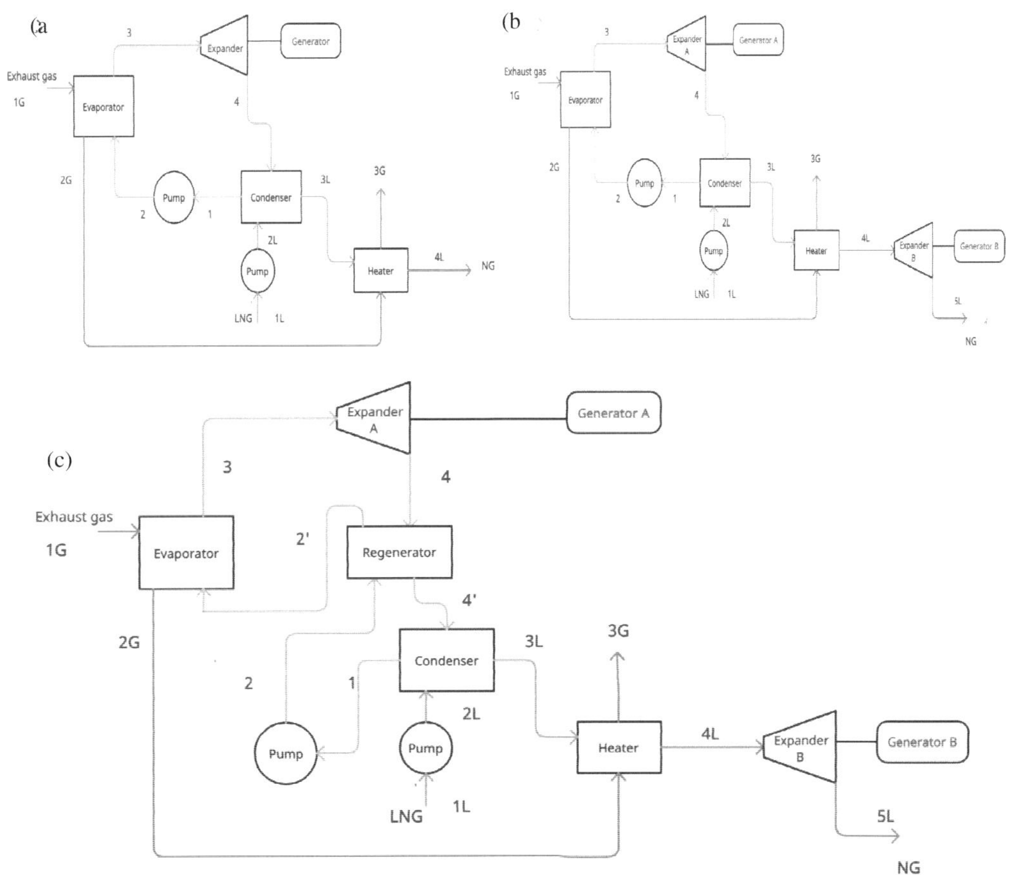

Figure 1: Schematic of the ORC system configuration: a) Simple ORC, b) ORC + DE, c) Regenerative ORC

Reformed Exhaust Gas Recirculation (REGR), which recirculates the exhaust gas back through the LNG fueled engine reducing the temperature of the final exhaust gas [16]. The maximum pressure in the cycle is set to the recommendation of Ref. [7] at 15 bar, and the minimum pressure is set at atmospheric pressure to avoid vacuum conditions.

The working fluid starts at 1 bar pressure and is pumped to the evaporator at 15 bars. The exhaust gas enters the evaporator at 150 ̊C and heats the working fluid up to 130○ ̊C. The high-pressure vapor is then expanded through the expander, producing power. The expanded vapor is then passed through the condenser, where it is condensed by the LNG cryogenic energy. The LNG start at 5 bar and -138 ̊C and is pumped to the condenser at 15 bars. After the condenser, the LNG is passed through the heater, where it is heated up to 25 ̊C by the residual exhaust gas and sent to the engine. In fig. 1(b), ORC+DE configuration, the LNG is heated to 95 ̊C (State 4L) and is then expanded by expander B to produce power. The low-pressure NG at 25 ̊C is then sent to the engine. In fig.1(c), a regenerator is added that absorbs the excess heat at the expander outlet and heats the WF before the evaporator inlet. The mass flow rate of the WF, exhaust gas and LNG is set to 1 kg/s, 10 kg/s and 0.875 kg/s, respectively.

2.2. THERMODYNAMIC ANALYSIS

The following assumptions have been made for the thermodynamic analysis based on recommendations from Ref. [17][18][19]:

1. The pressure drop in the heat exchangers and pipes is ignored.
2. The efficiencies of the heat exchangers are assumed to be 0.8 [19].
3. The isentropic efficiencies of the pump and expander are assumed to be 0.65 and 0.8, respectively [18].
4. The pinch point in the evaporator and regenerator is set to 20 ̊C and 30 ̊C, respectively [7].
5. It is assumed that the WF starts as saturated liquid at state 1.
6. Heat loss in the pump and expander is ignored.
7. For modelling purposes, LNG is assumed to be 100 % methane.

Based on the first and second laws of thermodynamics, the theoretical formulas used for the thermodynamic analysis have been shown below. Eq. (1) shows the energy balance in the evaporator, and Eq. (2) shows the heat absorbed by the WF.

$$n_{evap} = \frac{\dot{m}_{WF}.(h_{out} - h_{in})}{\dot{m}_{eg}.C_p.(T_{1G} - T_{2G})} \quad (1)$$

$$Q_{in,ORC} = \dot{m}_{WF}.(h_{out} - h_{in}) \ (2)$$

Here, ṁ is the mass flow rate, h is the enthalpy, n is the efficiency, Q is the total heat transfer, and C_p is the specific heat of the exhaust gas. Eq. (3) shows the energy balance in the condenser, and Eq. (4) shows the energy absorbed by the LNG before being heat for the DE cycle. The subscript WF, eg, evap, and cond stand for the working fluid, exhaust gas, evaporator, and condenser, respectively.

$$n_{cond} = \frac{\dot{m}_{LNG}.(h_{out} - h_{in})}{\dot{m}_{WF}.(h_{in} - h_{out})} \quad (3)$$

$$Q_{in,heater} = \dot{m}_{LNG}.(h_{4L} - h_{3L}) \ (4)$$

The overall heat energy input into the system is the total energy absorbed by the WF and LNG from the exhaust gas, as shown in Eq. (5).

$$Q_{in,total} = Q_{in,ORC} + Q_{in,heater} \ (5)$$

The work input required by the pumps and the work output by the turbines is shown in Eq. (6).

$$W_{P,T} = \dot{m}_{WF}.(h_{out,P,T} - h_{in}) \ (6)$$

In Eq. (6), the calculation of h_{out} is done based on the isentropic efficiencies of the pump and expander, shown in equations (6A) and (6B).

$$h_{out,P} = \frac{h_{out,isentropic} - h_{in}}{n_{pump}} + h_{in} \ (6A)$$

$$h_{out,T} = h_{in} - n_{expander}(h_{in} - h_{out,isentropic}) \ (6B)$$

Here, the isentropic enthalpy is calculated using REFPROP based on the entropy at the inlet of the pump or expander. For the regenerator, the energy balance is shown in Eq. (7).

$$n_{regen} = \frac{\dot{m}_{WF}.(h_{2'} - h_2)}{\dot{m}_{WF}.(h_4 - h_{4'})} \quad (7)$$

The subscripts 2' and 4' represent States 2' and 4', respectively, as shown in fig. 1. The objective functions being prioritized in this study are the net power output, thermal efficiency, and exergy efficiency, which can be calculated using Eq. (8) – (10).

$$W_{net} = \sum W_T - \sum W_P \ (8)$$

$$n_{th} = \frac{W_{net}}{Q_{in,total}} \quad (9)$$

$$n_{ex} = \frac{W_{net}}{E_{xin,total}} \quad (10)$$

Exergy is based on the second law of thermodynamics and is a better performance indicator than the first law efficiency. The exergy at any point 'n' in the system can be calculated using Eq. (11), and the exhaust gas exergy can be calculated using Eq. (12). The dead state is set as T_o=15 ̊C and P_o=1 bar.

$$E_{xn} = \dot{m}_n(h_n - h_o) - T_o.(s_n - s_o) \ (11)$$

$$E_{xeg,xheater} = \dot{m}_{eg}.C_p.(T_{1G,2G} - T_o - T_o.ln\frac{T_{1G}}{T_o}) \ (12)$$

Here, s is the entropy at the given state, and E_x is the exergy. The exhaust gas exergy is calculated at the inlet of the evaporator and the inlet of the heat at the corresponding temperatures. The total exergy input in the system is given by:

$$E_{xin,total} = E_{x,eg} + E_{x,LNG} + E_{x,heater} \ (13)$$

It is pertinent to note that other studies in the literature [7][13] do not include the exergy input at the heat sink side of the DE configuration. This results in a lower exergy efficiency in this study as compared to other studies. However, the addition of the input exergy before the DE cycle turbine provides a better thermodynamic overview of the overall system performance.

The system was simulated in MATLAB connected with REFPROP 9.1, which was used to obtain the thermos physical properties of the fluids. The pure fluid properties are based on the Helmholtz equation of state, whereas the mixture properties are based on the KW1 Kunz and Wagner model for hydrocarbon mixtures, as defined in REFPROP [20].

2.3. WORKING FLUID SELECTION

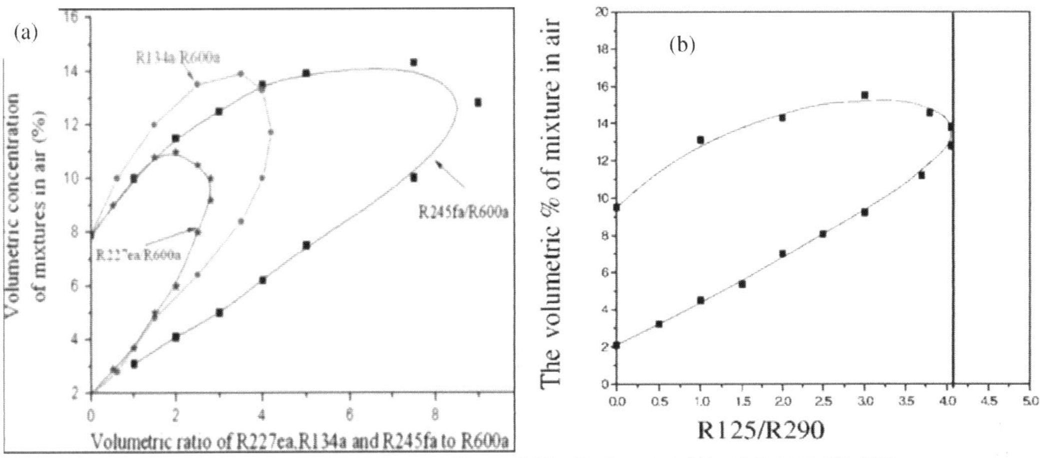

Figure 2: Volumetric ratios for flammability limits: a) R600a [21], b) R290 [22].

Table 1: Working fluid properties and compositions:

Fluid	T_{crit} (°C)	P_{crit}(bars)	Melting point (°C)	Normal Boiling point (°C)	ODP	Toxic	GWP	Flammable
Ethylene (R1150)	9.2	50.418	-169.16	-103.77	0	No	4	Yes
Ethane (R170)	32.172	48.722	-182.78	-88.581	0	No	6	Yes
Propane (R290)	96.74	42.512	-187.63	-42.114	0	No	3	Yes
Ethylene/R23 (0.1/0.9)	21.533	57.83	-161	-86.149	0	No	14800	No
Ethane/R23 (0.1/0.9)	20.177	51.4	-163	-82.9	0	No	14800	No
Propane/R22 (0.1/0.9)	92.4	48.6	-165	-41.9	0	No	1810	No
Propane/R134a (0.1/0.9)	95.89	41.68	-124	-28.49	0	No	1430	No

The critical criterion for working fluids for ORCs utilizing cryogenic energy is based on low melting point and high thermal stability above 150 °C. Additionally, environmental factors such as the Ozone Depletion Potential (ODP), Global Warming Potential (GWP), flammability, critical temperature and pressure, and normal boiling point are all crucial factors when deciding the working fluid. Based on this, the pure working fluids chosen for this analysis have been shown in table 1.

One of the main things to note is that the pure working fluids chosen are highly flammable. To counter the issue of flammability, fire-retardants will be mixed with the working fluids, and the effect on system performance will be measured. From the experimental analysis conducted by Ref. [21] [22] on the flame retardant tendencies of hydrofluorocarbons, it was found that there is a critical flammability ratio for each fire retardant and hydrocarbon mixture, based on the number of fluorine and hydrogen atoms in the compound and the molecular weight of the fire retardant. The results of their studies are shown in fig. 2.

From fig. 2, it is apparent that as the number of fluorine atoms and molecular mass of the fire-retardant compound increases, relative to R600a and R290, the volumetric ratio of the retardant required for inflammability decreases. These conclusions were the verification of the results achieved by Ref. [23] [24]. For most hydrofluorocarbons, the volumetric ratio of 4:1 is satisfactory for producing a fire-retardant compound. However, a conservative approach has been taken in this paper, and the fire-retardant volume percentage has been kept to 90%. Fig. 2 shows the addition of fire-retardants to common refrigerants, R600a and R290. However, these fire-retardants do not match the boiling points of the working fluids used in this study. The choice of the fire retardants was based on their boiling points relative to the hydrocarbon fluid to reduce the temperature glide. A high temperature glide would cause increased exergy destruction in the condenser, hence reducing the efficiency. Accordingly, fire-retardants with similar boiling temperatures to the working fluids are selected. R23 was chosen to be mixed with R170 and R1150, and R22 and R134a were chosen to be mixed with R290. R23, R22 and R134a have boiling points within 20 ℃ of the pure working fluids they are being mixed with, which would ultimately decrease the temperature glide in the evaporator and condenser. The properties of the chosen fire-retardant mixtures are shown in table 1.

From table 1, it is evident that the environmental impact of the chosen working fluids is very high. R23 and R22 have been banned as of 2020 [25], and the GWP of R134a is also very high. The work presented in this paper is preliminary, and a search is continued for finding better fire-retardants with similar properties but with a smaller environmental impact.

3. RESULTS AND DISCUSSIONS:

3.1. PURE HYDROCARBON WORKING FLUIDS:

Parametric analysis on the evaporation and condensation pressures was conducted on the three ORC configurations, heated up to 130 ℃. In the case of the condensation pressure, it is apparent in fig. 3 that the thermal and exergy efficiencies of each cycle configuration decreases with the increase in condensation pressure. The opposite trend is seen for the evaporation pressure, wherein the performance of the system increases with increasing evaporation pressure.

In the case of ethane, for the simple ORC, increasing the condensation pressure from 100-500 kPa decreases the thermal and exergy efficiency from 19.54 % to 9.39 % and 24.51 % to 10.67 %, respectively, decreasing the performance by more than 50 %. Alternatively, increasing the evaporation pressure increases the system performance by approximately 10-12%, as shown in fig. 3 (b). In the case of the ORC and DE cycle integration, it is evident from fig. 3(c,d) that the introduction of the DE cycle increases the system performance by utilizing the mechanical exergy of the LNG, which is wasted in the simple ORC. An interesting observation from fig. 3 is that ethylene performs only slightly better than ethane, even though its boiling point is much lower than ethane. This is because the evaporation temperature has been kept constant, which allows for a higher degree of superheat for ethylene as compared to ethane. The same is true for propane, which shows a considerably poor system performance when compared to the other working fluids. In the case of different heat source temperatures, a higher heat source temperature for propane would have increased system performance.

From fig. 3(c,d,e,f), it can be seen that the performance of the ORC+DE system is not improved with the addition of the regenerator. Rather, it exhibits almost the same thermal and exergy efficiencies. However, this should not be the case, as studied by Ref. [7] [26]. The reason for the similar performance in this study, as compared to other studies, is because this model considers all the heat and exergy input to the system when calculating system performance, as shown in Eq. 5, 9, 10 and 13. This shows a better overall view of the system exergy input and output. In the case where only the ORC system performance may be analyzed, neglecting heat energy input to the heater to heat the LNG before the DE turbine inlet and DE power output, the addition of the regenerator improves the system performance as shown in fig. 4. In each case, the system performance increases at least by 1-3 %.

3.2. FIRE RETARDANT MIXTURES:

The previous section discussed the implications of varying the condensation and evaporation pressures on the system performance, as well as the viability of adding a regenerator in the system. It was found that while the regenerator increases the ORC efficiency, the overall system efficiency remains unchanged due to the heating of LNG before the DE turbine inlet. Hence, for a marine ship, it is better to use the ORC+DE cycle since the regenerator would increase system complexity and cost without any gain in performance.

Extrapolating from the results of the parametric analysis, it was assumed that the mixed fluids would showcase similar trends at varying condensation and evaporation pressures. Hence, it was decided that the condensation and evaporation pressure for the mixed working fluid analysis would be fixed at 1 bar and 15 bar, respectively. The results of the mixed working fluid analysis are shown in table 2. The addition of the fire-retardants to the hydrocarbons reduces the system performance, especially the net power output. This is because the mole % of fire-retardants in the WF mixture is very high. Hence, the performance of the system is only slightly better than using pure fire-retardants as the working fluid. However, the addition of the fire-retardants can be deemed as necessary since the added safety is essential in closed spaces (such as marine ships) with a limited chance of survival in case of an explosion or fire. In the case of ethane/R23 and ethylene/R23, they achieved similar results for the objective parameters. However, the temperature glide of ethane/R23 is lower than ethylene/R23 (due to a smaller difference in normal boiling points of the individual components). Hence, ethane/R23 is preferred over ethylene/R23.

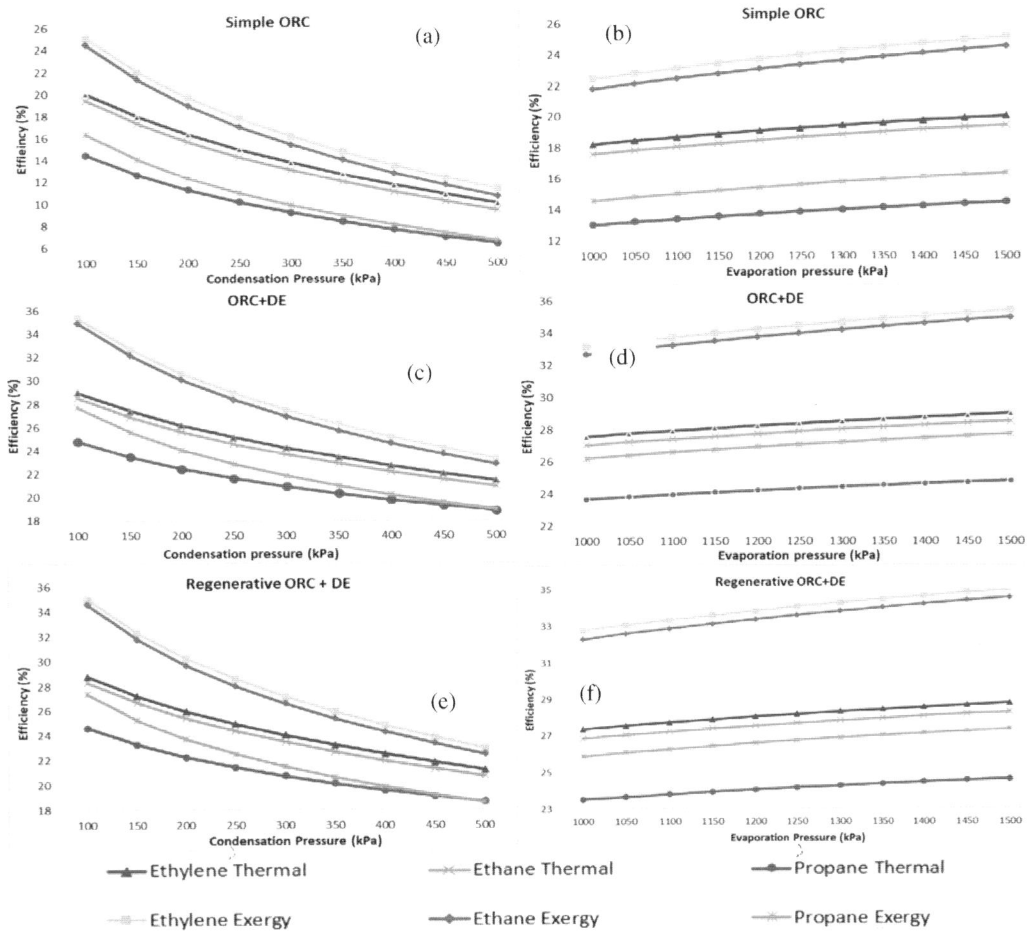

Figure 3: Results of the parametric analysis: a,b) Simple ORC, c,d) ORC+DE, e,f) Regenerative ORC+DE

Table 2: Fire-retardant hydrocarbon mixtures:

Configuration	Performance Parameter	Ethylene	Ethylene /R23 (0.1/0.9)	Ethane	Ethane /R23 (0.1/0.9)	Propane	Propane / R22 (0.1/0.9)	Propane /R134a (0.1/0.9)
Simple ORC	n_{th} (%)	20.02	10.04	19.45	10.09	14.47	8.09	7.26
	n_{Ex} (%)	25.12	9.53	24.51	9.58	16.36	7.35	6.54
	W_{net} (kW)	193.55	82.9	187.56	83.307	129.88	65.09	57.94
ORC+DE	n_{th} (%)	29.07	21.81	28.6	21.81	24.86	20.41	19.78
	n_{Ex} (%)	35.19	28.19	34.73	28.18	27.49	25.25	24.33
	W_{net} (kW)	320.25	209.64	314.25	209.64	256.58	191.82	184.67

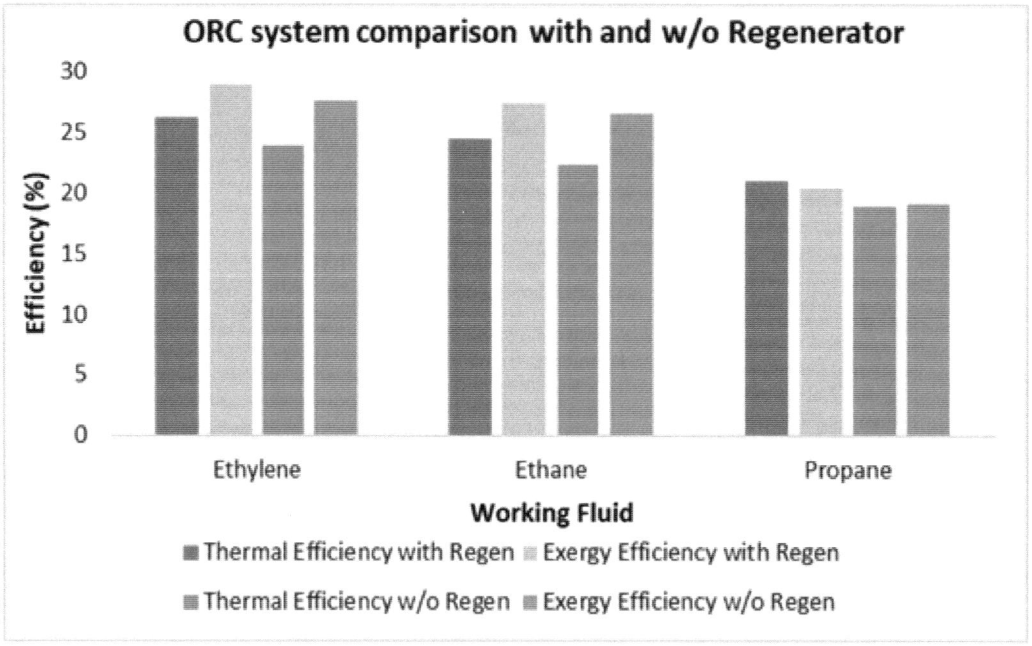

Figure 4: Effect of regenerator on ORC + DE configuration performance at 1 bar condensation pressure and 15 bar evaporation pressure.

4. CONCLUSIONS

This paper performed a comparative analysis on three different ORC configurations for LNG cryogenic energy recovery onboard an LNG powered sea vessel. A parametric analysis was performed with thermal and exergy efficiency as the objective functions and the evaporation and condensation pressures as the variables. Three pure hydrocarbons, namely ethylene, ethane, and propane, were used as the working fluids, and four fire retardants were added to the hydrocarbons with mole fraction of 0.1 hydrocarbon and 0.9 fire-retardant fluid. The choice of the fire-retardant was based on the boiling temperatures of the pure hydrocarbons to reduce the temperature glide in the condenser and evaporator. The following conclusions can be drawn from this study:

- The integration of the DE cycle must always be incorporated with an ORC to fully utilize the mechanical and thermal exergy of the LNG.
- Increasing the evaporation temperature increases the system performance, whereas increasing the condensation temperature decreases system performance. Hence, the maximum possible pressure ratio should be considered for an ORC.
- The ORC+DE configuration is best suited for dual waste thermal and cryogenic energy recovery onboard marine ships.
- The addition of fire-retardants to pure hydrocarbons is necessary, even though it means sacrificing a lot of the system performance. In this study, a conservative approach was taken when deciding the mole fractions of the mixtures. In future studies, an extensive search of hydrocarbons and fire-retardants needs to be conducted.
- The best performing pure fluid was ethylene, achieving 29.07 % energy and 35.19 % exergy efficiency and producing 320.25 kW of net power output at 1 bar and 15 bar condensing and evaporating pressures, respectively.
- The best performing fire-retardant mixture was ethane/R23, achieving 21.81% energy and 28.18 % exergy efficiency and producing 209.64 kW of power. The reason for choosing ethane/R23 mixture over ethylene/R23 mixture, even though they achieved the same performance, is that the temperature glide of ethane/R23 is lower than ethylene/R23. Additionally, ethane is cheaper than ethylene, making it a more economically viable option.

5. REFERENCES

1. Badami M, Bruno JC, Coronas A, Fambri G. Analysis of different combined cycles and working fluids for LNG exergy recovery during regasification. Energy. 2018.

2. Bao J, Yuan T, Zhang L, Zhang N, Zhang X, He G. Comparative study of liquefied natural gas (LNG) cold energy power generation systems in series and parallel. Energy Conversion & Management. 2019.

3. Watanabe KJME. High operation capable marine dual fuel engine with LNG. 2015.

4. Liu Y, Guo K. A novel cryogenic power cycle for LNG cold energy recovery. Energy. 2011.

5. Griepentrog H, Sackarendt P, editors. Vaporization of LNG with closed-cycle gas turbines. Turbo Expo: Power for Land, Sea, and Air; 1976: American Society of Mechanical Engineers.

6. Ng C, Tam IC, Wu D. System Modelling of Organic Rankine Cycle for Waste Energy Recovery System in Marine Applications. Energy Procedia. 2019.

7. Tsougranis EL, Wu D. A feasibility study of Organic Rankine Cycle (ORC) power generation using thermal and cryogenic waste energy on board an LNG passenger vessel. International Journal of Energy Research. 2018.

8. Liu Y, Guo K, editors. Efficiency of power generation by LNG cold energy. 2010 Asia-Pacific Power and Energy Engineering Conference; 2010: IEEE.

9. Wu D, Tsougranis E-L. LNG Cryogenic Energy Technologies. Advances in Energy Research: Nova Science Publishers; 2020.

10. Bao J, Zhao L. A review of working fluid and expander selections for organic Rankine cycle. Renewable and sustainable energy reviews. 2013.

11. Sun Z, Lai J, Wang S, Wang T. Thermodynamic optimization and comparative study of different ORC configurations utilizing the exergies of LNG and low grade heat of different temperatures. Energy. 2018.

12. Emadi MA, Chitgar N, Oyewunmi OA, Markides CN. Working-fluid selection and thermoeconomic optimisation of a combined cycle cogeneration dual-loop organic Rankine cycle (ORC) system for solid oxide fuel cell (SOFC) waste-heat recovery. Applied Energy. 2020.

13. Sun H, Zhu H, Liu F, Ding H. Simulation and optimization of a novel Rankine power cycle for recovering cold energy from liquefied natural gas using a mixed working fluid. Energy. 2014.

14. Kang Z, Zhu J, Lu X, Li T, Wu X. Parametric optimization and performance analysis of zeotropic mixtures for an organic Rankine cycle driven by low-medium temperature geothermal fluids. Applied Thermal Engineering. 2015.

15. Song J, Gu C-w. Analysis of ORC (Organic Rankine Cycle) systems with pure hydrocarbons and mixtures of hydrocarbon and retardant for engine waste heat recovery. Applied Thermal Engineering. 2015.

16. Du Y, Yu X, Liu L, Li R, Zuo X, Sun Y. Effect of addition of hydrogen and exhaust gas recirculation on characteristics of hydrogen gasoline engine. International Journal of Hydrogen Energy. 2017.

17. Kim K, Lee U, Kim C, Han C. Design and optimization of cascade organic Rankine cycle for recovering cryogenic energy from liquefied natural gas using binary working fluid. Energy. 2015.

18. Xue X, Guo C, Du X, Yang L, Yang Y. Thermodynamic analysis and optimization of a two-stage organic Rankine cycle for liquefied natural gas cryogenic exergy recovery. Energy. 2015.

19. Sun Z, Zhang H, Zhang T, Lin L, Lin K. Optimizations and comparison of three two-stage Rankine cycles under different heat source temperatures and NG distribution pressures. Energy Conversion & Management. 2020.

20. Lemmon EW, Huber ML, McLinden MOJNsrd. NIST reference fluid thermodynamic and transport properties–REFPROP. 2002.

21. Yang Z, Wu X, Peng J. Theoretical and experimental investigation on the flame-retarding characteristic of R245fa. Experimental thermal and fluid science. 2013.

22. Zhao Y, Ting W, Xihong L. Experimental studies and estimates of the explosion limit of some environmentally friendly refrigerants. Combustion science and technology. 2005.

23. Zabetakis MG. Flammability characteristics of combustible gases and vapors. Bureau of Mines Washington DC; 1965.

24. Oellrich LR, Leisenheimer B, Srinivasan K. Flammability behavior of (octafluoropropane plus propane) and (octafluoropropane plus methane). Chemie Ingenieur Technik. 1997.

25. United States Environmental Protection Agency. URL:http://www.epa.gov/ozone/snap/subsgwps.html . 2018.

26. Zhang J, Zhang H, Yang K, Yang F, Wang Z, Zhao G, et al. Performance analysis of regenerative organic Rankine cycle (RORC) using the pure working fluid and the zeotropic mixture over the whole operating range of a diesel engine. Energy conversion and management. 2014.

6. AUTHOR BIOGRAPHIES

Salman Farrukh is a PhD student in the department of Mechanical Engineering at the University of Birmingham. His research interests include cryogenics, refrigeration, power generation and decarbonization.

Dr. Dawei Wu is a senior lecturer in Mechanical Engineering at the University of Birmingham and an EPSRC (Engineering and Physical Sciences Research Council) Fellow. Dawei has published over 50 peer reviewed papers and book chapters in the field of Mechanical and Marine Engineering, mainly on power generation systems, decarbonization, cryogenic energy applications, novel linear engine generators and zero carbon propulsions.

Dr. Raya Al-Dadah is a Reader in Sustainable Energy Technologies and Head of the Sustainable Energy Technology Laboratory at the University of Birmingham. She has co-authored four book chapters, presented several keynote speeches and published around 200 papers in peer reviewed international journals and conference proceedings.

SAFER ON BOARD ENVIRONMENTS FOR INDONESIAN SEAFARERS AND FISHERS IN THE TIME OF COVID-19 AND BEYOND

G Thomas, L Huang and **C Ryan,** Department of Mechanical Engineering, University College London, UK
I K A P Utama, S Riyadi, and **D Setyawan,** Sepuluh Nopember Institute of Technology, Indonesia
W Hetharia, Department of Naval Architecture, University of Pattimura, Indonesia

SUMMARY

The seafaring and fishing industries in Indonesia have been severely affected by the COVID-19 pandemic. There is an urgent need to research the best technical solutions to ensure the safety of seafarers and fishers so that they can keep working safely at sea. It is highly unlikely that the Indonesian fleet will be able to rapidly return to the pre-COVID-19 situation within the next few years. Therefore, long-term solutions are required. This research aims to provide guidance for owners, operators and masters to reconfigure their vessels and operations to ensure a safe working environment to counter the potential spread of the virus, as well as maintaining the integral safety of the vessels. The present paper reports several primary activities, including analysis of existing guidelines,questionnaire survey, computational fluid dynamics, remote monitoring and machine learning. These works are expected to provide indications on virus-hazardous areas for ships, inform guidelines on hygienic and reconfiguring means to counter COVID-19 and potentially the spread of similar viruses, and further establish a long-term monitoring and optimising system for ship safety and performance.

1. INTRODUCTION

Commercial shipping and fisheries in Indonesia have severe short-term and long-term social and economic issues. In the short term, COVID-19 has decimated the marine industry by reducing operations with many vessels not able to leave port or being severely restricted[1]. There is a dilemma between(a) theneeds of food and income for the Indonesian people that in majority come from the fishing activities, and (b) the health protocolsand the fear of fishermen to the danger of continues work at sea. For example, in the Maluku fishing fleet only 30% of purse seiner vessels are currently able to operate; whilst crew boats servicing offshore platforms are operating at 1/2 capacity and are not economically viable. Restrictions due to COVID-19 have major impact on operations, for example during fishing operations hauling a purse seine net is completely manual with between 20 to 25 crew members participating at once with physical distancing nigh on impossible to implement.

In the longer term the design and fit out of vessels is hampering their ability to be economically viable; for example, in many sectors of the fishing fleet there is no preservation (ice hold) to maintain fish quality of fish and increase the activity. The low level of vessel technology results in poor maritime industry economic viability and safety [2,3]. There are approximately 2.2M fishers in Indonesia, operating more than 600,000 fishing vessels, over 90% of which are small (up to 5GT); equipping low levels of technology is a huge problem [4]. The development of vessels that can support safe and sustainable fishing practices is an enormous challenge for Indonesia. Wibawa et al. [3] proposed that the starting point is to use traditional fishing boats and evolving solutions for local problems is likely to increase local fishers' acceptance. Fisher surveys were conducted to understand the actual issues around economics and safety in East Java [5] and Ambon by Hetharia [6], showing that stability characteristics of the vessels are unsatisfactory and very often there is a complete lack of stability data.

2. OBJECTIVES

In order to address the short-term and long-term issues, the present research has two clear timelines in mind. Firstly, with the focus on mitigating the influence of COVID-19 from a safety and economic perspective, the timeline is short with the impact ambition in terms of 6 to 18 months. The second, longer-term focus, is on new cost-effective remote monitoring techniques, using sensors, machine learning techniques and post-processing algorithms; here the pathway to impact will be longer, 3 to 8 years.

The ultimate vision of this research is to enable Indonesia to have technologically-advanced fleets of safe commercial and fishing vessels that are operating in a profitable and safe way. This will contribute to sustainable development, a reduction in poverty, and improved social welfare with fewer fatalities in the maritime industry.

The impact will be ensured by the active participation of key stakeholders beyond the project consortium, such as the Directorate of Shipbuilding and Seafarers, Ministry of Transportation, Department of Marine and Fisheries - Maluku Province, which controls the operation of commercial and fishing vessels in Moluccas waters, the fishing company CV Jora, which operates a fleet of fishing vessels and the Orela shipyard which builds a range of commercial vessels.

A list of specific objectives is as follows:

- Develop clear guidance for reconfiguring commercial and fishing vessels and their operations to ensure a safe working environment given the impact of COVID-19, as well as maintaining the integral safety of the vessels.
- Provide vessel design and operational adjustments to ensure the future economic

215

viability of commercial and fishing vessels given the impact of COVID-19.
- Demonstrate the long-term safety and economic viability of the proposed solutions through a series of case studies.
- Develop new remote monitoring techniques, using sensors, machine learning techniques and post-processing algorithms to establish key indicators of ship safety and performance remotely.

The work towards the above objectives is still ongoing, but some preliminary activities have been conducted, which will be reported in the next section.

3. ACTIVITIES

In order to achieve the objectives, a range of activities have been and are being conducted. Work for short-term COVID-19 safety includesanalysis of existing guidelines, questionnaire survey, computational fluid dynamics.Work for long-term ship performance optimisation includes remote monitoring and machine learning. They are reported separately in this Section.

3.1 ANALYSIS OF EXISTING GUIDELINES

A comprehensive survey of current guidelines and regulations pertinent to dealing with COVID-19 in the maritime industries around the world was carried out. The main guidelines currently available are those from the IMO, Interferry, US Passenger Vessel Association, Canadian Government and the Indonesian General Director of Sea Transportation [7,8].

The similarities and differences between these guidelines have been established and compared with the Indonesian guideline [8]. Whilst the international guidelines have comprehensive advice available for passenger ships and cargo ships, the Indonesian guideline is lacking advice for fishing vessels particularly for small vessels which account for the vast majority of the Indonesian fleet. Therefore, intended future work is to fill the gap by developing a supplementary guideline for Indonesian small vessels.

3.2 QUESTIONNAIRE SURVEY

Before COVID-19 guideline can be developed for Indonesian vessels, the first step is to understand the current situation and anti-pandemic measurements there, for which, questionnaires were distributed to Indonesian ship and boat operators.
The scope of the questions covers vessel and voyage situation, social distancing, hygiene measurements and crew training. A series of vessel types were studied, including crew boats, fishing vessels, ferries and cargo vessels. A version of the questionnaire for passenger vessels is shown Table 1.
The collected results suggest that currently the crew are required to provide recent COVID-19 negative results before boarding, which means the main potential source of COVID-19 would be the passengers. Masks are required to wear but people may not strictly follow the rule, which exposures a significant risk. In addition, it lacks guideline on what areas should be cleaned more often or how to reconfigure the ships during the pandemic.

Table 1: COVID-19 questionnaire for passenger ships

1. General
How many ships do you operate?
What is the size of the ships that you operate?
How many people are on the ships that you operate?
How long is a normal voyage on the ships that you operate?
Do you have any ships that you know are not COVID safe? How many?
Have you had any positive COVID cases onboard previously? If so, how was this handled?
Are you aware of the national/international rules and recommendations for COVID safety onboard ships?
How many of the crew have been vaccinated?
2. Pre-departure
Do you carry out routine COVID testing before boarding? (Are crew required to take the RT-PCR Test / Rapid Test Antigen before boarding? Maximum time of sample taken)
Do you carry out body temperature measurements before boarding?
Do you ask about possible COVID exposure before boarding?
Do you ask about possible COVID symptoms before boarding?
3. Social distancing
Are there social distancing measures in place onboard? If yes, what are they?
Do you implement a minimum distance for crew/passengers? If so, what distance and in what areas of the ship?
Do you implement maximum capacity of crew/passengers on board? If so, what capacity and in what areas of the ship?
Do you split the crew into separate isolated groups to stop spread through the entire crew?
Are physical distancing markers placed on the floor/different areas of the ship?
Has overall passenger capacity been reduced? If so, what was the original capacity and what is the new capacity?
Are one-way systems used in any part of the ship?
If you do not require social distancing, what is the reason for this?
4. Masks

216

Do you require crew to wear fabric masks?
Do you require passengers to wear fabric masks?
Are there any physical barriers/screens installed between crew and passengers?
Do you experience any difficulty in buying masks?
If you do not require mask use, what is the reason for this?
5. Personal hygiene
Do you have hand sanitiser placed throughout the ship? If so, where is the hand sanitiser located?
If you do not have any hand hygiene stations, what is the reason for this?
6. Environmental Hygiene
Is the ship cleaned between voyages?
How often is the ship cleaned?
What cleaning products are used to clean the ship?
Are any areas cleaned more or less frequently?
Is shared equipment sanitised between uses e.g. lifejackets?
Is there a dedicated area to isolate possible COVID cases?
What ventilation/air conditioning is used onboard the vessel? Has this been altered to reduce the risk of COVID?
If you do not have any environmental hygiene measures, what is the reason for this?
7. Training
Are crew trained on COVID prevention?
Are crew trained on recognising symptoms of COVID?
Are passengers given information on COVID prevention?
Are passengers given information on recognising symptoms of COVID?
Is safety information regularly announced throughout the voyage?
8. Open questions
Do you need any resources/training to improve COVID safety of your ships?
Do you have any future plans to implement COVID safety measures?
If you do not have any training or COVID information measures, what is the reason for this?

3.3 COMPUTATIONAL FLUID DYNAMICS

During ship operations, winds can enter a ship and induce potential airborne transmission of virus particles. According to the survey results, there is a risk that people do not strictly wear masks on board. The compact space inside a ship implies a high risk of COVID outbreak. To investigate this problem, a model was developed combining computational fluid dynamic (CFD) and Lagrangian particles, in which, CFD is used to simulate the airflow and the transmission of virus particles is tracked by the Lagrangian approach.

The model was first validated against experimental measurements of particle diffusion in a standard room. As shown in Figure 1, a rectangular room is applied, where an air inlet is near the left top side and an outlet is near the right bottom side. From the inlet, air is coming into the room at a constant velocity, carrying particles that have a similar size to flu viruses.

Figure 1: A standard room case to validate the computational model

Chen et al. [9] measured the velocity profile and particle distribution inside the room, and the developed CFD-Lagrangian model was used to replicate their experiment. Figure 2 compares the experimental and computational results are in the central section (profile direction) of the room. It can be seen that excellent agreement is achieved, which demonstrates the present computational model is suitable to study the airborne transfer of COVID virus that is similar to a flu virus.

Upon validation of the computational model, the standard room geometry is replaced by the internal space of a crew boat, as shown in Figure 3. This example shows the velocity profile inside the ship advancing at 6 knots, and the distribution of virus particles following an import event, e.g. a passenger cough. Further analyses are to study different scenarios, such as when the front door is closed and when a passenger is speaking other than coughing. The results are expected to provide indications on virus-hazardous areas for ships, and further inform guidelines on hygienic and reconfiguring means to counter COVID-19 and potentially the spread of similar viruses in the future.

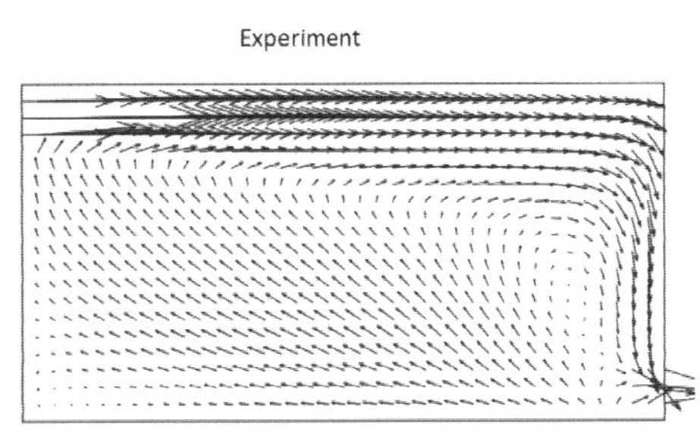

217

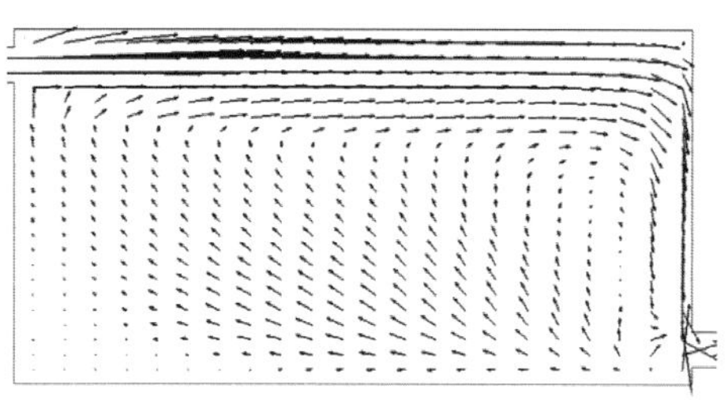

(a) Velocity profile

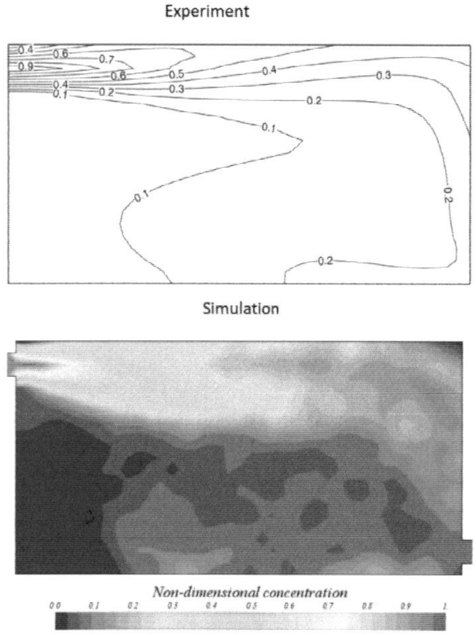

(b) Particle concentration

Figure 2: Experimental and computational comparison, showing results in the central section (profile direction) of the room

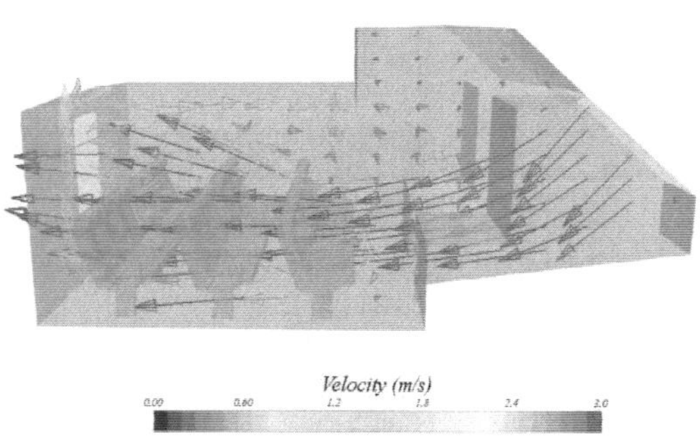

(a) Velocity profile

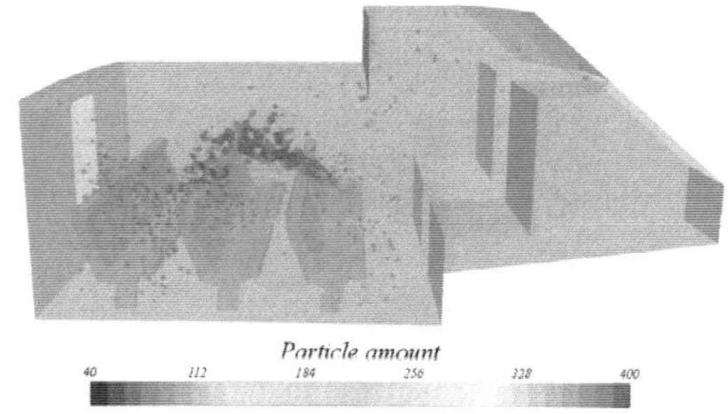

(b) Virus distribution

Figure 3: Simulation of virus-laden flow inside a ship

3.4 REMOTE MONITORING AND MACHINE LEARNING

Beyond COVID-19, the Indonesian fleet will also benefit significantly from improved ship safety, stability and efficiency. To facilitate this development, an onboard data monitoring system, named SHIMOS®, has been developed by the Orela shipyard.
Data recorded by the system are shown below:

- Date & Time
- Vessel location (latitude and longitude)
- Speed (knot)
- Heading (degree)
- RPM 1 (engine starboard) & RPM 2 (engine portside)
- Fuel consumption

The recording is done every 2 minutes and the user interface is shown in Figure 4.

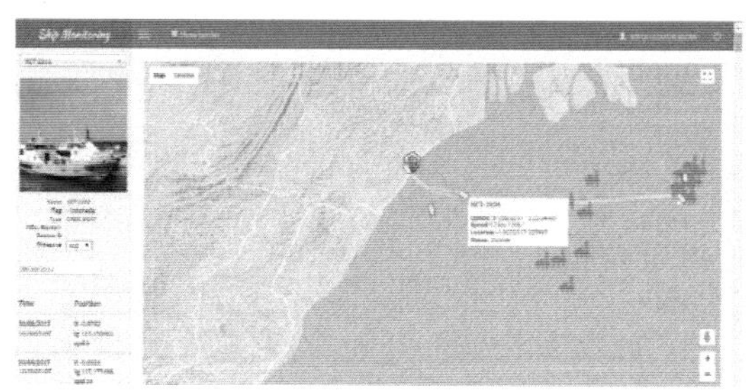

Figure 4: User interface of the SHIMOS® monitoring system

With the data collected by the monitoring system, machine learning techniques are used to indicate ship safety and performance on a real-time basis. The Artificial Neural Network (ANN) method is used for this purpose. ANN was inspired by the biological brain and comprise multiple artificial neurons which receive a signal, process it and send it to neurons connected to it [10]. In recent years, ANN has been successfully applied to predict ship performance with high accuracy, as demonstrated by Laurie et al. [11].

218

An example is given here, where ANN training was performed using the data recorded from three sister ships with the same design between 31 December 2016 to 5 March 2020. The total data obtained for each ship is 834,040 points, so the total data for 3 vessels is 2,502,120 points. The data were used to generate a function for predicting shaft power, dictated by heading angle and speed (Figure 5). The results were compared with measurement data that were not used for training the ANN model, and the average deviation is fairly low, 6.2% [12]. In the future, the ANN model may be further trained for other prediction purposes, such as optimising routes based on real-time environmental data, and suggesting potential threats to shipping safety [13].

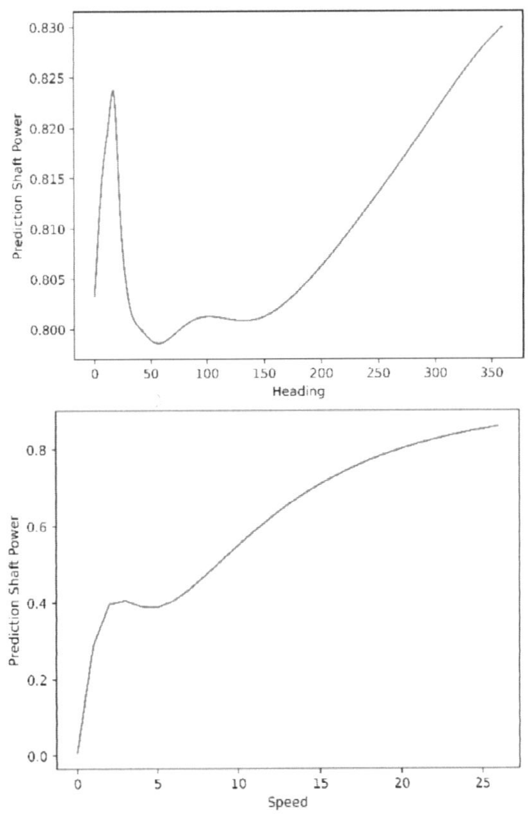

Figure 5: Machine learning function of shaft power versus heading angle and ship speed

4. CONCLUSIONS

In order to improve the short-term COVID-19 safety and long-term ship performance in Indonesia, a series of tasks is performed in this research. Addressing COVID-19, analysis of current regulations, questionnaire survey and CFD work are conducted to inform onboard anti-virus guideline. For facilitating ship performance, remote monitoring and machine learning technique are developed to establish long-term monitoring and optimising. The presented work is still ongoing – more results will be reported in the near future.

5. ACKNOWLEDGEMENTS

This work is part of a project that has received funding from the British Council under the Newton Institutional Links Grants - Ensuring the safety of Indonesian seafarers and fishers in the time of COVID-19 and beyond (agreement No. 623457938)

6. REFERENCES

1 I. Utama, W. Hetharia, COVID-19 and the Problem with Fishing Activities, in: Various Perspectives of Covid-19 Impacts, Contribution of AIPI Fellows to the Country, 2021.

2 D. Attwood, Insight report on safety in the fishing industry., Lloyd's Register Foundation, n.d. https://www.lrfoundation.org.uk/en/publications (accessed September 26, 2021).

3 P.A. Wibawa, R.W. Birmingham, M.D. Woodward, Design of sustainable fishing vessels, future challenges for the Indonesian fisheries, in: 12th International Marine Design Conference, 2015: pp. 357–367.

4 CEA, Indonesia Fisheries: 2015 Review, The David and Lucile Packard Foundation, 2016.

5 I.P.A. Wibawa, R. Birmingham, Improving Safety Working Environment on Indonesian Fishing Fleet: A Case Study on Local Fishing Communities in East Java, Mastic2018 Maritime Safety International Confrence. (2018).

6 W. Hetharia, Interviews with Ambon Fishers, University of Pattimura, 2019.

7 IMO, Coronavirus (COVID 19) – WHO guidance to promote public health measures on cargo ships and fishing vessels, 2020.

8 KEMENTERIAN PERHUBUNGAN, Petunjuk Pelaksanaan Perjalanan Orang Dalam Negeri Dengan Transportasi Laut Pada Masa Pandemi Corona Virus Disease (Indonesian Language), General Director of Sea Transportation SE25, 2021.

9 F. Chen, C.M. Simon, A.C. Lai, Modeling particle distribution and deposition in indoor

219

environments with a new drift–flux model, Atmospheric Environment. 40 (2006) 357–367.

10 B. Yegnanarayana, Artificial neural networks, PHI Learning Pvt. Ltd., 2009.

11 A. Laurie, E. Anderlini, J. Dietz, G. Thomas, Machine learning for shaft power prediction and analysis of fouling related performance deterioration, Ocean Engineering. (2021) 108886.

12 S. Riyadi, I. Utama, W.D. Aryawan, R. Rulaningtyas, G.A. Thomas, An Investigation into the Operational Characteristics of High-Speed Crew Boat Based on Artificial Neural Network, in: IOP Conference Series: Earth and Environmental Science, IOP Publishing, 2020: p. 012054.

13 L. Huang, Z. Li, C. Ryan, J.W. Ringsberg, B. Pena, M. Li, L. Ding, G. Thomas, Ship resistance when operating in floating ice floes: Derivation, validation, and application of an empirical equation, Marine Structures. 79 (2021) 103057.

7. AUTHORS BIOGRAPHY

ProfessorGiles Thomas BEng, MPhil, PhD, CEng, FRINA, currently holds the position of BMT chair of maritime engineering at University College London. Anaval architect, his research focusses on the performanceof ships, boats and offshore structures. Specific interestsinclude fluid-structure interaction, hydrodynamics, fullscale measurements, model testing and design. He is the project leader for thisNewton Institutional Links project.

Dr Luofeng Huang BEng, MSc, PhD, AFHEA is a research fellow of naval architecture at University College London. His work has a focus on computational fluid dynamics, which is used to simulate various marine engineering and science problems. In particular, he has been studying the modelling of sea ice and its interactions with ships and ocean waves.

Mr Christopher Ryan BEng, MSc, is a PhD student of naval architecture at University College London. He was a research assistant in the SEDNAproject (https://www.sedna-project.eu/), working on ship performance modelling, strategic research planning, accident analysis, and identifying future trends in Arctic shipping.

Professor I Ketut Aria Pria Utama BEng, MSc, PhD, CEng, FRINA, FAIPI, is a Professor of Ship Hydrodynamics at the Department of Naval Architecture, Sepuluh Nopember Institute of Technology. He is the Vice President of the Asian Region of the Royal Institution of Naval Architects (RINA). His research interests include ship resistance and propulsion, experimental and computational fluid dynamics, marine renewable energy, and maritime education.

Mr Soegeng Riyadi BEng, MEng, is PhD student at the Department of Naval Architecture, Sepuluh Nopember Institute of Technology. He has experience designing ships in several shipyards, including reducing ship resistance with foil at the back of the ship and investigating the operational mode of the fleet to achieve optimum fuel consumption.

Mr Dony Setyawan BEng, MEng, is PhD student at the Department of Naval Architecture, Sepuluh Nopember Institute of Technology. He is also a lecturer in the department. His research interests include ship structure and strength, advanced ship materials, and Ship stability.

Dr Wolter Hetharia BEng, MEng, PhD, FRINA, currently holds the position of lecturer at the Department of Naval Architecture, Pattimura University, where he is responsible for the courses of ship design and fishing vessels. His research focusses on small boats and fishing vessels. Before coming back to Indonesia, He obtained his PhD degree from the University of Liege, Belgium, in 2015.

220

EFFECT OF OCCUPATIONAL, NON OCCUPATIONAL, AND SOCIODEMOGRAPHICS CHARACTERISTICS FACTORS TO THE FATIGUE RISK OF SHIP STEEL SMELTING PRODUCTION WORKERS

A M Disrinama, F Rachman, and **S Y P Mahardika,** Politeknik Perkapalan Negeri Surabaya, Indonesia

SUMMARY

Work fatigue is a process of decreasing efficiency, work performance, and reduced physical strength or endurance to continue the activities that must be done. This study was conducted to determine the effect of occupational factors (work climate, physical workload), non-occupational factors (sleep quality, disease history, exercise habits, drinking water consumption), and socio-demographic characteristics (age, marital status, nutritional status, childhood). work) against work fatigue. Based on the results of research using the Industrial Fatigue Research Committee questionnaire, as many as 36% of workers experienced moderate fatigue and 22% of workers experienced severe fatigue. The research method used ordinal logistic regression, Mann Whitney and Kruskall Wallis. The results of the partial test showed that sleep quality (p-value 0.021), age (p-value 0.009), and years of service (p-value 0.000) had an effect on work fatigue. The results of the different test indicate that there are no variables that have differences in work fatigue. The recommendations given are doing progressive muscle relaxation, education about important aspects of sleep quality, changing work positions, providing job rotation and training planning, conducting regular and periodic medical check-ups.

1. INTRODUCTION

Health is a very important factor for productivity and increased productivity of the workforce as human resources. Jobs that require high work productivity can only be carried out by workers with excellent health conditions. To work productively, work must be carried out in a working manner and in a work environment that meets health requirements so that health problems and work fatigue do not occur (Suma'mur, 2013). The steel smelting industry as a construction industry has a strategic role in the service business of public transportation machines, along with the increasing needs and desires of the community, the company is required to be able to provide machine components in adequate types, quantities, quality and in accordance with consumer demand.

This industry has several parts (units) of work, one of which is the production department. The processes that occur in the production area include the initial sorting of scrap as raw material to the production process to process steel into billets. To process it requires high temperatures to heat the steel back above a temperature of 1800 °C. As a result, it creates heat in the work area so that workers are exposed to heat during work. The presence of high temperatures can cause the workload to increase due to uncomfortable working environment temperatures so that workers will quickly experience fatigue.

Work fatigue is one of the health and safety problems accompanied by a decrease in efficiency and resilience at work, this is one of the occupational health and safety problems. Fatigue is caused by factors that are divided into work-related and non-work related factors (Hendra & Nasri, 2019). Work-related factors are factors that cause fatigue that come from a person's conditions and activities during work, including workload, physical work environment, work duration, work shifts, psychosocial at work, and work stress. Non-work related is a factor that causes fatigue that is not related to work, including sleep quality, health problems, family conflicts, travel time, and socio-demographic characteristics.

This study was conducted to determine the risk factors for worker fatigue. The risk factors for fatigue in this study were occupational factors (work climate, physical workload), non-work factors (sleep quality, disease history, exercise habits, drinking water consumption), and socio-demographic characteristics (age, marital status, nutritional status, years of service).

Research field based contain a lot of factor within maritime background are rarely described. Beside that, researcher use three methods simultaneously and describe each level with brief analysis.

By knowing the factors causing work fatigue obtained from the results of this study, it can be used to provide recommendations for improvement in order to minimize the level of fatigue risk felt by production workers in the steel smelting industry.

2. METHODS

The method used for data analysis is ordinal logistic regression test to find the effect of independent variables on the dependent variable by focusing on simultaneous and individual tests. Ordinal logistic regression is a good statistical method to analyze the effect of predictor variables on response variables. The response variable in ordinal logistic regression has an ordinal scale with three or more variables (Hosmer & Lemeshow, 2000). The previous research that used this method was the Analysis of the Effect of Work and Characteristics on Fatigue (Fahrezy & Rachman, 2017) and the influence of the influence, work climate and individual factors of industrial workers on work fatigue (Nisa, K., & Rachman, F, 2018).

The next test is a different test using the Mann Whitney test and the Kruskall Wallis test. The basis for using this method is because the test results with Kolmogorof Smirnov show that the data is not normally distributed, so that the different test method used is non-parametric, namely the Mann Whitney U and Kruskall Wallis tests. Within all variables, only those that have an effect are tested. There are ordinal logistic regression, Mann Whitney test, andthe Kruskall Wallis test above.

This study is a quantitative study using analytical methods that use data analysis to prove the influence between two variables, namely the independent and dependent variables. The sampling technique in this study is total sampling, namely the entire population of 50 workers. Primary data collection in this study was obtained from the results of structured questionnaires, direct measurements and field observations. The primary data to be collected are:

Fatigue (Variable Y)

The level of worker fatigue was measured using the Industrial Fatigue Research Committee (IFRC) questionnaire which contained 30 questions related to fatigue symptoms. Each question has the lowest scale of 1 which is never felt up to the highest scale of 4 which is very often felt. The total score will show the level of fatigue experienced in categories which are divided into low (score 30 – 52), moderate (score 53 – 75), high (score 76 -98) and very high (score 99 – 120) (Ningsih & Nilamsari, 2018).

Job Factor

Work factors are factors that cause fatigue that comes from a person's conditions and activities during work. In this study, work factors include work climate and physical workload.

Work Climate (X1)

ISBB measurement according to Minister of Health Regulation No. 70 of 2016 and SNI 16-7061-2004 conducted 3 times (beginning, middle, end of work shift) for 8 hours. The results of the ISBB calculation are corrected with the correction value of work clothes, then determine the workload based on the metabolic rate in accordance with PMK No. 70 of 2016. The ISBB value and workload based on the metabolic rate are used to determine the allowable NAV based on the Minister of Manpower Regulation No. 5 of 2018.

Physical Workload (X2)

Physical workload is obtained by measuring the pulse using a pulse meter. Measurement of the physical workload is obtained by calculating the value of the workload based on the CVL percentage. Furthermore, the calculation results are classified based on the workload category according to %CVL (Tarwaka, 2015).

Non Employment Factor

Non-work related factors are factors that cause fatigue that have nothing to do with work. In research (Pelders & Nelson, 2019), non-work related factors are factors related to lifestyle. In this study, non-occupational factors included sleep quality, disease history, exercise habits, and drinking water consumption.

Sleep Quality (X3)

Sleep quality was obtained with primary data in the form of the distribution of the Pittsburgh Sleep Quality Index (PSQI) questionnaire. This questionnaire consists of 19 lists of questions related to the characteristics of a person's sleep within a period of 1 month subjectively. Sleep quality was categorized into Good (score <5) and Poor (score >5) (Gaol et al., 2018).

History of Disease (X4)

Disease history is obtained from primary data by distributing questionnaires to workers whether the worker has a history of illness that is being suffered. Then the categories are divided into: There is disease and No disease (Gaol et al., 2018).

Sports Habits (X5)

Exercise habits were obtained from primary data by distributing questionnaires to workers. Exercise habits are categorized into 4 groups, namely never, sometimes (1-2 a week), often (3-5 a week), and every day (Prawira et al., 2017).

Drinking Water Consumption (X6)

Water consumption is obtained from primary data in the form of distributing questionnaires to workers in the conversion of how many degrees of water per day, then assuming glass capacity = 0.2 liters, the value of water consumption is converted into liters. Categories are divided into 3, namely < 1.9 liters (light), 1.9 – 2.8 liters (medium), and > 2.8 liters (high) (Faturahman and Setiyono, 2014) in (Syah, 2020).

Socio-demographic Characteristics

Age (X7)

Age is obtained from secondary data of Steel Smelting Industry companies. Then categorized into <40 years and 40 years (Nisa, 2018).

Marital Status (X8)

Marital status was obtained from primary data in the form of distributing questionnaires to workers. Then categorized into Married and Unmarried (Agustin, 2018).

Nutritional Status (X9)

Measurement of Nutritional Status by measuring the weight of workers using a scale measuring tool for the weight of workers. After that, the worker's height was measured using a stature meter. Then the data that has been obtained is then entered into the Body Mass Index (BMI) formula based on the Regulation of the Minister of Health of the Republic of Indonesia No. 41 in 2014.

Working Period (X10)

The working period is obtained from secondary data of the Steel Smelting Industry company. Then they are categorized into 4 groups, namely tenure <3 years, 3-6 years, 7-10 years, >10 years (Puspitasari, 2018).

After the research data has been collected, the next technique is to analyze the data using SPSS statistical analysis.

Data analysis

The method used for data analysis is ordinal logistic regression test to find the effect of independent variables on the dependent variable by focusing on simultaneous and individual tests. The next test is a different test using the Mann Whitney test and the Kruskall Wallis test. The basis for decision making in this study is based on the significance value, H0 is accepted if p-value > 0.05 or H0 is rejected if p-value 0.05.

3. RESULT

The results of the use of fatigue using a subjective fatigue questionnaire (Industrial Fatigue Research Committee) which divides fatigue into 4 levels of feeling tired, namely mild, moderate, severe, and very heavy, and also obtained results from filling out a questionnaire for the distribution of work factors (work climate, physical workload), non-occupational factors (sleep quality, disease history, habits, drinking water consumption), and sociodemographic characteristics (age, marital status, nutritional status, years of service). Thus, researchers can find out the distribution of each variable when viewed from a predetermined category. It can be seen in the following table.

Table 1. Characteristics of Respondents

Dependent	**Category**	**Frequency**	
		N	**%**
Fatigue	Low	21	42
	Medium	18	36
	Weight	11	22
Job Factor	**Category**	**Frequency**	
		N	**%**
Work Climate	≤ NAB	11	22
	> NAB	39	78
Physical Workload	Low	13	26
	Medium	35	70
	Weight	2	4
Non Employment Factor	**Category**	**Frequency**	
		N	**%**
Sleem Quality	Good	17	34
	Poor	33	66
Disease History	Yes	11	22
	No	39	78
Sport Habits	Never	10	20
	1-2 times a week	25	50
	3-5 times a week	15	30
Drinking Water Consumption	< 1.9 liters / day	8	16
	1,9 - 2,8 liters / day	24	48
	> 2,8 liters / day	18	36
Socio Demographic Characteristics	**Category**	**Frequency**	
		N	**%**
Age	< 40 Years Old	23	46
	≥ 40 Years Old	27	54
Marital status.	married.	48	96
	Not married yet.	2	4
Nutritional status	Thin	1	2
	Normal	21	42
	Fat	18	36
	Obesity	10	20
Length of working	< 3 Years	13	26
	3 - 6 years	5	10
	7 - 10 years	1	2
	> 10 years	31	62

Based on Table 1, the results of the frequency of respondents' characteristics are based on categories. The data is then analyzed for the effect and differences in the effect on work fatigue as an independent variable.

Effect of work climate on work fatigue.

Based on the results of the test using ordinal logistic regression, it did not show any effect of work climate on work fatigue because the p-value was 0.882 (> 0.05). The results of the Mann Whitney test showed no significant difference in work climate on work fatigue with a p value of 0.411 (> 0.05). This is because the results of measurements of the working climate at natural wet temperatures and bulb temperatures do not show results that differ much from measurements in one work location so that they cannot be identified by statistics.

Another factor is the workers who have acclimatized to the work climate in the work location. Workforce acclimatization with a work climate is an adaptation of the workforce to a new work environment and new or different working climatic conditions. Work and workplaces are generally hot working climates, where the heat pressure usually exceeds everyday conditions in general. This causes new workers who start working in the work environment to experience a process of acclimatization to the intensity of heat exposure that has never been experienced before (Suma'mur, 2009) in (Mustofani & Dwiyanti, 2019).

The results of the hypothesis obtained in this study are in accordance with research conducted by (Suryaningtyas, 2017) explaining that the work climate does not affect work fatigue. However, the results of this test are different from the results of research (Eka et al., 2019) that the work climate affects work fatigue.

Effect of physical workload on work fatigue.

Based on the test results using ordinal logistic regression, it did not show any effect of physical workload on the risk of work fatigue because the p-value was 0.063 (> 0.05). The results of the Kruskall Wallis test showed that there was no significant difference in physical workload on the level of risk of work fatigue with a p-value of 0.118 (> 0.05).

This is because the results of research in the field, workers who experience heavy physical workloads actually have a lower percentage of only 4% or as many as 2 workers compared to workers who experience work fatigue with light and moderate workload levels. In this study, the steel processing production workers worked no more than 8 hours or approximately 40 to 48 hours a week by getting enough rest time to rest and drink. According to (Tarwaka, 2015) in (Sinaga, 2020), giving rest time can prevent fatigue, give the body an opportunity to recover or refresh, and provide time for social contact.

The results of this study are in line with research conducted by (Nisa, 2018) that physical workload has no effect on work fatigue.

Effect of sleep quality on work fatigue.

Based on the results of the test using ordinal logistic regression, it showed that there was an effect of sleep quality on the risk of work fatigue with a p-value of 0.021 (<0.05). Meanwhile, the results of the Mann Whitney test showed no significant difference in sleep quality to the level of risk of work fatigue with a p-value of 0.071 (> 0.05). In this study, it was concluded from the results of research data that the worse the quality of sleep experienced by workers, the higher the level of perceived work fatigue. Workers with poor sleep quality will get tired easily compared to workers who have good sleep quality.

The results of this study are in accordance with the theory of Kroemer and Grandjean in (Gaol et al., 2018) that circadian rhythms are one of the factors causing work fatigue in the industry, which if the circadian rhythms are disturbed it will cause disturbances in a person's sleep patterns and can affect sleep quality. . Poor sleep quality for employees can be seen from the length of time employees sleep at night on average only 4-5 hours, problems that are often felt that interfere with their sleep such as not being able to fall asleep for 30 minutes since lying down, waking up in the middle of the night, waking up to sleep for 30 minutes. going to the bathroom, getting cold or hot at night, and some mention other reasons.

Another study conducted by (Nidia Nuraini, Winarko, 2018) stated that sleep quality affects the level of work fatigue. However, the results of this study differ from research by (Hauteas et al., 2019) that sleep quality has no effect on work fatigue.

Effect of disease history on work fatigue.

Based on the results of the test using ordinal logistic regression, it did not show that there was an effect of medical history on the risk of work fatigue with a p-value of 0.102 (> 0.05). While the results of the Mann Whitney test showed no significant difference in the history of the disease to the level of risk of work fatigue with a p-value of 0.511 (> 0.05). This is due to other factors such as health conditions at the time of the study. Even though the worker has a history of illness, the disease may have been experienced for quite a long time and at the time of the study his health condition was in good condition.

So that it does not affect work fatigue. In addition, there are other factors, namely the small number of samples and the number of respondents who are not balanced between respondents who have a history of disease and those who do not have a history of disease. The number of respondents who have a history of disease is smaller than the number of respondents who do not have a history of disease.

The results of this study are in line with the results of research (Usman & Yuliani, 2019) which states that work fatigue is associated with a history of illness. This is because health status can affect work fatigue which can be seen from the history of the illness. While the influence of a history of illness on work fatigue in this study is in accordance with the research conducted (Fahrezy, 2017).

The effect of exercise habits on work fatigue.

Based on the results of the test using ordinal logistic regression, it did not show any influence of exercise habits on the risk of work fatigue with a p-value of 0.161 (> 0.05). While the results of the Kruskall Wallis test showed that there was no significant difference in exercise habits on the level of risk of work fatigue with a p-value of 0.284 (> 0.05). This can be caused by the data obtained, workers who still do exercise 1-2 times a week, there are also those who experience moderate and severe work fatigue, even workers who exercise up to 3-5 times a week there are still those who experience high work fatigue..

So, even though the workers had done enough exercise, they still experienced moderate and severe fatigue. This can be caused by exercise that is not carried out regularly or exercise that is carried out regularly but the time for workers to rest is still lacking. In addition, the types of exercise that each worker does are different and the length of time they exercise is also different.

The results of this study are in line with research conducted by (Fahrezy, M. Fahri, 2017) explaining that exercise habits do not affect work fatigue. Research (Sudira & Indarwati, 2018) states that there is no difference in the level of work fatigue based on exercise habits. However, the results of this study are different from research conducted (Arini & Dwiyanti, 2017) that exercise habits are associated with work fatigue.

The effect of drinking water consumption on work fatigue.

Based on the test results using ordinal logistic regression, it did not show any influence of exercise habits on the risk of work fatigue with a p-value of 0.090 (> 0.05). While the results of the Kruskall Wallis test showed no significant difference in sleep quality on the level of risk of work fatigue with a p-value of 0.080 (> 0.05). This can be caused by an unbalanced distribution of categories between workers who consume drinking water in the light, medium, and heavy categories.

Based on the research data that has been obtained, it shows that workers who already have a habit of drinking 1.9-2.8 liters of water per day still experience moderate and high levels of work fatigue. Based on the conditions and conditions in the field, there is sufficient drinking water available around the work area and appropriate rest and drinking time is provided. In addition, the place for drinking water in the form of gallons of water is close to the production site so that workers can easily reach their drinking water needs to avoid dehydration and get tired quickly.

This study is in line with research conducted by (Ferusgel, 2018) explaining that drinking water consumption does not affect work fatigue. However, the results obtained are contrary to research (Zahra & Kurniawidjaja, 2018) that drinking water consumption can affect work fatigue.

Effect of age on work fatigue.

Based on the results of the test using ordinal logistic regression showed that there was an effect of age on the risk of work fatigue with a p-value of 0.009 (<0.05). While the results of the Mann Whitney test on the level of risk of work fatigue with a p-value of 0.111 (> 0.05). With increasing age, the level of fatigue will occur more quickly, and a person's age will affect the condition, ability and capacity of the body in carrying out activities. According to the theory in research (Wahyuni & Indriyani, 2019), workers who are more than 45 years old experience more fatigue.

This is because human work capacity reaches its peak at the age of 25-30 years and declines at the age of more than 30 years. In addition, older workers will experience a decrease in muscle strength which has an impact on fatigue in doing their work and a decrease in muscle strength will cause muscle fatigue that occurs due to the accumulation of lactic acid in the muscles (Wahyuni & Indriyani, 2019).

The results of this study are in line with research conducted by (Nidia Nuraini, Winarko, 2018) explaining that age affects work fatigue. Another study conducted by (Rosanti et al., 2016) stated that there was no difference in the level of work fatigue based on age. However, another conflicting study, namely the results of research by (Ferusgel, 2018) that age does not affect work fatigue.

Effect of marital status on work burnout.

Based on the results of the test using ordinal logistic regression, there was no effect of marital status on the risk of work burnout with a p-value of 0.755 (> 0.05). Meanwhile, the results of the Mann Whitney test showed no significant difference in marital status on the level of risk of work burnout with a p-value of 0.487 (> 0.05). This could be due to the small number of samples and the unbalanced distribution of the unmarried and married status groups. In this study, married workers were more motivated and enthusiastic to carry out their work.

The results of this study are in line with research conducted by (Bangkit Rizky P, 2019). However, the results obtained are contrary to research (Zahra & Kurniawidjaja, 2018) that marital status can affect work fatigue. Married workers have a burden of responsibility in providing for them. So that this triggers bad conditions for workers, namely experiencing work fatigue.

Effect of nutritional status on work fatigue.

Based on the test results using ordinal logistic regression, it did not show any influence of nutritional status on the risk of work fatigue with a p-value of 0.374 (> 0.05). While the results of the Kruskall Wallis test showed no significant difference in nutritional status to the level of risk of work fatigue with a p-value of 0.578 (> 0.05). This can be caused by the conditions and conditions in the field, the workers have received good nutrition. Prior to work, the workers have been given nutritional intake in the form of milk and vitamins, while during the day the workers have had enough lunch. In addition, from the research data, there were 21 workers (42%) showing normal nutritional status, and only 1 worker (2%) showing underweight nutritional status. So it can be concluded that the workers have received good and sufficient nutrition during work.

The results of this study are in line with research conducted by (Fahrezy, 2017) which states that nutritional status does not affect work fatigue. Research (Nurullita & Joko, 2017) states that there is no difference in the level of work fatigue based on nutritional status. However, the results obtained are contrary to research (Mehta & Cavuoto, 2017) that nutritional status can affect work fatigue. Another study conducted by (Sudira & Indarwati, 2018) that there are differences in the level of work fatigue based on nutritional status. According to (Wahyuni & Indriyani, 2019), a worker with good nutritional status will have better body resistance and work capacity, while a worker with poor nutritional status will have poor body resistance and work capacity.

The effect of working period on work fatigue.

Based on the test results using ordinal logistic regression, it shows that there is an effect of age on the risk of work fatigue with a p-value of 0.000 (<0.05). While the results of the Mann Whitney test on the level of risk of work fatigue with a p-value of 0.191 (> 0.05). According to (Tarwaka, 2015) in (Prastuti & Martiana, 2017), a long working period will result in feelings of boredom and boredom if the work done is monotonous. The length of the work period is determined by the level of fatigue and muscle complaints felt by workers. The longer a person works, the longer it will take to do repetitive activities. This can increase the risk of fatigue. Based on the theory put forward by (Asriyani & Karimuna, 2017) that a higher level of fatigue is experienced by workers with a longer working period because the longer they work, the feeling of boredom due to the monotonous work will affect the level of fatigue they experience. Fatigue that occurs continuously results in chronic fatigue.

The results of this study are in line with research conducted by (Mulyadi and Arminah, 2018) that working period affects work fatigue. Another study conducted by (Rosanti et al., 2016) stated that there was no difference in the level of work fatigue based on years of service. However, the results obtained are contrary to research (Lahay, et al, 2018) that the length of work cannot affect work fatigue. The results of research conducted by (Fahrezy, 2017) stated that working period did not affect work fatigue.

4. RECOMMENDATIONS

Recommendations are given to control the fatigue experienced by workers so that the level can be decreased. Recommendations focus on overcoming the independent variables that have a significant effect on fatigue, namely sleep quality, age, and years of service. In the control hierarchy there are 5 levels that can be done, namely elimination, substitution, engineering, administration and personal protective equipment (PPE). However, the three initial stages mentioned, cannot be applied because of the 3 variables that affect fatigue, none of which can be eliminated, substituted and technical engineering is carried out on materials, materials or work tools. The most appropriate administrative control to control variables that affect fatigue include, progressive muscle relaxation, education on important aspects of sleep quality, shifting work positions, providing job rotation and training planning, conducting regular and periodic medical check-ups by company doctors.

5. CONCLUSIONS

Based on the results of the partial effect test using ordinal logistic regression, it was found that the non-occupational factor that affects work fatigue is sleep quality with p-value = 0.021, and socio-demographic characteristics that affect work fatigue are age with p-value = 0.009 and years of service p-value=0.000. The results of different tests on the variables of work climate, sleep quality, disease history, age, and years of service using the Mann Whitney test showed that there was no difference in the level of work fatigue. Meanwhile, for the variables of physical workload, exercise habits, drinking water consumption, nutritional status, and years of service using the Kruskall Wallis test, the results showed that there was no difference in the level of work fatigue. It is hoped that the company can implement the results of this study in the form of improving administrative control and medical check-ups.

6. REFERENCES

1 Agustin, N. (2018). Hubungan Faktor Individu Dengan Kelelahan Kerja Pada Karyawan Di PT. Adhi Persada Gedung Bekasi Tahun 2018. *Jurnal Persada Husada Indonesia*, *5*(19), 18–30.

2. Arini, S. Y., & Dwiyanti, E. (2017). ANAlISIS FAKTOR YANG BERHUBUNGAN DENGAN TERJADINYA KELELAHAN KERJA PADA PENGUMPUl TOL DI PERUSAHAAN PENGEMBANG JALAN TOL SURABAYA. *The Indonesian Journal of Occupational Safety and Health*, *4*(2), 113. https://doi.org/10.20473/ijosh.v4i2.2015.113-122

3. Asriyani, N., & Karimuna, S. (2017). Faktor Yang Berhubungan Dengan Terjadinya Kelelahan Kerja Pada Pekerja Pt. Kalla Kakao Industri Tahun 2017. *Jurnal Ilmiah Mahasiswa Kesehatan Masyarakat Unsyiah*, *2*(6), 198202.

4. BANGKIT RIZKY PRATOMO, N. B. P. (2019). ANALISIS PENYEBAB KELELAHAN OPERATOR Haul Dumptruck (HD) (Studi Kasus di PT X Rantau Nangka Kalimantan Selatan). *Applied Sciences (Switzerland)*, *9*(13), 1–11. https://doi.org/10.3390/app9132584

5. Eka, L., . D., & DN, A. T. (2019). Pengaruh Iklim Kerja Terhadap Kelelahan Tenaga Kerja. *Gema Lingkungan Kesehatan*, *17*(2), 100–104. https://doi.org/10.36568/kesling.v17i2.1061

6. Fahrezy, M. F. (2017). *Analisis Pengaruh Beban Kerja Dan Karakteristik Individu Terhadap Kelelahan Pada Operator Ctt*. *2581*, 575–580.

7. Ferusgel, A. (2018). Faktor Yang Mempengaruhi Kelelahan Fisik Kerja Pada Pekerja Molding Plant Di Pt. Pacific Medan Industri Tahun 2016. *Jurnal Kesehatan Masyarakat Dan Lingkungan Hidup*, *1*(1), 32–38.

8. Gaol, M. J. L., Camelia, A., & Rahmiwati, A. (2018). ANALISIS FAKTOR RISIKO KELELAHAN KERJA PADA KARYAWAN BAGIAN PRODUKSI PT. ARWANA ANUGRAH KERAMIK, Tbk. *Jurnal Ilmu Kesehatan Masyarakat*, *9*(1), 53–63. https://doi.org/10.26553/jikm.2018.9.1.53-63

9. Hauteas, N. R., Ruliati, L. P., & Doke, S. (2019). Analisis Faktor yang Berpengaruh terhadap Kelelahan Kerja pada Pekerja Rambut Palsu di CV. Dona Mandiri Cabang Lasiana Kota Kupang. *Lontar : Journal of Community Health*, *1*(4), 127–139. https://doi.org/10.35508/ljch.v1i4.2178

10. Hendra, N. M. A., & Nasri, S. M. (2019). Analysis of work related and non-work related factors relationship with fatigue among stamping workers in automotive industry. *Indian Journal of Public Health Research and Development*, *10*(5), 1504–1510. https://doi.org/10.5958/0976-5506.2019.01227.0

11. Hosmer, D. W., & Lemeshow, S. (2000). Applied Logistic Regression.

12. Lahay, Idham Halid., Eduart wolok, H. dan H. U. (2018). Pengaruh Usia Dan Lama Kerja Terhadap Kelelahan Kerja Pada Pekerja Pembuat Batako Di. *Sentra*, 64–67.

13. Mehta, R. K., & Cavuoto, L. A. (2017). Relationship between BMI and Fatigability Is Task Dependent. *Human Factors*, *59*(5), 722–733. https://doi.org/10.1177/0018720817695194

14. Mulyadi dan Nurhajja Arminah. (2018). *ANALISIS FAKTOR-FAKTOR YANG MEMENGARUHI KELELAHAN KERJA PADA PEKERJA MEBEL DI KECAMATAN MANGGALA KOTA MAKASSAR*. *18*(2), 184–188.

15. Mustofani, M., & Dwiyanti, E. (2019). Relationship between Work Climate and Physical Workload with Work-Related Fatigue. *The Indonesian Journal of Occupational Safety and Health*, *8*(2), 150. https://doi.org/10.20473/ijosh.v8i2.2019.150-157

16. Nidia Nuraini, Winarko, S. B. E. W. (2018). FAKTOR YANG BERPENGARUH TERHADAP KELELAHAN (Studi Pada Pekerja Proyek Kapal Perang di PT. X Tahun 2018). *Gema Lingkungan Kesehatan*, *2*(1), 1–13. 3%0Ahttp://dx.doi.org/10.1016/j.jff.2015.06.018 %0Ahttp://dx.doi.org/10.1038/s41559-019-0877-3%0Aht

17. Ningsih, S. N. P., & Nilamsari, N. (2018). Hubungan kelelahan kerja dengan produktivitas kerja karyawan di fakultas dakwah dan komunikasi UIN Sunan Kaligaga Yogyakarta. *Journal of Industrial Hygiene and Occupational Health*, *3*(1), 69–82.

18. Nisa, K. (2018). PENGARUH KEBISINGAN, IKLIM KERJA DAN FAKTOR INDIVIDU PEKERJA INDUSTRI PENGECORAN LOGAM TERHADAP KELELAHAN KERJA. *Proceeding 2nd Conference on Safety Engineering and Its Application*, *2581*, 541–548.

19. Nurullita, U., & Joko, T. (2017). Pengaruh Beban Kerja dan Faktor Lingkungan Fisik Terhadap Tekanan Darah, Denyut Nadi dan Tingkat Kelelahan Pekerja Bagian ARC FURNACE dan ROLLING MILL PT. Inti General Yaja Steel Semarang. *Jurnal Kesehatan Lingkungan Indonesia*, *6*(1), 13–16.

20. Pelders, J., & Nelson, G. (2019). Contributors to Fatigue of Mine Workers in the South African Gold and Platinum Sector. *Safety and Health at Work*, *10*(2), 188–195. https://doi.org/10.1016/j.shaw.2018.12.002

21. Prastuti, T. N., & Martiana, T. (2017). Analisis Karakteristik Individu Dengan Keluhan Kelelahan Kerja Pada Pengemudi Taksi Di Rungkut Surabaya. *The Indonesian Journal of Public Health*, *12*(1), 64. https://doi.org/10.20473/ijph.v12i1.2017.64-74

22. Prawira, M. A., Yanti, N. P. N., Kurniawan, E., & Artha, L. P. W. (2017). Factors Related Musculoskeletal Disorders on Students of Udayana University on 2016. *Journal of Industrial Hygiene and Occupational Health*, *1*(2), 101. https://doi.org/10.21111/jihoh.v1i2.888

23. Puspitasari, W. (2018). Pengaruh Faktor Pekerja Dan Pekerjaan Terhadap Kelelahan Kerja Pada Area Cetak Pabrik Pengecoran. *Proceeding 2nd Conference on Safety Engineering and Its Application*, *2581*, 2–5.

24. Rosanti, E., Dan, H., Budiastuti, S., Kunci, K., & Kebisingan, : (2016). Perbedaan Tingkat Kelelahan Tenaga Kerja Yang Mengalami Kecemasan Akibat Kebisingan Pada Bagian Weaving, Riching, Dan Administrasi Di PT. Iskandar Indah Printing Textile Surakarta 1. *Jurnal EKOSAINS*, *VI VI*(2), 39–45.

25. Sinaga, N. (2020). Analisis Hubungan Beban Kerja dengan Kelelahan Kerja Pada Pekerja Bangunan Kota Medan. *Seminar Nasional Teknik Industri Universitas Gadjah Mada*, 61–65.

26. Sudira, P. G., & Indarwati, S. (2018). *Hubungan tingkat kantuk terhadap derajat kelelahan dokter residen di RSUP Dr . Sardjito*.

27. Suma'mur. (2009). *Higene Perusahaan dan Kesehatan Kerja (Hiperkes)*. Sagung Set.

28. Suma'mur. (2013). *Higiene Perusahaan dan Kesehatan Kerja (Hiperkes) (2 ed.)*. CV Sagung Seto.

29. Suryaningtyas, Y. (2017). Iklim Kerja Dan Status Gizi Dengan Kelelahan Kerja Pada Pekerja Di Ballast Tank Bagian Reparasi Kapal Pt. X Surabaya. *Jurnal Manajemen Kesehatan Yayasan RS.Dr. Soetomo*, *3*(1), 17. https://doi.org/10.29241/jmk.v3i1.87

30. Suwignyo, & Ningsih, R. R. A. (2021). Pengaruh Iklim Kerja Panas dan Pencahayaan Terhadap Tingkat Kelelahan pada Operator Di PLTD Karang Asam Samarinda Tahun 2019. *Jurnal Penelitian Kesehatan Suara Forikes*, *12*, 195–198.

31. Syah, M. L. (2020). Pengaruh Faktor Beban Kerja Mental , Non-Working Cause s dan Sosioeconomic Causes terhadap Kelelahan Pekerja pada Proyek Konstruksi Pipa Gas. *Proceeding 4nd Conference on Safety Engineering and Its Application*, *2581*, 283–287.

32. Syuhada, A. D., & Widodo, R. S. (2019). *Faktor Dominan Yang Mempengaruhi Kelelahan Kerja Pada Pekerja Bagian Tiang Pancang Di Bandung Barat*. *8*(3), 192–202.

33. Tarwaka. (2015). *Ergonomi Industri : Dasar-dasar Pengetahuan Ergonomi dan Aplikasi di Tempat Kerja*. Harapan Press.

34. Usman, S., & Yuliani, I. (2019). Faktor-Faktor Yang Berhubungan Dengan Kelelahan Kerja Pada Karyawan Produksi PT Gerbang Sarana Baja Jakarta Utara. *Journal Educational of Nursing(Jen)*, *2*(1), 141–146. https://doi.org/10.37430/jen.v2i1.18

35. Wahyuni, D., & Indriyani, I. (2019). FAKTOR-FAKTOR YANG BERHUBUNGAN DENGAN

KELELAHAN KERJA PADA PEKERJA BAGIAN PRODUKSI DI PT. ANTAM Tbk. UBPP LOGAM MULIA. *Jurnal Ilmiah Kesehatan*, *11*(1), 73–79. https://doi.org/10.37012/jik.v11i1.70

36. Zahra, A. N., & Kurniawidjaja, L. M. (2018). Construction Workers' Fatigue Conditions at PT. X Construction Contractor Apartment Development in the 2017 Work Year. *KnE Life Sciences*, *4*(5), 46. https://doi.org/10.18502/kls.v4i5.2538

7. AUTHOR BIOGRAPHY

Am Maisarah Disrinama, dr., M.Kes, medical doctor that pass Assistant Professor last years, has 6 six years experience on safety engineering. Within on Politeknik Perkapalan Negeri Surabaya (Shipbuilding Institute of Polytechnic Surabaya), the field of research based on safety, include paper and prototype. Many papers focus on workers that several of them work on maritime field. There are characteristic maritim workers, risk identification, and others.

Farizi Rachman, S.Si., M.Si is currently a Lecturer at Politeknik Perkapalan Negeri Surabaya (Shipbuilding Institute of Polytechnic Surabaya). He is a graduate of the Department of Statistics, Sepuluh Nopember Institute of Technology. Currently his area of expertise is Biostatistics and Quality Control. He does a lot of research on Health and occupational safety in various industries with statistical analysis. In addition, he also does a lot of research on quality control the maritime industry.

Shafira Yuana Putri Mahardika, S.ST is post graduate of bachelor safety engineering inPoliteknik Perkapalan Negeri Surabaya (Shipbuilding Institute of Polytechnic Surabaya). Nowadays, she work on HSE with occupied her specialty on safety.

THE ANALYSIS OF ELECTRICAL SYSTEM REQUIREMENT OF JAVA COASTAL FISHING VESSELS

P Asri, H A Widodo, A T Nugraha, J E Poetro, M B Rahmat, I Rachman, I A Septiadi, H A Basyari, and **D R Ruwahida**, Politeknik Perkapalan Negeri Surabaya, Indonesia

SUMMARY

Traditional ships are dependent on diesel which functions not only as a propulsion, but also to supply power to the tools they use. This causes a decrease in the profits obtained by fishermen considering that the price of diesel fuel is increasingly expensive. This problem is also supported by the availability of diesel which is one of the non-renewable fossil fuels on earth which is getting thinner. The design and calculation of electrical power to supply tools that can make fishermen more effective in catching fish where the electrical energy obtained come from solar energy and is assisted by a portable generator as an emergency supply. The innovation of using solar energy as an effort to reduce the use of solar which will indirectly save fishermen's expenses. The 3GT fishing boat uses 4 solar panels with 450Wp specifications. This fishing boat design is equipped with radar, AIS, and navigation lights.

1. INTRODUCTION

Sea transportation is one of the technological sectors that continues to develop [1]. This can be seen through the functions and uses of sea transportation itself, especially for fishermen on the Java Coast where boats are the primary infrastructure for their daily livelihood needs. Ship is a category of marine transportation that uses diesel fuel. Solar is one of the preparations that are sourced from fossil energy, where fossil energy is energy that is unrenewable resources which means it cannot be renewed and one day it will definitely run out [2].

Based on the identification of data on several energy sub-sectors, such as oil and gas, renewable energy and electricity, the trend of investment realization in the energy sector throughout the 2010-2017 period tends to be below the target. Based on the calculation and analysis of the potential and challenges of investment in the fossil energy sector according to the RUEN, several important things need to be considered, one of which is to meet the RUEN target in the fossil energy sector, it is projected that Indonesia requires investment of more than USD 500 billion until 2050. the upstream oil and gas sector fails to meet, then the government has the potential to spend more than USD 400 billion by 2050 [3]

With the limited availability of fossil energy, especially diesel fuel, this is what demands fishermen as consumers to use diesel as well and efficiently as possible to conduct research to reduce ship fuel that was previously unrenewable to renewable material. This is done so that if at any time the diesel fuel runs out, the fishermen as consumers will have no trouble because there are other fuels to replace diesel.

On the other hand, solar energy is energy obtained by converting solar thermal energy (the sun) through certain equipment into other forms of power. In addition, solar energy is categorized as energy that is currently being actively developed by the Government of Indonesia, where Indonesia as a tropical country has great solar energy potential [4].

From the two problem topics above, the author aims to conduct an analysis by replacing the role of diesel which was previously the main fuel for ships into the role of solar energy through solar panels. The advantage of using solar panels is that apart from being an innovation from the development of renewable technology, this solar panel can also ease the operational needs of fishermen, improve fishermen's welfare because it can save on ship fuel costs, this technology is also environmentally friendly, and reduces dependence on fuel.

In this study, the electric power requirements of fishing boats will be designed and analyzed with the specifications of the boat as a small fishing fleet. This study will discuss the design and calculation of electrical power to supply tools that can make fishermen more effective in catching fish where the electrical energy obtained comes from solar energy and is assisted by a portable generator as an emergency supply.

The basic concept of this research is the need for the suitability of solar panels that replace the role of diesel as the main fuel for fishing boats. The solar panels are designed in such a way with the addition of several components and supporting sensors such as PLTS Solar module 450 Wp, Off grid inverter 5000W, 12V 120Ah battery, ESP 32, DC voltage sensor control system, PZEM-004T Sensor, ACS712 Sensor, Portable Generator, Fish Finder Garmin 585 plus, AIS class B Xinuo XF11069B plus promarine VHF AIS antenna, Furuno 1623 radar, navigation light, ceiling light, bilge pump 1100GPH, etc.

2. METHOD

A. Solar Energy

Solar energy can be converted directly into other forms of energy by three separate processes, namely the heliochemical process, the helioelectrical process and the heliothermal process. Meanwhile, the conversion of solar energy into electricity is a helioelectrical process. This process can take place if photovoltaic or solar cells are used. Photovoltaic is a device used to convert or convert solar energy into direct electrical energy, which is made of semi-conductor material [5].

In its utilization, solar panels require several other components, namely solar charge control and batteries.

Solar Charge Control (SCC) or also known as Battery Controller Unit (BCU) is an electronic device that functions to control the voltage and current from the solar cell to the battery and from the battery to the load. A battery that functions as a storage for power generated by solar cells [6].

B. Solar Charge Controller

The main controller of the solar charge functions to maintain the amount of charge coming from the solar PV module that flows into the battery bank to prevent the battery from being overcharged. It performs three basic functions: (i) It limits and regulates the voltage from the solar panels to avoid overcharging the battery. (ii) While the dc load is in use, the controller does not let the battery discharge. (iii) Allows use of different dc loads [7].

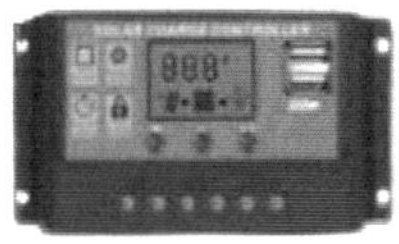

Figure 2.1 Solar Charge Controller (SCC)**[8]**

C. Sensor PZEM-004T

PZEM-004T is a sensor that can be used to measure rms voltage, rms current and active power that can be connected via Arduino or other opensource platforms. The physical dimensions of the PZEM-004T board are 3.1 × 7.4 cm. The PZEM-004t module is bundled with a 3mm diameter current transformer coil which can be used to measure a maximum current of 100A [9].

Figure 2. 2 Sensor PZEM-004T**[10]**

D. Sensor ACS712

ACS712 is a packaged IC which is useful as a current sensor replacing a current transformer which is relatively large in terms of size. In principle, the ACS712 is the same as other hall effect sensors, namely by utilizing the magnetic field around the current and then converting it to a linear voltage with changes in current [11].

Figure 2. 3 Sensor ACS712**[12]**

E. Generator Set

Genset (generator set) is a device that functions to generate electrical power. Referred to as a generator set with the understanding is a set of equipment combined from two different devices, namely the engine and generator or alternator. The engine as a player device while the generator or alternator as a generating device. In a generator set system, the drive or engine greatly influences the generator's working system. Because a stable generator rotation can make the generator output a maximum [13].

Figure 2. 4 Generator Set 1 Phase**[14]**

F. Battery VRLA

The type of battery used in this study is a battery with a voltage of 12 Volts and 120 Ah. The battery was chosen because it has a high number of hours of charging and discharging [15]. The batteries analyzed in this study have the specifications shown in Table 2.4 as follows:

Table 3.1 Battery Specifications

Parameter	Description
Capacity	12 V 120 Ah
Long	405 mm
Wide	170 mm
Tall	235 mm
Heavy	33 kg

3. RESULT AND DISCUSSION

3.1 SYSTEM CONCEPT

This 3gt fishing boat used a DC power source as the main supply to suppled almost all the equipment used by fishermen. Using batteries as a medium for stored power generated by solar panels so that it could be used even at night when fishermen went to sea.

The load supplied by the DC electric current source is divided into 2 types, namely crucial loads and non-critical loads. The crucial load is the electrical load that had to be lit continuously when the fishing vessel is operated, these tools are the bilge pump, radar, AIS, and navigation lights which are further divided into mashhead lights, green side lights, side red lights, and side lights. side. stern light.

Non-crucial loads are electrical loads that do not have to be always on and can be turned on/off according to the needs of fishermen, these tools are lighting lamps and fish finder.

The working principle of the electrical system on this 3 gt fishing boat is that all DC loads would be connected to the battery as the main supplied, when in an emergency the battery cannot supplied the load because the battery capacity runs out or there is a problem with the battery, the supply voltage for the DC load would transferred to the adapter while disconnecting the battery to the load with the help of a relay. However, not all DC loads would be transferred to the adapter, only crucial loads would turned on in an emergency.

The AC power source itself is basically only used to supply cool storage and later to supplied an adapter during an emergency which comes from a portable generator when the ship is operated or from a shore connection when the ship is in port.

3.2 SYSTEM DESIGN

This 3 GT fishing boat system had 2 power supplies, namely AC (Alternating Current) and DC (Direct Current). The DC source is obtained from 4 x 450 Wp solar panels which will be stored in a 24 Vdc battery / battery used to supply loads that do require DC current. While the AC source is obtained from a portable generator with a voltage of 220 Vac and can also be obtained from the Shore Connection at the port.

Loads connected to the supply are divided into 2, namely crucial loads and non-crucial loads. Crucial loads are tools that are important and shold be always on when fishing boats are operated, these tools are navigation lights, radar, AIS, and bilge pumps. Meanwhile, non-crucial loads are loads that are considered less important so that they can be turned on or off according to the user's wishes, these tools are lighting lamps and fish finder as well as cool storage. Figure 3.1 shows a load block diagram supplied by a DC current source from the battery.

The DC current source generated by solar panels is used as the main supply for almost all the tools used in this 3 gt fishing boat so that dependence on fossil energy sources can be reduced. The power generated by the solar panels will be processed first in the control batterycharger before being stored in a battery with a voltage of 24 volts and then connected to a DC load. Figure 3.2 shows a load block diagram supplied by an AC current source from a portable generator and shore connection.

The AC current source is used as an emergency supply and turn on the 220 V AC load. The AC current source comes from a portable generator that can generate 220 V AC voltage which is used to turn on cool storage which is useful for cooling fish caught by fishermen and supplying an adapter that can be used in an emergency as a DC current supply at crucial loads so that important tools can still be used when fishermen are looking for fish. This 3 gt fishing boat is also equipped with a shore connection as a 220 V AC current supply when the ship rests in port. Figure 3.3 shows a block diagram of the crucial loads that can be supplied by the adapter.

Although the main supply in a 3 gt fishing boat is the battery/battery, in an emergency such as when a problem occurs in the solar panel system or when the battery or battery capacity runs out, the DC load supply can be taken over by the adapter. However, not all loads will be covered by the adapter, only crucial loads that should be always on when the fisherman is operated

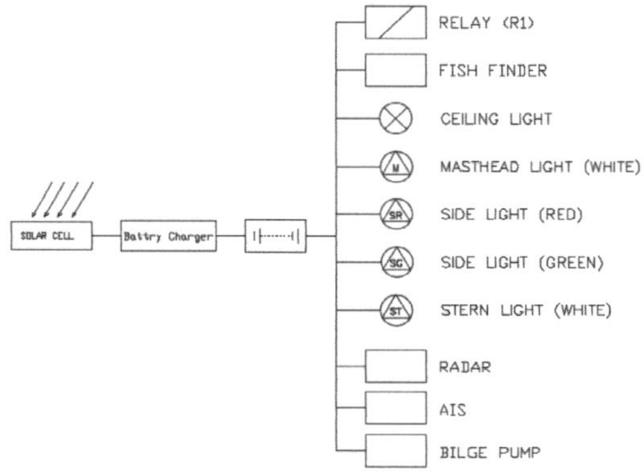

Figure 3. 1DC Source to Load Block Diagram

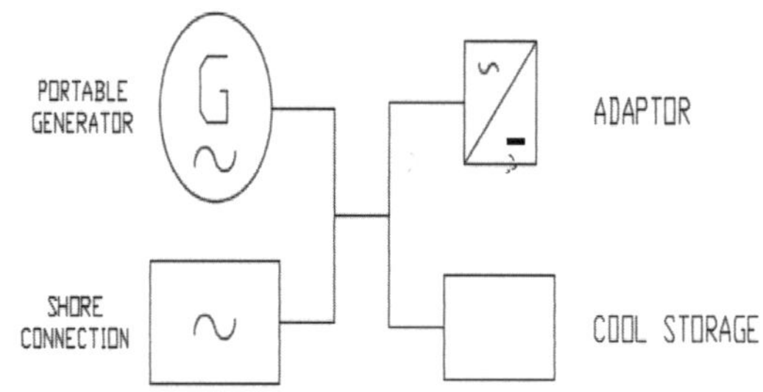

Figure 3. 2AC Source to Load Block Diagram

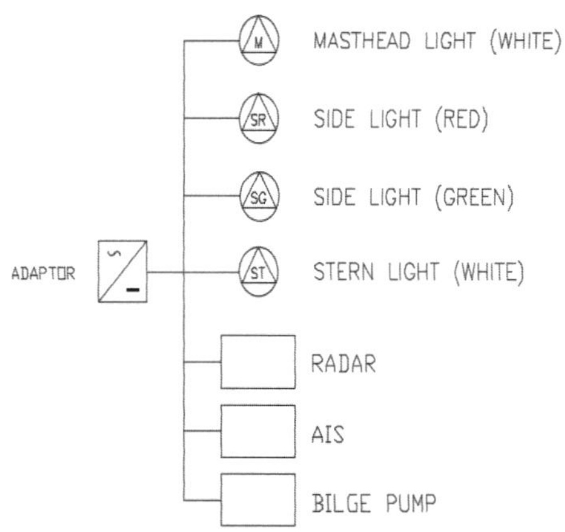

Figure 3. 3 Block Diagram from Adapter to Load

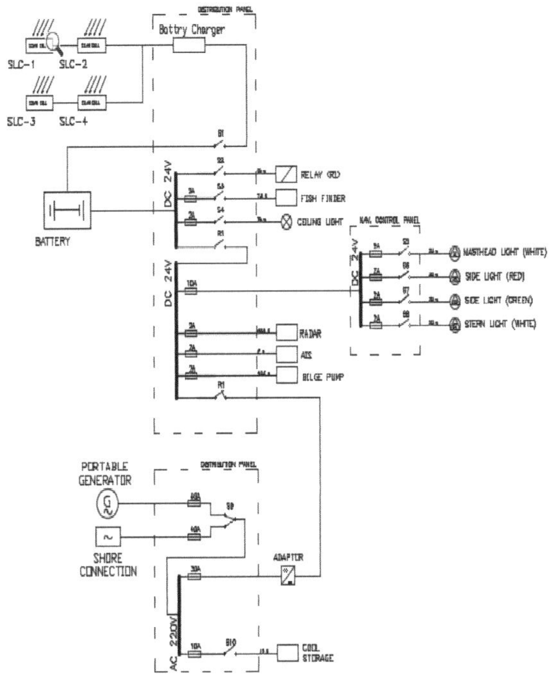

No.	SYMBOL	DESCRIPTION
1		SOLAR PANEL 450 WP
2		BATERAI / AKI
3		SHORE CONNECTION
4		PORTABLE GENERATOR
5		ADAPTOR 220 Vac TO 24 Vdc
6		FUSE WITH AMPERE CAPACITY
7		LAMPU PENERANGAN 20W
8		MASTHEAD LIGHT 25W
9		SIDE LIGHT RED 25 W
10		SIDE LIGHT GREEN 25 W
11		STERN LIGHT 25 W
12		SPST SWITCH
13		SPDT SWITCH
14		KONTAK NO RELAY
15		KOTAK NC RELAY

Figure 3. 4 Description Symbol Design One Line Overall System Diagram

3.3 SHIP BODY DESIGN

In this research, several stages of work have been carried out, namely the design of the ship's hull, the calculation of the analysis of electricity needs, the assembly of ship components and their installation, then testing and analysis of the ship's electrical power.

This boat has been designed with a capacity of 2-3 people, 9.46 meters long and 1.25 meters wide and can go at a speed of 12-15 knots. This vessel has been designed for a cruising radius of 50 miles, for operations in the coastal areas of Java where the southern coast of Java has extreme currents and waves. This fishing vessel has a fish capacity of 50 kg and is equipped with cool storage that can maintain the freshness of the caught fish.

The design of this ship has a total length (Loa) of 9.46 meters which is the overall length of the ship measured from the end of the rear of the ship to the end of the forehand of the hull, the length of the waterline (Lwl) is 8.05 meters measured from the intersection the height of the stern with the waterline at draft up to the intersection of the height of the bow with the waterline, the width of the ship (B) 1.25 meters, the height of the main deck of the ship (H) with a height of 0.84 meters measured in the midship plane from above keel (keel) to the top of the deck on the side of the ship. Then the ship's draft height (T) is 0.35 meters measured from the top of the keel to the waterline. From these specifications the design of this fishing boat is included in the type of 3GT fishing boat.

Determination of the specifications of solar panels as a source of ship electrical energy

Determination of solar panel specifications is very necessary as a basic design for determining energy that can be used as driving energy. The solar panels used are mono-crystalline types with a capacity of 450 Wp. The specifications of the solar panels used are as follows:

Table 3.2 Solar Panel Specification

Maximum Power (Pmax)	:	**450WP**
Maximum Power Current (Imp)	:	13.15A
Maximum Power Volatge (Vmp)	:	34,20V
Maximum Circuit Voltage (Voc)	:	41.04V
Short Circuit Current (Isc)	:	13.95A
Maximum System Voltage	:	1000V
Operating Temperature	:	-40°C to +85°C
Dimensions	:	1900 x 1134 x 35mm

Determination of fish finder specifications as a tool for fishermen to find fish

Determination of fish finder specifications is very necessary as a useful tool to help fishermen in finding the location of the fish. The fish finder used is the Garmin 585 Plus type.

Physical dimensions: 6.93" x 7.48" x 2.60" (17.6 cm x 19.0 cm x 6.6 mm)

Power consumption: 7.6 watts

Current : 0.32 A

Voltage range: 10 - 36 VDC

Determination of AIS specifications as ship traffic communication

AIS is needed as a useful tool to convey data via VHF Data Link (VDL) to send and receive information automatically to other ships. The AIS used is the AIS class B Xinuo XF1069B.
Physical dimensions: 6.93" x 7.48" x 2.60" (17.6 cm x 19.0 cm x 6.6 mm)
Power consumption: 7.6 watts
Current : 0.32 A
Voltage range: 10 - 36 VDC

Determination of Radar specifications as a ship navigation tool
Radar is one of the navigations in the ship that serves to provide information related to the distance of the ship to the mainland, other ships, and possible obstacles that will be faced in order to avoid a collision. The radar used is the Furuno 1623 Radar Silver LCD Small Ship which is equipped with an antenna.
Power Supply : 12-24 VDC
Current consumption: 3.2-1.4 A

Other supporting equipment, among others:
1. Ship Navigation Lights (mast light / M, side light (red) / SR, side light (green) / SG stern light / ST)
- Voltage : 24 V
- Power : 25 W
- Current : 1.05 A

2. Ceiling Lights
- Voltage : 24 VDC
- Power : 15 W
- Current : 0.625 A

3. Bilge Pump 12V 24V DC 1100GPH Submersible Pump Submersible Pump 4100L/hour
- Voltage : 24 V
- Current : 1.8 A

4. Cool Storage
- Length : 3 meters
- Width : 0.8 meters
- Height : 0.5 meters
- Power : 100 Watt
- Fish capacity: 50kg

Table3.3 Equipment Current and Voltage Specifications

Tool's name	Voltage (Volt)	Ampere/ hour (Ah)	Power (Watt)
Fish Finder	24	0,32	7,6
Bilge Pump	24	1,8	43,2
Radar	24	1,4	33,6
AIS	24	0,08	2
Lighting	24	0,625	15
Masthed Light (white)	24	1,05	25
Side Light (red)	24	1,05	25
Side Light (green)	24	1,05	25
Stern Light (white)	24	1,05	25
TOTAL			201.4

From the exposure of the tools above, the power used in one hour is 201.4 Watt. And if the fishing boat operates for 12 hours then:
Power in 1 hour x operating time
201.4 watts x 12 Hours = 2416.8 Watts or 2,417 Kw

To find out how much battery capacity is used then:
Battery capacity = 2416.8 watts / 24volt = 100.7 Ah
Because in this study the voltage used is 24 VDC, the batteries must be connected in series so that the batteries used in this fishing boat are 2 12 Volt batteries with a capacity of 100.7 Ah.

If you use a portable generator with the same load, you will spend as much diesel as :

S = 0.2 x P x T
= 0,2 x 2,014 x 12
= 5,07 liter

The description
- K = 0.2 (Factor of determination of diesel consumption per kilowatt per hour)
- P = Generator power
- T = Time

When compared to fishing boats that use a solar panel system that is designed in a hybrid way with the use of a portable generator with fishing boats that use a full portable generator system, it can be said that these fishing boats are much more efficient than fishing boats. which uses a fully portable generator system.

4. CONCLUSION

The location of Indonesia which is on the equator with a tropical climate makes Indonesia have sunlight available throughout the year, this point makes the use of solar panels have enormous potential. Especially in the mode of transportation of ships which have advantages where the sun's rays are not blocked by anything, in contrast to land buildings.

The design of this 3 GT fishing boat is equipped with supporting equipment that is able to make fishermen effective in catching fish and keeping them fresh on the way to the mainland, these tools are bilge pump, fish finder and cool storage.

The design of this fishing boat is also equipped with radar, AIS, and navigation lights in the form of a masthead light, 2 side lights on the right and left of the ship, and a stern light so that it is deemed to have met the standards for operating on the high seas.

This 3 gt fishing boat system has 2 power supplies, namely AC (Alternating Current) power and DC (Direct Current) power where DC electric power is the main supply and for AC power sources it is only used to turn on cool storage and in an emergency. just. Equipped with a shore connection so that when you lean back you can use the resources available on land.

Based on the analysis carried out in this study,the ship use 4 solar panels with 450Wp specifications. The electrical power generated by the solar panels will be stored in a battery with a capacity of 120 Ah and used to supply loads with a total power of 201.4 watts as long as this fishing boat operates for approximately 12 hours.

If you compare this fishing boat that uses a solar panel system designed in a hybrid manner with the use of a portable generator with a fishing boat that uses a full portable generator system, it can be said that this fishing boat is much more efficient than fishing boats that use a full portable generator system. This fishing boat is able to save 5 liters of diesel which is replaced with free energy obtained from solar panels installed on the roof of the fishing boat.

5. REFERENCES

1 M. T. S. U. Rifki Firdaus, "ESTIMASI KEBUTUHAN BAHAN BAKAR KAPAL PENUMPANG PELABUHAN TANJUNG EMAS SEMARANG SAMPAI TAHUN 2040 MENGGUNAKAN," *Jurnal Teknik Mesin S-1,* vol. 4, no. 1, p. 10, 2016.

2 D. P. M. T. M. F. Elinur, "PERKEMBANGAN KONSUMSI DAN PENYEDIAAN ENERGI DALAM PEREKONOMIAN INDONESIA," *Indonesian Journal of Agricultural Economics (IJAE),* vol. 2, no. 1, p. 98, 2010.

3 P. S. P. I. G. R. P. a. P. F. S. R. N. S. Julius Christian Adiatma, Kebutuhan Investasi Energi di Indonesia dengan Studi Kasus Rencana Umum Energi Nasional, Jakarta Selatan: Institute for Essential Services Reform (IESR), 2019.

4 G. Widayana, "PEMANFAATAN ENERGI SURYA," *Jurnal Pendidikan Teknologi dan Kejuruan,* vol. 9, no. 1, p. 37, 2019.

5 B. A. Sudiyono, "Perancangan dan Pembuatan Kapal Wisata dengan Motor Generator Listrik Tenaga Surya Sebagai Energi Alternatif Penggerak Propeler," *Jurnal Keilmuan dan Terapan Teknik Mesin,* vol. 10, no. 1, pp.52-62, 2008.

6 B. Y. D. Iradiratu D.P.K, "Perhitungan Kebutuhan Daya Listrik untuk Penggerak Perahu Nelayan Bertenaga Surya," *CYCLOTRON,* vol. 3, no. 1, p. 19, 2020.

7 Nugraha, Anggara Trisna, and Dadang Priyambodo. "Design of Hybrid Portable Underwater Turbine Hydro and Solar Energy Power Plants: Innovation to Use Underwater and Solar Current as Alternative Electricity in Dusun Dongol Sidoarjo." Journal of Electronics, Electromedical Engineering, and Medical Informatics 3.2 (2021): 93-98.

8 S. S. Pamor Gunoto, "PERANCANGAN PEMBANGKIT LISTRIK TENAGA SURYA 100 Wp UNTUK PENERANGAN LAMPU DI RUANG SELASAR FAKULTAS TEKNIK UNIVERSITAS RIAU KEPULAUAN," *Sigma Teknika,,* vol. 3, no. 2, p. 99, 2020.

9 T. A. N. D. Salwin Anwar, "Pengukuran Energi Listrik Berbasis PZEM-004T," in *Proceeding Seminar Nasional Politeknik Negeri Lhokseumawe,* Aceh, 2019.

10 D. N. D. B. F. Asep Muhamad Alipudin, "RANCANG BANGUN ALAT MONITORING BIAYA LISTRIK TERPAKAI BERBASIS INTERNET OF THINGS (IOT)," *JURNAL ONLINE MAHASISWA (JOM) UNIVERSITAS PAKUAN,* vol. 1, no. 1, p. 3, 2018.

11 M. Y. H. A. M. Muhammad Taif, "Penggunaan Sensor ACS712 dan Sensor Tegangan untuk Pengukuran Jatuh Tegangan Tiga Fasa Berbasis Mikrokontroler dan Modul GSM shield," *Jurnal PROtek,* vol. 6, no. 1, p. 43, 2019.

12 Nugraha, Anggara Trisna, and Dadang Priyambodo. "Prototype Hybrid Power Plant of Solar Panel and Vertical Wind Turbine as a Provider of Alternative Electrical Energy at Kenjeran Beach Surabaya." Journal of Electronics, Electromedical Engineering, and Medical Informatics 2.3 (2020): 108-113.

13 B. Saputro, "ANALISIS KEANDALAN GENERATOR SET SEBAGAI POWER SUPPLY DARURAT APABILA POWER SUPPLY DARI PLN MENDADAK PADA MDI MORODADI POULTRY SHOP BLITAR," *Jurnal Qua Teknika,* vol. 7, no. 2, p. 18, 2017.

14 M. R. Amir Hamzah Pohan, "GENERATOR MINI PORTABLE SATU PHASA 2500 VA UNTUK PENGATURAN FREKUENSI INVERTER SATU PHASA," *Jurnal Sain dan Teknologi TEKNIK UTAMA,* vol. XVI, no. 3, p. 14, 2021.

15 W. J. Z. Liang Song, "Residual Capacity Estimation of Valve-regulated Lead-acid (VRLA) Batteries for Second-use," Innovation Smart Grid Technologies - Asia (ISGT Asia), 2019. [Online]. Available: http://doi.org/10.1109/ISGT-Asia.2019.8881200. [Accessed 25 September 2021].

6. AUTHORS BIOGRAPHY

Purwidi Asri is aLecturer of Marine Electrical Engineering, Politeknik Perkapalan Negeri Surabaya (Shipbuilding Institute of Polytechnic Surabaya). Her Previous Research is aboutAnalysis of the Electricity

Needs of a Tourist Boat with a Capacity of 10 People Based on Solar Power

Hendro Agus Widodo is aLecturer of Marine Electrical Engineering, Politeknik Perkapalan Negeri Surabaya (Shipbuilding Institute of Polytechnic Surabaya). His Previous Research is aboutAnalysis of the Electricity Needs of a Tourist Boat with a Capacity of 10 People Based on Solar Power

Jossianto Eko Poetro is a Lecturer of Marine Electrical Engineering, Politeknik Perkapalan Negeri Surabaya (Shipbuilding Institute of Polytechnic Surabaya). His Previous Research is aboutAnalysis of the Electricity Needs of a Tourist Boat with a Capacity of 10 People Based on Solar Power

Anggara Trisna Nugraha is aLecturer of Marine Electrical Engineering, Politeknik Perkapalan Negeri Surabaya (Shipbuilding Institute of Polytechnic Surabaya). His Previous Research is aboutAngga, Anggara Trisna Nugraha, et al. "Solutions For Growing the Power Factor Prevent A Reactive Electricity Tariff And Decrease Warmth On Installation With Bank Capacitors." Applied Technology and Computing Science Journal 4.1 (2021): 35-46.

Mohammad Basuki Rahmat is aLecturer of Marine Electrical Engineering, Politeknik Perkapalan Negeri Surabaya (Shipbuilding Institute of Polytechnic Surabaya). His Previous Research is aboutAnalysis of the Electricity Needs of a Tourist Boat with a Capacity of 10 People Based on Solar Power

Isa Rachman is aLecturer of Marine Electrical Engineering, Politeknik Perkapalan Negeri Surabaya (Shipbuilding Institute of Polytechnic Surabaya). His Previous Research is aboutSmart Docking Ship Based on AIS (Automatic Identification System) to streamline the process of docking ships at the seaport

Ivan Azwar Septiadi is an Undergraduate Student of Otomation Engineering, Politeknik Perkapalan Negeri Surabaya (Shipbuilding Institute of Polytechnic Surabaya).

Habibi Ahmad Basyari is an Undergraduate Student of Marine Electrical Engineering, Politeknik Perkapalan Negeri Surabaya (Shipbuilding Institute of Polytechnic Surabaya).

Dewi Rizani Ruwahida is an Undergraduate Student of Otomation Engineering, Politeknik Perkapalan Negeri Surabaya (Shipbuilding Institute of Polytechnic Surabaya).

MONITORING AND FAULT DETECTION OF MAIN ENGINE COOLING WATER BASED ON MODBUS COMMUNICATION AND INTERFACE

A T Nugraha and **M Z A Tiwana**, Politeknik Perkapalan Negeri Surabaya, Indonesia

SUMMARY

The main engine failure can damage the engine system, reduce navigation safety, and lead to severe maritime accidents. The cooling system is essential to avoid engine mechanical efficiency reduction and engine failure due to overheating. Thus, a monitoring and fault detection system is needed. The system made is based on RS485 Modbus Communication and Interface. The acquisition device uses existing components in the 7S-50MC Diesel Engine. PLC is used as a processing device and Logic Panel Autonics S070 as an interface. From the test, fault detection in this study can provide error indicators when abnormal conditions occur. In addition, data monitoring and system error indication can be displayed clearly on the Interface. Testing this system proves compliance with Biro Klasifikasi Indonesia (BKI) rules regarding the machinery alarm system. The developed research can be the basis for a more complex and reliable monitoring and fault detection system for application on ships.

1. INTRODUCTION

Main engine failure is the most expensive damage, reaching 34.3% of the total machining cost. The Swedish Club noted that the average price of this damage was almost USD 634,000 from 2005 to 2011 [1]. In addition to being expensive, damage to the engine system can reduce navigation safety and result in severe maritime accidents [2]. One of the causes is the failure of the engine cooling system.

The cooling system is essential to maintain the operation of the ship's main engine because the combustion temperature ranges from 600°C to 800°C. If this temperature is not cooled each time the combustion process is complete, a decrease in engine mechanical efficiency and engine failure due to overheating can occur [3]. Thus, a monitoring and fault detection system is needed.

In addition to improving the limits of maintenance, this system regularly collects information from machines. The information obtained is processed to prevent unexpected control actions that affect the plant and early restoration by the backup system [4]. So, the safety and reliability of the ship become more effective. Biro Klasifikasi Indonesia (BKI) has established rules for monitoring and fault detection in the Rules for Automation System. The Machinery Alarm System must provide an optical and audible signal for fault identification in giving a warning. The alarm indication is from the text or symbol, and the optical signal must remain visible until the fault is resolved [5]. In addition, the system must also include capabilities in various operating conditions, cycles, and operations [6].

So much research related to shipping engine controlin academic journals. Yemao et al. [7] take a practical application design of a marine engine room monitoring system. This research provides an engine monitoring system design through the S7-200PLC and an alarm board. Serial communication with RS232 and Ethernet is sending data to the FameView application. Unfortunately, this research focuses only on the system's architectural design without clearly explaining the field device and error determination algorithm.

Sanadhya et al. [8] provide a monitoring solution with a specific field device and predict fault, but the scope of the discussion is limited. They usean MPU6050 accelerometer and piezo-type microphone to acquire vibration and noise data from the fuel oil separator. They are predicting fault using Crest factor calculation from the data obtained to detect impulsiveness in the signal. More extensive parameter monitoring has been investigated by Hu et al. [9] in the design of monitoring and diagnostic systems for marine diesel engines integrated with multiple methods and parameters. The system made consists of six sub-monitoring systems.

This study uses the integration of research in[7], [8], [9], and the provisions of the machinery alarm system in [5]. The parameters monitored are different from previous studies, namely the temperature of the main engine cooling water cylinder on the ship's main engine. The fault detection is functioning asan early warning of abnormal conditions in each cylinder. The Modbus RS485 communication protocol on the PLC sends field device data to the Logic Panel S070 as a human-machine interface. The system created can display data through the Interface and can detect errors.

2. BASIC RULES REQUIREMENT

The basic rule of this research is the machinery alarm requirement on the cylinder cooling water temperature at the engine outlet. BKI sets alarms in the form of indicators and upper high limit alarms[10]. The temperature indicator is from the thermocouple, and the upper high limit alarm is from the temperature switch on the engine. In addition, the monitoring and fault detection systems also follow BKI rules. These rules are summarized as follows [5]:

- The machinery alarm system shall provide an optical and an audible signal of unacceptable deviations from operating Figures

- Alarm delays shall be kept within time limits to prevent any risk to the monitored system in the event of exceeding the limit value
- The meaning of the individual indications shall be identifiable by text or symbols. Also, the optical signal shall remain visible until the fault disappear.
- The acknowledgment of an alarm shall not inhibit an alarm that new causes have generated.
- Transient faults which are self-correcting without intervention shall be memorized and indicated by optical signals.

3. METHODOLOGY

3.1. SYSTEM ARCHITECTURE

This system consists of input, process, and output. The input consists of analog and digital signals. The Analog signal is from the thermocouple, and the digital signal is from the temperature switch. The voltage change data from the thermocouple is equivalent to temperature change. The thermocouple sends the changes of voltage to the temperature controller.While in this research use two temperature controller becausethere are only four thermocouple channels in one device. Figure 1 shows the block diagram of the thermocouple input reading.

Digital signal depends on the state of the temperature switch. The principle of this switch is contact change due to the expansion of the bimetallic strip. Then, these changes will provide digital data to the programmable logic control (PLC) as the fault detection. There is also a push-button, an acknowledgment button, which provides information that the engineer is aware of the alarm condition. Figure 2 shows the digital input reading architecture.

The PLC processes all input. Different from the digital input, the temperature controller sends the thermocouple temperature readings via the RS485 Modbus communication. Modbus communication protocol is a standard communication protocol to transmit signals from the control device to the controller or data collecting system. Figure 3 shows the communication architecture between devices. The output of this system is buzzer and Interface—the buzzer asthe alarm indication with audible and optical alerts. Meanwhile, the Interface monitoring will display data and warnings of abnormal conditions by the Logic Panel Autonics S070. Figure 4 shows the hardware architecture.

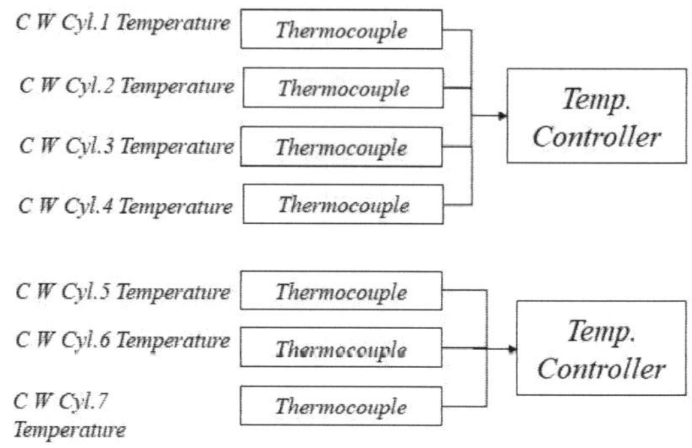

Figure 1: Analog Input Block Diagram

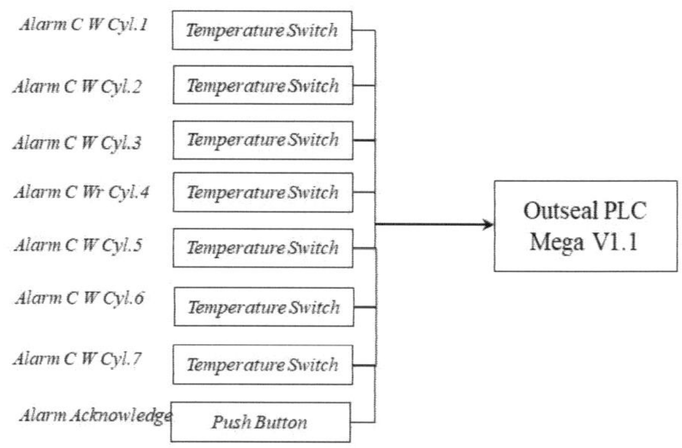

Figure2: Digital Input Block Diagram

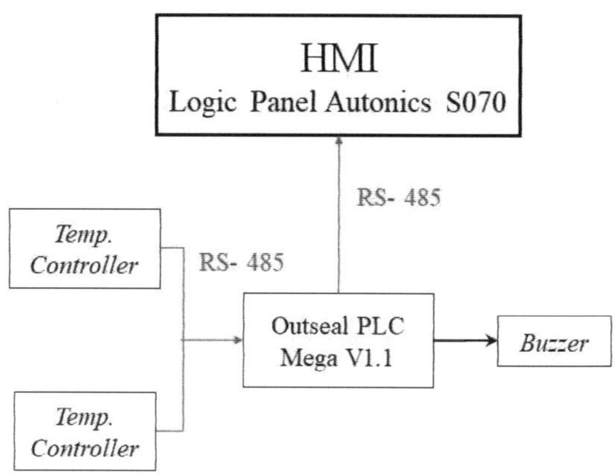

Figure3: Devices Communication Diagram

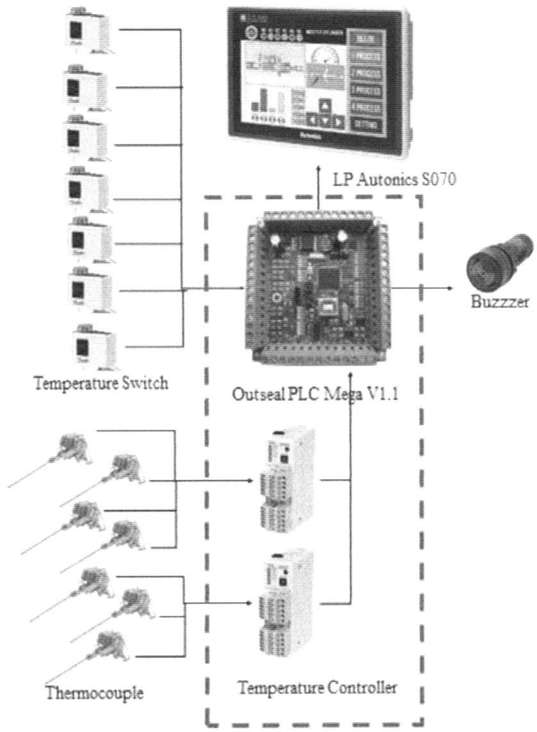

Figure 4: Hardware Architecture

3.2. SYSTEM CONCEPT

Figure 5 shows the process flow of the monitoring and fault detection system. The PLC reads the cooling water temperature data in real-time. The logic panel reads and displays the temperature value, and the digital signal is becoming the fault detection.
The fault is detected when the cooling water temperature exceeds the setting temperature, 80°C [11]. Once the PLC receives the fault data, the processing fault will begin, and the buzzer will be active.
The fault message was sent to the Logic Panel S070 to inform the engineer. The Logic Panel also shows clear alarm information by the alarm list on display. With the interlock function, the optical signal of the fault condition will be visible until the fault acknowledge. If another error appears, the alarm will be active again, and the alarm list will display new errors that occurred. The alarm list only disappears when the engineer has acknowledged the fault.

3.3. SYSTEM DESIGN

GP Editor is the interface design application. The display contains two pages. Figure 6(a) displaysthe engine cooling water, and Figure 6(b) displays alarm monitoring.

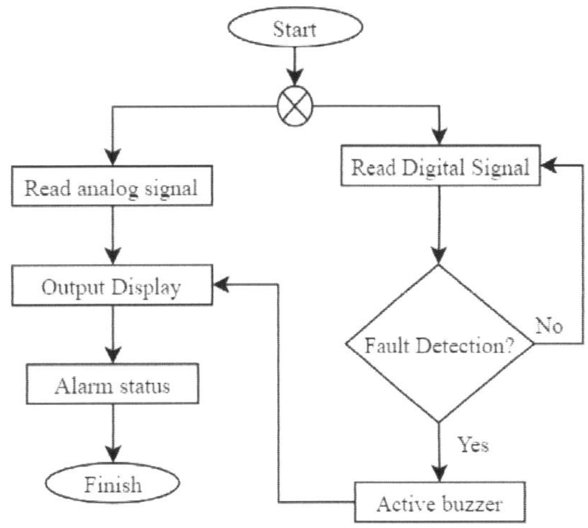

Figure 5: The Process Flow

Changes in cooling water temperature will be displayed graphically and dynamically with a bar graph in Figure 6(a).For easy monitoring, each graph uses CW Cyl x as the name of the thermocouple location in the outlet of cylinder x. In addition, the temperature display setting isan intuitive format with an unsigned data type that can adapt to the source value. The temperature value setting required precision and a read-only value. Each value has been assigned a unit of °C following the temperature unit.
The second display is the Alarm Monitoring in Figure6(b). This display is essential to warn engine errors: the Alarm & Safety List and the Warning section. Alarm & Safety List section to record faults and display their Interface according to the date and time they occurred. A flash button gives a signal to the engineer in case of a problem. The "WARNING" text also gives an indicator signal with the color turning red when an error is detected. Changing the position from the first section to the second section can be done with the left-right arrow buttons at the bottom right of the screen.

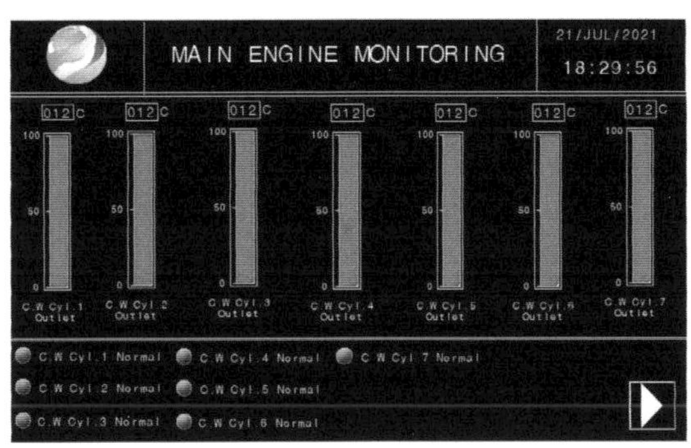

(a)

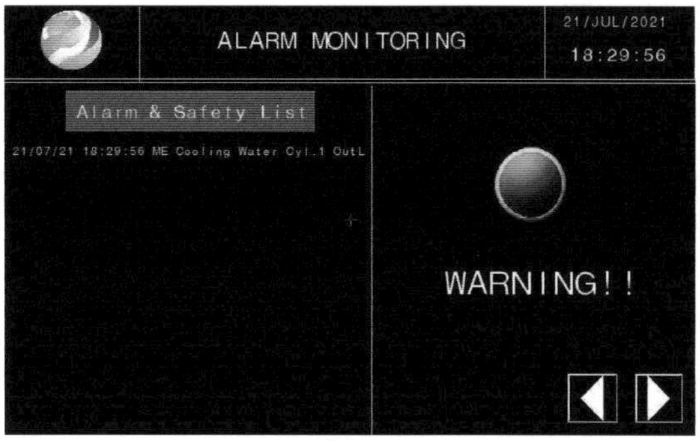

(b)

Figure 6: Main Engine and Alarm Monitoring Display

6. RESULT AND DISCUSSION

6.1 RESULT

The PLC was programmed in Outseal Studio and tested to determine the response of digital input, digital output, and communication with Logic Panel S070. The digital input test uses a normally open (NO) switch to trigger the PLC's internal memory. Meanwhile, the Digital Output test is to know the condition of the PLC relay.Real-time testing monitoring is done through Outseal Studio. The testing of the digital input is shown in Figure 7, represented by S1 with internal memory B1, and the test result is in Table 1. The Digital Output test is shown in Figure 8, represented by S1 with relay R1, and the result is in Table 2.

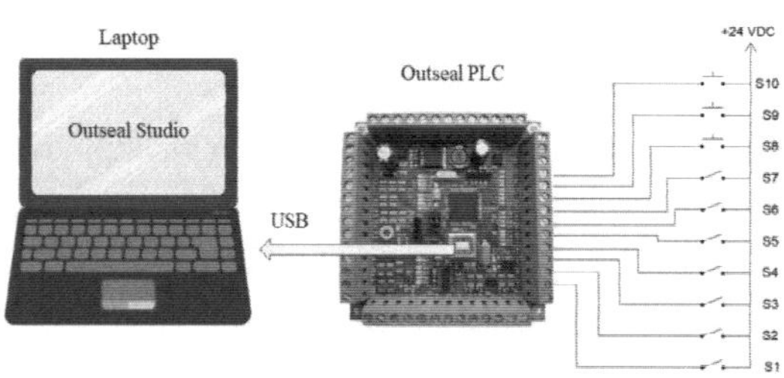

(a)

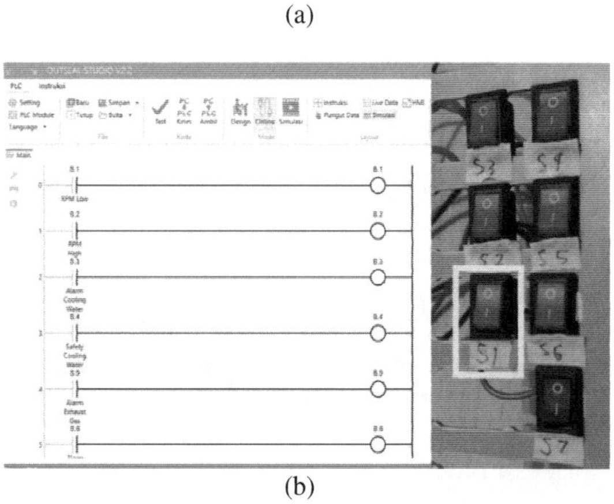

(b)

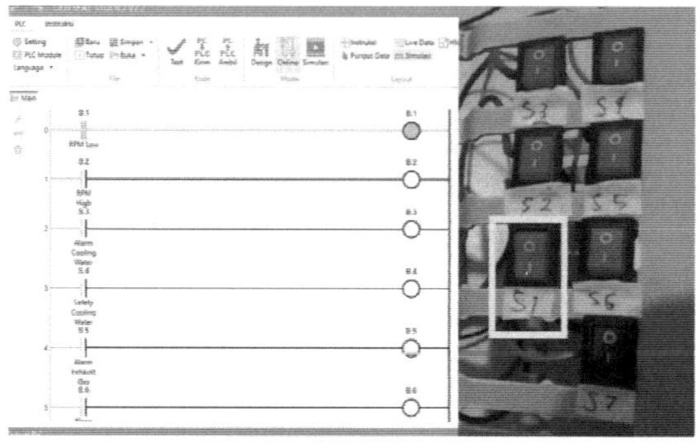

(c)

Figure 7: Digital Input Testing

The testing of the digital input is in Figure 7. Figure 7(a) shows the wiring diagram of digital input testing. This study only uses toggle switches on S1 to S7. All switches connect with the digital input port of the PLC. The responses of the PLC input port shows in PC through Outseal Studio.

Figures 7(b) and 7(c) show the switches and monitoring display. Figure 7(b) is "off" condition, and the S1 contact in the monitor will not emit green light. Figure 7(c) is the"on" condition. So, the S1 contact in the monitor emits green light. The test result is in Table 1.

Figure 8 shows the digital output testing. Figure 8(a) is the wiring diagram. The switches connect the PLC digital input, and the relay connects the PLC digital output. Each relay will respond to every state of the switches. If the switch is on, so does the relay and vice versa. Same with the digital input testing, the digital output testing also uses Outseal Studio to monitor the relay responses.

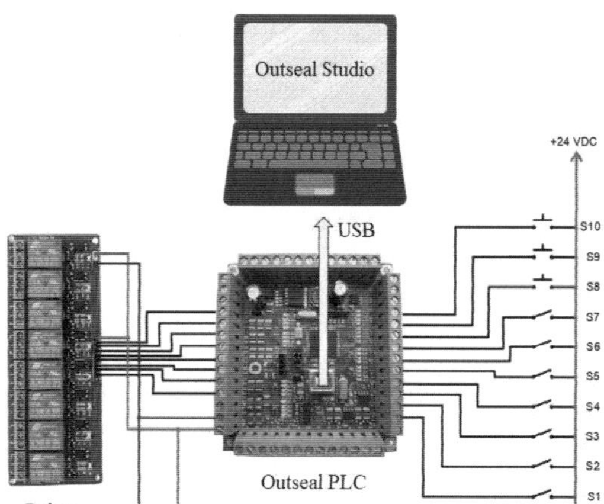

(a)

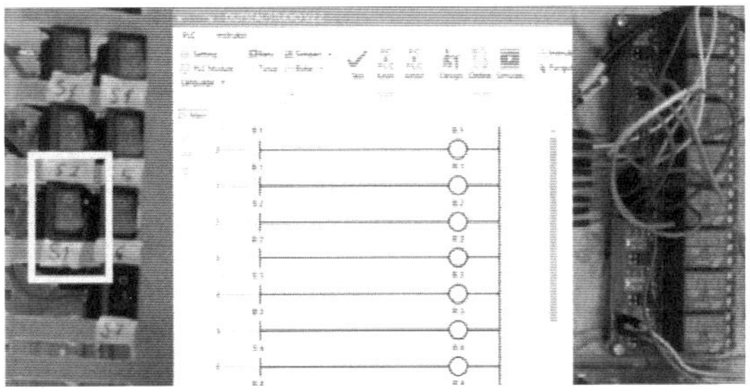

(b)

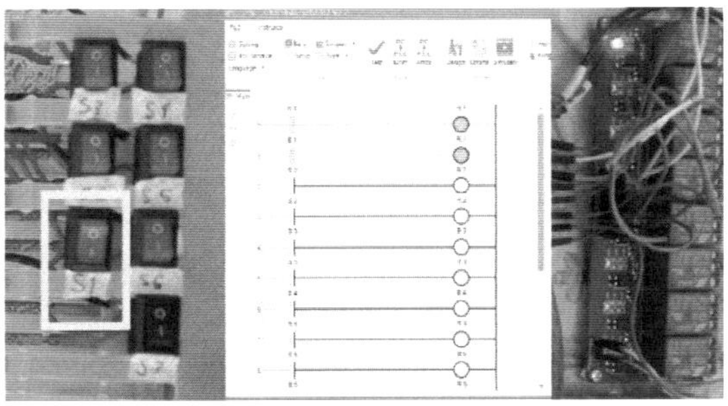

(c)

Gigure 8:Digital Output Testing

Figures 8(b) dan 8(c) show the switches, monitoring display, and relays. Figure 8(b) shows the switch in off condition. So, the S1 dan R1 contact in the Outseal Studio will not emit the green light. As a consequence, the relay will not work. Figure 8(c} shows that the switch S1 is on. Then, the S1 and R1 contact in Outseal studio will emit the green light. So, the relay R1 works. Table 2 shows the digital output testing result.

Table 1. Digital Input Test Result

No	Digital Input	Switch State	Internal Memory Response
1	S1	0	0
		1	1
2	S2	0	0
		1	1
3	S3	0	0
		1	1
4	S4	0	0
		1	1
5	S5	0	0
		1	1
6	S6	0	0
		1	1
7	S7	0	0
		1	1

Table 2: Digital Output Test Result

No	Digital Input	Switch State	Output Respon	Digital Output
1	S1	0	0	R1
		1	1	
2	S2	0	0	R2
		1	1	
3	S3	0	0	R3
		1	1	
4	S4	0	0	R4
		1	1	
5	S5	0	0	R5
		1	1	
6	S6	0	0	R6
		1	1	
7	S7	0	0	R7
		1	1	

Communication test is for Modbus communication test between temperature controller, PLC, and Logic Panel. The thermocouple sends the temperature value to the temperature controller. Then the PLC receives the value and forwards it to the Logic Panel. Figure8 shows the testing process. Figure 9(a) is the wiring diagram, Figure 9(b) shows the testing process, and Figure 9(c) shows the temperature from the thermocouple. Table 3 shows the result of the digital output testing.

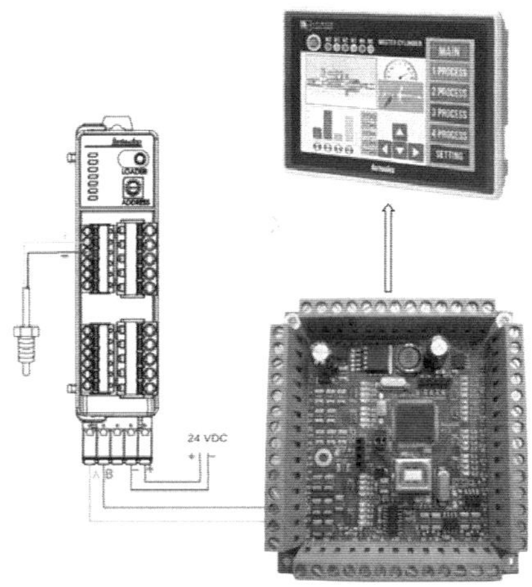

(a)

(b)

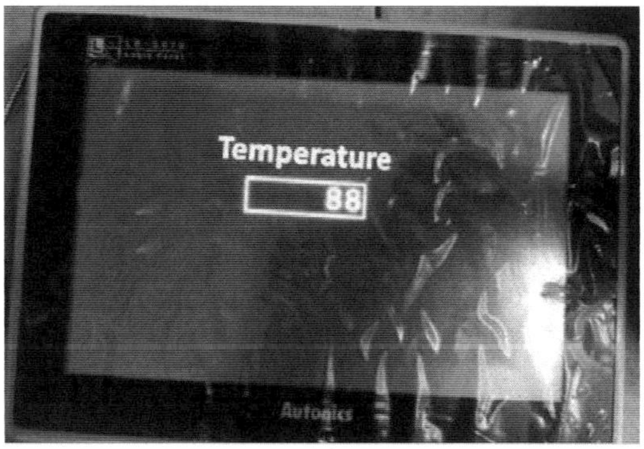

(c)

Figure9: Communication Testing

Table 3: Communication Test Result

Data order	Temperature(°C)
1	27
2	30
3	36
4	40
5	42
6	48
7	53
8	59
9	66
10	71
11	79
12	83
13	85
14	88
15	90

Experiments on the ship were also carried out by connecting the data acquisition on the ship with the prototype. The seven existing temperature switches on the 7S-50MC Diesel Engine connect the digital input of the Outseal PLC. The systemswitches are in a normally closed (NC) condition. Thus, the buzzer as an audible signal will not be active. However, a fault occurs if the switch state changes from NC to NO. The results of the standard engine condition test are in Table 4.

Table 4: Normal Engine Condition Test Result

No	Switch Fault	Input Switch	Output Response	Fault Indication
1	Alarm C.W Cyl.1	1	0	Normal
2	Alarm C.W Cyl.2	1	0	Normal
3	Alarm C.W Cyl.3	1	0	Normal
4	Alarm C.W Cyl.4	1	0	Normal
5	Alarm C.W Cyl.5	1	0	Normal
6	Alarm C.W Cyl.6	1	0	Normal
7	Alarm C.W Cyl.7	1	0	Normal

Obtaining the fault condition is by removing the temperature switch cable on the digital input PLC. So,the contact conditions change from normally open(NO) to normally close(NC). The active buzzer proves the working response of the fault detection, and an alarm indication is on the Interface. Table 5 shows the fault condition testing. Analog data from the thermocouple is in Table 6. The average cooling water temperature at the cylinder outlet when the main engine is not operating is 72.2°C. Figure10 shows the process of taking temperature data from one of the main engine cooling water cylinders.

Table 5: Fault Engine Condition Test Result

No	Switch Fault	Input Switch	Output Response	Logic Panel Fault Indication
1	Alarm C.W Cyl.1	0	1	Warning
2	Alarm C.W Cyl.2	0	1	Warning
3	Alarm C.W Cyl.3	0	1	Warning
4	Alarm C.W Cyl.4	0	1	Warning
5	Alarm C.W Cyl.5	0	1	Warning
6	Alarm C.W Cyl.6	0	1	Warning
7	Alarm C.W Cyl.7	0	1	Warning

Table 6: Cooling Water Temperature Test Result

No	Thermocouple Location	Cooling Water Temperature (°C)
1	Cylinder 1 Outlet	72.3
2	Cylinder 2 Outlet	72.3
3	Cylinder 3 Outlet	72.1
4	Cylinder 4 Outlet	72.2
5	Cylinder 5 Outlet	72.3
6	Cylinder 6 Outlet	72.1
7	Cylinder 7 Outlet	72.0

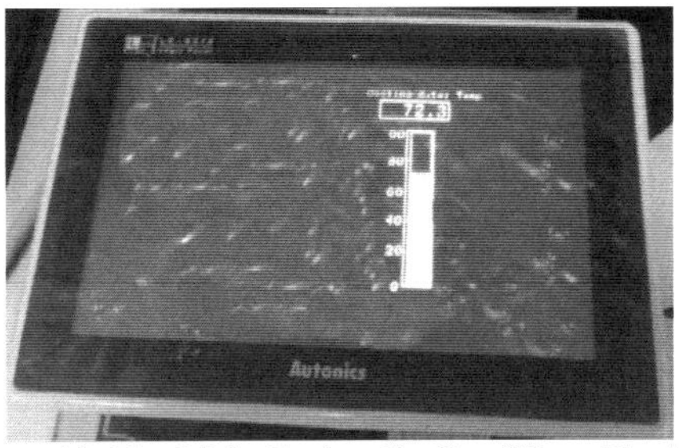

Figure10: The Main Engine Cooling Water Testing

6.2 DISCUSSION

A clear explanation of the system is system architecture, hardware architecture, system concepts, system interface design, and system testing process. System architecture to hardware architecture provides clear connections between devices. System concepts provide a directed system workflow. Moreover, most importantly, an informative and precise interface design is also provided for monitoring the condition of the main engine.

The digital input test proves that the internal memory changes according to the switch conditions. With an input voltage of +24VDC, the PLC can read the input signal and forward it to the internal memory according to its contacts. The digital output test produces a relay response that matches the switch conditions and input voltage. The results of this test prove that PLC Outseal can work according to the program created.

The test result of communication between temperature controller, PLC, and Logic Panel shows that the Temperature controller can send data to the PLC for processing. Furthermore, the data received by the Logic Panel is graphically displayed.

Furthermore, the test was carried out on the 7S-50MC Diesel Engine by connecting the temperature switch of each cylinder and the temperature controller to the PLC. The test produces data on the normal condition of the engine because the engine is in a non-operating condition. The fault condition is from disconnecting the switch to the PLC input to change the normally close (NC) contact to normally open (NO). From the test, fault detection in this study can provide error indicators when abnormal conditions occur.

This test produces cooling water temperature at a high-temperature freshwater outlet with an average of 72.2°C. Analog data only for temperature monitoring without error detection to avoid errors if the thermocouple is damaged and cannot read the temperature value according to the actual value.

7. CONCLUSIONS

This study provides a monitoring and fault detection system for the7S-50MC cooling water. During the test, the engine condition was not active. The system made is based on RS485 Modbus Communication and Interface. Different from previous studies, the acquisition device uses existing components in the main engine.

This study gives the system architecture, hardware architecture, system concept with a clear workflow, interface design, and the fulfillment of the BKI Alarm requirement during testing. First, the system can display optical and audible signals through the Interface and buzzer. Second, the alarm delay ranges from 1 to 2 seconds from the abnormal condition appearing. Third, fault indications can be displayed clearly through the safety list section on the Interface. Fourth, acknowledging an error does not block a new fault signal. Moreover, fifth, the alarm list function will memorize the alarm condition until the engineer acknowledges the condition.

8. REFERENCES

1 The Swedish Club, "Main Engine Damage Study," 2015. [Online]. Available: https://www.swedishclub.com/media_upload/files/Publications/Loss%20Prevention/Main%20Engine%20damage%202015%20The%20Swedish%20Club.pdf. [Accessed 17 July 2021].

2 X. Xu, X. Yan, K. Yang, J. Zhao, C. Sheng and C. Yuan, "Review of Condition Monitoring and Fault Diagnosis for Marine Power System,"*Transportation Safety and Environment,* vol. 3, no. 2, pp. 85-102, 2021.

3 C. Jiale, L. Tie and Z.Xinyi, "A study of smart thermal insulation coating on improving thermal efficiency in a marine two-stroke low-speed diesel engine," *Fuel,* vol. 304, 2021.

4 M. Adel, S. Farhang, A. Mohammad, Q. Ali Nouri, B. Abbas, "Resilience estimation of critical infrastructure systems: Application of expert judgment,"*Reliability Engineering & System Safety,* vol. 215,2021.

5 Biro Klasifikasi Indonesia, "Rules for Automation,"*Rules for Classification and Construction,* vol. VII, pp. 2-1, 2018.

6 Biro Klasifikasi Indonesia, "Rules for Automations,"*Rules For Classification and Construction,* Vol. VII, pp. 1-2, 2018.

7 M. Yemao, L. Monica, and M. Scott N, "Managing unruly technologies in the engine control room: from problem patching to an architectural thinking and standardization,"*WMU Journal of Maritime Affairs,* pp. 497-519, 2018.

8 D. P. Sanadhya and R. P. Sharma, "Condition Monitoring of Marine Fuel Oil Separator System,"*International Journal for Research in*

Applied Science & Engineering Technology (IJRASET), vol. 8, no. 2, pp. 260-2264, 2020.

9 Y. J. Y. Y. Hu N., "Design and Implementation of Integrated Monitoring and Diagnosis System for Marine Diesel Engine,"*engineering Asset Management - Systems, Professional Practices and Certification.,* pp. 647-659, 2015.

10 Nugraha, Anggara Trisna, and Dadang Priyambodo. "Design of Pond Water Turbidity Monitoring System in Arduino-based Catfish Cultivation to Support Sustainable Development Goals 2030 No. 9 Industry, Innovation, and Infrastructure." Journal of Electronics, Electromedical Engineering, and Medical Informatics 2.3 (2020): 119-124.

11 Biro Klasifikasi Indonesia, "Rules for Machinery Installations,"*Rules for Classification and Construction,* Vol. III, pp. 2-26, 2019.

12 G. Theotokatos, K. Sfakianakis and D. Vassalos, "Investigation of ship cooling system operation for improving energy efficiency,"*J Mar Sci Technol,* p. 38–50,2017.

9. AUTHORS BIOGRAPHY

Anggara Trisna Nugraha Lecturer of Marine Electrical Engineering, Politeknik Perkapalan Negeri Surabaya (Shipbuilding Institute of Polytechnic Surabaya). His Previous Research is about the Design of a Monitoring System for Hydroganics based on Arduino Uno R3 to Realize Sustainable Development Goal's number 2 Zero Hunger.

Mayda Zita Aliem Tiwana, Undergraduate Student of Marine Electrical Engineering, Politeknik Perkapalan Negeri Surabaya (Shipbuilding Institute of Polytechnic Surabaya).

CAUSATIVE ANALYSIS FOR A SUNK FERRY RO-RO IN INDONESIA: SEAKEEPING, WAVE HEIGHT AND LASHING CORRELATION

W Mutmainnah, S. Anggara, T. Firmanda, M. R. Fitra, S. Komariyah, R. Syarif and **M. A. Kurniawan,** PT. Biro Klasifikasi Indonesia, Indonesia

SUMMARY

In this research, one sunk case is reanalysed based on official investigation report from Indonesian National Transportation Safety Committee. At first, causative factors contributing to the accident are identified utilizing MOP Model. Second, the technical analysis is carried out to support the previous qualitative analysis by MOP Model, such as Seakeeping and minimum angle analysis for vehicle being slide on car deck. The analysis result shows that the minimum angle causing the vehicle being slide on vehicle deck is 20 degrees for unlashed vehicle. That angle number is reached when the ship was hit by wave height of 3 m, which was the wave height at the time of the accident with several additional condition that is discussed in this paper. The aim of this research is finding any correlation between seakeeping, wave height, and unlashed vehicle on Ferry Roro safety that happen in the case study.

NOMENCLATURE

H_s	Significant wave height (m)
$T_{average}$	Average wave period (s)
θ	Inclination angle (deg)
μ_s	Static friction coefficient

1. INTRODUCTION

The International Maritime Organization (IMO) hold a meeting every year. Domestic ferry safety is one of 10 topics that is discussed on 103rd and 104th session of Maritime Safety Committee in 2021, from now on it will be called as MSC 103 and MSC 104 [1]. Even though IMO has SOLAS and ISM Code already, the IMO meeting session is now trying to establish an effective global safety standard which may be applied to domestic ferries in order to reduce the number of domestic ferry accidents.

In Indonesia, the National Transportation Safety Committee (NTSC) is still conducting any investigation on ferry accidents. From the statistical analysis of ship accidents in Indonesia, from 2018-2019 there were 763 accidents that involved 10 types of ships and 7 types of accidents. The highest number of accidents was passenger ships, including HSC, which count for 29% then followed by fishing vessel for 25% and dry cargo ships for 18%. In total, there were 714 people dead or missed and 1052 people injured. Almost 30% of the people died/missed caused by ship accidents was from passenger ship accident. Type of accident that was common in that year was Capsizing or sunk, counted for 43%, then followed by fire or explosion for 18% [2].

As many ship accident news are still released, several government agencies make some efforts to reduce the accidents. Indonesian Meteorological, Climatological, and Geophysical Agency ('BMKG') issues an advice related to the high risk for ship sailing in a certain wave height, called as Safety Voyage Advice. The advice defines 4 levels of wave height that will cause high risk voyage for certain types of ships. For ferry ships, the possible high risk voyage will be at wind speed of more than 21 knots and wave heights above 2,5 m [3]. The advice released because there are still many capsized cases happen every year in Indonesian Waterways caused by high wave and wind speed.There are several investigation reports published by NTSC that declare any wave height contribution to ship accidents, especially sunk or capsized. Even though NTSC did not investigate all accidents happen in Indonesian waterways, there are several investigation reports that people can have lesson learned from it. In 2009, there are 2 sinking ships that was investigated by NTSC [4] [5].

One of the cases was involving a ship departed from Cappa Ujung, Pare-Pare Port headed to Samarinda, East Kalimantan with 365 passengers and mixed cargo up to 443 tonnage. In the middle of its voyage, the ship started unstable because of the high wave and windy. The main engine was not good enough to keep the ship stable, thus the ship sunk quickly [4]. In other years, similar accidents happen where the ship departed in a good weather but in the middle of its voyage the weather changed become windy with high wave. On that moment, the ship become unstable with different conditions [5][6][7][8]. The wave heights that hit the ships were higher than the advice from BMKG. However, the ship condition on the moment contributed how the accidents happen where the conditions are different for each ship. The condition that contributed to accident can be analysed to be mitigated so than the accident can be avoided.

In this paper, all the ship condition, got from its investigation report, will be analysed in order to find the causative factors (CF) of the accidents using MOP Model [2]. The case studied in this paper was in 2018.

A ship was heeled due to high waves reaching 3,0 meters from the portside. The seawatergot into main deck through openings on the ship's side and also from non-tight ramp doors. Incoming seawater could not be discharged outof the shipquickly because the ship does not have a freeing port on the main deck, causing seawater being accumulated on the portside so that the ship heeled in this direction. When the ship heeled, vehicles without correct lashing were shifted to the portside of the ship. This conditionleads the ship heeled to the portside quickly. When the ship heeled 15 degrees to the portside and the ocean waves had reached the top deck, the captain decided to run the ship to the nearest shore to avoid the ship capsized and sinking [8].

After getting the CFs, two technical analysis are conducted to verify how the accident happen, namely seakeeping analysis and minimum heeling angle calculation for sliding vehicle. The aim of this research is finding any correlation between seakeeping, wave height, and unlashed vehicle on Ferry Roro safety that happen in the case study.

2. CASE STUDY: A SUNK FERRY

The case that is studied in this paper is a sunk ferry on July 3rd 2018 from Port of Bira to Port of Pamatata. As stated before, the facts of the accidents can be read from investigation reports by NTSC [8]. The ship's principal dimension is LOA 47,75 m, B 16,5 m, D 2,45 m, and draught 1,85 m. On 3rd July the manifest of ships carried passengers, car, and truck. Number of passengers is 190 people, the number of car/truck carriage 49 units (Gol. II 19 unit, Gol IV 14 unit, Gol V 8 unit, and Gol VI 8 unit). From 190 people onboard, 34 people were dead and 1 person lost.

When the ship is proceeded, the ship is heeled due to high waves that reach 3.0 meters from towards the portside. Seawater from the wave enters to the main deck through the openings on the side ship's hull and also from its non-tight ramps. The incoming seawater cannot be drained out quickly off the ship because the ship did not have freeing port in the main deck, causing seawater is being accumulated on the left so that the ship heels to the portside.

When the ship heel to the portside, there is also a shift in the vehicle that is not tied to the ship, so that the speed of the ship's heeling becomes faster. When the ship is heeled 15 degrees to left and the seawater have reached the upper deck, the captain decided to run aground the ship to the nearest shore to avoid capsizing and sinking.

3. MOP MODEL

MOP (4M Overturned Pyramid) Model is a new model developed by Mutmainnah and Furusho since 2014 [9] that can be utilized to describe characteristics of Maritime Transportation System (MTS) which is a socio-technical environment system. The model is based on the epidemiological model that consists of the latent condition, barriers, and active condition.

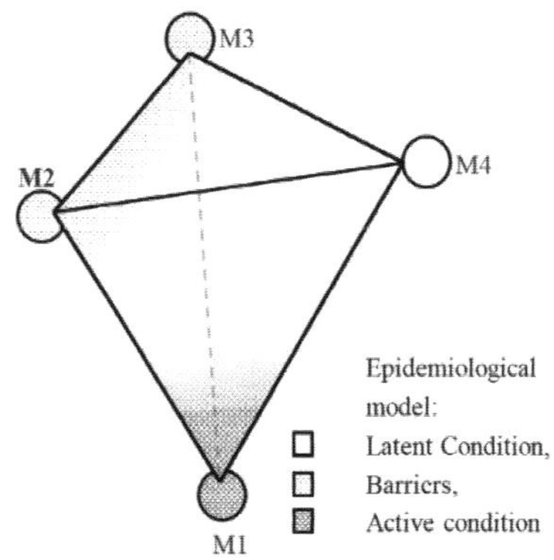

Figure 1: MOP Model

Figure 1 shows an image of the MOP model that looks like three sided overturned pyramid in 3D. Each corner of the model represents each M factor, namely Man, Machine, Media, and Management. In the model, all the M factors contribute to the balance of the MTS. Some mistakes or errors happen in the system, whether in each M factor or the interaction between or among the M factors, will cause the system become unstable then the accident happen.

MOP Model utilized in this paper is the newest development of MOP model that was modified by making a fix classification of CFs in each category of 4M, adopting several classifications of Human Factors Analysis and Classification System (HFACS) method [2], for the Corner Analysis as the first step of utilizing MOP Model. However, the second step of MOP Model does not carry out in this paper. The new model can represent how the accident happen by categorizing the CFs into a fix classification in the first step, called as Corner Analysis (CA). The next step is re-arranging or ordering all CFs based on causation condition and/or timeline.

Here is how the accident, studied in this paper using MOP Model, happen:

A. Seawater come into main deck/ vehicle deck through openings

M2010202 Front ramp door is not tightly closed
M2010101 There is no freeing port
M3010901 Wave height up to 3 m
M1010403 The mean operational draft is over (2.02m) compare to the approved draft (1,75m)

B. The ship is unstable (heeled to portside) and it cannot be back at the upright position

M3030402 There is shifted vehicle
M3030403 Not all vehicles are correctly lashed

M1010404 Vehicle arrangement is based on the ships condition
M1010403 The ship stability is not calculated correctly before departure
M3030401 The vehicle weight is not known

C. The emergency condition is more uncontrollable
M1010301 Some passengers jump into sea without wearing life jacket
M1020101 Passengers were panic
M1020301 Seafarer and ship's crew do not conduct crisis and crowd management
M2010501 There is no muster station
M1010405 Seafarers do not carry out safety drill
M4010101 Seafarer's qualification is not complied

Those are the CFs gotten from MOP Model analysis that are categorized based on the sequence of the accidents. There are three sequences that happen to the accident consecutively from point A to B to C. The CFs in each sequences are the failures/errors/facts that cause the sequences.To complete the analysis, two technical analysis carried out in this paper. Those technical analysis will be complements to sequences A and B which are related to the issue of wave height limitation issued by BMKG [3] and the conditions happen after getting the high wave.

4. TECHNICAL ANALYSIS

The concept of this technical analysis is to know the maximum wave height that becomes the safety operation limit for the ship. It is understood that there are multiple factors leading to safety. However, this study only captures the effect of wave height to vehicle sliding on board which induces many accidents for ferry roro.

4.1 SEAKEEPING ANALYSIS

In order to know the minimum angle for vehicles to slide onboard, the response of the ship above the wave shall be calculated at first. There are conditions to be set up to approach the simulation, as follow:

1. H_s (Significant wave height) was varied from 0.5 to 8.5 m with the average wave period data ($T_{average}$) taken from North Atlantic wave scattering [10] as seen in Table 1. Period T is the center of gravity of the wave energy distribution curve which is the weibull distribution.

Table 1: H_s and $T_{average}$ variation

H_s	0.5	1.5	2.5	3.5	4.5	5.5	6.5	7.5	8.5
$T_{average}$	6.50	7.78	8.53	9.09	9.55	9.95	10.31	10.63	10.92

2. The wave energy spectrum is based on Jonswap formulation with a peakness coefficient of 3.3 which is the average value for the characteristics of North-Atlantic waters
3. Wave heading was variated from 0 - 360 degree direction, 0 degree is coming from stern of the ship
4. The ship is moving forward at a speed of 12 knots and Full load draft condition
5. Strip theory is used as methodology

Seakeeping analysis shows that the most significant rolling amplitude due to waves occurred when waves coming from beam sea, as seen in Figure 2. It is captured that with 3 m significant wave height, the ship has been heeling up to 18 degrees. Therefore, this angle is used as a reference angle to see if the vehicle without lashing has shifted at that angle, which in turn causes the ship will experience stability failure.

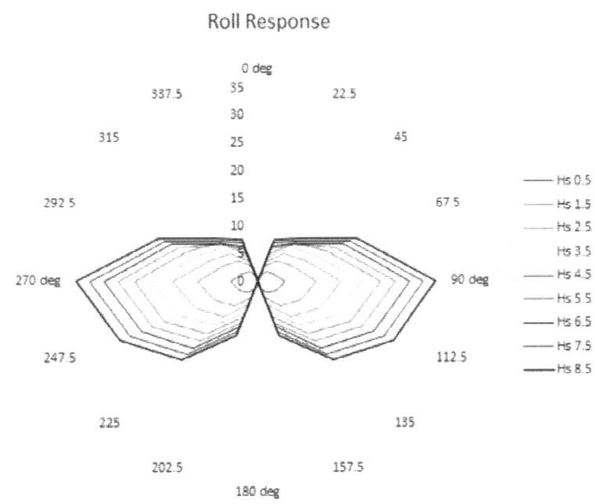

Figure 2: Rolling response to wave height variation

4.2 MINIMUM HEELING ANGLE CALCULATION FOR SLIDING VEHICLE

Literature study was conducted to determine the inclination angle of the ship that is able to move the vehicle without ties. This angular limitation can be achieved when the vehicle will start to shift, so the static friction coefficient will play a role in this phenomenon, the formula can be seen as follows:

$$\theta = \tan^{-1}(\mu_s) \quad (1)$$

where θ[deg] is the inclination angle of the ship and μs is the static friction coefficient of the vehicle wheel with the deck plate. Utilizing equation (1), $\theta$is calculated using known μ_s from several literatures. Table 2 shows a summary of the criteria for limiting the ship's angle to vehicle displacement.

Table 2: Summaries of inclination angle and static friction coefficient

Literature	μ_s	θ (deg)	Note
KNKT [8]	-	10 -15	-
IMO CSS [11]	0,3	16,7	Steel-Rubber
J.Jia [12]	0,65	33	Dry
	0,37	**20**	**Wet***

* If the condition of the deck is wet then the coefficient of static friction is reduced by 57% (Jones & Children report)

5. DISCUSSION

MOP Model help researcher to figure out the causation condition of the accident as a qualitative approach. There are three sequences that happen in the accident. First is seawater come into main deck/vehicle deck through openings.The front ramp door was not closed tightly and the mean operational draft was over (2.02m) compare to the approved draft (1,75m) when the ship was operated. In the middle of the voyage, the weather changed, 3 m wave height hit the ship thus the seawater come into main deck through the non tightly closed ramp door. Since the ship had no freeing port, the incoming seawater could not be coming out by itself and was accumulated on the main deck.

As the seawater was accumulated, the ship heeled to portside and the main deck become wet. Then, some vehicles that were not lashed correctly were shifted to portside quickly and can not be back at upright position. Other contributing factors were the vehicle arrangement was based on the ships condition, the ship stability was not calculated correctly before departure, and the vehicle weights were not known. All of those condition leads to second sequence of the accident, namely The ship is unstable (heeled to portside) and it cannot be back at the upright position.

Final Sequence of the accident is The emergency condition is more uncontrollable. Since the ship condition became worst, passengers were panic and some of them jumped into sea without wearing life jacket. Were seafarer and ship's crew conducted crisis and crowd management and there were muster station, passengers would be less panic. Other contributing condition were the seafarer did not carry out safety drill periodically and some of their qualification were not complied.

To verify the MOP Model results, two technical analysis is conducted for first and second sequences of the accidents as the quantitative approach. According to investigation report from NTSC, the wave height at the time of the accident was 3 meters, The seakeeping analysis and its relationship with friction coefficient shows that the ship's heeling angle of 18 degrees due to 3 m wave height is able to shift the unlashed vehicle (according to the NTSC criteria and the IMO CSS Code), while the J.Jia criteria requires approximately 2 degrees or the TCG of the ship shifts 0.5 m from the center line position. There are several possibility explaining those condition since J.Jia state that minimum degree that will make unlashed vehicle being slide on board on wet condition is 20 degree. However, the seakeeping analysis show that the rolling angle of the ship caused by 3m wave height is 18 degrees.

Firstly, It should be noted that the calculation of vehicle shifting above is done statically, not considering the acceleration due to the ship's motion which can increase the magnitude of the force to shift the cargo on deck. Secondly, considering that this accident was short-lived and progressive, which meant that the ship's heeling was not only influenced by waves, but the incremental shifting of the vehicle itself also contributed to the ship heeling. It leads the ship to heel further even though there is no additional external load applied. The two things above become limitations of technical analysis and at the same time become compensation for the value of 2 degrees heel angle until the ship has an accident.

From the explanation above, there is a match correlation between the actual conditions while incident occurred and the reconstruction/simulation result.

6. CONCLUSION

The qualitative approach using MOP Model and quantitative approach in technical analysis have been carried out. All the results show that they are correlated. The technical analysis can explain how the ship heeled caused by 3m wave height quantitatively. The ship heeled 18 degrees after getting 3m wave height of beam seas. While the minimum heeling angle if the ship that can slide the unlashed vehicle on wet deck based on J.Jia criteria is 20 degrees. On dry deck, the vehicles will be slide if the ship heels up to 33 degree.

Qualitative approach explained that the seawater come into main/vehicle deck through non tight front ramp door and there were no freeing port. Thus the seawater was accumulated on vehicle deck making possibility of shifting TGC causing any additional heeling degree and wet floor condition. These conditions are very correlated to the quantitative result causing the ship heeled then the unlashed vehicles were slide then the ship could not be back at upright position. The analysis carried out in this paper shows that there is correlation between seakeeping, wave height and unlashed vehicle on Ferry Ro-Ro safety.

7. REFERENCES

1. IMO, 'Maritime Safety Committee, 103rd session', https://www.imo.org/en/MediaCentre/MeetingSummaries/Pages/MSC-103rd-session.aspx, 2021.

2. Mutmainnah, W., Bowo, L.P., Nurwahyudy, A., Prasetyo, F.A., and Furusho, M., 'Causative Factor Analysis of Passenger Ship Accident (Fire/Explosion) in Indonesia', *Maritime Safety International Conference- IOP Conf. Series: Earth and Environmental Science, 557, 012037*, 2021.

3. BMKG, 'Saran Keselamatan Berlayar', *https://maritim.bmkg.go.id/info/37/Saran-Keselamatan-Berlayar,* 2018. (Indonesian Language)

4. NTSC, 'Laporan Investigasi Kecelakaan Pelayaran', *http://knkt.dephub.go.id/knkt/ntsc_maritime/Laut/2009/Laporan%20Akhir%20KM%20Teratai%20Prima.pdf,* 2009.

5. NTSC, 'Laporan Investigasi Kecelakaan Pelayaran', *http://knkt.dephub.go.id/knkt/ntsc_maritime/Laut/2009/Laporan%20Final%20KM.%20Dumai%20Express%2010.pdf,* 2010.

6. NTSC, 'Laporan Investigasi Kecelakaan Pelayaran', *http://knkt.dephub.go.id/knkt/ntsc_maritime/Laut/2013/Laporan%20Final%20KM.%20Irama%20Nusantara.pdf,* 2013.

7. NTSC, 'Laporan Investigasi Kecelakaan Pelayaran', *http://knkt.dephub.go.id/knkt/ntsc_maritime/Laut/2015/FINAL%20LAUNCHED_Marina%20Baru%202B.pdf*, 2015.

8. NTSC, 'Laporan Investigasi Kecelakaan Pelayaran', *http://knkt.dephub.go.id/knkt/ntsc_maritime/Laut/2018/FINAL%20KNKT-18-07-22-03%20Lestari_Maju.pdf*, 2018.

9. Mutmainnah W., 'The 4M Overturned Pyramid(MOP) Model Development to Characterize Accidents in Maritime Transportation System', *Kobe University - Thesis Disertation*, 2014.

10. IACS, 'Standard Wave Data', *IACS* rec. 34, 2000.

11. IMO, 'Code of Safe Practice for Cargo Stowage and Securing', *IMO Publication*, 1992.

12. Jia, J., 'Investigations of vehicle securing without lashings for Ro-Ro ships', *Journal of Marine Science and Technology*, 2007.

9. AUTHORS BIOGRAPHY

Wanginingastuti Mutmainnah holds the current position as a researcher at Biro Klasifikasi Indonesia. She is responsible for analysing accident causation. She is interested in Human Factors, Accident Analysis, and Safety.

Sony Anggara holds the current position as a researcher at Biro Klasifikasi Indonesia,interested in Offshore technology.

Topan Firmandaholds the current position as a researcher at Biro Klasifikasi Indonesia, interested in ship structure and numerical method.

M. Rizqi Fitra holds the current position as a researcher at Biro Klasifikasi Indonesia.He is responsible for review accident report from NTSC. His major research interests are machinery, Autonomous, green technology, and safety

Siti Komariyahholds the current position as a researcher at Biro Klasifikasi Indonesia, interested in materials and ship structure.

Rosihan Syarif holds the current position as a senior researcher at Biro Klasifikasi Indonesia. His major research interests is ship stability.

M. Arif Kurniawan holds the current position as a Senior Manager of Technical Research at Biro Klasifikasi Indonesia. He is responsible for coordinating and supervising all the technical research carried out in his team.

PERFORMANCE EVALUATION ON LOCAL CULTURE-BASED PASSIVE RADAR REFLECTOR FOR TRADITIONAL FISHING BOATS IN INDONESIA

I P A Wibawa, M B Rahmat, D A Utari, and **E Julianto,** Politeknik Perkapalan Negeri Surabaya, Surabaya, Indonesia.
E Setijadi, Institut Teknologi Sepuluh Nopember, Surabaya, Indonesia.
R W Birmingham, Newcastle University, Newcastle upon Tyne, United Kingdom

SUMMARY

The accident of a fishing vessel hit by a larger merchant ship has been quite common in Indonesia. It is because the large vessels failed to detect the fishing boats clearly, which was exacerbated by the absence of passive radar reflector on-board. The previous study found that the attempt to introduce passive radar reflector to local fishing communities need to consider the design of the radar reflector in harmony with the ornamentation found on local fishing boats. Based on the above issue, the passive radar reflector that adopts the ornamentation on local traditional fishing vessels has been designed and tested. The standard octahedral passive radar reflector was covered by casing in the form ofMosque dome made of fiberglass reinforced plastic as an adaptation of local culture. The test showed that there is a slight difference on the performance of passive radar reflector with and without FRP casing.

1. INTRODUCTION

Fishing or being a fisherman is among the most dangerous and riskiest occupations in the world. The International Labor Organization (ILO) estimated that around 24,000 work accidents happen every year in the fisheries worldwide [1]. The occupational health and safety on board is also a major problem in capture fisheries in Indonesia. One type of frequently happened accident is a collision between a fishing boat and a much larger steel ship. An example of a case is the collision between a fishing boat with a container ship in the waters of Masalembu, Madura in November 2019, which nearly killed 12 fishermen on the boat [2]

Collisions between traditional fishing boats and large ships happen because the presence of traditional fishing boat is often not detected by the captain of large ships. This applies especially at night or in foggy areas as the captain's view is relatively limited. A non-fulfilled safety requirement by traditional fishing vessels such as the absence of standard lights and navigation equipment is also one of the aspects that increase the risk of a traditional fishing vessel collided with a large ship [3, 4].

In addition to navigation lights, according to FAO and IMO recommendations, navigation equipment that is also required to be installed on fishing vessels is radar reflector. Radar reflector serves to reflect radar waves from large ships, especially on small non-ferrous vessels such as wooden and Fiberglass Reinforced Plastic (FRP) boats, so the position of small vessels equipped with radar reflectors can be detected and identified by large ships.

Currently, most of the small vessels with wood and fiberglass hull and slim design (curved or sloping surface reflectsless), even those with masts and engines donot possess enough reflective qualities. This results in insufficientreflective qualities to make it highly visible on radar screens. A Container ships and Tankers require long down time and a wide turning radius. Therefore, ensuring detection of other ships or boats in a safe distance under all weather and sea conditions is very important to avoid collision.

In the meantime, the introduction of a new technology to the fishing community should pay attention to the fishing community's concern about local culture in order to ensure the fishing community's acceptance of the new technology or new design on fishing boats [5]. By considering the fishing community embrace of new designs and technologies influenced by local culture, the design of the radar reflector to be installed on local fishing boats should give room for the existing design or ornamentation on local fishing vessels.

A study of social and technical aspects was carried out to ensure that the alternative design of the radar reflector, which was based on ornamentation commonly found on traditional fishing vessels, could be accepted by the fishing community as well as work effectively as a radar reflector. Analysis of the fishing community potential acceptance of radar reflector showed that the fishing community was highly interested in installing a radar reflector on their fishing boats after knowing the function of the radar reflector. However, they were more enthusiastic in using a radar reflector that resembled the ornamental design on local fishing boats compare to the IMO standard [4].

The conventional passive radar reflector is made of three planarcircles or squares of metal intersecting at right angles, formingeight trihedral reflectors, oranoctahedral. Anessential design feature of reflector which greatly enhances the reflective amount of signal returned is the accuracy of right angles formed between the plates. In the research that was carried outin the coastal area of Banyuwangi, this designwas considered less attractiveto fishermen since they preferedto applya more local cultural design of ornament into their ships.

Therefore, a modification of the passive radar reflector

design was made in which the dome shape which is based on ornamentation found in local fishing boat is applied as a casing for the octahedral passive radar reflector. Meanwhile, theoctahedral radar reflector is made of 2 milimeters Alumnium plate with dimension: 20cmx20cmx 20 cm.

The test wascarried out by placing theoctahedral radar reflector covered by the dome in a controlled room, an air-tightroom called the anechoic chamber.The measurements were made to obtain the value of the transmission coefficient (S21) thattells us how the transmission line transmits the signal. This is related to the insertion loss in the transmission. By knowing this value, the information will be obtained whether the object is good enough to reflect back the signal that hit it, and whether the FRP dome casing have any effect to the performance of radar reflector in reflecting the signal.

2. EXPERIMENTAL AND EVALUATION

The first stage of the study was designing the prototype of passive radar reflector. As the results of the initial study, the design of the local culture-based passive radar reflector was referred to the forms of ornamentation that exist on traditional fishing boatsfound in local fishing community in Muncar, East Java. The ornamentation that was used as reference based on the form of Mosque doom model which can be found in top of the local fishing boat mast, as seen in Fig.1

The prototype consists of standard octahedral passive radar reflector that is placed in the "catch rain" position and covered by casing in the form of Mosque dome made of Fiberglass Reinforced Plastic. The use of mosque dome is based on ornamentation that is mostly found on the traditional fishing boat in East Java.

Figure 1. The ornamentation of traditional fishing boats

To evaluate the performance of the proposed passive radar reflector, Indoor measurements were carried out in the anechoic chamber at the National Institute of Aeronautics and Space (LAPAN). The anechoic chamber is a simulation of free space and far-field conditions. The equipment used in the measurement of the Radar CrossSectionconsistsof:

1) AnechoicRoomlocatedattheEMCLaboratoryofthe LAPANBRINSatelliteTechnologyCenter.Thefrequency used is adjusted to the frequency between 2-13GHz.
2) VectorNetworkAnalyzertypeVNAAgilentN5221 AE
3) Antenna3-12GHz
4) Radar Reflector
5) Dome

The measurement method is as shown in Figure2. [6]

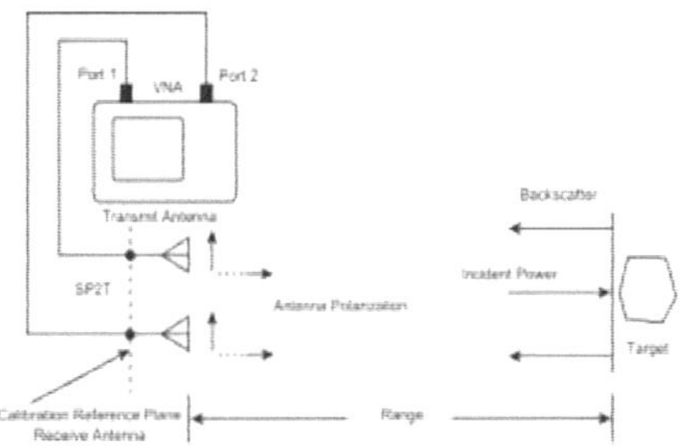

Figure 2. Measurement Setup.

The transmitting antenna (connected to port 1 VNA) and receiving antenna (connected to port 2 VNA) are positioned in the same plane position, as shown in figure 2. The measurement target is a radar reflector mounted on a low reflectance base. The measurements as shown in the figure 2, represent the VNA as radar, where S21 is measured. The coaxial cable output port 1 is connected to the coaxial to rectangular waveguide transition (E plane in the vertical direction). The output of port 2 is connected to the output of the wave receiving antenna. The two antennas are placed as close as possible [6, 7, 8, 9, 10]. Where the distance between the two antennas is determined to be 1.6 cm. Figure 3 shown how to setup the antenna.

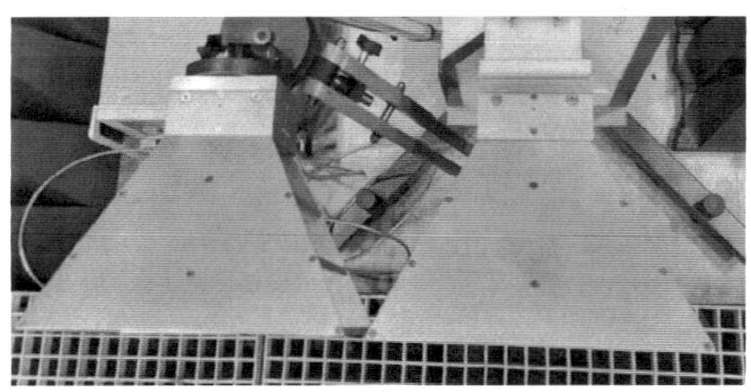

(a)

(b)

Figure 3.The Antenna Setup

The distance between the passive radar reflector as the test object and the sending and receiving antennasis adjusted to the settings that already available at the test location, which is 10 meters. The measurement setup in the anechoic chamber looks like Figure 4..

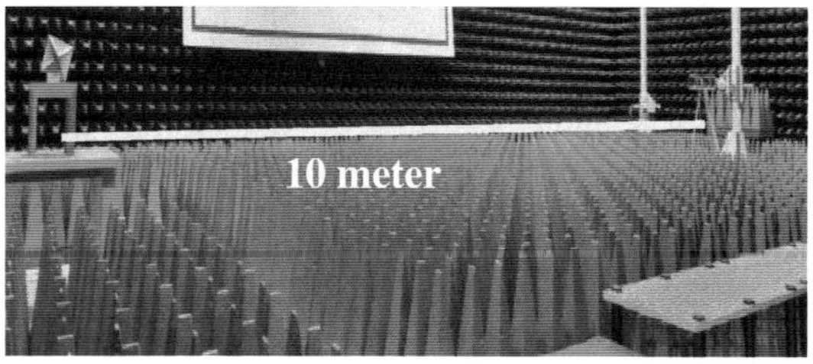

Figure 4. The measurement setup in anechoic chamber

3. RESULT AND DISCUSSION

The measurements were conducted by placing the test object, which is the radar reflector, as far as 10 meters from the transmitting and receiving antennas. The position of the radar reflector is shown on Figure 4. The first measurement was at 0 degrees. Afterward, the test object was rotated horizontally every 15-degree. Therefore, from an angle of 0-degree to 90-degree, 7 measurement data were obtained from the test. The next step was to rotate the radar reflector in the opposite direction from the initial position by the same degree which was 15 degrees. As a result, another6-measurement data was obtained. The data resulted from this first stage is measurement data with various positions for radar reflector without dome casing.

The next step was to cover the passive radar reflectorwith a dome casing made of 2 layers (600 gram/m^2) of Fiberglass Reinforcement Plastic as seen in Figure 5.

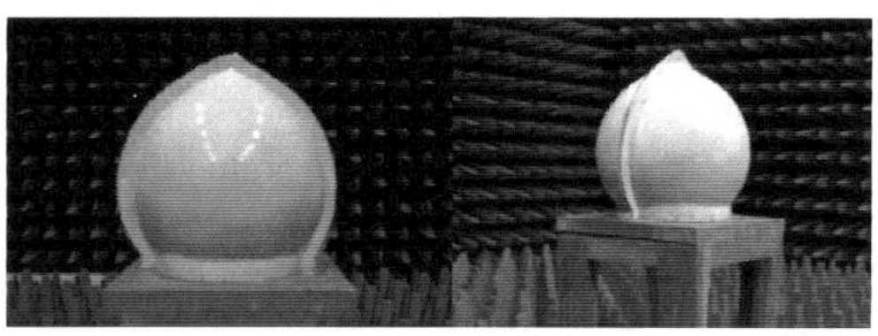

Figure 5 Passive Radar Reflector casing in the form of Dome

The next measurement process was carried out as in the first stage. The measurements were made in the frequency range of 3 GHz to 12 GHz. The results of the measurement of the radar reflector without dome with a shift in the value of the positive degree are shown in Figure 6 and 7 below sequentially.

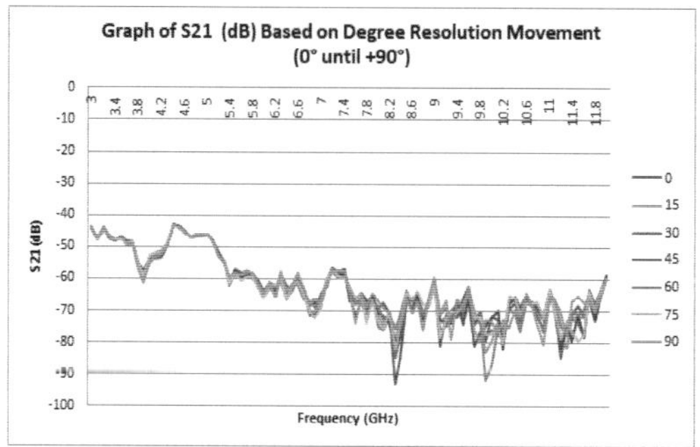

Figure 6.Graph of S21 (dB) Based on Degree Resolution Movement ($0^0 - 90^0$)

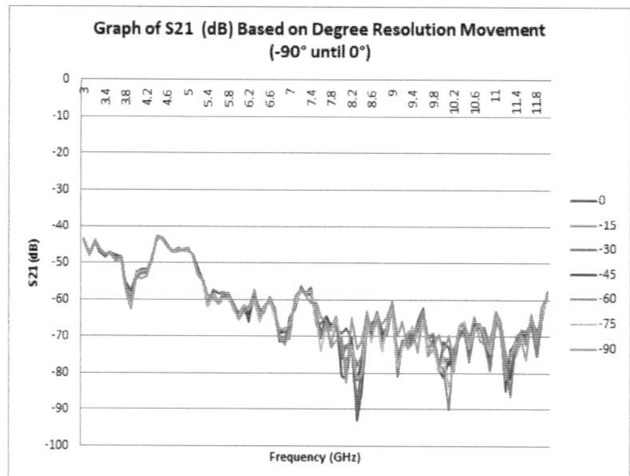

Figure 7Graph of S21 (dB) Based on Degree Resolution Movement (-90^0 - 0^0)

The measurement results of the passive radarreflector covered by FRP dome casing with positive and negative degree rotated values are shown in Figure 8 and 9

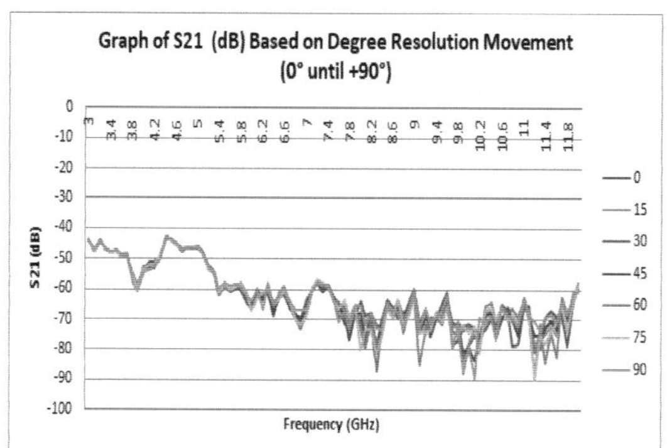

Figure 8. Graph of S21 (dB) Based on Degree Resolution Movement ($0^0 - 90^0$) with Dome

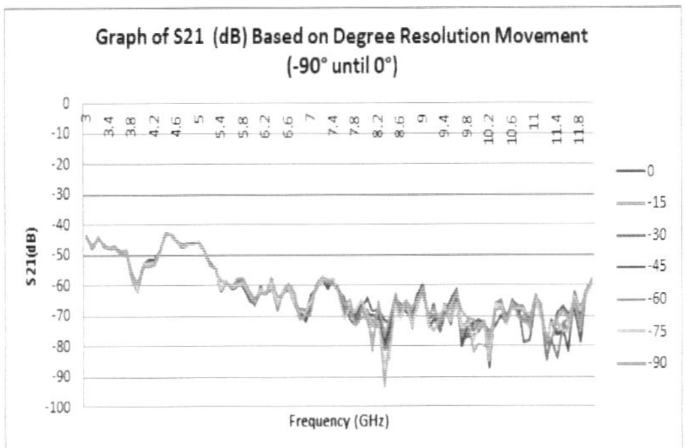

Figure 9. Graph of S21 (dB) Based on Degree Resolution Movement (−90^0 - 0^0) with Dome

In order to identify the effect of FRP casing in the form of mosque dome, data from previous results have been compared. As seen in Figure 10, the measurementsare resulted from 2 conditions: radar reflector without dome casing, and radar reflector using dome casing. The graph shows that there is no significant difference between the presence of DOME and the absence of DOME.

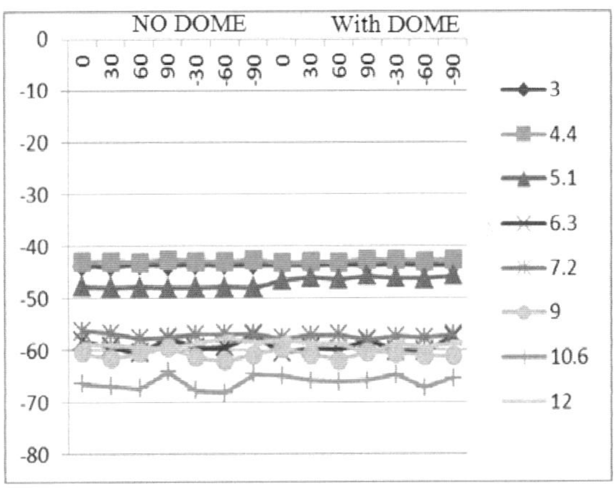

Figure 10. S21(Transmission Coefficient) Result for both passive radar reflector with and without FRP casing

7. CONCLUSIONS

The result of the test on the performance of passive radar reflector, which was placed in a "catch rain" position, showed that there are no significant difference of S21(Transmission Coefficient) value on all position (0^0, 15^0, 30^0, 45^0, 60^0, 75^0, 90^0). Furthermore, on average, there is no difference S21 value for passive radar reflector that was covered by FRP Dome casing and those without FRP Dome casing. It can be concluded that the design of passive radar reflector that adopted the design of ornamentation on local fishing boat in the form of mosque dome made of FRP do not affect the performance of passive radar reflector itself.

8. ACKNOWLEDGEMENTS

Our high appreciation goes to Center of Research and Community Service (P3M) PPNS for supporting this research through the DIPA 2021.

9. REFERENCES

1 GUDMUNDSSON, A.,'International instrument on the safety of fishing vessels and fishermen', Bay of Bengal News, 2006

2. RIVAI, M., dan SURYA, 'Perahu hancur ditabrak kapal besar, di Perairan Masalembu, 12 nelayan sempat tenggelam'. TribunNews.com. 2019

3 WIBAWA, P.A. and BIRMINGHAM, R.W. 'Improving Safety Working Environment on Indonesian Fishing Fleet: A Case Study on Local Fishing Communities in East Java', *Mastic2018 Maritime Safety International Confrence*, Indonesia, 2018

4 WIBAWA, P.A.,RAHMAT, M.B.,UTARI,D.A., "Penggunaan Radar Reflektor Berbasis Budaya Lokal Pada Kapal Ikan Tradi-sional",Seminar Nasional Terapan Riset Inovatif (SENTRINOV) Ke-6,ISAS Publishing, Vol. 6 No. 1 (2020) E-ISSN: 2621-9794, P-ISSN:2477-2097, 2020.

5 BIRMINGHAM, R.W., WIBAWA, P.A.,'The role of aesthetics inengineering design – insights gained from cross-cultural research intotraditional fishing vessels in indonesia'. *Marine Design XIII*, Volume 1:Proceedings of the 13th International Marine Design Conference (IMDC2018),June10-14,Finland, 2018.

6 ANRITSU, 'Measurement of Radar Cross Section using the VNA Master Handheld VNA',Application Note, Anritsu Company,2011

7 KNOTT, E.F., SCHAEFFER,J.F.;TULEY, M.T. 'Radar Cross Section' Artech House, INC, New Jersey,1985

8 YIM, J.B and KIM, W.S. 'Design of Passive Type Radar Reflector',International Journal of Navigation and Port Research,Vol.27,No.3 pp.267-27,ISSN-1598-5725, 2003

9 CURRIE, N. C., 'Radar Reflectivity Measurement', Technics Applications, Artech House Inc., 1989

10 CERQUEIRA R, MARTIN, I.M.,FAEZ, R. "Radar Cross Section Measurement (8-12GHz) of Magnetic and Dielectric Microwave

Absorbing Thin sheets", Revisa de Fesica Aplicda e instrumentacao,Vol.15,No.1,Dezembro, 2002.

9. AUTHORS BIOGRAPHY

I Putu Arta Wibawa got his Ph.D on Naval Architecture from Newcastle University, United Kingdom. His research interests include design and production of small craft, especially fishing vessels; the materials for boats, Occasional Health and Safety aspects of fishing vessels and the Life Cycle Assessment. He is a member of the Royal Institution of Naval Architects (MRINA) since started his PhD study in 2012.

Muhammad Basuki Rahmat is a senior lecturer and head of Research and Community Service at Politeknik Perkapalan Negeri Surabaya (Shipbuilding Institute of Polytechnic Surabaya) He got his bachelor and master degree from Institut Teknologi Sepuluh Nopember (ITS) Surabaya. His research interests covering the Navigation, Metamaterial, Radar Absorber and Wireless Sensor Network.

Dian Asa Utari gained her Bachelor degree from the State University of Surabaya (UNESA) and Master degree from State University of Malang (UM). She teaches English for Marine Engineering (English for Specific Purposes) at Politeknik Perkapalan Negeri Surabaya (Shipbuilding Institute of Polytechnic Surabaya). She is interested in exploring the use of ICT on teaching and ESP teaching methods and approaches.

Eko Setijadi received Ph.D from Kumoto University Japan in 2010 focusing on Computational Electromagnetics (CEM). He completed graduate and postgraduate program at Institut Teknologi Sepuluh Nopember (ITS) Surabaya on Telecommunication Engineering. He is a lecturer as well as researcher at the Department of Electronic Engineering, ITS. His research mainly focuses on Computational Electromagnetics (CEM), Antenna and microwave devices.

Eko Julianto is the Director of Politeknik Perkapalan Negeri Surabaya (Shipbuilding Institute of Polytechnic Surabaya). He is honoured as a fellow of the Royal Institution of Naval Architects (RINA). He received his M.Sc degree from Newcastle University, UK. Among his research interests are in marine engineering and piping engineering.

Richard W Birmingham is a Professor of Small Craft Design, De Newcastle University. His research interests include design methodologies applied in the marine environment, design of small craft, including working craft (such as tugs and fishing vessels), high speed craft and recreational craft, ecologically sustainable recreational craft, design for safety applied in the marine environment, design and production of traditional working craft in developing economies, sustainable operation of working craft in developing countries and stability of marine vehicles. He was President of the RINA (the Royal Institution of naval Architects) from 2018 until 2020.

The Royal Institution of Naval Architects
8-9 Northumberland Street
London, WC2N 5DA
United Kingdom

ISBN 978-1-7138-8303-6

CURRAN ASSOCIATES INC.
proceedings.com